THE JAHN–TELLER EFFECT

A Bibliographic Review

THE JAHN–TELLER EFFECT

A Bibliographic Review

I. B. Bersuker

Institute of Chemistry
Modavian Academy of Sciences
Kishinev, USSR

IFI/PLENUM • NEW YORK—WASHINGTON—LONDON

Library of Congress Cataloging in Publication Data

Bersuker, I. B. (Isaak Borisovich)
The Jahn–Teller effect.

Bibliography: p.
Includes index.
1. Jahn–Teller effect—Bibliography. I. Title.
Z5524.P6B47 1983 016.5412′2 83-17635
[QD461]
ISBN 0-306-65206-4

This book is published under an agreement with VAAP, the copyright agency of the USSR

A Division of Plenum Publishing Corporation
233 Spring Street, New York, N.Y. 10013

Printed in the United States of America

FOREWORD

This bibliography is the first in the world literature compiling and systematizing 3264 publications which appeared during the first 50 years on the Jahn-Teller effect and vibronic interactions - the rapidly developing modern trend in the physics and chemistry of molecules and crystals. These works are classified and categorized into 5 sections and 34 subsections, and briefly reviewed in the section and subsection introductions. The information was compiled from all possible sources, including direct contacts with the main specialists in the field.

This fractional rubrication contains the theory of vibronic interactions, including ideal, multimode, and cooperative Jahn-Teller, pseudo Jahn-Teller, and Renner effects with applications to electron, photoelectron, vibrational, rotational, ESR, NMR, NGR, electron and neutron scattering, and acoustic spectroscopy of molecules and crystals, magnetic properties, structural and related phase transitions in solids, ferroelectricity, impurity crystal properties, crystal chemistry, stereochemistry, chemical reactivity, and chemical activation in catalysis and biology. The titles of the publications contain short codified abstracts, and these their titles are translated into English if published in another language. Also included are cross-references and a subject, formula, substance, and author index.

This book is intended for physicists, chemists, and biologists involved in research, teaching, engineering, and bibliographic search.

CONTENTS

INTRODUCTION

The term Jahn-Teller Effect is used to denote all the effects and regularities which arise due to the mixing of the (fixed nuclei) electronic states of molecules and crystals by nontotally symmetric nuclear displacements. Research in this field constitutes a new trend in modern physics and chemistry and employs a new approach to the general problem of correlating the structure and properties of polyatomic systems based on the concept of electron-vibrational (vibronic) interactions. It incorporates the ideal, multimode, and cooperative Jahn-Teller, pseudo Jahn-Teller and Renner effects in spectroscopy (in the whole range of electromagnetic measurements, electron and neutron scattering, and acoustic losses), solid state physics (including impurity crystal properties and, especially, structural and ferroelectric phase transitions), stereochemistry and crystal chemistry, mechanisms of chemical reactions, and chemical activation in catalysis and biology.

The vibronic effects are most important in the case of electronic degeneracy (Jahn-Teller and Renner effects) and of near-lying electronic states - pseudodegeneracy (pseudo Jahn-Teller effect), and the concept as a whole is to a certain degree concerned with all kinds of molecular systems and solids.*

Due to its diverse applications, the literature on vibronic interactions appears in many different sources, in journals of various scopes and scientific levels (all kind of theoretical and/or applied physics, chemistry, biology, and others). As a result, some of the publications under consideration remain unknown and are practically inaccessible to specialists. The use of abstract journals and special computer programs is also hindered by the same reasons (note that the sections on the Jahn-Teller effect were introduced

---

*For a review of this sugject, see: (1) R. Englman, The Jahn-Teller Effect in Molecules and Crystals, Wiley, New York (1972). (2) I. B. Bersuker and V. Z. Polinger, Vibronic Interactions in Molecules and Crystals [in Russian], Nauka, Moscow (1983). (3) I. B. Bersuker, The Jahn-Teller Effect and Vibronic Interactions in Modern Chemistry, Plenum, New York (1984).

in the abstracts journals only in the late sixties and early seventies and do not include all the works mentioned above).

These circumstances induced the editor and his co-workers to try to collect and systematize all the publications on vibronic interactions and to prepare an appropriate bibliographic review. Being engaged in these topics for more than 23 years, I thought that it might not be too difficult to prepare such a bibliography. Actually, it turned out to be very time and labor consuming. First, it was difficult to find all the publications on the subject; the usually available sources, including the main journals and books, contain only about 80% of these works. The remainder was obtained by means of special searches, including direct contacts with the main specialists in the area.

But the most laborious was the classification of the 3264 publications into the 34 subsections and elucidating all the other characteristics needed for the short abstract, cross-references, and subject, formula, and substance indexes. The work was performed by the staff of the Department of Quantum Chemistry of the Institute of Chemistry (V. Z. Polinger, G. I. Bersuker, V. P. Zenchenko, S. S. Stavrov, B. G. Vekhter, I. Ya. Ogurtsov, and S. S. Budnikov) together with the Central Scientific Library (A. S. Moldavskaya, M. M. Kogan, and E. A. Faverman) of the Academy of Sciences of Moldavian SSR.

In this bibliographic review the publications of the first 50 years (1929-1979) on vibronic interactions are collected, briefly reviewed, and systematized. Concerning the scope of this bibliography, the following comments should be made.

First, the literature on vibronic effects begins not from the publication of Jahn and Teller in 1937, which contains the so-called Jahn-Teller theorem, but from the earlier work of Wigner and von Neumann, published in 1929, in which the mixing of two near-lying electronic states by nuclear displacements was considered for the first time.

Second, the boundary between the topics under consideration and the related problems is somewhat uncertain. In fact, the mixing of near-lying electronic states by nuclear displacements (in our terminology, the pseudo Jahn-Teller effect) is considered in many physical and chemical phenomena without independently of the Jahn-Teller effect. In some cases the authors approach the vibronic problem starting from other conceptions, and sometimes use another terminology for the same effects. Works on related problems, in which the authors do not recognize the vibronic effect, are as a rule, not included in this bibliography.

Some of the topics related to the vibronic effect develop into their own independent fields. These include phase transitions in metals resulting from two-band and multiband interactions, nonadiabatic chemical reactions, nonradiative transitions and relaxation phenomena, mixing of impurity states by vibrations in metals, etc. All these and similar problems are not covered fully in this bibliography. Nevertheless, in order to maintain the link between the vibronic effect and the related topics, we have included some of these works in the appropriate sections.

Third, abstracts of papers presented at conferences which duplicate publications in journals and which do not contain additional information are, as a rule, not included.

Finally, there may be some minor omissions, as revealed by spot checking, and some works of vibronic interactions may have been excluded due to oversight. I cordially beg the author's and reader's pardon for these and other shortcomings of this book.

For the reader's convenience, this bibliography is presented in such a way that one can easily find all the papers on a certain subject collected in the appropriate subsection of the five sections. Each of the sections and subsections is preceded by a brief introduction written by a specialist in the area. It must be emphasized that in some of the works several vibronic problems are discussed simultaneously, and therefore their inclusion in a certain section (subsection) is rather arbitrary. This inconvenience is compensated by additional cross-references, that are given at the end of each subsection, as well as by additional information in the codified abstract to each of the publications. The titles of papers published in other langugages are translated into English. We have also included a detailed formula, substance, and subject index (the latter serves as a convenient supplementary means of orientation and does not contain references to all the papers on the subject), as well as an author index.

I am most grateful to my co-workers for their patient participation in compiling this book.

I am also greatly indebted to the authors of the works compiled in this bibliography and for their help and cooperation in sending the titles and reprints of their works: J. H. Ammeter, M. Abou-Ghantous, V. I. Avvakumov, L. S. Bartell, M. Bacci, C. A. Bates, H. Barentzen, D. Bertini, J. Brickmann, H. Bilz, B. D. Bhattacharyya, S. Brühl, L. Boatner, L. S. Cederbaum, R. E. Coffman, L. L. Chase, R. S. Dagis, E. Duval, A. M. De Goer, T. Dunn, R. Englman, T. L. Estle, J. P. Fackler, Jr., J. R. Fletcher, A. Fukuda, J. Feder, E. M. Ganapol'skiy, D. Gatteschi, N. D. Glinchuk, B. Halperin, B. Hoffman, J. R. Herrington, P. C. Jaussand, B. R. Judd, S. A. Kazanskiy, O. Kahn, Y. Kamishina, J. K. Kjems, D. I.Khomskiy, V. A. Krylov,

N. N. Kristofel, Le Si Dang, R. Lacroix, S. Leach, A. A. Levin, V. D. Lipatov, A. V. Malakhovskiy, A. A. Maradudin, F. Mehran, R. L. Melcher, K. A. Müller, Y. Merle d'Aubigne, E. Mulazzi, S. Muramatsu, K. Nasu, A. N. Nikiforov, M. C. M. O'Brien, V. I. Osherov, V. V. Ovsyankin, V. S. Ostrovskiy, R. Pearson, O. E. Polansky, D. R. Pooler, E. Pytte, M. Ratner, A. Ranfagni, D. Reinen, M. Rueff, E. G. Reut, E. Sigmund, K. W. H. Stevens, N. F. Stepanov, R. B. Stinchcombe, M. D. Sturge, W. H. E. Schwarz, J. C. Slonczewski, L. S. Sochava, V. A. Shutilov, N. Terzi, H. Thomas, Y. Toyozawa, M. F. Trinkler, V. S. Vikhnin, G. Viliani, M. Wagner, G. Wenzel, Yu. V. Yablokov, H. Yamatera, S. G. Zazubovich, M. Z. Zgierski, and A. I. Zvyagin.

I. B. Bersuker, Editor

HOW TO USE THE BIBLIOGRAPHY

1. MODE OF PRESENTATION

All the 3204 bibliographic descriptions are divided into 5 sections and 34 subsections. Each section and subsection is preceded by a brief introduction serving as a supplementary means of orientation. The descriptions in each subsection are presented chronologically according to year of publication, and within the given year they are ordered alphabetically after the name of the first author. For papers published in proceedings of conferences, symposia, etc., the year of publication of the proceedings (and not the year of the conference) is used in the chronological ordering. The names of authors in their original non-Latin spelling are transliterated according to the British-American Standard System.

At the end of each subsection cross-references are given. These give important additional information on the topics under consideration, since the classification of some of the papers in certain subsections is rather arbitrary. Important additional information is also given in the subject, formula, substance, and author indexes.

2. THE BIBLIOGRAPHY DESCRIPTION

Each bibliography entry consists of the number of the publication, the bibliographic description (the author, title, and source), the abstracting or indexing source, and a short codified abstract.

For example,

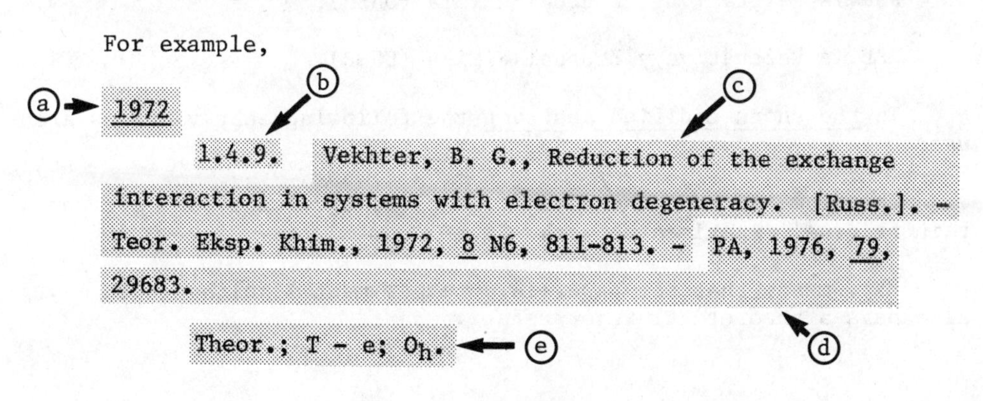

(a) Year of publication.

(b) Number of publication.

(c) Bibliographic description (author, title, and source).

(d) Abstracting or indexing source.

(e) Short codified abstract.

The number of publication consists of three numbers separated by points: the section number, the subsection number, and the ordering number inside the given subsection, respectively. If several publications of the same subsection are cited (e.g., in the introductory reviews, author and subject indexes), their ordering numbers are preceded by a hyphen. For example, 3.1.-7, 15, 16, 20 refer to the following numbers: 3.1.7, 3.1.15, 3.1.16, and 3.1.20.

The bibliographic description is presented in the following order: (1) the author(s) of the publication; (2) the title, and (3) the source, including the first and last pages of the paper. The titles of articles are given in English, the original language of publication (if not English) being indicated in brackets. The book titles are given in their original language (transliterated if in non-Latin alphabet), and the English translation is given in the comments. The abbreviations of titles of journals and other serial publications follow the standard used in Physics Abstracts.

The abstracting or indexing source refers to bibliographic publications that abstract or index the article or book, or that give the English translation of nonEnglish, nonserial publications. These bibliographic publications are abbreviated as follows:

CA is Chemical Abstract,

PA is Physics Abstract,

RZK is Referativnyy Zhurnal Khimia (USSR),

RZP is Referativnyy Zhurnal Fizika (USSR).

In the short codified abstract the following abbreviations are used:

Theor. means that the paper is a theoretical and does not contain original experimental work.

Exp. means that the paper is an experimental, although it may also have a theoretical interpretative part.

Rev. means that the paper is a review article.

E, T, $A_1 + E$, etc., mean that electronic Jahn-Teller terms E, T, $A_1 + E$, etc., are considered. The irreducible representations used for electronic terms and nuclear displacements are given after Mulliken, with the exception of double representations, which are given after Bethe.

E - e, T- t_2, T - (e + t_2), etc., refer to the particular vibronic problem considered. Small letters denote (as usual in Jahn-Teller effect theory) the irreducible representations for nuclear displacements active in the vibronic problem under consideration.

D_{2d}, D_{3h}, C_{3v}, O_h, T_d, etc., refer to symmetry of the polyatomic system studied. Symmetries are denoted after Schoenflies. If the Jahn-Teller center is a translatory invariant element of the proper crystal lattice, the local symmetry is indicated, and not the space group of the lattice, since the former determines the nature of the vibronic interaction.

d^1, d^2, sp, p , etc., indicate the electronic configuration of the Jahn-Teller center.

The numbers 1.1, 1.2, etc., given in the codified abstract mean that the paper under consideration may also be classified under Subsections 1.1, 1.2, etc.

Below we show how to use the system, using publication 2.1.228 as an example.

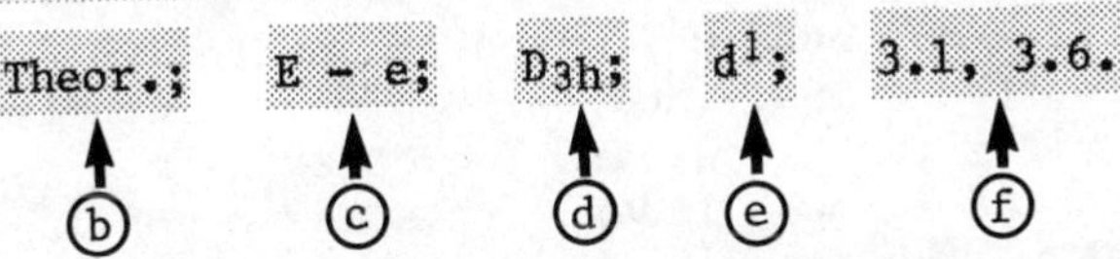

(a) Additional information is available in Chemical Abstracts, Vol. 83 (1975), No. 199596, and in Physics Abstracts, Vol. 79 (1976), No. 5761.

(b) The paper is theoretical.

(c) The E - e problem is considered.

(d) The symmetry of the Jahn-Teller system under consideration is D_{3h}.

(e) The electronic configuration of the Jahn-Teller system is d^1.

(f) The publication can also be classified under Subsections 3.1 and 3.6.

3. INDEXES

In the Author Index there are also cross-references to names which may take different forms, e.g.: Van der Waals J. N. is given also as Waals J. N., van der, with an appropriate cross-reference.

The Subject Index covers terms of interest in vibronic theory, including those which are given in the Contents, but the list of papers attributed to a given term does not claim to be exhaustive.

The Formula Index contains information on the substances mentioned in the appropriate papers (not necessarily Jahn-Teller systems).

The chemical formula ordering corresponds to the standard used in Chemical Abstracts. Exceptions are those where atoms are grouped in a structural unit. In these cases we use generally accepted notation. The ordering of these groups follows the alphabetic-numeric system. For example, $Mn(cp)_2$ is placed after $MnCl_2$, but before MnF_2; the formula $(MnO)(Cr_2O_3)$ is placed after MnO, but before $(MnO)(Fe_2O_3)$.

A list of chemical abbreviations for some structural units and subunits in the chemical formulas is given at the end of the Substance Index.

The same chemical compound may often be written in different forms. For example, the spinel $(MnO)(Fe_2O_3)$ may be written $MnFe_2O_4$. In such cases the first form is used. The other form $(Fe_2O_3)(MnO)$, is also included in the Formula Index with a corresponding cross-reference. Similarly, the formula $Mn_xFe_{2-2x}O_{3-2x}$ is repaced by $(MnO)_x(Fe_2O_3)_{1-x}$ and $(Fe_2O_3)_{1-x}(MnO)_x$ with cross-references. The formulas Cu^{2+} in ..., Zn^{2+} in ..., etc., mean the impurity ion in the corresponding crystal matrix.

The Substance Index is given as an appendix to the Formula Index. It contains a list of chemical compounds that are not included in the Formula Index since their chemical formulas are too complex and less informative. The whole class of chemical compounds such as spinels, perovskites, porphyrins, hemoproteins, metallocenes, etc., is also listed.

1. GENERAL THEORY OF VIBRONIC INTERACTIONS*

The concept of intrinsic instability of high-symmetry nuclear configurations of molecular systems in orbital degenerate electronic states was first suggested by L. D. Landau in 1934. The circumstances which led Jahn and Teller to the proof of the so-called Jahn-Teller theorem [1.1.-3, 5] are described by Teller in the "Historical Note" given as an introduction to the book [1.1.37].

The mixing of two Born-Oppenheimer (degenerate and pseudodegenerate) states by nuclear displacements was first used in 1929 by Von Neumann and Wigner [1.1.1] for the investigation of adiabatic potential intersections. Similar considerations were used by Herzberg and Teller in 1933 in the analysis of selection rules for the electronic transitions to pseudodegenerate electronic states [1.5.1]. In 1934 Renner employed the limited basis of degenerate Born-Oppenheimer electronic states in his work on adiabatic potentials of linear molecules [1.6.1].

The above papers contain the first principles on which the Jahn-Teller effect is based. For a review of the subject the articles [1.1.55] and [3.1.151] may be useful. More detailed information can be found in the reviews [1.1.22], [3.1.-144, 253], and [1.1.37].

1.1. General. The Jahn-Teller Theorem. Vibronic Hamiltonian. Group-Theoretical Aspects

In the papers [1.1.-3, 5] the proof of the Jahn-Teller theorem is reduced to the question whether the matrix elements of vibronic interactions equal zero or not. The answer is found in group-theoretical selection rules (see, e.g., [1.1.12]) preceded by a group-theoretic classification of the nuclear displacements by irreducible representations of the appropriate symmetry group. Thus the paper [1.1.3] contains also useful data on irreducible representations for rotations, translations, and vibrations of the framework of any symmetry-type molecular system.

*The introductory reviews for this section were written by V. Z. Pollinger.

A more constructive proof of the Jahn-Teller theorem was given in 1965 in the work [1.1.20]. In 1971 a somewhat similar proof was suggested by Blount [1.1.31], while in 1973 the analytical proof based on the Frobenius theorem appeared [1.1.-42, 43, 44]. Note that the proof in [1.1.44] covers also the case of space groups. In 1976 the algebraic proof of the Jahn-Teller theorem including accidental degeneracy was presented [1.1.61].

The lack of extremum of the adiabatic potential at the point of degeneracy is often interpreted as the instability of the corresponding nuclear framework. This conclusion, taken literally, is not true, at least not always; the physical meaning of the Jahn-Teller theorem is discussed in the work [1.1.21].

The solution of the vibronic equations is very difficult. Some simplification may be made by means of group-theoretic methods. It was shown that the approximations usually used to simplify the vibronic Hamiltonian result in an increase of its symmetry, the latter being sometimes one of the infinitesimal groups. The appropriate mathematical formalism is given in the review article [1.1.50].

1929

1.1.1. Von Neumann, J., Wigner, E., Behavior of eigenvalues in adiabatic processes. - Phys. Z., 1929, 30, Aug., 467-470.

Theor.; 1.2

1936

1.1.2. Jahn H., Teller E., Stability of degenerate electronic states in polyatomic molecules. - Phys. Rev., 1936, 49, N11, 874-880. - CA, 1937, 31, 5633[7].

Theor.

1937

1.1.3. Jahn, H. A., Teller, E., Stability of polyatomic molecules in degenerate electronic states. I. Orbital degeneracy. - Proc. R. Soc. London. Ser. A., 1937, 161, N905, 220-235. - CA, 1938, 32, 11[6].

Theor.

1.1.4. Teller, E., Crossing of potential surfaces. - J. Phys. Chem., 1937, 41, N1, 109-116. - CA, 1937, 31, 4861[6].

1938

1.1.5. Jahn, H. A., Stability of polyatomic molecules in degenerate electronic states. II. Spin degeneracy. - Proc. R. Soc. London. Ser. A., 1938, 164, 117-131. - CA, 1938, 32, 3221[5].

Theor.

1939

1.1.6. Van Vleck, J. H., The Jahn-Teller effect and crystalline Stark splitting for clusters of the Form XY_6. - J. Chem. Phys., 1939, 7, 72-84. - CA, 1939, 33, 1561[5].

Theor.

1948

1.1.7. Landau, L., Lifshits, E., Kvantovaja Mekhanika: Ch. I. Nerelyativistskaya Teoriya. [Russ.]. - Moscow; Leningrad: Gostekhizdat, 1948.

§100. The stability of symmetric configuration of molecules, p. 423-426.

Engl. ed.: Landau, L. D., Lifshitz, E. M., Quantum Mechanics: Non-Relativistic Theory. - London: Pergamon Press, 1958. - 515 p. - (Addison-Wesley Ser. in Adv. Phys.; Course of Theor. Phys. Vol. 3).

1957

1.1.8. Liehr, A. D., On the use of the Born-Oppenheimer approximation in molecular problems. - Ann. Phys. (USA), 1957, 1, N3, 221-232.

Theor.

1.1.9. Ruch, E., Symmetry relations and valence phenomena [Germ.]. - Z. Elektrochem., 1957, 61, N8, 913-923. - CA, 1958, 52, 4262a. RZK, 1958, N23, 76252.

Theor.

1959

1.1.10. Clinton, W. L., Rice, B., Reformulation of the Jahn-Teller theorem. - J. Chem. Phys., 1959, 30, N2, 542-546. - RZF, 1960, N1, 723.

Theor.

1960

1.1.11. Clinton, W. L., Dynamical Jahn-Teller theorem. - J. Chem. Phys., 1960, 32, N2, 626-627. - CA, 1960, 54, 14918a.

Theor.; E - e

1.1.12. Heine, V., Group Theory in Quantum Mechanics: An Introduction to Its Present Usage. - London, etc.: Pergamon Press, 1960. - IX, 468 p.

1.1.13. Orgel, L. E., An introduction to Transition-Metal Chemistry: Ligand-Field Theory. - London: Methuen; New York: Wiley, 1960. - 180 p.

Ch. 4. Sec. 2. The Jahn-Teller Effect in Octahedral Complexes.

Ch. 4. Sec. 4. Distortions in Tetrahedral Complexes.

1962

1.1.14. Balhausen, C. J., Introduction to Ligand Field Theory. - New York: McGraw-Hill, 1962. - IX, 298 p.

1.1.15. Bersuker, I. B., Ablov, A. V., Khimicheskaya Svyaz' v Kompleksnykh Soedineniyakh. [Russ.]. - Kishinev: Shtiintsa, 1962. - 208 p.

Chemical Bond in Complex Compounds.

1.1.16. Birman, J. L., Configurational instability in solids: diamond and zinc blende. - Phys. Rev., 1962, 125, N6, 1959-1961. - RZF, 1962, 12E30.

Theor.

1.1.17. Ham, N. S., The Jahn-Teller theorem. - Spectrochim. Acta, 1962, 18, N3, 775-789. - CA, 1962, 57, 8076a.

Theor.

1963

1.1.18. Birman, J. L., Jahn-Teller effect and valence-bond method. - J. Chem. Phys., 1963, 39, N11, 3147-3148. - CA, 1964, 60, 6227e.

Theor.

1964

1.1.19. Bersuker, I. B., Issledovaniya v Oblasti Kvantovoy Teorii Kompleksov Perekhodnykh Metallov: Avtoref. Dis. ... Dokt. Fiz.-Mat. Nauk. [Russ.]. - Leningrad, 1964. - 24 p. - Leningr. Univ. Fiz. Fak.

Investigation in the Field of Quantum Theory of Transition Metal Complexes.

Theor.; E - e; T; 1.2; 3.1.

1965

1.1.20. Ruch, E., Schoenhofer, A., Proof of the Jahn-Teller theorem with the help of a theorem on the induction of representations of finite groups. [Germ.]. - Theor. Chim. Acta, 1965, 3, N4, 291-304. - CA, 1965, 63, 12334d.

Theor.

1966

1.1.21. Bersuker, I. B., Formulation and meaning of the Jahn-Teller theorem. [Russ.]. - Teor. & Eksp. Khim., 1966, 2, N4, 518-21. - CA, 1967, 66, 22403. PA, 1968, 71, 38793.

1.1.21a. Dagis, R., Levinson, I. B., Group theoretical properties of the adiabatic potential in molecules. [Russ.]. - In: Opt. Spectrosk.: Sb. Statei. Leningrad, 1967, 3, 3-8. - CA, 1967, 67, 36453.

Theor.

1967

1.1.22. Sturge, M. D., The Jahn-Teller effect in solids. - In: Solid State Phys.: Adv. Res. & Appl. New York; London, 1967, Vol. 20, p. 91-211. - RZF, 1968, 9E339.

Rev.

1.1.23. Yoshito Amako, Yoshiichi Sato, Hiroshi Azumi, The irreducible representations of spin functions in molecular system. - Sci. Rep. Tohoku Univ. Ser. I, 1967, 50, N2, 100-107. - CA, 1968, 68, 16271.

Theor.

1968

1.1.24. Yäger, E., Jahn-Teller effect at lattice sites of symmetry D_{3d}. - Phys. Status Solidi, 1968, 25, N1, K43-K46. - CA, 1968, 68, 73196. PA, 1968, 71, 17308.

1.1.24a. Martinenas, B. P., Dagis, R. S., The application of the symmetry theory for the investigation of the electron-vibrational interaction in molecules. [Russ.]. - Litov. Fiz. Sb., 1968, 8, N1-2, 97-106.

Theor.; E - e; D_{3d}.

1969

1.1.25. Martinenas, B. P., Dagis, R. S., On the classification of inversion levels in symmetric molecules. [Russ.]. - Theor. & Eksp. Khim., 1969, 5, N1, 123-125.

Theor.

1970

1.1.26. Butucelea, A., Jahn-Teller effect in some fluorine compounds with octahedral symmetry. [Rom.]. - Stud. & Cercet. Chim., 1970, 18, N8, 803-813. - CA, 1971, 74, 36336.

Rev.; O_h.

1.1.27. Wybourne, B. G., Jones, G. D., Interaction effects in solids using transition metal and rare earth ions as probes. - Christchurch (N. Z.), 1970. - 9 p. - US Clearinghouse Fed. Sci. Tech. Inform. A. D. N719868. - CA, 1971, 75, 69088. PA, 1972, 75, 20012.

Theor.

1971

1.1.28. Bates, C. A., Chandler, P. E., Stevens, K. W. H., Isomorphisms and spin-phonon interaction for $(3d)^n$ ions. - J. Phys. C, 1971, 74, N14, 2017-2023. - PA, 1971, 74, 83620.

Theor.; O_h; T_d.

1.1.29. Bersuker, I. B., Vekhter, B. G., Ogurtsov, I. Ya., Electronic degeneracy and quasi-degeneracy effects in coordination systems with strong electron-vibrational coupling. [Russ.]. - In: Theory of Electronic Shells of Atoms and Molecules: Rep. Int. Symp., 1969. Vilnius, 1971, p. 281-289. - RZF, 1972, 2E493.

Theor.; 3.1; 3.4.

1.1.30. Bersuker, I. B., Stroenie i Svoystva Koordinatsonnykh Soedineniy: Vvedeniye v Teoriyu. [Russ.]. - Leningrad: Khimiya, 1971. - 312 p.

Structure and Properties of Coordination Compounds: Introduction to the Theory.

1.1.31. Blount, E. I., The Jahn-Teller theory. - J. Math. Phys., 1971, 12, N9, 1890-1896. - CA, 1971, 75, 114339. PA, 1971, 74, 68228. RZF, 1972, 3D104.

Theor.

1.1.32. Creighton, J. A., Jahn-Teller effect in transition element compounds. - In: Essays in Chem. London; New York, 1971, Vol. 2, p. 45-59. - CA, 1971, 75, 133083.

Rev.

1.1.33. Ham, F. S., The Jahn-Teller effect. - Int. J. Quantum Chem. Sympos., 1971, N5, 191-199. - CA, 1972, 76, 39341.

Rev.

1.1.34. Jothan, R. W., Kettle, S. F. A., Geometrical consequences of the Jahn-Teller effect. - Inorg. Chim. Acta, 1971, 5, N2, 183-187. - CA, 1971, 75, 121590.

Theor.

1.1.35. Kiselev, A. A. Petelin, A. N., Born-Oppenheimer method for normal and linear molecules. [Russ.]. - In: Theory of Electronic Shells of Atoms and Molecules: Rep. Int. Symp., 1969. Vilnius, 1971, p. 211-213. RZF, 1972, 2D134.

Theor.; 1.6.

1972

1.1.36. Ballhausen, C. T., Hansen, A. E., Electronic spectra. - In: Ann. Rev. Phys. Chem. Palo Alto, 1972, 23, 15-38. - RZK, 1973, 9B166.

Theor.

1.1.37. Englman, R., The Jahn-Teller Effect in Molecules and Crystals. - New York: Wiley, 1972. - XVI, 350 p. - (Monogr. Chem. Phys.) - CA, 1972, 76, 160021. PA, 1972, 75, 48118.

1.1.38. Kamimura, Hiroshi, Dynamical Jahn-Teller effect. [Jap.]. - Kotai Butsuri, Solid State Phys., 1972, 7, N7, 365-375, 388. - CA, 1972, 132560. PA, 1972, 75, 77809.

Rev.

1.1.39. Sturge, M. D., Jahn-Teller effect. - In: McGraw-Hill Yearbook of Science and Technology. New York, 1972, p. 263-264.

Rev.

1.1.40. Tsukerblat, B. S., Rosenfeld, Yu. B., Vekhter, B. G., On the possibility of a photon-induced Jahn-Teller effect. - Phys. Lett. A, 1972, 38, N5, 333-334. - CA, 1972, 76, 119174. PA, 1972, 75, 26089. RZF, 1972, 7E376.

Theor.

1973

1.1.41. Orlandi, G., Seibrand, W., Theory of vibronic intensity borrowing: Comparison of Herzberg-Teller and Born-Oppenheimer coupling. - J. Chem. Phys., 1973, 58, N10, 4513-4523.

Theor.

1.1.42. Raghavacharyulu, I. V. V., Simple proof of the Jahn-Teller theorem. - In: Proc. Nucl. Phys. & Solid State Phys. Symp., Bangalore, New Delhe, 1973, p. 296-298. - PA, 1975, 78, 46578.

Theor.

1.1.43. Raghavacharyulu, I. V. V., The Jahn-Teller theorem for space groups. - In: Proc. Nucl. Phys. & Solid State Phys. Symp., Bangalore, New Delhi, 1973, p. 299. - PA, 1975, 78, 46579.

Theor.

1.1.44. Raghavacharyulu, I. V. V., Simple proof of the Jahn-Teller theorem. - J. Phys. C, 1973, 6, N24, L455-L457. - CA, 1974, 80, 137374. PA, 1974, 77, 16712. RZF, 1974, 6E567.

Theor.

1.1.45. Stepanov, N. F., Zhilinskiy, B. I., The methods of separation of variables in the molecular quantum mechanical problems. [Russ.]. - In: Sovrem. Probl. Fiz. Khim. Moscow, 1973, Vol. 7, p. 3-31.

Theor.; 1.5; 1.6.

1974

1.1.46. Aktins, P. W., Quanta: A Handbook of Concepts. - Oxford: Clarendon Press, 1974. - VI, 309 p. - (Chem. Ser. 21).

1.1.47. Bersuker, I. B., The Jahn-Teller effect, its real meaning and applications. - In: New Developments in Coord. Chem.: Sect. Lect. 13th Int. Conf. Coord. Chem., Cracow - Zakopane, 1970. Wroclae; Warszawa, 1974, p. 9-18.

Rev.; 1.2; 1.5; 2.1; 3.1.

1.1.48. Cracknell, A. P., Group theory in solid-state physics is not dead yet, alias: Some recent developments in the use of group theory in solid-state physics. - Adv. in Phys., 1974, 23, N5, 673-866. - Bibliogr.: p. 860-866.

Theor.

1.1.49. Epstein, S. T., Some theorems satisfied and some theorems not satisfied by Born-Oppenheimer type wave functions. - J. Chem. Phys., 1974, 60, N1, 147.

Theor.

1.1.50. Judd, B. R., Lie groups and the Jahn-Teller effects. - Can. J. Phys., 1974, 52, N11, 999-1044. - CA, 1974, 81, 140911. PA, 1974, 77, 60745. RZF, 1974, 12E427.

Theor.; T - (t + e); O_h.

1.1.51. Landau, L. D., Lifshitz, E. M., Teoreticheskaya Fizika. T.3.Kvantovaya Mekhanika: Nerel'ativistakaya Teoriya. [Russ.]. - Izd. 3-e. Pererab. i Dop. - Moscow: Nauka, 1974. - 752 p.

§102. Ustoychivost' Simmetrichnykh Konfiguratsiy Molekuly, p. 464-470.

Engl. ed.: Landau, L. D., Lifshits, E. M., Quantum Mechanics: Non-Relativistic Theory. - 3rd ed., rev. & enlarged. - Oxford; New York: Pergamon Press, 1977. - 673 p. - (Course Theor. Phys.; Vol. 3).

1.1.52. Leung Chi-hong, Symmetry studies of the Jahn-Teller effect. - 1974. Diss. Abstr. Int. B, 1975, 35, N9, 4595. - CA, 1975, 83, 34866.

Rev.; 1.4.

1975

1.1.53. Ballhausen, C. J., Basic theory of the electronic states of molecules. - In: Electronic States in Inorg. Compounds: New Exp. Techn. Dordrecht, 1975, p. 1-25. - PA, 1976, 79, 12908.

Theor.

1.1.54. Bersuker, I. B., Vekhter, B. G., Ogurstov, I. Ya., Tunneling effects in polyatomic systems with electronic degeneracy and pseudodegeneracy. [Russ.]. - Usp. Fiz. Nauk, 1975, 116, N4, 605-641. - PA, 1976, 79, 87253. RZF, 1975, 12E45.

Rev.

1.1.55. Bersuker, I. B., The Jahn-Teller effect in crystal chemistry and spectroscopy. - Coord. Chem. Rev., 1975, 14, N4, 357-412. - CA, 1975, 82, 161931.

Rev.

1.1.56. Burrows, E. L., Clark, M. J., Kettle, S. F. A., A simple rule for the prediction of Jahn-Teller active vibrational modes. - J. Chem. Phys., 1975, 63, N9, 4086-4087. - CA, 1975, 83, 211087. PA, 1976, 79, 17133. RZK, 1976, 6B67.

Theor.; I_h; D_{4d}; C_{8v}; C_{nv}.

1.1.57. Cianchi, L., Mancini, M., Moretti, P., The Jahn-Teller effect in localized centers: a review of the theory. - Riv. Nuovo Cimento, 1975, 5, N2, 187-254. - Bibliogr.: 121 ref. CA, 1976, 84, 10961. PA, 1976, 79, 31067.

Rev.

1.1.58. Pearson, R. G., Concerning Jahn-Teller effects. - Proc. Natl. Acad. Sci. USA, 1975, 72, N6, 2104-2106. - CA, 1975, 83, 87550. PA, 1975, 78, 84640.

Theor.; 1.5.

1.1.59. Stevens, K. W. H., Bates, C. A., Crystal field theory (and magnetic oxides). - In: Magn. Oxides. Pt. I. Chichester, 1975, p. 141-180. - PA, 1975, 78, 62177.

Rev.

1.1.60. Zhilinskiy, B. I., Stepanov, N. F., The singular perturbation of the two-level electron equation with degeneracy. [Russ.]. - in: Mat. Probl. Khim. Novosibirsk, 1975, Pt. 2, p. 211-216. - RZF, 1975, 9D98.

Theor.

1976

1.1.61. Aronowitz, S., Adiabatic approximation and the Jahn-Teller theorem. - Phys. Rev. A, 1976, 14, N4, 1319-1325. - CA, 1976, 85, 182655. PA, 1977, 80, 10575.

Theor.

1.1.62. Baranowski, J., Uba, S., The Jahn-Teller effect. [Pol.]. - Postepy Fiz., 1976, 27, N5, 437-448. - CA, 1977, 86, 98307.

Theor.; d^7.

1.1.63. Bersuker, I. B., Elektronnoye Stroeniyi i Svoystva Koordinatsionnykh Soedineniy: Vvedeniye v Teoriyu. [Russ.]. - Izd. 2, Pererab. i Dop. - Leningrad: Khimiya, 1976. - 349 p. - RZK, 1976, 15B35K.

Electronic Structure and Properties of Coordination Compounds: Introduction to the Theory.

1.1.64. Hughes, A. E., Optical techniques and an introduction to the symmetry properties of point defects. - in: Defects. Their Struct. Nonmet. Solids, 1976, p. 133-154. (NATO Adv. Study Inst. Ser., Ser. B, 1975, 19). - Bibliogr.: 56 ref. CA, 1977, 86, 162850.

Rev.

1.1.65. Judd, B. R., Lie groups and the Jahn-Teller effect for a color center. - In: Group Theor. Methods in Phys.: 4th Int. Colloq. Nijmegen, 1975. Berlin, 1976, p. 312-321. (Lect. Notes in Phys.; Vol. 50.) - CA, 1976, 85, 151095. PA, 1977, 80, 20279.

Theor.; O_h.

1.1.66. Kiselev, A. A., Adiabaticheskaya Teorya Vozmushcheniya v Zadachakh Molekulyarnoy Spektroskopii: Avtoref. Dis. ... Dokt. Fiz.-Mat. Nauk. [Russ.]. - Leningrad, 1976, - 22 p. - Leningr. Univ..

The Adiabatic Perturbation Theory in Molecular Spectroscopy Problems.

1977

1.1.67. Azumi Tohru, Matsuzaki Kazuo, What does the term "vibronic coupling" mean? - Photochem. & Photobiol., 1977, 25, N3, 315-326. - CA, 1977, 87, 73428.

Theor.; 1.5.

1.1.68. Bersuker, I. B., Problems of quantum chemistry of coordination compounds. [Russ.]. - In" Fizika Molekul: Resp. Mezhved. Sb. Kiev, 1977, N5, p. 58-69. - CA, 1978, 89, 80299.

Theor.; 1.5; 2.1; 3.1; 4.6; 5.1.

1.1.69. Judd, B. R., The Jahn-Teller effect for degenerate systems. - In: Spectroscopie des elements ourds dans less solides. - Paris, 1977, p. 127-132. (Colloq. Int. CNRS, N255).

Theor.; $\Gamma_8 - e$; $\Gamma_8 - t$; $\Gamma_8 - (e + t)$; O_h; 1.3.

1.1.70. Kiselev, A. A., Lyuders, K., On the symmetry group of the nonrigid impurity center. [Russ.]. - In: Fiz. Mat. Metody Koord. Khim.: Tez. Dokl. 6 Vsesoyuz. Soveshch. Kishinev, 1977, p. 167-168.

Theor.

1.1.71. Sturge, M. D., The Jahn-Teller effect. - In: McGraw-Hill Encyclopedia of Science and Technology. New York, 1977, p. 336a-336b.

1.1.72. Zhilinskiy, B. I., Stepanov, N. F., Stability of polyatomic molecules in degenerated electronic states with respect to the redistribution of effective charge. [Russ.]. - Koord. Khim., 1977, 3, N3, 321-326. - CA, 1977, 87, 11863.

Theor.

1978

1.1.73. Kiselev, A. A., Adiabatic perturbation theory in molecular spectroscopy. - Can. J. Phys., 1978, 56, N6, 615-647. - RZF, 1978, 11D139.

Rev.; 1.6.

1.1.74. Kristofel', N. N., On the adiabatic approximation at finite temperatures. [Russ.]. - Fiz. Tverd. Tela, 1978, 20, N2, 618-620.

Theor.

1.1.75. Pooler, D. R., Continuous group invariances of linear Jahn-Teller systems. - J. Phys. A., 1978, 11, N6, 1045-1055. - PA, 1978, 81, 60092.

Theor.

1.1.76. Zhilinskii, B., The Jahn-Teller theorem. [Fr.]. - C.R. Hebd. Seances Acad. Sci. Ser. B, 1978, 287, N10, 249-251. - CA, 1979, 90, 94540. PA, 1979, 82, 31421.

Theor.

1979

1.1.77. Ogawa Tohru, Nara Shigetoshi, The electronic viewpoint on the morphology of the microclusters. II. The conflict of the Hund tendency and the Jahn-Teller tendency. - Z. Phys. B, 1979, 33, N1, 69-77. - CA, 1979, 90, 157360. PA, 1979, 82, 46996.

Theor.

See also: 1.3-56; 1.4-14; 1.5-1, 46, 50, 51.

1.2. Adiabatic Potentials

The first investigations of adiabatic potentials of Jahn-Teller degenerate electronic terms were made in 1939 [1.1.6, 5.2.1]. The review article [2.1.13] published in 1941 also contains interesting results. It was shown that the Jahn-Teller adiabatic potentials possess a complicated many-valley shape. Its qualitative features may be obtained by investigating the extremum points of the adiabatic potentials, their coordinates and energy depths. In 1957 the important paper [1.2.1] appeared where, among other interesting results, a special method of extremum point investigaton for Jahn-Teller adiabatic potentials was developed.

At the same time, at the end of the fifties, an interesting series of works was published by Liehr [1.2.-2, 3, 4, 5, 7, 8, 9, 10, 12], who considered a large number of different Jahn-Teller and pseudo-Jahn-Teller molecular situations with different multiplicity of electronic (pseudo) degeneracy and for different symmetries. We note the papers of Moffitt and Liehr [1.3.1] and of Ballhausen and Liehr [1.2.5] where the quadratic vibronic coupling is shown to be the possible cause of the tetragonal minima that occur on the lowest sheet of the cubic (or trigonal) E term adiabatic potential. The adiabatic potentials of the electronic E term of tetragonal molecules (possessing one four-fold symmetry axis) coupled to B_1 and B_2 nuclear vibrations has been investigated in Refs. [1.6.8] and [1.3.18].

The extremum points of the adiabatic potential of an electronic T term coupled to E and T_2 vibrations were studied by Öpik and Pryce [1.2.1]. O'Brien [1.3.33] considered the special case of this problem where the Jahn-Teller stabilization energy due to vibronic coupling to E and to T_2 vibrations, respectively, has the same value. In this case the five-dimensional adiabatic potential has been shown to possess a two-dimensional trough.

The adiabatic potential in the case of a T term coupled to T_2 vibrations only, taking into account linear and quadratic vibronic interaction, was studied by Muramatsu and Iida [1.2.19]. It has been shown that, for some relations between the vibronic constants, as distinguished from the linear coupling case, orthorhombic-type minima on the lowest sheet of the adiabatic potential may occur. More complicated situations, in particular the coexistence of different types of minima, were shown to be possible in the papers [1.2.-30, 35, 36], provided other possible interactions for a cubic T term (e.g., linear, quadratic and higher-order vibronic coupling to both E and T_2 vibrations, spin-orbital interaction, etc.) are taken into account. Similar results for a tetragonal E term have been obtained in Ref. [1.2.45].

1957

1.2.1. Öpik, U., Pryce, M. H. L., Studies of the Jahn-Teller effect. I. A survey of the static problem. - Proc. R. Soc. London. Ser. A, 1957, 238, N1215, 425-447. - CA, 1957, 51, 15247.

Theor.; E - e; T - (e + t);.1.5.

1958

1.2.2. Liehr, A. D., Ballhausen, C. J., Inherent configurational instability of octahedral inorganic complexes in E_g electronic states. - Ann. Phys. (USA), 1958, 3, N2, 304-319.

Theor.; E - e; O_h; d^1; d^4; d^6; d^8; 3.1.

1960

1.2.3. Liehr, A. D., "Ionic radii," spin-orbit coupling and the geometrical stability of inorganic complexes. - Bell. Syst. Tech. J., 1960, 39, N6, 1617-1626. - RZF, 1961, 6V41.

Theor.; O_h.

1.2.4. Liehr, A. D., Semiempirical theory of vibronic interactions in some simple conjugated hydrocarbons. - Rev. Mod. Phys., 1960, 32, N2, 436-439. - CA, 1961, 55, 420e. - RZF, 1961, 3V35.

Theor.; $^2E_{1g}$; D_{6h}.

1961

1.2.5. Ballhausen, C. J., Liehr, A. D., An application of the Jahn-Teller theorem to the VCl_4 molecule. - Acta Chem. Scand., 1961, 15, N4, 775-781. - CA, 1962, 56, 20076. RZF, 1962, 4V28.

Theor.; E - e, T_d.

1962

1.2.6. Avvakumov, V. I., Distortion of the complexes $Cr^{3+}(H_2O)_6$ and $Ni^{2+}(H_2O)_6$ and the splitting of the spin levels of Cr^{3+} and Ni^{2+} due to the Jahn-Teller phenomenon. [Russ.]. - Opt. & Spektrosk., 1962, 13, N4, 588-591. - CA, 1963, 58, 5158a. RZF, 1963, 2E330.

Theor.; T - (e + t); O_h; d^3; d^8.

1.2.7. Liehr, A. D., An essay on higher-order vibronic interactions. - In: Ann. Rev. Phys. Chem. Stanford, 1962, Vol. 13, p. 41-76.

Rev.; 1.5; 1.6.

1.2.8. Liehr, A. D., A comparison of the theories of molecular orbitals, valence bond and ligand field. - J. Chem. Educ., 1962, 39, N3, 135-139.

Theor.

1.2.9. Liehr, A. D., The coupling of vibrational and electronic motions in degenerate electronic states of inorganic complexes. Part I. States of double degeneracy. - In: Prog. Inorg. Chem., New York; London, 1962, Vol. 3, p. 281-314.

Rev.; E - $(b_1 + b_2)$; E - e.

1.2.10. Liehr, A. D., The coupling of vibrational and electronic motions in degenerate electronic states of inorganic complexes. Part II. States of triple degeneracy and systems of lower symmetry. - In: Prog. Inorg. Chem., New York; London, 1962, Vol. 4, p. 455-540.

Rev.; T; T_d; O_h; I; 2.1; 2.7; 2.8; 3.1; 3.6; 5.2.

1963

1.2.11. Herzberg, G., Longuet-Higgins, H. C., Intersection of potential energy surfaces in polyatomic molecules. - Discuss. Faraday Soc., 1963, 35, N7, 77-82.

Theor.

1.2.12. Liehr, A. D., Topological aspects of the conformational stability problem.

Pt. 1. Degenerate electronic states. ; J. Phys. Chem., 1963, 67, N2, 389-471.

Pt. 2. Non-degenerate electronic states. - Ibid., 471-494.

Rev.; 1.6.

1964

1.2.13. Hatfield, W. E., Piper, T. S., Distortions about six-coordinated ferrous ion. - Inorg. Chem., 1964, 3, N9, 1295-1298. - CA, 1964, 61, 9053c.

Theor.; $^5T_2$; O_h; d^6.

1.2.14. Liehr, A. D., Conformational instability of non-cubosymmetric inorganic compounds in degenerate electronic states. - In: Theory and Struct. Complex Compounds: Papers Symp., Wroclaw, 1962. London; New York, 1964, p. 95-135. - CA, 1965, 63, 12508e.

Rev.; 1.5.

1966

1.2.15. Bir, G. L., The Jahn-Teller effect for impurity centers in semiconductors. [Russ.]. - Zh. Eksp. & Teor. Fiz., 1966, 51, N2, 556-569. - CA, 1967, 66, 15023. RZF, 1967, 1E456.

Theor.; E; T_1; T_2; Γ_8.

1.2.16. Englman, R., Covalency effects in trigonal distortion. - J. Chem. Phys., 1966, 45, N10, 3862-3869. - CA, 1967, 66, 23748.

Theor.

1968

1.2.17. Wysling, P., Muller, K. A., Stable potential minima and the Jahn-Teller interaction. - Phys. Rev., 1968, 173, N2, 327-332. - PA, 1969, 72, 751.

Theor.

1969

1.2.18. Stoneham, A. M., Lannoo, M., The Jahn-Teller instability with accidental degeneracy. - J. Phys. & Chem. Solids, 1969, 30, N7, 1769-1777. - CA, 1969, 71, 74190. PA, 1969, 72, 45215. RZF, 1970, 1D109.

Theor.; E; T; O_h; T_d.

1970

1.2.19. Muramatsu Shinji, Iida Takeshi, Possibility of orthorhombic Jahn-Teller distortion. - J. Phys. & Chem. Solids, 1970, 31, N10, 2209-2216. - CA, 1970, 73, 135442. PA, 1970, 73, 79217.

Theor.; T - t.

1.2.20. Zgierski, M. Z., Effective potential curves for degenerate vibrations in the Jahn-Teller effect. - Acta Phys. Pol. A, 1970, 38, N2, 185-188. - CA, 1971, 74, 34811. PA, 1970, 73, 70843. RZF, 1971, 5D77.

Theor.; E; C_{3v}.

1972

1.2.21. Dixon, J. M., Smith, R. M., Anharmonicity and the Jahn-Teller effect. - J. Phys. C, 1972, 5, N20, 2941-2948. - PA, 1972, 75, 80957. RZF, 1973, 3E438.

Theor.; E - e.

1.2.22. Leonardi, C., Messina, A., Persico, F., Ground state of the spin-phonon interaction. - J. Phys. C, 1972, 5, N16, L218-L222.

Theor.

1.2.23. Ranfagni, A., Jahn-Teller effect in the emission of thallium-doped potassium-iodide-type phosphors, coexistence of two kinds of minima on the $^3T_{1u}$ adiabatic potential-energy surfaces. - Phys. Rev. Lett., 1972, 28, N12, 743-746. - CA, 1972, 77, 158178. PA, 1972, 75, 33314.

Theor.; T - (e + t).

<u>1973</u>

1.2.24. Bersuker, I. B., Vekhter, B. G., Rafalovich, M. L., Stable configurations of the bioctahedral complex with Jahn-Teller centers. [Russ.]. - Kristallografiya, 1973, <u>18</u>, N1, 11-18. - CA, 1973, <u>78</u>, 128590. PA, 1973, <u>76</u>, 72035. RZF, 1973, 5E856.

Theor.; E.

1.2.25. Bersuker, I. B., Polinger, V. Z., The second-order T - (e - t_2) problem in the Jahn-Teller effect theory: A new type of adiabatic potential minima and inversion (tunneling) splitting. - Phys. Lett. A, 1973, <u>44</u>, N7, 495-496. - PA, 1973, <u>76</u>, 68551. RZF, 1974, 1E437.

Theor.; T - (e + t_2); 1.4.

1.2.26. Dushin, R. B., Shcherba, L. D., Parameters of electron-vibrational interaction in the static Jahn-Teller effect. [Russ.]. - In: Abstr. 11th Eur. Congr. Mol. Spectrosc. Tallin, 1973, 58. - PA, 1973, <u>76</u>, 67765.

Theor.

1.2.27. Ranfagni, A., Pazzi, G. P., Fabeni, P., Viliani, G., Fontana, M. P., A possible interpretation of the double luminescence of KI:Tl-type phosphors excited in the A band. - In: Abstr. 11th Eur. Cong. Mol. Spectrosc. Tallin, 1973, 44(B2).

Theor.; T - (a + e + t_2); O_h; 1.3.

1.2.28. Tachiya, M., Tabata, Y., Oshima, K., Configuration coordinate model for the hydrated electron. II. Jahn-Teller splitting of the excited state. - J. Phys. Chem., 1973, <u>77</u>, N19, 2286-2290. - CA, 1973, <u>79</u>, 129298. PA, 1974, <u>77</u>, 29365. RZF, 1974, 2D107.

Theor.

<u>1974</u>

1.2.29. Bacci, M., Fabeni, P., Fontana, M. P., Pazzi, G. P., Ranfagni, A., Viliani, G., Possibility for the coexistence of tetragonal with orthorhombic or trigonal Jahn-Teller distortions in alkali halide phosphors. - In: Extended Abst. Int. Conf. Color Centers in Ionic Crystals. Sendai, 1974, G133, 2 p. - PA, 1975, <u>78</u>, 18567.

Theor.; T - (e + t).

1.2.30. Bersuker, I. B., Polinger, V. Z., Jahn-Teller effect for the T-term. [Russ.]. - Zh. Eksp. & Teor. Fiz., 1974, <u>66</u>, N6, 2078-2091. - CA, 1974, <u>81</u>, 112963. PA, 1975, <u>78</u>, 12934. RZF, 1974, 11E596.

Theor.; T - (e + t); 1.4; 4.1.

1.2.3. Ranfagni, A., Viliani, G., Quadratic Jahn-Teller effect in the emision of KI:Tl-type phosphors. - Phys. Rev. B, 1974, 9, N10, 4448-4454. - CA, 1974, 81, 56117. PA, 1974, 77, 70355. RZF, 1974, 11E968.

Theor.; ($^3T_{1u}$ + $^3A_{1u}$) - e_g; 1.5.

1.2.32. Ranfagni, A., Viliani, G., Two kinds of Jahn-Teller distortions in an octahedral complex with the $^3T_1$ and $^1T_1(a_{1g}t_{1u})$ electronic states coupled to E_g vibration modes. - J. Phys. & Chem. Solids, 1974, 35, N1, 25-31. - CA, 1974, 80, 126341. PA, 1974, 77, 24703. RZF, 1974, 7E909.

Theor.; T - e; 2.2.

1.2.33. Ranfagni, A., Pazzi, G. P., Fabeni, P., Viliani, G., Fontana, M. P., Large quadratic interactions for the coexistence of two kinds of Jahn-Teller distortions in KI:Tl-type phosphors. - Solid State Commun., 1974, 14, N11, 1169-1172. - PA, 1974, 77, 57041. RZF, 1974, 12E850.

Theor.; T - (e + t).

1975

1.2.34. Bacci, M., Fontana, M. P., Ranfagni, A., Viliani, G., Coexistence of tetragonal with orthorhombic or trigonal Jahn-Teller distortions in an O_h complex. - Phys. Lett. A, 1975, 50, N6, 405-406. - PA, 1975, 78, 25613. RZF, 1975, 7E66.

Theor.; T - (e + t); O_h.

1.2.35. Bacci, M., Ranfagni, A., Fontana, M. P., Viliani, G., Coexistence of tetragonal with orthorhombic or trigonal Jahn-Teller distortions in an O_h complex: Plausible interpretation of alkali-halide phosphor luminescence. - Phys. Rev. B, 1975, 11, N8, 3052-3059. - CA, 1975, 83, 18183. PA, 1975, 78, 51025. RZF, 1975, 10D541.

Theor.; T - (e + t); 2.2.

1.2.36. Bacci, M., Ranfagni, A., Cetica, M., Viliani, G., Coexistence of tetragonal with orthorhombic or trigonal Jahn-Teller distortions in an O_h complex. II. Effect of anharmonicity. - Phys. Rev. B, 1975, 12, N12, 5907-5911. - CA, 1976, 84, 67035. PA, 1976, 79, 35785. RZK, 1976, 13B31.

Theor.; T - (e + t); O_h.

1.2.37. Bhattacharyya, B. D., Dynamic Jahn-Teller effect in nickel (1+):I_B-III-VI_2 chalcopyrite semiconductors. - In: Proc. Nucl. Phys. & Solid State Phys. Symp., 1975, 18c, p. 186-188. - CA, 1977, 87, 93154.

Theor.; E - e; T_d; d^9.

1.2.38. Khlopin, V. P., Polinger, V. Z., Bersuker, I. B., The Jahn-Teller effect in icosahedral molecules and complexes. [Russ.]. - In: Kvant. Khim.: Tez. Dokl. 6 Vsesoyuz. Soveshch. Kishinev, 1975, X-232. - CA, 1976, 84, 128139. RZK, 1975, 24B83.

Theor.; T - v; I.

1976

1.2.39. Bacci, M., Ranfagni, A., Cetica, M., Viliani, G., Effect of anharmonicity on the T × (ε_g + τ_{2g}) Jahn-Teller problem. - In: Proc. 12th Eur. Congr. Mol. Spectrosc. Amsterdam, 1976, p. 503-506. - CA, 1977, 86, 23712.

Theor.; T - (e + t_2).

1.2.40. Kapelewski, J., Jahn-Teller effect for surface impurities in crystals of perovskite structure. [Pol.]. - Biul. Wojsk. Akad. Tech., 1976, 25, N11, 115-129. - CA, 1977, 86, 63757.

Theor.

1.2.41. Kayanuma Yosuke, Toyozawa Yutaka, Vibronic problem for the relaxed excited state of the F-center in alkali halides. I. Vibronic energy schemes. - J. Phys. Soc. Jpn., 1976, 40, N2, 355-362.

Theor.; (s + p) - (e_g + t_{2g}); O_h.

1.2.42. Stepanov, N. F., Zhilinskii, B. I., Random degeneracy of electron terms: Model nonadiabatic equation. [Russ.]. - In: Fizika Molekul: Resp. Mezhved. Sb. Kiev, 1976, Vol. 2, p. 25-41. - CA, 1977, 86, 146180.

Theor.

1977

1.2.43. Kapelewski, J., The Jahn-Teller effect for surface impurities in crystals of perovskite structure. - J. Tech. Phys., 1977, 18, N2, 239-253. - CA, 1978, 88, 112622.

Theor.

1.2.44. Khlopin, V. P., Polinger, V. Z., Bersuker, I. B., The Jahn-Teller effect in icosahedral molecules and complexes. [Russ.]. - In: Fizika Molekul: Resp. Mezhved. Sb. Kiev, 1977, Vol. 5, p. 70-76, 103. - CA, 1978, 89, 82217.

Theor.; T - v; U - v.

1978

1.2.45. Bacci, M., Coexistence of two kinds of stable Jahn-Teller distortions in molecules with a fourfold symmetry axis. - Phys. Rev. B, 1978, 17, N12, 4495-4498. - CA, 1978, 89, 153081. PA, 1978, 81, 84176.

Theor.; E - $(a + b_1 + b_2)$; D_{4h}.

1.2.46. Judd, B. R., Ligand trajectories for a degenerate Jahn-Teller system. - J. Chem. Phys., 1978, 68, N12, 5643-5646. - CA, 1978, 89, 119852. PA, 1978, 81, 76028.

Theor.; T - $(e + t_2)$; O_h.

1.2.47. Khlopin, V. P., Polinger, V. Z., Bersuker, I. B., The Jahn-Teller effect in icosahedral molecules and complexes. - Theor. Chim. Acta, 1978, 48, N2, 87-101. - CA, 1978, 89, 33418. PA, 1978, 81, 78642.

Theor.; T - v; U - v; V - v; I.

1979

1.2.48. Askarov, B., Oksengendler, B. L., Yunusov, M. S., The atomic effects of vibronic interaction in semiconductors. [Russ.]. - in: Vlyaniye Nesovreshenstv Struktury na Svoystva Kristallov. Tashkent, 1979, p. 3-10. - CA, 1980, 92, 119923. RZK, 1979, 23B614.

Theor.

1.2.49. Iogensen, L. V., Malov, V. V., Competition between Jahn-Teller and Zeeman splitting for π-electron singlet terms of cyclic molecules. [Russ.]. - Zh. Fiz. Khim., 1979, 53, N7, 1883-1885. - CA, 1979, 91, 165789.

Theor.; D_{4h}.

1.2.50. Judd, B. R., The F^+ center in CaO. - J. Lumin., 1979, 18/19, N2, 868-870. - CA, 1979, 90, 129656. PA, 1979, 82, 42499.

Theor.; T - $(e + t_2)$; O_h.

1.2.51. Ranfagni, A., Mugnai, D., Bacci, M., Montagna, M., Pilla, O., Viliani, G., Coexistence of tetragonal with orthorhombic or trigonal Jahn-Teller distortions in an O_h complex. III. Effect of totally symmetrical vibrational mode. - Phys. Rev. B, 1979, 20, N12, 5358-5365. - PA, 1980, 83, 49680.

Theor.; T - (e + t); O_h; 5.3.

1.2.52. Vikhnin, V. S., Sochava, L. S., The off-center Jahn-Teller ion: coupled polar and tetragonal deformation. [Russ.]. - Fiz. Tverd. Tela, 1979, 21, N17, 2083-2090. - CA, 1979, 91, 165986. PA, 1980, 83, 44555. RZF, 1979, 10E320.

Theor.; E - e; O_h; d^9.

See also: 1.1.-1, 4, 6, 47, 74; 1.3.-1; 1.5.-2, 9, 12, 43, 45; 1.6.-2, 19; 1.7.-51; 1.8.-6, 11; 2.1.-3, 4, 291; 2.2.-81, 82, 106, 115, 157; 2.4.-88; 2.5.-28; 2.7.-20; 4.2.-3, 10; 4.6.-79; 5.1.-1, 34; 5.2.-1, 93; 5.3.-8, 25; and Section 5.4.

1.3. Solutions of Vibronic Equations

The eigenvalue problem for the matrix vibronic Jahn-Teller effect Hamiltonian, as mentioned above, results in a complicated system of coupled differential equations, usually called the vibronic equations. The analytical solution of these equations is very difficult. However, for the limiting cases of weak and strong vibronic coupling, a qualitative understanding of the dynamics in the coupled electron-nuclear system can be obtained by means of approximate solutions (e.g., using perturbation theory methods).

In 1957 Moffitt and Liehr [1.3.1] considered the weak coupling limit for an E term linearly coupled to E vibrations. The role of quadratic vibronic coupling was also qualitatively discussed. At the same time, Moffitt and Thorson [1.3.2] employed a similar perturbation approach for both the E term coupled to E vibrations and the T term coupled to E and T_2 vibrations.

In 1958 Longuet-Higgins et al. published an important work [1.3.3] where the detailed numerical solution of the linear E - e problem is given. Analytical solutions for the limiting cases of weak and strong coupling are also given. In the same year Moffitt and Thorson published their paper [1.3.4] containing rather similar results in a little-known publication, and this work is rarely cited.

Another possibility for a simple analytical solution and clear physical interpretation is the case of several equivalent deep-enough minima of the adiabatic potential causing the localization of the nuclear motion in the neighborhood of the minima. In 1960 it was firstly noted that tunneling through the potential barriers dividing the minima is manifested in a characteristic splitting of the vibronic levels [1.3.7] (the splitting was then named "inversion splitting"). In 1962 this inversion (tunneling) splitting was calculated more precisely for several typical Jahn-Teller situations [1.3.10]. In this work an universal analytical method of tunneling-splitting calculations was developed. Somewhat latter in Ref. [3.1.45] the $E - e$ problem was considered taking into account linear vibronic interaction and small to moderate anharmonicity, the latter causing the warping of the adiabatic potential energy surface. By means of adiabatic separation of the angular (slow) nuclear motion from the radial (fast) one, the lowest vibronic energy levels were obtained, among which the tunneling splitting in the strong coupling limit is also present.

A general group-theoretic classification of tunneling energy levels was developed in Ref. [1.1.25].

For the case of intermediate coupling the numerical computation approach is inevitable. The works [1.3.-3, 4] were the first to introduce numerical methods in the $E - e$ problem; a similar calculation for the $T - t_2$ problem is reported in Ref. [1.3.22]. We note that numerical calculations in Jahn-Teller problems require very large sets of basis functions. It leads to some special difficulties in numerical diagonalization of Jahn-Teller Hamiltonians. One way of overcoming these difficulties is to use symmetry considerations in order to divide the basis space into irreducible subspaces. As mentioned above, Jahn-Teller problems sometimes possess infinitesimal symmetries. A detailed review of symmetry properties of Jahn-Teller systems and group-theoretic methods in the numerical computation of vibronic problems is given in Ref. [1.1.50].

Another way is to use more efficient algorithms. One of them is the so-called Lanzosh method employed in Ref. [1.3.95] for the linear $E - (b_1 + b_2)$ problem. The more complicated $T - (e + t_2)$ case has been investigated within the framework of different simplifications. O'Brien [1.3.-33, 40] assumed that both the linear vibronic constants and the frequencies have the same values for E and T_2 vibrations, respectively (the so-called d mode model). In Ref. [1.3.33] the difference between the vibronic constants V_E and V_T is considered as a small perturbation, the latter being diagonalized within a large basis set of strong-coupling wave functions. In Ref. [1.3.97] the same problem was considered using the Lanzosh method within the d mode approximation with appropriate values of coupling constants.

A numerical solution of vibronic equations for the electronic term linearly coupled to T_2 vibrations is given in Ref. [1.3.30]. An alternative quasi-classical approach to the vibronic problems, valid for large values of quantum numbers, is developed in Refs. [1.3.-63, 67].

<u>1957</u>

1.3.1. Moffitt, W., Liehr, A. D., Configurational instability of degenerate electronic states. - Phys. Rev., 1957, <u>106</u>, N6, 1195-1200.

Theor.; E - e; D_{6h}; 1.2; 1.6.

1.3.2. Moffitt, W., Thorson, W., Vibronic states of octahedral complexes. - Phys. Rev., 1957, <u>108</u>, N5, 1251-1255. - PA, 1958, <u>61</u>, 2005.

Theor.; E - e; T - (e + t); O_h.

<u>1958</u>

1.3.3. Longuet-Higgins, H. C., Öpik, U., Pryce, M. H. L., Sack, R. A., Studies of the Jahn-Teller effect. II. The dynamical problem. - Proc. R. Soc. London. Ser. A, 1958, <u>244</u>, N1, 1-16. - CA, 1958, <u>52</u>, 9745.

Theor.; E - e; 2.1.

1.3.4. Moffitt, W., Thorson, W., Some calculations related to Jahn-Teller effect. [Fr.]. - In: Calcul des fonctions d'onde moleculaire. Paris, 1958, p. 141-148. - RZK, 1959, 84810.

Theor.; E - e.

<u>1960</u>

1.3.5. Child, M. S., Dynamical Jahn-Teller effect in molecules possessing one four-fold symmetry axis. - Mol. Phys., 1960, <u>3</u>, N6, 601-603. - CA, 1961, <u>55</u>, 8039e. RZF, 1961, 9V14.

Theor.; E - (b_1 + b_2); 1.7.

1.3.6. Child, M. S., Vibrational-electronic coupling in IrF_6, $OsCl_6^{2-}$, and $IrCl_6^{2-}$. - Mol. Phys., 1960, <u>3</u>, N6, 605-607. - CA, 1961, <u>55</u>, 10061f. RZF, 1961, 9V37.

Theor.; O_h.

<u>1961</u>

1.3.7. Bersuker, I. B., Hindered motions in the transition metal complexes. [Russ.]. - Opt. & Spektrosk., 1961, <u>11</u>, N3, 319-324.

Theor.; E - e; O_h.

1.3.8. Longuet-Higgins, H. C., Some recent developments in the theory of molecular energy levels. - In: Adv. Spectrosc., New York, 1961, Vol. 2, p. 429-472. - RZF, 1962, 6V30.

Rev.; E - e; 2.7; 2.8.

1.3.9. McLachlan, A. D., The wave-functions of electronically degenerate states. - Mol. Phys., 1961, 4, N5, 417-423. - RZF, 1962, 4V18.

Theor.

1962

1.3.10. Bersuker, I. B., Inversion of splitting of energy levels in free transiton metal complexes. [Russ.]. - Zh. Eksp. & Teor. Fiz., 1962, 43, N4, 1315-1322. - RZF, 1963, 3D68.

Theor.; E - e; T - e; T - t.

1.3.11. Child, M. S., The rotational levels of electronically degenerate dihedral molecules. - Mol. Phys., 1962, 5, N4, 391-396. - RZF, 1963, 2D82.

Theor.; D_{3h}.

1.3.12. Longuet-Higgins, H. C., The vibrational spectra of Jahn-Teller molecules. - In: Proc. Int. Symp. Mol. Structure Spectr. Tokyo, 1962, B101, 4 p. - CA, 1964, 61, 7839d.

Theor.; E - e; 2.7; 2.8.

1.3.13. McLachlan, A. D., Snyder, L. C., Spin density fluctuations in the cyclooctatetrane negative ion. - J. Chem. Phys., 1962, 36, N5, 1159-1162. - RZF, 1962, 11V437.

Theor.; E - $(b_1 + b_2)$; D_{8h}.

1963

1.3.14. Bersuker, I. B., Vekhter, B. G., Titova, Yu. G., The effect of inversion splitting in tetrahedral complexes. [Russ.]. - Izv. Akad. Nauk Mold.SSR, 1963, N9, 16-23.

Theor.; T_d.

1.3.15. Bersuker, I. B., The spin-electronic vibrational problem for free transition metal complexes. [Russ.]. - Litov. Fiz. Sb., 1963, 3, N1-2, 327-340. - CA, 1964, 61, 169f.

Theor.; E - e; T - e; T - t; O_h.

1.3.16. Bersuker, I. B., The effect of inversion splitting in transition metal complexes. [Russ.]. - In: Tez. Dokl. 3 Soveshch. Kvant. Khim. Kishinev, 1963, p. 8.

Theor.

1.3.17. Ellis, D. E., Englman, R., The quadratic Jahn-Teller effect. - In: Quart. Progr. Rept. Solid State and Mol. Theory Group. Mass. Inst. Technol., 1963, N48, p. 61-66. - RZF, 1966, 2D82.

Theor.

1965

1.3.18. Ballhausen, C. J., Jahn-Teller configuratinal instability in square-planar complexes. - Theor. Chim. Acta, 1965, 3, N4, 368-374. - CA, 1965, 63, 12509b. RZF, 1966, 7D98.

Theor.; E - $(b_1 + b_2)$; E - b_1; 2.1.

1.3.19. Child, M. S., Strauss, H. L., Causes of 1-type doubling in the 3p (E") Rydberg state of ammonia. - J. Chem. Phys., 1965, 42, N7, 2293-2292. - CA, 1965, 62, 12605h.

Theor.; E - e; T_d.

1.3.20. Slonczewski, J. C., Resonant excited states in the dynamic Jahn-Teller effect. - Helv. Phys. Acta, 1965, 38, N4, 359. - RZF, 1966, 1D506.

Theor.; E.

1966

1.3.21. Bersuker, I. B., Vekhter, B. G., Jahn-Teller distortions and inversion splitting for transition metal complexes and local centers in crystals with degenerate electronic T-terms. - Phys. Status Solidi, 1966, 16, N1, 63-68. - CA, 1966, 65, 9887e. RZF, 1966, 12E562.

Theor.; T - t.

1.3.22. Caner, M., Englman, R., Jahn-Teller effect on a triplet due to threefold degenerate vibrations. - J. Chem. Phys., 1966, 44, N10, 4054-4055. - CA, 1966, 65, 9925f. PA, 1966, 69, 23249. RZF, 1966, 12E298.

Theor.; T - t; 1.4.

1.3.23. Coffman, R. E., Operator form for the linear and quadratic Jahn-Teller coupling in octahedrally coordinated E electronic states. - J. Chem. Phys., 1966, 44, N6, 2305-2306. - CA, 1966, 64, 13553d. RZF, 1966, 8E382.

Theor.; E - e; O_h; d^9.

1.3.24. Lepard, D. W., The Coriolis ζ sum rules. - Can. J. Phys., 1966, 44, N3, 461-466. - PA, 1966, 69, 17545.

Theor.; D_{nd}; S_{2n}.

1.3.25. Struck, C. W., Herzfeld, F., Vibrational levels of the ^2D - E state of Ce^{3+} in CaF_2. - J. Chem. Phys., 1966, 44, N2, 464-468. - CA, 1966, 64, 7530d. RZF, 1966, 7D127.

Exp.; E - e; O_h; 2.1.

1.3.26. Uehara Hiromichi, Vibronic effect on the mean-square amplitudes of internuclear distances in a tetrahedral molecule. - J. Chem. Phys., 1966, 45, N12, 4536-4542. - CA, 1967, 66, 32094. PA, 1967, 70, 6605.

Theor.; E - e; T_d.

1967

1.3.27. Perrin, M. H., Gouterman, M., Vibronic coupling. IV. Trimers and trigonal molecules. - J. Chem. Phys., 1967, 46, N3, 1019-1022. - PA,1967, 70, 13757.

Theor.

1.3.28. Van Eekelen, H. A.M., Stevens, K. W. H., On the use of Green functions for the calculation of Jahn-Teller effects. - Proc. Phys. Soc. London, 1967, 90, Pt. I, 199-205. - CA, 1967, 66, 42046. RZF, 1967, 7D507.

Theor.; 3.1.

1.3.29. Waals, J. H., van der, Berghuis, A. M. D., Groot, M. S., de, Vibronic interaction in the lower electronic states of benzene. I. Review of the static problem and solution of the pseudocylindrical vibronic equations. - Mol. Phys., 1967, 13, N4, 301-321. - CA, 1968, 68, 109453.

Theor.; $(^3E_{1u} + {}^3B_{1u}) - e_{2g}$; D_{6h}; 1.5.

1968

1.3.30. Thorson, W., Moffitt, W., Some calculations on the Jahn-Teller effect in octahedral systems. - Phys. Rev., 1968, 168, N2, 362-365. - PA, 1968, 71, 29687.

Theor.; $\Gamma_8 - t_2$.

1969

1.3.31. Alper, J. S., Silbey, R., On the Jahn-Teller and pseudo-Jahn-Teller effect. - J. Chem. Phys., 1969, 51, N7, 3129-3130. - CA, 1969, 71, 128835. PA, 1970, 73, 20109. RZF, 1970, 5D146.

Theor.; E - e; 1.5.

1.3.32. Bates, C. A., Dixon, J. M., Jahn-Teller theory of Cr^{2+} in corundum. - J. Phys. C, 1969, 2, N12, 2209-2224. - CA, 1970, 72, 26020. PA, 1970, 73, 31575.

Theor.; E - e; C_{3v}; D_4.

1.3.33. O'Brien, M. C. M., Dynamic Jahn-Teller effect in an orbital triplet state coupled to both E_g and T_{2g} vibrations. - Phys. Rev., 1969, 187, N2, 407-418. - CA, 1970, 72, 61074. PA, 1970, 73, 25994.

Theor.; T - (e + t); 1.2; 1.4.

1.3.34. Purins, D., Karplus, M., Spin delocalization and vibrational-electronic interaction in the toluene ion-radicals. - J. Chem. Phys., 1969, 50, N1, 214-218.

Theor.; D_{6h}.

1970

1.3.35. Alper, J. S., Silbey, R., Vibronic interactions in the anions of benzene and substituted benzenes. - J. Chem. Phys., 1970, 52, N2, 569-579. - CA, 1970, 72, 70742. PA, 1970, 73, 25118.

Theor.; E - e; D_{6h}; 1.5.

1.3.36. Boese, F. K., Wagner, M., Influence of the Jahn-Teller effect on the U-center local mode absorption. - Z. Phys., 1970, 235, N2, 140-147. - PA,1970, 73, 58828. RZF, 1970, 11D632.

Theor.; T - t.

1.3.37. Mizuhashi Seiji, Numerical analysis of dynamical Jahn-Teller effect. - Rep. Comput. Cent. Univ. Tokyo, 1970, 3, NI-2, 7-18. - PA, 1972, 75, 78056. RZF, 1972, 2E494.

Theor.

1.3.38. Vallin, J. T., Dynamic Jahn-Teller effect in the orbital ^5E state of Fe^{2+} in CdTe. - Phys. Rev. B, 1970, 2, N7, 2390-2397. - PA, 1971, 74, 5767.

Theor.; ^5E - e; d^6.

1971

1.3.39. Brown, J. M., Rotational energy levels of symmetric top molecules in ^2E states. - Mol. Phys., 1971, 20, N5, 817-834. - CA, 1971, 75, 27655. PA, 1971, 74, 32564.

Theor.; E - e.

1.3.40. O'Brien, M. C. M., The Jahn-Teller effect in a p state equally coupled to E_g and T_{2g} vibrations. - J. Phys. C, 1971, 4, N16, 2524-2536. - CA, 1972, 76, 39683. PA, 1972, 75, 7012. RZF, 1972, 4E405.

Theor.; T - (e + t); 1.4; 2.1.

1.3.41. Sloane, C. S., Silbey, R., Many-body approach to the Jahn-Teller problem. - J. Chem. Phys., 1971, 55, N6, 3053-3054. - PA, 1971, 74, 72283.

Theor.; E - e.

1972

1.3.42. Bersuker, I. B., Tunneling effects in electronically degenerate E-term systems. [Russ.]. - In: Phys. Impurity Cent. Cryst.: Proc. Int. Semin. "Selected Probl. Theory Impurity Cent. Cryst.," 1970. Tallin, 1972, p. 479-482. - RZF, 1973, 1D425.

Theor.; E - e.

1.3.43. Dixon, J. M., Strong Jahn-Teller coupling of ionic E_g orbitals to a cluster in the harmonic approximation. - Phys. Kondens. Mater., 1972, 15, N3, 237-246. - PA, 1973, 76, 8115. RZF, 1973, 5E912.

Theor.; E - e.

1.3.44. Fletcher, J. R., A variational Jahn-Teller ground state: comparison with exact results. - Solid State Commun., 1972, 11, N5, 601-603. - PA, 1972, 75, 72234. RZF, 1973, 2E454.

Theor.; E - e; 1.4.

1.3.45. Wagner, M., Nonlinear canonical transformations in the coupling between degenerate high- and low-energy excitations: Dynamical resonances. - Z. Phys., 1972, 256, N4, 291-308. - PA, 1973, 76, 4014.

Theor.; E - e.

1973

1.3.46. Babcenco, A., Canonical transformation approach to the intermediate-coupling T - t Jahn-Teller effect. - Solid State Commun., 1973, 12, N8, 799-801. - CA, 1973, 78, 166253. PA, 1973, 76, 33811. RZF, 1973, 9E348.

Theor.; T - t.

1.3.47. Bersuker, I. B., Polinger, V. Z., The linear Jahn-Teller effect for an orbital triplet. - Phys. Status Solid b, 1973, 60, N1, 85-96. - CA, 1974, 80, 8520. PA, 1974, 77, 1204. RZF, 1974, 4D110.

Theor.; T - (e + t); 1.4.

1.3.48. Bhattacharyya, B. D., Jahn-Teller manifestation in an orbital triplet coupled to E_g and T_{2g} modes. - Phys. Status Solidi b, 1973, 57, N2, K149-K153. - CA, 1973, 79, 71589. PA, 1973, 76, 46952. RZF, 1973, 12E1369.

Theor.; T - (e + t); O_h; d^1.

1.3.49. Grevsmuhl, U., Wagner, M., A nonlinear canonical transformation for the dynamical Jahn-Teller problem in cubic symmetry (optical resonance effect). - Phys. Status Solidi b, 1973, 58, N1, 139-148. - PA, 1973, 76, 55608. RZF, 1974, 2E402.

Theor.; T - t; 2.1.

1.3.50. Halperin, B., Englman, R., Absence of isotopic stabilization in Jahn-Teller systems. - Solid State Commun., 1973, 13, N8, 1185-1187. - CA, 1974, 80, 31991. PA, 1974, 77, 1207. RZF, 1974, 4E545.

Theor.; E - e; O_h; d^7.

1.3.51. Kristoffel, N. N., Sigmund, E., Wagner, M., Utilization of canonical transformation for the vibronic self-energies of coupled impurity centers. - Z. Naturforsch. a, 1973, 28, N11, 1782-1786. - CA, 1974, 80, 75194.

Theor.

1974

1.3.52. Polinger, V. Z., Elektronno-Kolebatel'noye Vzaimodeystviye v Mnogoatomnykh Sistemakh s Elektronno-Vyrozhdennym T-Termom: Avtoref. Dis. ... Kand. Fiz.-Mat. Nauk. [Russ.]. - Kishinev, 1974. - 20 p. - Akad. Nauk Mold.SSR.

Electron-Vibrational Interaction in Polyatomic Systems with Electronic Degenerate T-Terms.

Theor.; T - (e + t); 1.4.

1.3.53. Polinger, V. Z., Tunneling splitting in tetragonally and trigonally distorted Jahn-Teller systems. [Russ.]. - Fiz. Tverd. Tela, 1974, 16, N9, 2578-2583. - PA, 1975, 78, 46576. RZF, 1975, 1E171.

Theor.; E - e; T - (e + t_2).

1.3.54. Purins, D., Feeley, H. F., Jahn-Teller effect in the cyclopentadienyl radical: an example of three-mode coupling. - J. Mol. Struct., 1974, 22, N1, 11-27. - CA, 1974, 81, 119662. RZF, 1974, 12D157.

Theor.; E - e; D_{5h}; 2.1.

1.3.55. Ray, T., Regnard, J. R., Dynamical Jahn-Teller effects in the ground $^4T_{1g}$ and the excited $^4T_{2g}$ orbital triplets of cobalt (2+) ion in a magnesium oxide crystal. - Phys. Rev. B, 1974, 9, N5, 2110-2121. - CA, 1974, 80, 150508. RZF, 1975, 6E216.

Theor.; $^4T_{1g}$ - e_g; O_h; d^7; 1.4.

1.3.56. Roche, M., Jaffé, H. H., A modification of the Herzberg-Teller expansion for vibronic coupling. - J. Chem. Phys., 1974, 60, N4, 1193.

Theor.

1.3.57. Shultz, M. J., Silbey, R., T × t Jahn-Teller problem: Energy levels for strong coupling. - J. Phys. C., 1974, 7, N17, L325-L326. - CA, 1974, 81, 143572. PA, 1974, 77, 79297. RZF, 1975, 2E194.

Theor.; T - t.

1.3.58. Sigmund, E., Wagner, M., U-matrix as a simple tool to derive canonical transformations. - Z. Phys., 1974, 268, N2, 245-249. - PA, 1974, 77, 51433. RZF, 1974, 11E511.

Theor.; E - e; T - t.

1975

1.3.59. Judd, B. R., Vogel, E. E., Coherent states and Jahn-Teller effect. - Phys. Rev. B, 1975, 11, N7, 2427-2435. - CA, 1975, 83, 34868. PA, 1975, 78, 46573. RZF, 1975, 10E68.

Theor.; E - e; T - (e + t); 1.4.

1.3.60. Lacroix, R., Duval, E., Champagnon, B., Louat, R., Effect of the trigonal field and uniaxial stress on the vibronic triplet states of $(3d)^n$ ions in α-Al_2O_3. - Phys. Status Solidi b, 1975, 68, N2, 473-484.

Theor.; T - (e + t); C_{3v}; d^3.

1.3.61. Purins, D., Karplus, M., Isotope effect on the hyperfine splittings of the deuteriobenzene anions: Example of the dynamic Jahn-Teller effect. - J. Chem. Phys., 1975, 62, N2, 320-332. - CA, 1975, 83, 8586. PA, 1975, 78, 32620. RZF, 1975, 7D172.

Theor.; D_{6h}.

1.3.62. Vogel, E. E., Ionic systems: the Jahn-Teller effect and electric multipole interaction. - 1975. - 113 p. - [Order N76-1508.] Diss. Abstr. Int. B, 1976, 36, N7, 3464. CA, 1976, 84, 128115.

Theor.; E - e; 1.4.

1.3.63. Voronin, A. I., Osherov, V. I., Karkach, S. P., Quasi-classical vibronic levels in Jahn-Teller systems. [Russ.]. - In: Kvant. Khim.: Tez. Dokl. 6 Vsesoyuz. Soveshch. Kishinev, 1975, B-47. - CA, 1976, 84, 142682. RZK, 1975, 24B94.

Theor.; E - e.

1976

1.3.64. Bacci, M., Dynamic Jahn-Teller effect and tunneling splitting. - Lett. Nuovo Cimento, 1976, 16, N11, 340-342. - CA, 1976, 85, 101400. PA, 1976, 79, 72534.

Theor.; T.

1.3.65. Bersuker, I. B., Boldyrev, S. I., Polinger, V. Z., The T - (e + t_2) problem in the Jahn-Teller effect taking into account spin-orbital interaction and low-symmetry crystal fields. [Russ.]. - In: Tez. Dokl. V Vsesoyuz. Simp. Spektrosk. Kristallov Aktivir. Redkimi Zeml'ami Elementami Gruppy Zheleza. Kazan', 1976.

Theor.; ${}^3T$ - (e + t).

1.3.66. Duval, E., Champagnon, B., Lacroix, R., Jahn-Teller effect in trigonal symmetry: $3d^n$ ions in corundum. - J. Phys. (France), 1976, 37, N12, 1391-1407. - CA, 1977, 86, 62819. PA, 1977, 80, 3584. RZF, 1977, 6E71.

Theor.; T - e; C_{3v}.

1.3.67. O'Brien, M. C. M., Wave functions for dynamic Jahn-Teller systems. - J. Phys. C, 1976, 9, N12, 2375-2382. - CA, 1976, 85, 168965. PA, 1976, 79, 72533.

Theor.; E - e; T - (e + t).

1.3.68. O'Brien, M. C. M., Some fractional parentage coefficients for R_5 and their application to T × (ε + τ_2) Jahn-Teller systems such as F and F^+ centers. - J. Phys. C, 1976, 9, N17, 3153-3164. - CA, 1977, 86, 35909. PA, 1976, 79, 87246.

Theor.; T - (e - t); 2.4.

1.3.69. Ranfagni, A., Viliani, G., The tunneling splitting in the T × (ε_g + τ_{2g}) Jahn-Teller problem. - Solid State Commun., 1976, 20, N10, 1005-1008. - CA, 1977, 86, 48659. PA, 1977, 80, 8059.

Theor.; T - (e + t).

1.3.70. Shultz, M. J., Silbey, R., A theoretical study of the strongly coupling T × t Jahn-Teller system. - J. Chem. Phys., 1976, 65, N11, 4375-4383. - CA, 1977, 86, 62816. PA, 1977, 80, 28147.

Theor.; T - t.

1.3.71. Sigmund, E., Jahn-Teller and "Fano" systems: on the approximation of resonance structures. - J. Phys. (France), 1976, 37, Colloq. N7,117-122. - CA, 1977, 86, 179688.

Theor.; E - e; T - t.

1.3.72. Voronin, A. I., Karkach, S. P., Osherov, V. I., Ushakov, V. G., Quasiclassical dynamics of symmetric molecules. [Russ.]. - Zh. Eksp. & Teor. Fiz., 1976, 71, N3, 884-895. - CA, 1976, 85, 184161. PA, 1977, 80, 3342. RZF, 1977, 1D87.

Theor.; E - e.

1977

1.3.73. Bacci, M., Quadratic Jahn-Teller effect and tunneling splitting in octahedral clusters. - Phys. Status Solidi b, 1977, 82, N1, 169-177. - CA, 1977, 87, 92737. PA, 1977, 80, 72840.

Theor.; T - (e + t); O_h; d^3; 2.1.

1.3.74. Barentzen, H., Polansky, O. E., Variational approach to the linear Jahn-Teller effect E × e. - Chem. Phys. Lett., 1977, 49, N1, 121-124. - CA, 1977, 87, 108640. PA, 1977, 80, 71759.

Theor.; 1.4.

1.3.75. Bates, C. A., Steggles, P., Weak and intermediate-strength Jahn-Teller effects. - Physica B + C, 1977, 86-88, N3, 1130-1131. - CA, 1977, 86, 197028. PA, 1977, 80, 66143.

Theor.; T - e.

1.3.76. Chau, F. T., Karlsson, L., Vibronic interaction in molecules and ions. I. The first order Jahn-Teller interaction in doubly degenerate electronic states of C_{3v} type molecules. - Phys. Scr., 1977, 16, N5-6, 248-257. - CA, 1978, 88, 179513.

Theor.; E - e; C_{3v}; 2.1.

1.3.77. Chau, F. T., Karlsson, L., Vibronic interaction in molecules and ions. II. The first order Jahn-Teller interaction and spin-orbit coupling in doubly degenerate electronic states of C_{3v} type molecules. - Phys. Scr., 1977, 16, N5-6, 258-267. - CA, 1978, 88, 200283.

Theor.; E - e; C_{3v}; 2.1.

1.3.78. Cianchi, L., Mancini, M., Moretti, P., The role of nonadiabatic operators in the dynamic Jahn-Teller effect in Al_2O_3: Ti^{3+}. - Lett. Nuovo Cimento, 1977, 19, N4, 129-133. - CA, 1977, 87, 31272. PA, 1977, 80, 59270.

Theor.; C_{3v}; d^1.

1.3.79. Cianchi, L., Manchini, M., Moretti, P., On the existence of tunneling splitting in the dynamic Jahn-Teller effect. - Lett. Nuovo Cimento, 1977, 19, N10, 381-382. - PA, 1977, 80, 69275.

Theor.; T.

1.3.80. Dixon, J. M., Lacroix, R., Weber, J., Jahn-Teller effect of a triplet state on a trigonally distorted site. - J. Phys. C, 1977, 10, N19, 3793-3802. - CA, 1978, 88, 143574. PA, 1977, 80, 89739.

Theor.; T - t.

1.3.81. Gregory, A. R., Henneker, W. H., Siebrand, W., Zgierski, M. Z., Exactly solvable models for vibronic coupling in molecular spectroscopy. II. Totally symmetric harmonic mode. - J. Chem. Phys., 1977, 67, N7, 3175-3180. - RZF, 1978, 6D407.

Theor.; 1.5; 2.1.

1.3.82. Judd, B. R., Jahn-Teller degeneracies of Thorson and Moffitt. - J. Chem. Phys., 1977, 67, N3, 1174-1179. - CA, 1977, 87, 124745. PA, 1977, 80, 88402.

Theor.; E - e; Γ_8 - t_2.

1.3.83. Judd, B. R., The Jahn-Teller effect in the actinides. - In: Proc. 2nd Int. Conf. Electron. Struct. Actinides. Wroclaw, 1977, p. 93-103. - CA, 1977, 87, 92735.

Theor.; Γ_8; 1.4.

1.3.84. Lacroix, R., Jahn-Teller effect in trigonal symmetry. [Fr.]. - In: Spectroscopie des elements lourds dans les solids. Paris, 1977, p. 133-135. (Colloq. Int. CNRS, N255.)

Theor.; (E + A) - e; E - e; C_{3v}; 1.4.

1.3.85. Pooler, D. R., O'Brien, M. C. M., The Jahn-Teller effect in a Γ_8 quartet: equal coupling to ε and τ_2 vibrations - J. Phys. C, 1977, 10, N19, 3769-3791. - CA, 1978, 88, 143573. PA, 1978, 81, 2454.

Theor.; Γ_8 - (e + t); 2.1.

1.3.86. Ranfagni, A., WKB approximation in multidimensional problems. - Phys. Lett. A, 1977, 62, N6, 395-396.

Theor.

1.3.87. Rueff, M., Wagner, M., E - e Jahn-Teller system: Calculations of the low-lying energy levels. - J. Chem. Phys., 1977, 67, N1, 169-172. - CA, 1977, 87, 92745. PA, 1977, 80, 76780.

Theor.; E - e.

1.3.88. Shultz, M. J., The nature of the eigenfunctions in a strongly coupled Jahn-Teller problem. - In: Electron-Phonon Interactions and Phase Transitions. New York; London, 1977, p. 331-336. (NATO Adv. Study Inst. Ser., Ser. B., 1977, N29.) - CA, 1979, 90, 143632.

Theor.; T - t; 2.1.

1.3.89. Sigmund, E., Occupation dynamics of an E - e Jahn-Teller system. - Z. Phys. B, 1977, 26, N3, 239-244. - PA, 1977, 80, 35900.

Theor.; E - e.

1978

1.3.90. Barentzen, H., Polansky, O. E., Canonical transformation approach to the linear Jahn-Teller effect E - e. - J. Chem. Phys., 1978, 68, N10, 4398-4405. - CA, 1978, 89, 67696. PA, 1978, 81, 67993.

Theor.; E - e; 1.4.

1.3.91. Bernstein, E. R., Webb, J. D., Quadratic Jahn-Teller coupling in octahedral systems. - Mol. Phys., 1978, 36, N4, 1113-1118. - CA, 1979, 90, 212415. PA, 1979, 82, 6881.

Theor.; Γ_8 - e; O_h.

1.3.92. Bernstein, E. R., Webb, J. D., Quadratic Jahn-Teller coupling in octahedral systems. - 1978. - 18 p. - US NTIS, AD Rep., AD - A052755. - CA, 1978, 89, 97250.

Theor.; Γ_8 - e; O_h.

1.3.93. Bersuker, I. B., The phenomenon of tunneling splitting of energy levels of polyaromatic systems in electronically degenerate states. Discovery, Diploma N202. Priority from September 12, 1960. [Russ.]. - Bul. Izobret., 1978, N40, p. 3.

See also in: Otkrytiya v SSSR, 1978, Moscow, 1979, p. 10-11.

Theor.; E - e; T - t; T - (e + t_2); 2.3; 2.7; 3.1; 3.2; 4.2; 4.6; 5.1; 5.2.

1.3.94. Karkach, S. P., Osherov, V. I., Quasiclassical dynamics of Jahn-Teller molecules. - Mol. Phys., 1978, 36, N4, 1069-1084. - CA, 1979, 90, 212414. PA, 1979, 82, 6880.

Theor.; E - e.

1.3.95. Muramatsu Shinji, Sakamoto Nobuhiko, Dynamic Jahn-Teller vibronic coupling in tetragonal symmetry. - J. Phys. Soc. Jpn., 1978, 44, N5, 1640-1646. - CA, 1978, 89, 50683. PA, 1978, 81, 60091.

Theor.; E - (b_1 + b_2); T_d; 1.4.

1.3.96. Polinger, V. Z., Boldyrev, S. I., Bersuker, I. B., Electron-vibrational interaction in Jahn-Teller systems with a ^3T-term. [Russ.]. - In: Tez. Dokl. Vsesoyuz. Soveshch. Kvant. Khim. Novosibirsk, 1978, p. 183.

Theor.; ^3T - (e + t_2).

1.3.97. Sakamoto, N., Muramatsu, S., Dynamic Jahn-Teller vibronic coupling for an orbital triplet $T \times \tau_2$ and $T \times (\tau_2 + \varepsilon)$ Phys. Rev. B., 1978, 17, N2, 868-875. - CA, 1978, 88, 160808.

Theor.; T - t; T - (e + t_2); 1.4; 2.1.

1.3.98. Sigmund, E., Brühl, S., Ground-state calculation for the T - t Jahn-Teller system by an exponential transformation approach. - Solid State Commun., 1978, 27, N8, 789-792. - CA, 1978, 89, 223303. PA, 1979, 82, 4540. RZF, 1979, 4D109.

Theor.; T - t.

1.3.99. Sviridov, D. T., Smirnov, Yu. E., Tolstoy, V. N., The dynamic Jahn-Teller effect in the T (a_{1g} + e_g + t_{2g}) system. [Russ.]. - Opt. & Spektrosk., 1978, 44, N5, 932-937. - CA, 1978, 89, 82246. PA, 1979, 82, 28327. RZF, 1978, 9E71.

Theor.; T - (a_1 + e + t_2); 2.1.

1979

1.3.100. Barentzen, H., Analytical treatment of the linear Jahn-Teller effect E × e. - Solid State Commun., 1979, 32, N12, 1285-1288. - CA, 1980, 92, 99807. PA, 1980, 83, 34404.

Theor.; E - e; 1.4.

1.3.101. Birkhold, M., Wagner, M., Jahn-Teller E - e problem: diagonalization by means of a transformed basis. - Z. Naturforsch. a, 1979, 34, N6, 667-671. - CA, 1979, 91, 79143. PA, 1979, 82, 82566.

Theor.; E - e.

1.3.102. Calles, A., Goscinski, O., The use of Green functions for the calculation of dynamic Jahn-Teller effect in electron spin resonance spectra. - Rev. Mex. Fis., 1979, 26, N1, 23-32. - PA, 1980, 83, 84643.

Theor.; T - (e + t).

1.3.103. Judd, B. R., Exact solutions to a class of Jahn-Teller systems. - J. Phys. C, 1979, 12, N9, 1685-1692. - CA, 1979, 91, 199156.

Theor.; E - e; Γ_8 - t_2.

1.3.104. Lacroix, R., Weber, J., Duval, E., Jahn-Teller effect in trigonal symmetry. - J. Phys. C, 1979, 12, N11, 2065-2080. - CA, 1979, 91, 219466. PA,1979, 82, 82563.

Theor.; E - e; (E + A) - e; C_{3v}; D_6: 1.5.

1.3.105. Muramatsu Shinji, Sakamoto Nobuhiko. A method for solving combined spin-orbit and Jahn-Teller coupling and application to $^4T_2$ of KMgF: V^{2+}. - J. Phys. Soc. Jpn., 1979, 46, N4, 1273-1279. - CA, 1979, 90, 212660.

Theor.; T - e; d^3; 2.2.

1.3.106. Schmutz, M., Explicit representation of a single para-Bose operator. - Phys. Lett. A, 1979, 72, N4/5, 301-312. - PA 1979, 82, 85064.

Theor.

See also: 1.2.-27, 30, 45; 1.4.-5, 10, 21, 22, 23; 1.5.-5, 13, 28, 30, 41, 44, 45, 47, 48; 1.6.-5, 7, 10, 25, 27, 28; 1.7.-4, 5, 6, 7, 8, 11, 13, 15, 16, 19, 20, 21, 26, 28, 30, 32, 35, 37, 38, 39, 40, 41, 44, 45, 50, 51; 2.1.-9, 219, 272, 291; 2.2.-175; 2.3.-90, 103, 110; 2.4.-25, 112; 2.7.-8, 14, 16, 46, 61; 2.8.-97; 3.1.-126, 253; 3.5.-12; 5.4.-92.

1.4. Vibronic Reduction Factors

In all the above cases the ground vibronic states possess the same transformation properties as the initial electronic states. For the E - e problem the ground vibronic energy level is an E term, and in the case of the T - (e + t_2) problem the ground vibronic term is T, etc. This allows one to simplify essentially the calculations of physical magnitudes determined by the ground state by just introducing the so-called vibronic reduction factors. First, the effect of vibronic reduction was noted in Ref. [3.1.25] where the first-order spin-orbital splitting of the ground vibronic energy level was shown to be multiplied by the overlap integral of vibrational wave functions located in different minima of the Jahn-Teller adiabatic potential. In 1965 Ham generalized this concept (Ref. [1.4.1]), demonstrating that such vibronic reduction takes place for matrix elements of every low-symmetry electronic operator. A general proof of the vibronic reduction theorem is given in Ref. [2.1.11] and [1.3.47] (see also the review article [3.1.114] and the monograph [1.1.37]).

Vibronic reduction factors for different Jahn-Teller systems were calculated soon after ground-state wave functions were obtained (analytically or numerically). Thus, for the E - e case in trigonal systems, the vibronic reduction of matrix elements of angular momentum transforming after the irreducible representation A_2 was calculated in Ref. [2.7.8]. A detailed investigation of the vibronic reduction factors for the Jahn-Teller E - e problem is given in Ref. [1.4.3] (see also the review article [3.1.151]. The quadratic vibronic interaction also affects the vibronic reduction. For the E - e problem this was considered in Refs. [1.4.-4, 5]. The multimode character of vibronic coupling also alters the vibronic reduction factor. In particular, some relations between the vibronic factors of the same electronic term (vibronic problem) for electronic operators of different symmetry, which exist in the case of an ideal vibronic system, are not valid in the multimode case. This conclusion was discussed in Refs. [1.7.-14, 15] and [1.4.14].

Vibronic reduction factors for the T term linearly coupled to E vibrations, as well as for that coupled to T_2 vibrations, are given in Ref. [1.4.1]. The numerical computation of the T - t_2 problem performed in Ref. [1.3.22] allows one to obtain the reduction factors more precisely.

The vibronic reduction for the case of the Jahn-Teller Γ_8 electronic term is considered in Ref. [1.4.6], where the special case of the d mode (i.e., the Γ_8 - d problem) is also considered.

In the case of additional infinitesimal symmetry, different vibronic reduction factors are not independent. A group-theoretic analysis of some relations between vibronic reduction factors is given in Ref. [1.4.14].

1965

1.4.1. Ham, F. S., Dynamical Jahn-Teller effect in paramagnetic resonance spectra: orbital reduction factors and partial quenching of spin-orbit interaction. - Phys. Rev. A, 1965, 138, N6, 1727-1740. - CA, 1965, 63, 1378 g. RZF, 1966, 2D479.

Theor.; E - e.

1967

1.4.2. McFarlane, K. M., Wong, J. Y., Sturge, M. D., Dynamic Jahn-Teller effect in octahedrally coordinated d^1 impurity systems. - Bull. Am. Phys. Soc., 1967, 12, N5, 709-710.

Theor.; T - e; d^1; 2.7; 3.1.

1968

1.4.3. Ham, F. S., Effect of linear Jahn-Teller coupling on paramagnetic resonance in a ^2E state. - Phys. Rev., 1968, 166, N2, 307-321. - CA, 1968, 68, 91740. PA, 1968, 71, 18402. RZF, 1968, 12D594.

Theor.; E - e; O_h; d^1.

1969

1.4.4. Halperin, B., Englman, R., Generalized reduction for the Jahn-Teller effect on a doublet. - Solid State Commun., 1969, 7, N21, 1579-1580. - CA, 1970, 72, 26544.

Theor.; E - e.

1.4.5. Slonczewski, J. C., Zeeman effect in dynamic Jahn-Teller impurities. - Solid State Commun., 1969, 7, N7, 519-520. - PA, 1969, 72, 31057. RZF, 1969, 11E627.

Theor.; E - e; 3.1.

1970

1.4.6. Morgan, T. N., Vibronic coupling in semiconductors: the dynamic Jahn-Teller effect. - Phys. Rev. Lett., 1970, 24, N16, 887-890. - CA, 1970, 73, 20228. PA, 1970, 73, 40762.

Theor.; Γ_8 - (e + t); O_h.

1971

1.4.7. Romestain, R., Merle d'Aubigne, Y., Jahn-Teller effect of an orbital triplet coupled to both E_g and T_{2g} modes of vibrations: Symmetry of the vibronic states. - Phys. Rev. B, 1971, 4, N12, 4611-4616. - CA, 1972, 76, 39369. PA, 1972, 75, 4418. RZF, 1972, 6E745.

Theor.; T - (e + t); O_h; p^1.

1.4.8. White, R. M., Reduction of orbital matrix elements in magnetic insulators. - Solid State Commun., 1971, 9, N4, 287-289. - PA, 1971, 74, 37463. RZF, 1971, 8E605.

Theor.; E - e; O_h; d^5.

1972

1.4.9. Vekhter, B. G., Reduction of the exchange interaction in systems with electron degeneracy. [Russ.]. - Theor. & Eksp. Khim., 1972, 8, N6, 811-813. - PA, 1976, 79, 29683.

Theor.; T - e; O_h.

1973

1.4.10. Bates, C. A., Jaussaud, P. C., Smith, W., The properties of $(3d)^4$ ions in corundum. III. The Jahn-Teller theory of the ^5E ground state and the parameters for Cr^{2+}. - J. Phys. C, 1973, 6, N5, 898-912. - CA, 1973, 78, 130009. PA, 1973, 76, 30688. RZF, 1973, 8E711.

Theor.; E - e; O_h; C_{3v}; d^4.

1.4.11. Halperin, B., Englman, E., Reduction factors for the Jahn-Teller effect in solids. - Phys. Rev. Lett., 1973, 31, N17, 1052-1055. - CA, 1973, 79, 151032. PA, 1973, 76, 75711. RZF, 1974, 3E477.

Theor.; E - e; 1.7.

1.4.12. Vekhter, B. G., Vibronic reduction factor for the second-order effects: Orbital doublet EPR spectrum. - Phys. Lett. A, 1973, 45, N2, 133-134. - CA, 1973, 79, 141287. PA, 1973, 76, 72507.

Theor.; E - e; O_h; 3.1.

1974

1.4.13. Bersuker, I. B., Polinger, V. Z., Vibronic reduction factors and EPR spectra of orthorhombically distorted Jahn-Teller systems. [Russ.]. - In: Fiz. Mat. Metody Koord. Khim.: Tez. Dokl. 5 Vsesoyuz. Sovesch. Kishinev, 1974, p. 72-73. - CA, 1975, 83, 17842.

Theor.; T - (e + t); O_h; T_d; 3.1.

1.4.14. Leung, C. H., Kleiner, W. H., Relations among reduction factors in Jahn-Teller systems. - Phys. Rev. B, 1974, 10, N10, 4434-4436. - CA, 1975, 82, 49386. PA, 1975, 78, 29597.

Theor.; E - e; T - (e + t); 1.7.

1.4.15. Passeggi, M. G., Stevens, K. W. H., Exchange interactions between Jahn-Teller ions. II. - Physica, 1974, 71, N1, 141-160. - CA, 1974, 80, 125031. PA, 1974, 77, 28355. RZF, 1974, 9E490.

Theor.; T - e.

1.4.16. Vekhter, B. G., Bersuker, I. B., Vibronic amplification of structural distortions in coordination compounds. [Russ.]. - In: Fiz. Mat. Metody Koord. Khim.: Tez. Dokl. 5 Vsesoyuz. Soveshch. Kishinev, 1974, p. 39-40.

Theor.; E - e; T - t.

1975

1.4.17. Kahn, O., Kettle, S. F. A., Vibronic coupling in cubic complexes. - Mol. Phys., 1975, 29, N1, 61-79. - PA, 1975, 78, 9971.

Theor.; T.

1976

1.4.18. Bersuker, I. B., Polinger, V. Z., Khlopin, V. P., Vibronic reduction and EPR spectra of orthorhombic distorted Jahn-Teller systems. - In: Magn. Resonance & Related Phenomena: Proc. 19th Congr. AMPERE, Heidelberg, 1976. Heidelberg: Geneva, 1976, p. 499-502. - CA, 1977, 86, 197415.

Theor.; T - (e + t); T - v; O_h; I_h.

1.4.19. Ham, F. S., Leung, C. H., Kleiner, W. H., Relationships among Jahn-Teller reduction factors for a Γ_8 state in cubic symmetry. - Solid State Commun., 1976, 18, N6, 757-759. - CA, 1976, 84, 171593. PA, 1976, 79, 45244.

Theor.; Γ_8 - (e + t); O_h.

1.4.20. Teodorescu, M., Jahn-Teller effects in the fundamental states of the triplet orbital. [Rom.]. - Rev. Fiz. si Chimie, 1976, 13, N7, 145-148. - CA, 1976, 85, 166719.

Theor.; T.

1977

1.4.21. Muramatsu Shinji, Numerical analysis of Jahn-Teller reduction factors in E × (e_1 + e_2). - Solid State Commun., 1977, 21, N1, 125-127. - PA,1977, 80, 15803.

Theor.; E - (e_1 + e_2); C_{4v}; D_{4h}; 1.7.

1.4.22. Payne, S. H., Stedman, G. E., A diagram approach to reduction factor calculations in Jahn-Teller systems. - J. Phys. C, 1977, 10, N23, L671-L675. - CA, 1978, 88, 161701. PA, 1978, 81, 19003.

Theor.; T - t; 1.7.

1.4.23. Sigmund, E., Wagner, M., Birkhold, M., Dynamical treatment of quenching factors (Ham-factors) for the linear E - e Jahn-Teller problem. - Solid State Commun., 1977, 22, N11, 719-720. - PA, 1977, 80, 59271.

Theor.; E - e.

1978

1.4.24. Leung, C. H., Comment on a recent calculation of the Ham's reduction factors by the use of the exponential transformation method. - Solid State Commun., 1978, 25, N10, 809-810. - CA, 1978, 88, 200088.

Theor.; E - e.

1979

1.4.25. O'Brien, M. C. M., Pooler D. R., Reduction factors for dynamic Jahn-Teller systems in strong coupling. - J. Phys. C, 1979, 12, N3, 311-320. - CA, 1979, 91, 129226. PA, 1979, 82, 42498.

Theor.; E - e; Γ_8 - (e + t); 1.3.

See also: 1.1.-52, 62; 1.2.-25, 30; 1.3.-22, 33, 40, 44, 47, 59, 62, 74, 83, 90, 95, 97, 100; 1.7.-17, 18, 22, 28, 29, 30, 42, 45; 1.8.-5, 6, 12; 2.1.-247; 2.2.-52, 132; 2.3.-8, 20, 53, 58, 76, 101, 112, 126, 127; 2.4.-19, 22, 28, 48, 52, 64, 93, 96, 97, 115, 119, 150, 181, 182, 188; 2.7.-24, 32, 68, 71, 72; 2.8.-21; 3.1.-44, 65, 115, 124, 146, 166, 182, 193, 215, 230, 254, 257, 266; 3.2.-5, 42, 69, 74, 89, 121; 3.4.-3, 36, 38; 3.5.-17, 25, 35, 46; 3.6.-14, 19, 27; 4.1.-82, 92, 157; 4.4.-72.

1.5. Pseudo Jahn-Teller Effect

In this subsection theoretical works concerning the vibronic effects resulting from the nonadiabatic mixing of near-lying (pseudo-degenerate) electronic states are listed. The adiabatic potential energy surfaces in these cases have been discussed in detail by Liehr in Ref. [1.5.9], whereas the main features of the corresponding vibronic wave functions and energy levels are given in Ref. [1.5.4].

1933

1.5.1. Herzberg, G., Teller, E., Vibrational structure of electronic transitions for polyatomic molecules. - Z. Phys. Chem. B (Leipzig), 1933, 21, N5-6, 410-446. - CA, 1933, 27, 4170.

Theor.

1955

1.5.2. Liehr, A. D., The interaction of vibrational and electronic motions in some simple conjugated hydrocarbons: Thesis. - Harvard Univ., 1955. - 127 p.

Theor.; $(E_{1u} + B_{1u} + B_{2u}) - (b_{1g} + b_{2g})$; $E - (e_1 + e_2)$; 1.7.

1960

1.5.3. Witkowski, A., Moffitt, W., Electronic spectra of dimers: derivation of the fundamental vibronic equation. - J. Chem. Phys., 1960, 33, N2, 872-875. - CA, 1961, 55, 7028c.

Theor.;

1961

1.5.4. Fulton, R. L., Gouterman, M., Vibronic coupling. I. Mathematical treatment for two electronic states. - J. Chem. Phys., 1961, 35, N3, 1059-1072. - PA, 1961, 64, 17334.

Theor.; 1.3.

1.5.5. McRae, E. G., Molecular vibrations in the exciton theory for molecular aggregates. II. Dimeric systems. - Austral. J. Chem., 1961, 14, N3-4, 344-353.

Theor.; $(A_u + B_g) - b_u$.

1.5.6. Witkowski, A., The nature of electronic-nuclear coupling in dimers. - Roczn. Chem., 1961, 35, N5, 1399-1408. - CA, 1962, 57, 4048i.

Theor.

1.5.7. Witkowski, A., Strong and weak coupling solutions of the vibronic equation for dimers. - Roczn. Chem., 1961, 35, N5, 1409-1418. - CA, 1962, 57, 4049a.

Theor.

1962

1.5.8. Englman, R. A., Suggested adiabatic approximation for vibronic problems with electronic near-degeneracy. - Phys. Lett., 1962, 2, N5, 227-228. - RZF, 1963, 4E159.

Theor.; 1.6.

1963

1.5.9. Liehr, A. D., The coupling of vibratonal and electronic motions in degenerate and nondegenerate electronic states of inorganic and organic molecules. Part III. Nondegenerate electronic states. - In: Prog. Inorg. Chem. New York; London, 1963, Vol. 5, p. 385-430.

Rev.; 1.2.

1964

1.5.10. Witkowski, A., Resonance interaction between degenerate electronic states: derivation of the vibronic equation. - J. Chem. Phys., 1964, 40, N2, 555-557. - CA, 1964, 60, 3611g.

Theor.

1.5.11. Witkowski, A., Trapping of the molecular exciton coupled with the nuclear vibrations. - J. Chem. Phys., 1964, 40, N5, 1453-1454. - CA, 1964, 60, 8739h.

Theor.

1965

1.5.12. Gouterman, M., Vibronic couling. III. Adiabatic potentials and vibronic functions. - J. Chem. Phys., 1965, 42, N1, 351-356. - CA, 1965, 62, 3526f. PA, 1965, 68, 11486.

Theor.; 1.2; 1.3.

1.5.13. Hobey, W. D., Vibronic interaction of nearly degenerate states in substituted benzene anions. - J. Chem. Phys., 1965, 43, N7, 2187-2199. - CA, 1965, 63, 12505b.

Theor.

1.5.14. Nicholson, B. J., Longuet-Higgins, H. C., Dipole moment of bis(hexamethylbenzene) cobalt. - Mol. Phys., 1965, 9, N5, 461-472. - CA, 1966, 64, 4390a.

Theor.

1.5.15. Witkowski, A., Coupling of the molecular exciton with the nuclear vibrations. - In: Mod. Quantum Chem. New York; London, 1965, Vol. 3, Sect. III.B.3, p. 161-175. - CA, 1968, 68, 7721.

Theor.

1966

1.5.16. Bierman, A., Vibronic states of dimers. - J. Chem. Phys., 1966, 45, N2, 647-653. - PA, 1966, 69, 33747.

Theor.

1.5.17. Witkowski, A., Exciton-exciton interaction resulting from phonon-exciton coupling. - Theor. Chim, Acta, 1966, 4, N4, 317-320. - CA, 1966, 65, 135g.

Theor.

1.5.18. Witkowski, A., Vibronic coupling in molecular crystals. - Acta Phys. Pol., 1966, 30, N3, 431-436. - CA, 1967, 66, 109718.

Theor.

1967

1.5.19. Bersuker, I. B., Vekhter, B. G., Rafalovich, M. L., Dipole moments of symmetrical molecular systems. [Russ.]. - In: Tez. Dokl. Vsesoyuz. Konf. Dipol'nym, Momentam Stroyeniyu Molekul. Rostov-na-Donu, 1967, p. 9.

Theor.; 1.2.

1.5.20. Fulton, R. L., Vibronic coupling. V. Several theorems concerning the eigenfunctions and spectra of the vibronic oscillator. - J. Chem. Phys., 1967, 46, N6, 2257-2261. - PA, 1967, 70, 23171.

Theor.

1.5.21. Kiselev, A. A., On the stability of molecules with nearly degenerate electronic states. [Russ.]. - Opt. & Spektrosk., 1967, 23, N3, 362-365.

Theor.; (A + E) - e; O_h; 1.2.

1.5.22. Witkowski, A. S., Exciton-phonon coupling in linear molecular crystals. - Acta Phys. Pol., 1967, 31, N1, 13-17. - CA, 1968, 68, 109016.

Theor.

1968

1.5.23. Kiselev, A. A., The stability of configurations with nearly degenerate electronic states. [Russ.]. - Izv. Akad. Nauk SSSR. Ser. Fiz., 1968, 32, N1, 14-15. - PA, 1969, 72, 39788. RZF, 1968, 8D72.

Theor.

1.5.24. Young, J. H., Vibronic coupling in the dimer. - J. Chem. Phys., 1968, 49, N6, 2566-2573. - CA, 1968, 69, 111956.

Theor.; 1.3; 2.3.

1969

1.5.25. Bersuker, I. B., Dipole moments of symmetrical molecular systems. [Russ.]. - Teor. & Eksp. Khim., 1969, 5, N3, 293-299.

Theor.; $(A_{1g} + T_{1u}) - t_{1u}$; 1.2.

1.5.26. Perrin, M. H., Gouterman, M., Perrin, C. L., Vibronic coupling. VI. Vibronic borrowing in cyclic polyenes and porphyrin. - J. Chem. Phys., 1969, 50, N10, 4137-4141.

Theor.; $(E_1 + E_2) - (a_1 + a_2 + b_1 + b_2)$; 2.4.

1.5.27. Stephens, P. J., Dynamic Jahn-Teller effect in trigonally distorted cubic systems. - J. Chem. Phys., 1969, 51, N5, 1995-2005. - CA, 1969, 71, 96626.

Theor.; $(A + E) - e$; O_h; C_{3v}.

1.5.28. Zgierski, M. Z., A note about the possibility of WKB solutions of the vibronic equation. - Acta Phys. Pol., 1969, 36, N2, 273-275.

Theor.; 1.3.

1970

1.5.29. Fedder, R. C., Model for Jahn-Teller distortions of Ag^{2+} ions by odd modes in fluorite cyrstals. - Phys. Rev. B, 1970, 2, N1, 40-48. - CA, 1971, 74, 25132. PA, 1970, 73, 76213. RZF, 1971, 4D664.

Theor.; O_h; d^9.

1.5.30. Witkowski, A., Zgierski, M., Decoupling conditions for the vibronic equation in dimers. - Int. J. Quantum Chem., 1970, 4, N4, 427-429. - CA, 1970, 73, 18631.

Theor.

<u>1971</u>

1.5.31. Orlandi, G., Siebrand, W., Vibronic coupling and line broadening in polyatomic molecules. - Chem. Phys. Lett., 1971, <u>8</u>, N6, 473-476. - CA, 1971, <u>75</u>, 12554.

Theor.

1.5.32. Waals, J. H., van der, Berghuis, A. M. D., Groot, M. S., de, Vibronic interactions in the lower electronic states of benzene. II. The hexagonal approximation for the free molecule and the effect of crystal field. - Mol. Phys., 1971, <u>21</u>, N3, 497-522. - CA, 1971, <u>75</u>, 103107.

Theor.; $(B_{1u} + E_{1u}) - e_g$; D_{6h}.

1.5.33. Witkowski, A., Zgierski, M. Z., Vibronic spectra of the perylene crystal. - Chem. Phys. Lett., 1971, <u>9</u>, N4, 336-338. - CA, 1971, <u>75</u>, 69077.

Theor.

<u>1972</u>

1.5.34. Azumi, T., Reinterpretation of the nonfluorescent properties of the η,π^* singlet state. - Chem. Phys. Lett., 1972, <u>17</u>, N2, 211-212. - CA, 1973, <u>78</u>, 49919. PA, 1973, <u>76</u>, 7216.

Exp.

1.5.35. Fulton, R. L., Vibronic interactions: The adiabatic approximation. - J. Chem. Phys., 1972, <u>56</u>, N3, 1210-1218.

Theor.; 2.3.

<u>1973</u>

1.5.36. Glinchuk, M. D., Deygen, M. F., Karmazin, A. A., Nature of the off-center position of impurity ions in lattices. [Russ.]. - Fiz. Tverd. Tela, 1973, <u>15</u>, N7, 2048-2052. - CA, 1973, <u>79</u>, 98782. PA, 1974, <u>77</u>, 37191. RZF, 1973, 11E381.

Theor.; $(A_{1g} + T_{1u}) - t_{1u}$; Oh.

1.5.37. Ham, F. S., Vibronic model for the relaxed excited state of the F center. I. General solution. - Phys. Rev. B, 1973, <u>8</u>, N6, 2926-2944. - RZF, 1974, 5E914.

Theor.; $(A_{1g} + T_{1u}) - t_{1u}$; O_h; 2s; 2p; 2.1; 2.4.

1.5.38. Ham, F. S., Grevsmuhl, U., Vibronic model for the relaxed excited state of the F center. II. Perturbation analysis for weak coupling limit. - Phys. Rev. B, 1973, 8, N6, 2945-2957. - RZF, 1974, 5E915.

Theor.; $(A_{1g} + T_{1u}) - t_{1u}$; O_h; 2s; 2p; 1.3; 2.4.

1.5.39. Koyanagi Motohiko, Goodman, L., Static and dynamical potential surface distortions in aromatic aldehyde $^3A''$ (η_π) states. - Chem. Phys. Lett., 1973, 21, N1, 1-8. - CA, 1973, 79, 11466. PA, 1973, 76, 63382.

Exp.; (A' + A") - a"; 2.7.

1.5.40. Zgierski, M. Z., Strong coupling in dimers. - Chem. Phys. Lett., 1973, 21, N3, 525-526. - CA, 1974, 80, 7318.

Theor.

1974

1.5.41. Kayanuma, Y., Toyozawa, Y., Vibronic interactions in the relaxed excited state of the F center. - In: Extended Abstr. Int. Conf. Color Centers in Ionic Crystals. Sendai, 1974, A4, 2 p. - PA, 1975, 78, 6739.

Theor.; $(A_{1g} + T_{1u}) - t_{1u}$; O_h; 2s; 2p.

1975

1.5.42. Bersuker, I. B., Vekhter, B. G., Rafalovich, M. L., Shaperev, Yu. V., The pseudo-Jahn-Teller effect in the adamantane molecule. [Russ.]. - In: Kvant. Khim.: Tez. Dokl. 6 Vsesoyuz. Soveshch. Kishinev, 1975, B-24. - CA, 1976, 84, 157417. RZK, 1975, 24B88.

1976

1.5.43. Dellinger, B., Kasha, M., Phenomenology of solvent matrix spectroscopic effects. - Chem. Phys. Lett., 1976, 38, N1, 9-14. - CA, 1976, 84, 157363. PA, 1976, 79, 34756.

Theor. $(T'_1 + T'_2)$.

1977

1.5.44. Feiner, L. F., Single ion and cooperative Jahn-Teller effect for a nearly degenerate E doublet. - In: Electron-Phonon Interactions and Phase Transitions. New York; London, 1977, p. 345-350. - CA, 1979, 90, 129700.

Theor.; $(A_1 + B_2) - (a_1 + b_2)$; C_{2v}; d^6; 4.1.

1.5.45. Henneker, W. H., Penner, A. P., Siebrand, W., Zgierski, M. Z., Anharmonic potential and vibronic energy levels for closely coupled molecular electronic states. - Chem. Phys. Lett., 1977, 45, N3, 407-410. - CA, 1977, 86, 130188. PA, 1977, 80, 23423.

Theor.; 1.2; 1.3; 1.6.

1.5.46. Siebrand, W., Scaled and unscaled crude-adiabatic Born-Oppenheimer wavefunctions. - Chem. Phys. Lett., 1977, 51, N1, 5-7.

Theor.

1.5.47. Zgierski, M. Z., Comparison of two linear approximatio-s to vibronic coupling. - Chem. Phys. Lett., 1977, 45, N1, 41-43.

Theor.

1978

1.5.48. Henneker, W. H., Penner, A. P., Siebrand, W., Zgierski, M. Z., Exactly solvable models for vibronic coupling in molecular spectroscopy. III. The pseudo Jahn-Teller effect. - J. Chem. Phys., 1978, 69, N5, 1884-1896. - CA, 1978, 89, 188170. PA, 1978, 81, 94234.

Theor.

1979

1.5.49. Kristoffel, N. N., Temperature effects in the dynamics of a center with nearly degenerate levels and localized phase transition. - Nuovo Cimento B, 1979, 49, N2, 307-313.

Theor.; 4.1; 4.2.

1.5.50. Pooler, D. R., Rotational invariances of the linear pseudo-Jahn-Teller effect on the (s × p) manifold. - J. Phys. C, 1979, 12, N21, 4483-4492. - CA, 1980, 92, 153381. PA, 1980, 83, 15482.

Theor.; $(A_{1g} + T_{1u}) - (a_{1g} + t_{1u} + e_g + t_{2g})$; O_h; 2s; 2p; 1.1.

1.5.51. Pooler, D. R., An equally coupled (s × p) pseudo-Jahn-Teller system. - J. Phys. C, 1979, 12, N21, 4493-4507.

Theor.; $(A_{1g} + T_{2u}) - (a_{1g} + t_{1u} + e_g + t_{2g})$; O_h; 2s; 2p; 1.1.

1.5.52. Seibrand, W., Zgierski, M., Interaction between vibronic coupling and spin-orbit coupling in pyrazine. - Chem. Phys. Lett., 1979, 67, N1, 13-16. - RZF, 1980, 3D129.

Theor.

1.5.53. Waals, J. H., van der, Conformational instability of excited states of symmetrical molecules: substituted benzenes and porphyrins. - Mol. Cryst. & Liq. Cryst., 1979, 50, N1/4, 301-304. - CA, 1979, 91, 19207. PA,1979, 82, 47062.

Theor.; D_4h; D_6h.

1.5.54. Zhilinskiy, B. I., Istomin, V. A., Stepanov, N. F., Vibration-rotational states of non-rigid molecules. [Russ.]. - In: Sovrem. Probl. Fiz. Khim. Moscow, 1979, Vol. 11, p. 259-304. - RZF, 1980, 5D179.

Rev.; 1.6.

See also: 1.1.-18, 30, 47, 58, 67, 68; 1.2.-1, 7, 14; 1.3.-29, 31, 81, 84, 104; 1.6.-25, 28; 1.7.-2; 1.8-16; 2.1.-7, 11, 25, 36, 48, 57, 83, 84, 101, 104, 121, 122, 126, 143, 144, 178, 191, 203, 207, 208, 219, 224, 226, 239, 245, 256, 262, 266, 267, 277, 289, 292; 2.2.-42, 62, 64, 65, 68, 73, 93, 97, 98, 110, 112, 118, 119, 130, 136, 137, 146, 149, 153, 158, 168; 2.3.-1, 5, 17, 19, 30, 56, 64, 83, 86, 100, 104, 112; 2.4.-7, 77, 81, 85, 94, 103, 112, 113, 125, 141, 145, 157, 189, 210; 2.5.-17, 20, 28, 35, 36, 44; 2.6.-2, 13, 16, 21; 2.7.-10, 16, 37, 43, 62, 65; 2.8.-2, 17, 22, 36, 53, 56, 61, 62, 63, 84, 87, 89, 90, 93, 95, 99, 100, 101, 102; 3.1.-43, 61, 67, 68, 103, 104, 156, 180, 208, 273, 284; 3.2.-45; 3.3.-21, 27, 31; 3.4.-19; 3.6.-29; 3.7.-7; 4.1.-49, 50, 115, 116, 118; 4.6.-114, 120, 147; 5.1.-3, 10, 12, 14, 15, 16, 19, 20, 21, 25, 32, 35, 44, 51, 52, 60; 5.2.-9, 21, 30, 33, 36, 41, 43, 51, 67, 68, 74, 75, 76, 77, 83, 84, 85, 92, 94, 95, 96, 97, 98, 99, 100, 101, 103, 106, 107, 108; 5.3.-1, 2, 3, 4, 6, 8, 12, 21, 26; 5.4.-57, 71, 75, 89, 100, 106, 107.

1.6. The Renner Effect

For a general review of the subject we refer the reader to the review article [1.6.22].

1934

1.6.1. Renner, R., Theory of the interaction between electronic motion and oscillation in linear triatomic molecules. [Germ.]. - Z. Phys., 1934, 92, N3-4, 172-193. - CA, 1935, 29, 2084^5.

Theor.; E - e; $D_{\infty}h$; 1.2.

1958

1.6.2. Pople, J. A., Longuet-Higgins, H. C., Theory of the Renner effect in the NH_2 radical. - Mol. Phys., 1958, 1, N4, 372-383. - RZK, 1959, 56059.

Theor.; E - e; $D_{\infty h}$; 5.4.

1960

1.6.3. Pople, J. A., Renner effect and spin-orbit coupling. - Mol. Phys., 1960, 3, N1, 16-22. - CA, 1960, 54, 14918h.

Theor.

1962

1.6.4. Halverson, F., Comments on potassium superoxide structure. - J. Phys. & Chem. Solids, 1962, 23, N3, 207-214.

Theor.

1.6.5. Hougen, J. T., Rotation energy levels of a linear triatomic molecule in a $^2\Pi$ electronic state. - J. Chem. Phys., 1962, 36, N2, 519-534.

Theor.; 1.3.

1.6.6. Hougen, J. T., Vibronic and rotational energy levels of a linear triatomic molecule in a $^3\Pi$ electronic state. - J. Chem. Phys., 1962, 36, N11, 1874-1881.

Theor.; 1.3.

1963

1.6.7. Hougen, J. T., Yesson, J. P., Anharmonic corrections in triatomic molecules exhibiting the Renner effect. - J. Chem. Phys., 1963, 38, N7, 1524-1525.

Theor.; 1.3.

1964

1.6.8. Hougen, J. T., Vibronic interactions in molecules with a fourfold symmetry axis. - J. Mol. Spectrosc., 1964, 13, N2, 149-167. - CA, 1964, 61, 2598g.

Theor.; E - e; D_{4h}; 1.3.

1965

1.6.9. Cosmo, C., Renner effect in the linear XY_2 molecule. - 1965. - 58 p. - AEC Accession N31943. Rept. N AEC-CONF-65-288-12. - CA, 1967, 66, 49375.

Theor.; E - e; $D_{\infty h}$.

1.6.10. Dixon, R. N., The Renner effect in a nearly linear molecule with application to NH_2. - Mol. Phys., 1965, 9, N4, 357-366. - CA, 1966, 64, 4460c. PA, 1965, 68, 32193.

Theor.; E - e; $D_{\infty h}$; 1.3.

1970

1.6.11. Chang Chen-fee, Vibrational magnetism and spin-vibronic interaction in linear triatomic molecules: Catholic Univ. America, Washington. Thesis. - 1970. - 110 p. - [Order N71-11070.] Diss. Abstr. Int. B, 1971, 31, N11, 6814. PA,1971, 74, 78812.

Theor.

1.6.12. Chang Chen-fee, Chiu Ying-nan, Magnetic Renner effect: direct orbital and spin interaction with "vibrational" rotation in linear triatomic molecules. - J. Chem. Phys., 1970, 53, N6, 2186-2195. - CA, 1970, 73, 93124.

Theor.

1.6.13. Kiselev, A. A., Degenerate electronic states in molecules and the Renner effect. [Russ.]. - Vestn. Leningr. Univ., 1970, N10. Fizika, Khim., N2, 7-9. - CA, 1970, 73, 125249. RZF, 1971, 1D134.

Theor.

1971

1.6.14. Kiselev, A. A., Ob'edkov, V. D., Yurova, I. Yu., Inelastic scattering of electrons by molecules with the Renner effect. [Russ.]. - Zh. Eksp. & Theor. Fiz., 1971, 61, N5, 1835-1840. - CA, 1972, 76, 52018. RZF, 1972, 4D244.

Theor.; 2.6.

1972

1.6.15. Jug, K., Weyssenhoff, H., von, Study of nonadiabatic transitions in triatomic molecules. - J. Chem. Phys., 1972, 56, N1, 517-523.

Theor.

1.6.16. Petelin, A. N., Teorija Effektov Neadiabatichnosti v Lineinykh Molekulakh_ Avtoref. Dis. ... Kand. Fiz.-Mat. Nauk. [Russ.]. - Leningrad, 1972. - 14 p. - Leningr. Univ.

The Theory of Nonadiabatic Effects in Linear Molecules.

Theor.; 2.1; 2.3.

1.6.17. Petelin, A.N., Kiselev, A. A., The Renner effect in four-atomic molecules. - Int. J. Quantum Chem., 1972, 6, N4, 701-716. - CA, 1972, 77, 81366.

Theor.; 2.1; 2.3.

1974

1.6.18. Barrow, T., Dixon, R. N., Duxbury, G., Renner effect in a bent triatomic molecule executing a large amplitude bending vibration. - Mol. Phys., 1974, 27, N5, 1217-1234. - CA, 1974, 81, 112960.

Theor.; E - e; $D_{\infty h}$.

1.6.19. Lacroix, R., Quadratic Jahn-Teller effect on an ion configuration d^9 in a fluorine structure. [Fr.]. - Helv. Phys. Acta, 1974, 47, N6, 689-704. - CA, 1975, 82, 175525. PA, 1975, 78, 65893. RZF, 1975, 10E207.

Theor.; T_{2g} - t_{1u}; O_h; d^9; 1.2; 5.3.

1.6.20. Spence, D., Schulz, G. J., Cross sections for production of O_2^- and C^- by dissociative electron attachment in CO_2: An observation of the Renner-Teller effect. - J. Chem. Phys., 1974, 60, N1, 216-220. - CA, 1974, 80, 112854.

Exp.; E - e; $D_{\infty h}$; 5.1.

1975

1.6.21. Brown, J. M., K-type doubling parameters for linear molecules in Π electronic states. - J. Mol. Spectrosc., 1975, 56, N1, 159-162. - CA, 1975, 83, 34881.

Theor.; E - e; $C_{\infty v}$; 1.3.

1976

1.6.22. Jungen, C., Merer, A. J., The Renner-Teller effect. - In: Mol. Spectrosc. Mod. Res. New York, 1976, Vol. 2, p. 127-164. - CA, 1977, 86, 197009.

Rev.

1.6.23. Herbst, E., Delos, J. B., Dynamic Renner effect in collisions of C^+ with H_2. - Chem. Phys. Lett., 1976, 42, N1, 54-58. - CA, 1976, 85, 151134.

Exp.; E - e; $D_{\infty h}$; 5.1.

1977

1.6.24. Brown, J. M., The effective Hamiltonian for the Renner-Teller effect. - J. Mol. Spectrosc., 1977, 68, N3, 412-422. - CA, 1977, 87, 208730.

Theor.

1978

1.6.25. Aarts, J. F. M., The Renner effect in $^2\Pi$ electronic states of linear triatomic molecules. I. Theory of vibronic interaction modified by spin-orbit coupling. - Mol. Phys., 1978, 35, N6, 1785-1803. - CA, 1978, 89, 223270.

Theor.; 1.3.

1.6.26. Pavlov, N. I., Kiselev, A. A., Lyaptsev, A. V., Stability of molecular configurations in the high field of a resonance light wave. [Russ.]. - Opt. & Spektrosk., 1978, 45, N2, 398-399. - CA, 1978, 89, 169420.

Theor.

1979

1.6.27. Kiselev, A. A., Lyaptsev, A. V., Pavlov, N. I., On the dynamics of the distortive vibration of a linear molecule in a resonance electromagnetic field. [Russ.]. - Opt. & Spektrosk., 1979, 46, N3, 460-466. - RZF, 1979, 7D176.

Theor.

1-6.28. Köppel, H., Cederbaum, L. S., Domcke, W., Von Niessen, W., Vibronic coupling in linear molecules and linear-to-bent transitions: HCN. - Chem. Phys., 1979, 37, N3, 303-317. - CA, 1979, 90, 194829.

Theor.; E - e; $C_{\infty v}$; 2.1; 2.3.

See also: 1.1.-35, 45, 66, 73; 1.2.-7, 12; 1.3.-1; 1.5.-54; 2.1.-6, 13, 15, 21, 61, 72, 124, 133, 138, 141, 149, 152, 154, 155, 158, 159, 176, 177, 229, 233, 251; 2.2.-42, 67; 2.3.-57; 2.5.-10; 2.6.-8, 9, 16; 2.7.-80; 2.8.-61, 98, 101; 3.1.-125; 5.2.-19; 5.4.-106.

1.7. The Multimode Jahn-Teller Effect

If there is only one Jahn-Teller active mode of each symmetry, the Jahn-Teller problem is called ideal (see, e.g., Ref. [1.1.37]). If the number of Jahn-Teller active vibrations of a given symmetry is more than one, the problem becomes a multimode one. The number of Jahn-Teller modes increases with the number of coordination spheres around the electronic degenerate center, the local centers in crystals being the limiting case for the multimode problem with an infinite number of Jahn-Teller active modes of the same symmetry.

The first interesting results in the multimode Jahn-Teller effect were obtained in 1963 by Slonczewski (Rev. [1.7.3]), who considered the strong coupling limit of the multimode E - (e_1 + e_2 +

e_3 + ...) problem. Narrow resonances were shown to appear in the vibronic density of states of the conical shaped upper sheat of the adiabatic potential. Similar concepts are used in Refs. [1.7.-4, 5] for the multimode Γ_8 - (e + t_2) problem.

In Ref. [1.7.11] the same E - e strong coupling case is considered, using the variational wave functions for the lowest rotational states. We note that the ground-state functions used in Ref. [1.7.11] do not contain the A_2 transforming part (see Ref. [1.4.14]). In Ref. [1.7.12] the so-called "simple approach" to the multimode Jahn-Teller effect (the cluster model) is suggested. The cluster model was further developed in Refs. [1.7.-14, 25, 26, 27, 31]. This trend in the theory of multimode Jahn-Teller effect is reviewed in Ref. [1.7.47].

Another approach, free of the cluster-model limitations, is given in Ref. [1.7.38], where the weak vibronic coupling in the multimode E - e problem is taken into account by perturbation theory.

1961

1.7.1. McConnell, H. H., MacLachlan, A. D., Nuclear hyperfine interactions in orbitally degenerate states of aromatic ions. - J. Chem. Phys., 1961, 34, N1, 1-12. - CA, 1961, 55, 21813g.

Theor.; E - e; D_{6h}; 5.3; 5.4.

1962

1.7.2. Aminov, L. K., Kochelaev, B. I., The influence of the spin-phonon interaction on the paramagnetic resonance spectra. [Russ.]. - Fiz. Tverd. Tela, 1962, 4, N6, 1604-1607. - RZF, 1962, 11V429.

Theor.; 1.5; 3.1.

1963

1.7.3. Slonczewski, J. C., Theory of the dynamic Jahn-Teller effect. - Phys. Rev., 1963, 131, N4, 1596-1610. - RZF, 1964, 4E131.

Theor.; E - e; 1.3.

1966

1.7.4. Fan Li-Chin, Ou Chin, The theory of the dynamic Jahn-Teller effect in the strong interaction approximation. [Chinese]. - Wuli xuebao, Acta Phys. Sin., 1966, 22, N4, 471-486. - PA, 1967, 70, 4058. RZF, 1966, 10E156.

Theor.; Γ_8 - (e + t); O_h; 2.7.

1.7.5. Fan Li-chin, Ou Chin, The theory of the kinetic Jahn-Teller effect in the case of strong coupling. [Chinese]. Kexue Tonbao, Chemistry, 1966, 17, N3, 101-104. - RZF, 1966, 10E155.

Theor.; Γ_8 - (e + t); O_h; 2.7.

1967

1.7.6. Slonczewski, J. C., Moruzzi, V. L., Excited states in the dynamic Jahn-Teller effect. - Physics, 1967, 3, N5, 237-254. - RZF, 1968, 5E195.

Theor.; E - e; 1.3; 2.2.

1969

1.7.7. Stevens, K. W. H., Effective Hamiltonian for the Jahn-Teller coupling of T_{1g} and T_{2g} ions. - J. Phys. C, 1969, 2, N11, 1934-1946. - CA, 1970, 72, 7351.

Theor.; T - e; O_h.

1970

1.7.8. Howgate, D. W., Coulter, C. A., Calculation of the internal energy state populations of a strongly coupled electron-phonon system in thermal equilibrium. - Phys. Rev. B, 1970, 2, N4, 1131-1139.

Theor.

1971

1.7.9. Bates, C. A., Chandler, P. F., Jahn-Teller effect in an orbital triplet state coupled to triplet modes of vibrations. - In: Magn. Resonance & Related Phenomena: Proc. 16th Congr. AMPERE, 1970. Bucharest, 1971, p. 249-257. - CA, 1973, 78, 9552.

Theor.; T - t; T_d; d^9; 3.1.

1972

1.7.10. Blumson, J., The effects of Jahn-Teller ions upon the dynamics of a cubic lattice: Thesis. - Univ. Nottingham., 1972. - PA, 1974, 77, 6594.

Theor.; E - e; O_h.

1.7.11. Fletcher, J. R., A variational ground state for the dynamic Jahn-Teller effect. - J. Phys. C, 1972, 5, N8, 852-862. - PA, 1972, 75, 40583. RZF, 1972, 10E448.

Theor.; E - e; O_h; 1.4.

1.7.12. O'Brien, M. C. M., The dynamic Jahn-Teller effect with many frequencies: a simple approach to a complicated problem. - J. Phys. C, 1972, 5, N15, 2045-2063. - PA, 1972, 75, 61565. RZF, 1973, 1E369.

Theor.; E - e; T - t; 2.3.

1.7.13. Sloane, C. S., Silbey, R., Vibronic interaction in doubly degenerate electronic states. - J. Chem. Phys., 1972, 56, N12, 6031-6043. - CA, 1972, 77, 26712. PA, 1972, 75, 49991. RZF, 1972, 11D70.

Theor.; E - (e_1 + e_2); C_{3v}.

1973

1.7.14. Englman, R., Halperin, B., Effective-Hamiltonian approach to Jahn-Teller coupling of impurities in crystals. - J. Phys. C, 1973, 6, N11, L219-L222. - PA, 1973, 76, 46798. RZF, 1973, 11E380.

Theor.

1.7.15. Gauthier, N., Walker, M. B., Dynamic Jahn-Teller effect for an electronic E state coupled to the phonon continuum. - Phys. Rev. Lett., 1973, 31, N19, 1211-1214. - PA, 1974, 77, 13142. RZF, 1974, 4D705.

Theor.; E - e; 1.4.

1974

1.7.16. Abou-Ghantous, M., Bates, C. A., Chandler, P. E., Stevens, K. W. H., Jahn-Teller effect in orbital triplets. I. Second-order contribution for D(T_2) in cubic crystals coupled to E-type lattice distortions. - J. Phys. C, 1974, 7, N2, 309-324. - PA, 1974, 77, 20549. RZF, 1974, 9E1025.

Theor.; T_2 - e; O.

1.7.17. Abou-Ghantous, M., Bates, C. A., Stevens, K. W. H., Jahn-Teller effect in orbital triplets. II. The theory of D(T_2) ions in trigonal symmetry. - J. Phys. C, 1974, 7, N2, 325-338. - CA, 1974, 80, 88981. PA, 1974, 77, 24700. RZF, 1974, 9E1026.

Theor.; T_2 - e; C_{3v}; d^6.

1.7.18. Abou-Ghantous, M., Jaussaud, P. C., Bates, C. A., Fletcher, J. R., Moore, W. S., Distinction between lattice and cluster models of the Jahn-Teller effect in an orbital doublet. - Phys. Rev. Lett., 1974, 33, N9, 530-533. - CA, 1974, 81, 112951.

Exp. ; E - e; C_{3v}; d^7; 1.4; 3.1; 3.5.

1.7.19. Bates, C. A., Chandler, P. E., Stevens, K. W. H., Jahn-Teller effects in orbital triplets. III. $D(T_2)$ ions coupled to T_2-type lattice distortions. - J. Phys. C, 1974, 7, N21, 3969-3980. - PA, 1975, 78, 6733. RZF, 1975, 4E224.

Theor.; T_2 - t_2; O_h; 3.2.

1.7.20. Halperin, B., Englman, R., Two-frequency description for the Jahn-Teller coupling of impurities in solids. - Phys. Rev. B, 1974, 9, N5, 2264-2272. - CA, 1974, 81, 7068. PA, 1974, 77, 53053. RZF, 1975, 6E86.

Theor.; E - (e_1 + e_2); 1.3; 1.4.

1.7.21. Vekhter, B. G., Polinger, V. Z., Rozenfel'd, Yu. B., Tsukerblat, B. S., Bound states with a Jahn-Teller impurity center. [Russ.]. - Pis'ma Zh. Eksp. & Teor. Fiz., 1974, 20, N2, 84-87, - PA, 1975, 78, 1291. RZF, 1974, 11E513.

Theor.

1975

1.7.22. Abou-Ghantous, M., Bates, C. A., Clark, I. A., Fletcher, J. R., Jaussaud, P. C., Moore, W. S., The multimode model of the Jahn-Teller effect in an orbital doublet. - In: Magn. Resonance & Related Phenomena: Proc. 18th Congr. AMPERE, 1974. Amsterdam, 1975, Vol. 2, p. 565-566. - CA, 1975, 83, 155307. PA, 1975, 78, 86080.

Exp.; E - e; O_h; d^7; 1.4; 2.1; 3.1.

1.7.23. Fischman, A. Ya., Ivanov, M. A., Mitrofanov, V. Ya., The spectrum of localized states of a magnetic insulator containing magnetic impurities with orbital degeneracy. [Russ.]. - Fiz. Tverd. Tela, 1975, 17, N10, 2961-2966. - PA, 1976, 79, 65727.

Theor.

1.7.24. Fletcher, J. R., Bates, C. A., Moore, W. S., Abou-Ghantous, M., Jaussaud, P. C., The multimode model of the Jahn-Teller effect in an orbital doublet. - In: Magn. Resonance & Related Phenomena: Proc. 18th Congr. AMPERE, 1974. Amsterdam, 1975, Vol. 2, p. 419-420.

Exp.; E - e; 1.4; 3.1.

1.7.25. Halperin, B., Englman, R., Lattice distortion induced by Jahn-Teller centers. - J. Phys. C, 1975, 8, N23, 3975-3987. - PA, 1976, 79, 13669.

Theor.; E - e; O_h; 1.2.

1.7.26. Halperin, B., Englman, R., Scattering of lattice waves due to Jahn-Teller interaction. - Phys. Rev. B, 1975, 12, N1, 388-399. - CA, 1975, 83, 106581. PA, 1975, 78, 81175.

Theor.; E - e; O_h; 3.7.

1.7.27. Halperin, B., Localized modes coupled to Jahn-Teller defects of intermediate coupling strength. - Phys. Lett. A, 1975, 55, N5, 301-302. - CA, 1976, 84, 81858. PA, 1976, 79, 26430.

Theor.; E - e.

1.7.28. Ray, T., Ray, D. K., Sangster, M. J. L., Comparison of the molecular cluster model with the phonon model for Jahn-Teller active impurities in crystals. - Solid State Commun., 1975, 17, N1, 93-96. - CA, 1975, 83, 124381. PA, 1975, 78, 61983. RZF, 1975, 12E84.

Theor.; T - e.

1976

1.7.29. Abou-Ghantous, M., Bates, C. A., Fletcher, J. R., Jaussaud, P. C., Multimode Jahn-Teller effects of ions in aluminum oxide. - In: Phonon Scattering in Solids: Proc. 2nd Int. Conf. New York; London, 1976, p. 169-171. - CA, 1977, 86, 179692.

Theor.; E - e; O_h; d^7; 1.4.

1.7.30. Bates, C. A., Brauns, P., Fletcher, J. R., Jaussaud, P. C., The Jahn-Teller theory of manganese(3+) ion-doped aluminum oxide. - J. Phys. (France), 1976, 37, N6, 763-767. - CA, 1976, 85, 84829. PA, 1976, 79, 68812.

Theor.; E - e; O_h; D_4; 1.4.

1.7.31. Englman, R., Halperin, B., Mukamel, S., Scattering from Jahn-Teller impurities in ionic solids. - In: Light Scattering in Solids: Proc. 3rd Int. Conf., 1975. Paris, 1976, p. 557-566. - CA, 1976, 85, 151140. PA, 1976, 79, 69935.

Theor.; 2.8.

1.7.32. Gauthier, N., Walker, M. B., Dynamic Jahn-Teller effect for an electronic E state coupled to the phonon continuum. II. - Can. J. Phys., 1976, 54, N1, 9-25. - CA, 1976, 84, 128420. PA, 1976, 79, 40749.

Theor.; E - e; 1.4; 3.2.

1.7.33. Gosar, P., Interaction of a two-level tunneling system with phonons. - Physica A, 1976, 85, N2, 374-388. - PA, 1977, 80, 15525.

Theor.

1.7.34. Halperin, B., Englman, R., Phonon scattering induced by Jahn-Teller impurities. - In: Phonon Scattering in Solids: Proc. 2nd Int. Conf. New York; London, 1976, p. 163-165. - CA, 1977, 86, 179691.

Theor.; E - e; 3.7.

1.7.35. Halperin, B., Quasi-molecular description of Jahn-Teller effects in solids at finite temperatures. - J. Phys. C, 1976, 9, N22, 4139-4149. - CA, 1977, 86, 147913. PA, 1977, 80, 7832.

Theor.

1.7.36. Ivanov, M. A., Mitrofanov, V. Ya., Falkovskaya, L. D., Fischman, A. Ya., The properties of a threefold degenerate orbital magnetic impurity in a ferromagnetic crystal. - Phys. Status Solidi b, 1976, 74, N1, 57-67. - PA, 1976, 79, 40632.

Theor.

1.7.37. Rashba, E. I., Optical spectra of phonons bound to impurity centers. [Russ.]. - Zh. Eksp. & Teor. Fiz., 1976, 71, N1, 319-329. - CA, 1976, 85, 133292.

Theor.; 2.7; 2.8.

1.7.38. Rozenfel'd, Yu. B., Polinger, V. Z., Dynamic Jahn-Teller effect for the E-term taking into account phonon dispersions. [Russ.]. - Zh. Eksp. & Teor. Fiz., 1976, 70, N2, 597-609. - PA, 1976, 79, 45012.

Theor.; E - e; 2.3; 2.7; 2.8.

1.7.39. Rueff, M., Sigmund, E., Wagner, M., Resonance scattering of phonons at trigonal Jahn-Teller centers. - In: Phonon Scattering in Solids: Proc. 2nd Int. Conf. New York; London, 1976, p. 166-168.

Theor.; E - e; 3.7.

1.7.40. Vasil'ev, A. V., Malkin, B. Z., Natadze, A. L., Ryskin, A. I., Microscopic determination of the Jahn-Teller interaction energy in a cobalt-doped zinc sulfide crystal. [Russ.]. - Zh. Eksp. & Teor. Fiz., 1976, 71, N3, 1192-1203. - CA, 1976, 85, 184367. PA, 1977, 80, 4623.

Exp.; T - (a_1 + e + t_2); T_d; d^7; 5.3.

1.7.41. Vekhter, B. G., Polinger, V. Z., Rozenfel'd, Yu. B., The Jahn-Teller effect for a T-term taking into account phonon dispersion. [Russ.]. - In: Tez. Dokl. 5 Vsesoyuz. Simp. Spektrosk. Kristallov Aktivir. Redkimi Zeml'ami Elementami Gruppy Zheleza. Kazan', 1976.

Theor.; T - t.

1977

1.7.42. Biernacki, S. W., The multimonde Jahn-Teller coupling for an orbital triplet in cubic crystals. - Phys. Status Solidi b, 1977, 84, N2, 699-708. - CA, 1978, 88, 56567. PA, 1978, 81, 23062.

Theor.; T - e; O_h.

1.7.43. Fal'kovskaya, L. D., Fishman, A. Ya., Ivanov, M. A., Mitrofanov, V. Ya., Electromagnetic absorption in magnetically ordered crystals with orbitally degenerate impurities. [Russ.]. - Fiz. Nizk. Temp., 1977, 3, N4, 488-496.

Theor.; E - e.

1.7.44. Rueff, M., Sigmund, E., Wagner, M., Resonance scattering of phonon at trigonal Jahn-Teller centers. - Phys. Status Solidi b, 1977, 81, N2, 511-520. - CA, 1977, 87, 60972. PA, 1977, 80, 66010.

Theor.; 3.7.

1.7.45. Steggles, P., Calculations on the multimode Jahn-Teller effect for T $\times$ e in a cubic lattice. - J. Phys. C, 1977, 10, N15, 2817-2830. - CA, 1978, 88, 14519. PA, 1977, 80, 76782.

Theor.; T - e; O_h; 1.4.

1978

1.7.46. Biernacki, S. W., Jahn-Teller coupling of chromium(2+) ion in zinc sulfide, zinc selenide, and zinc telluride. - Phys. Status Solidi b, 1978, 87, N2, 607-612. - CA, 1978, 89, 68019. PA, 1978, 81, 67997.

Theor.; T - e; T_d; D_4; 5.3.

1.7.47. Englman, R., Halperin, B., Cluster model in vibronically coupled systems. - Ann. Phys. (France), 1978, 3, N6, 453-478. - CA, 1979, 90, 127643. PA, 1979, 82, 19540.

Rev.

1.7.48. Polinger, V. Z., The Jahn-Teller effect of a T-term strongly coupled to E_g modes: An account of phonon dispersion. - In: Abstr. 3rd Symp. Jahn-Teller Effect, Levico Terme. Trento, 1978, 1 p.

Theor.; T - (e + t); O_h; 2.7; 2.8.

1979

1.7.49. Biernacki, S. W., An average phonon energy in the Jahn-Teller effect. - Phys. Lett. A, 1979, 69, N1, 61-62. - CA, 1979, 90, 31460. PA, 1979, 82, 15122.

Theor.

1.7.50. Polinger, V. Z., The multimode Jahn-Teller effect for a T-term: The case of prevailing coupling with E-vibrations. [Russ.]. - Zh. Eksp. & Teor. Fiz., 1979, 77, N4, 1503-1518. - CA, 1980, 92, 31169. PA, 1980, 83, 15490.

Theor.; T - (e + t); O_h; T_d; 2.7; 2.8.

1.7.51. Polinger, V. Z., Bersuker, G. I., Multimode Jahn-Teller effect for an E-term with strong vibronic coupling. I. Local and resonant states. - Phys. Status Solidi b, 1979, 95, N2, 403-411. - CA, 1979, 91, 219487. PA, 1980, 83, 24858.

Theor.; E - e; 1.4; 2.7; 2.8.

1.7.52. Polinger, V. Z., Bersuker, G. I., Multimode, Jahn-Teller effect for an E-term with strong vibronic coupling. II. Band shapes of the infrared and Raman spectra. - Phys. Status Solidi b, 1979, 96, N1, 153-161. - CA, 1980, 92, 31410. PA, 1980, 83, 20139.

Theor.; E - e; 1.4; 2.7; 2.8.

See also: 1.3.-5; 1.4.-11, 21, 22; 2.1.-94, 144, 160, 218, 243, 272; 2.2.-76; 2.3.-48; 2.8.-97; 3.1.-71, 193; 3.2.-89, 101; 3.5.-49, 60, 64; 4.1.-108.

1.8. The Jahn-Teller Effect in Polynuclear Metal Clusters

The first paper devoted to the Jahn-Teller effect in a binuclear copper(II) cluster was that by Lohr (Ref. [1.8.1]). The vibronic reduction of the Coulombic and exchange interactions in binuclear clusters was studied in Ref. [1.8.5]. In the paper [1.8.8] the role of superexchange and vibronic interaction in the formation of the ground state in binuclear clusters is discussed. The adiabatic potentials in binuclear Jahn-Teller clusters were reexamined in Ref. [1.8.12]. Note also Ref. [1.8.16] where the vibronic mechanism of intramolecular electron transfer is studied.

1968

1.8.1. Lohr, L. L., Jr., The Jahn-Teller effect in binuclear copper(II) complexes. - Proc. Natl. Acad. Sci. USA, 1968, 59, N3, 720-725. - CA, 1968, 69, 72704. PA, 1969, 72, 18181.

Theor.; E - e; O_h; D_{4h}; d^9; 1.2.

1969

1.8.2. Novak, P., Interactions between octahedrally coordinated E_g Jahn-Teller ions. - J. Phys. & Chem. Solids, 1969, 30, N10, 2357-2364. - CA, 1969, 71, 116608. PA, 1970, 73, 12326.

Theor.; E - e; O_h; 1.2.

1970

1.8.3. Novak, P., Stevens, K. W. H., The interaction between two $T_{1g}(T_{2g})$ Jahn-Teller ions. - J. Phys. C, 1970, 3, N8, 1703-1710. - CA, 1970, 73, 123590. PA, 1970, 73, 72293.

Theor.; T - e; O_h; 1.2.

1971

1.8.4. Baker, J. M., Interactions between ions with oribtal angular momentum in insulators. - Rep. Progr. Phys., 1971, 34, N2, 109-173.

Rev.

1972

1.8.5. Tsukerblat, B. S., Vekhter, B. G., Vibronic reduction of coulomb and exchange interactions in impurity pairs. [Russ.]. - Fiz. Tverd. Tela, 1972, 14, N9, 2544-2549. - CA, 1973, 78, 64750. PA, 1973, 76, 33813. RZF, 1972, 12E489.

Theor.; T - e; O_h; 1.4.

1973

1.8.6. Bersuker, I. B., Vekhter, B. G., Rafalovich, M. L., The Jahn-Teller effect in binuclear complexes: stable configurations, reduction of exchange interaction. - In: Proc. 15th Int. Conf. Cood. Chem., Moscow, 1973, p. 8-9.

Theor.; E - $(a_{1g} + a_{2u} + b_{2g} + b_{1u})$; D_{4h}; 1.2.

1.8.7. Fujiwara, Takeo, Two-exiton transitions in $RbMnF_3$. I. Jahn-Teller effects in a pair of excited ions. - J. Phys. Soc. Jpn., 1973, 34, N1, 36-43. - CA, 1973, 78, 50014. PA, 1973, 76, 11867. RZF, 1973, 6E764.

Theor.; T - $(a_{1g} + e_g)$; 1.2.

1.8.8. Kugel', K. I., Khomskiy, D. I., The exchange coupled pairs of Jahn-Teller impurities in crystals. [Russ.]. - Fiz. Tverd. Tela, 1973, 15, N7, 2230-2231. - PA, 1974, 77, 37408. RZF, 1973, 12E1370.

Theor.; 3.1.

1.8.9. Passeggi, M. C. G., Stevens, K. W. H., The direct exchange between two T_g Jahn-Teller ions. - J. Phys. C, 1973, 6, N1, 98-108. - CA, 1973, 78, 77434. PA, 1973, 76, 18088. RZF, 1973, 6E376.

Theor.; T - e; O_h; 3.1.

1974

1.8.10. Murao, T., Jahn-Teller effect in trinuclear complexes. - Phys. Lett. A, 1974, 49, N1, 33-35. - CA, 1974, 81, 143600. PA, 1974, 77, 75796. RZF, 1975, 2E195.

Theor.; E - e; C_{3v}.

1.8.11. Rafalovich, M. L., Vibronnye Vzaimodeistviya v Tetraedricheskikh, Bioktaedricheskikh i Tsepochechnykh Sistemakh: Avtoref. Dis. ... Kand. Fiz.-Mat. Nauk. [Russ.]. - Sverdlovsk, 1974. - 16 p. - Ural'sk. Univ.

The Vibronic Interaction in Tetrahedral, Bioctahedral, and Chain Systems.

Theor., E - e; T - e; T - t; 1.2; 2.1; 4.2.

1975

1.8.12. Bersuker, I. B., Vekhter, B. G., Rafalovich, M. L., Vibronic interaction in exchange-coupled impurity pairs. - Cryst. Lattice Defects, 1975, 6, N1/2, 1-6. - CA, 1976, 85, 70128. PA, 1976, 79, 13944.

Theor.; T; 1.2; 1.4.

1.8.13. Kugel', K. I., Khomskiy, D. I., The ground state of an exchange-coupled pair of Jahn-Teller ions. [Russ.]. - In: Kvant. Khim.: Tez. Dokl. 6 Vsesoyuz. Soveshch. Kishinev, 1975, K-136. - CA, 1976, 84, 157376. RZK, 1976, 1B61.

Theor.

1.8.14. Kurzynski, M., A simple theory of the temperature dependence of exchange interactions between E ions. - J. Phys. C, 1975, 8, N17, 2749-2759. - CA, 1975, 83, 171789. PA, 1976, 79, 2273. RZK, 1976, 6B631.

Theor.; E - e.

1978

1.8.15. Eremin, M. V., Kalinenkov, V. N., Rakitin, Yu. V., Exchange interaction of Cu^{2+} and Mn^{2+} ions in $KZnF_3$. [Russ.]. - Fiz. Tverd. Tela, 1978, 20, N9, 2832-2834. - CA, 1978, 89, 221139.

Theor.; E - e; O_h; d^9.

1.8.16. Ratner, M. A., Theoretical approaches to intramolecular electron transfer processes. - Int. J. Quant. Chem., 1978, 14, N5, 675-694. - CA, 1979, 90, 110259.

Theor.; E - $(b_1 + b_2)$; C_{4v}; 1.5.

See also: 2.5.-28; 3.1.-251; 4.4.-147; 5.4.-54, 67, 77, 96.

2. ELECTRONIC AND VIBRATIONAL SPECTROSCOPY*

This section is divided into 8 subsections. The greatest number of references are devoted to vibronic optical band shapes (Subsection 2.1.). A few current trends can be noted in this field of spectroscopy: i) the restoration of the vibronic optical bands by using numerical calculation of the vibronic energy levels; ii) development of approximate analytic approaches; and iii) theory of polarization dichroism. These questions are discussed in detail Ref. [2.1.-220].

The problems of luminescence and the related topic of nonradiative relaxation are listed in Subsection 2.2. The Jahn-Teller effect in the excited state is manifested in the polarization of emission and in peculiar temperature features. In the case of weak and moderate vibronic interaction the zero-phonon lines and phonon satellites occur. The zero-phonon splitting due to the influence of internal and external fields provides important information about the reduction factors (Subsection 2.3; see also Refs. [2.1.-22, 220]). Subsections 2.5 and 2.6 contain works on photo- and x-ray spectroscopy. Subsection 2.7 is devoted to the vibronic infrared spectra.

2.1. Optical Broad Bands of Electronic Transitions

The evaluation of band shapes for singlet-multiplet transitions dates back to Teller's paper [2.1.3]. The earliest calculation for the $A \rightarrow E$ type transitions was made by Longuet-Higgins et al. [1.3.3] using numerical computations of the energy levels for the $E - e$ problem. At $T = 0$ the optical band of the $A \rightarrow E$ transition represents a two-humped asymmetric curve, whereas the $E \rightarrow A$ transition results in a bell-shaped curve. When the vibronic coupling parameter increases, the asymmetry smooths out.

For a long time the paper [1.3.3] remained the only one in the field. The more recent work [2.1.63] based on the semiclassical approach (see also Ref. [2.1.50]) stimulated further developments of

---

*The short introductions to this section and subsections were prepared by B. S. Tsukerblat.

the theory of optical bands with the Jahn-Teller effect. In spite of its approximate nature, the semiclassic theory provides an understanding of the main features of the Jahn-Teller optical bands. Its advantage lies in the possibility of carrying out analytic calculations of band shape functions for most Jahn-Teller problems, of obtaining explicitly their temperature dependence, and of simplifying the multimode problem.

The case of strong spin-orbit interaction was considered in Ref. [2.1.50], the results of the theory being applied to the F band calculation in cesium halides. Spin-orbit interactions in the $^2E - b_1 + b_2$ problem for tetragonal centers and the optical $^2A \rightarrow {}^2E$ band were considered in Ref. [2.1.171]. A general consideration taking into account both the Jahn-Teller interactions with A_1, E, and T_2 vibrations and spin-orbit coupling in the $^2A \rightarrow {}^2T$ band shape was presented in Ref. [2.1.77]. Later the theory was applied to the Tl^+ type centers in alkali halides [2.1.-180, 181].

Some qualitative results on the band-shape theory are obtained directly from experimental data concerning the d-d spectra of transition metal ions ([2.1.-112, 125, 165, 167, 168]), Tl^+ type systems ([2.1.-113, 114, 121, 145, 161, 166, 180, 181, 211]), color centers ([2.1.-274, 135, 182]), and complex aggregate centers (R' centers) ([2.1.-117, 118]). For Jahn-Teller exitons, if the exiton levels form a continuous band, the vibronic interaction results in the pseudo-Jahn-Teller effect. In the case of narrow exiton bands the small-radius exiton is similar to the local center, and the dynamic Jahn-Teller interaction is manifested in the exiton absorption band splitting. In the opposite case of broad bands exceeding the energy of vibronic coupling, the autolocalization does not occur, and the Jahn-Teller splitting is smoothed out [2.1.-197, 198].

More recently, the semiclassic theory was extended to include forbidden transitions [2.1.82]. Renner-type systems (strong quadratic vibronic interaction) were considered in Refs. [2.1.-124, 138, 155, 159, 166].

The applicability condition of the semiclassic theory was used within the framework of quantum-statistical methods [2.1.-85, 86, 87, 169, 170]. Further refinement of this theory is given in Refs. [2.1.-170, 172, 220]. A more precise analytic approach is provided by the so-called "indendent-ordering approximation" [2.1.170]. The independent-ordering approximation was applied to $A \rightarrow E - e$ [2.1.240] and $A \rightarrow T - (a_1 + e + t_2)$ transitions [2.1.218]. The theory of multiplet-multiplet transitions is given in Refs. [2.1.-64, 65, 169, 87].

Both the semiclassic approach and the independent-ordering approximation allow one to obtain the envelope of optical bands, and no details of the vibronic band structure can be reproduced by these approaches. That is why numerical calculations were developed intensively in the sixties and seventies. The A → T - t_2 optical band was calculated in Ref. [2.1.111].

The important case of equal vibronic coupling with E and T_2 modes is considered in Refs. [1.3.-33, 40]. By means of magneto-optic measurements, was shown in Ref. [2.4.48] that the nontotally symmetric Jahn-Teller active modes give the main contribution to the second moment of the F^+ band of CaO, whereas the contribution of the total symmetric modes is less than 20% [2.4.48]. The observed band shape is close to the calculated one [1.3.40].

Numerical calculations allow one to reveal the resonance structure of the optical band. These features were first considered in Ref. [2.1.187] for the A → E band. The pseudo-Jahn-Teller effect was considered in Ref. [2.1.219].

1938

2.1.1. Seitz, F., Interpretation of the properties of alkali halide-thallium phosphors. - J. Chem. Phys., 1938, 6, N3, 150-162. - CA, 1938, 32, 3265[9].

Exp.

1941

2.1.2. Sponer, H., Teller, E., Electronic spectra of polyatomic molecules. - Rev. Mod. Phys., 1941, 13, N2, 76-170. - Bibliogr.: 226 ref.

Rev.

2.1.3. Teller, E., Asymmetric vibrations excited by electronic transitions. - Ann. N.Y. Acad. Sci., 1941, 41, 173-186.

Theor.; 1.2.

1942

2.1.4. Mulliken, R. S., Teller, E., Interpretation of the MeI absorption bands near 2000. - Phys. Rev., 1942, 61, N5, 283. - CA, 1942, 36, 2476[1].

Theor.; 1.2.

1953

2.1.5. Craig, D. P., The electronic spectra of the simple aromatic hydrocarbons. - Rev. Pure & Appl. Chem., 1953, 3, N4, 207-240. - CA, 1954, 48, 6229a. RZK, 1955, N5, 6982.

Theor.; E- e; D_{4h}.

1954

2.1.6. Innes, K. K., Analysis of the near-ultraviolet absorption spectrum of acetylene. - J. Chem. Phys., 1954, 22, N5, 863-876. - CA, 1954, 48, 11924c. RZK, 1955, N23, 54410.

Exp.; 1.6.

2.1.7. Moffitt, W. E., The electronic spectra of cata-condensed hydrocarbons. - J. Chem. Phys., 1954, 22, N2, 320-333. - CA, 1954, 48, 6233d. RZK, 1955, N17, 36634.

Theor.

2.1.8. Moffitt, W. E., Configuratonal interaction in simple molecular orbital theory. - J. Chem. Phys., 1954, 22, N11, 1820-1829. - CA, 1955, 49, 2176c. RZK, 1955, N17, 36631.

Theor.

1956

2.1.9. Liehr, A. D., Moffitt, W., Rydberg spectrum of benzene. - J. Chem. Phys., 1956, 25, N5, 1074.

Theor.; E - e; D_{6h}.

1957

2.1.10. Coulson, C. A., Kearsley, M. J., Color centers in irradiated diamonds. I. - Proc. R. Soc. London. Ser. A., 1957, 241, N1227, 433-454. - CA, 1958, 52, 7864e. RZK, 1958, N21, 69911.

Theor.; E; T; T_d.

2.1.11. Liehr, A. D., Ballhausen, C. J., Intensities in inorganic complexes. - Phys. Rev., 1957, 106, N6, 1161-1163. - RZF, 1958, N2, 3127.

Theor.; E; T; $(E_g + T_{1u}) - e_u$; $(^2T_{1u} + {}^2T_{2g}) - e_u$; O_h; d^1; d^9; 1.5.

1958

2.1.12. Pryce, M. H. L., Runciman, W. A., The absorption spectrum of vanadium corundum. - Discuss. Faraday Soc., 1958, N26, 34-42. - Discuss., 87-95. - RZK, 1960, N5, 16755.

Exp.; T; C_{3v}; d^2.

1959

2.1.13. Dressler, K., Ramsay, D. A., The electronic absorption spectra of NH_2 and ND_2. - Philos. Trans. R. Soc. London A, 1959, 251, N1002, 553-602.

Exp.; 1.6.

1960

2.1.14. Cotton, F. A., Meyers, M. D., Magnetic and spectral properties of the spin-free $3d^6$ systems iron(II) and cobalt(III) in cobalt(III) hexafluoride ion: probable observation of dynamic Jahn-Teller effects. - J. Am. Chem. Soc., 1960, 82, N19, 5023-5026. - CA, 1961, 55, 6135 i.

Exp.; E; O_h; d^6; 3.6.

2.1.15. Dixon, R. N., The absorption spectrum of the free NCO radical. - Philos. Trans. R. Soc. London A, 1960, 1960, 252, 165-192. - CA, 1961, 55, 10054g. PA, 1961, 64, 856.

Exp.; 1.6.

2.1.16. Eisenstein, J. C., Pryce, M. H. L., Theory of the magnetic and spectroscopic properties of neptunium hexafluoride. - Proc. R. Soc. London. Ser. A, 1960, 255, N1281, 181-198. - CA, 1961, 55, 16142f.

Theor.; Γ_8; O_h; 3.6.

2.1.17. Englman, R., Some temperature-dependent effects on the optical absorption line shape of paramagnetic ions. - Mol. Phys., 1960, 3, N1, 23-34. - RZF, 1961, 2V263.

Theor.; T - e.

1961

2.1.18. Bader, R. F. W., Westland, A. D., The electronic spectra of $MoCl_5$ and $NbCl_5$. - Can. J. Chem., 1961, 39, N12, 2306-2315. - CA, 1962, 56, 2089b.

Exp.; E - e; D_{3h}.

2.1.19. Douglas, A. E., Hollas, J. M., The 1600-Å band system of ammonia. - Can. J. Phys., 1961, 39, N4, 479-501. - RZF, 1962, 2V1400.

Exp.; E - e; D_{3h}.

2.1.20. Griffith, J. S., The Theory of Transition-Metal Ions. - Cambridge. Univ. Press, 1961. - 455 p.

3.1; 3.6.

2.1.21. Johns, J. W. C., The absorption spectrum of BO_2. - Can. J. Phys., 1961, 39, N12, 1738-1768. - CA, 1962, 57, 10662a. - RZF, 1962, 10V107.

Exp.; 1.6.

1962

2.1.22. Bersuker, I. B., Intrinsic asymmetry in coordination compounds. III. Influence on optical properties. [Russ.]. - Zh. Strukt. Khimii, 1962, 3, N1, 64-69.

Theor.; E - e.

2.1.23. Blankenship, F. A., Belford, R. L., VCl_4 vapor spectrum and Jahn-Teller splitting. - J. Chem. Phys., 1962, 36, N3, 633-639. - CA, 1962, 15056h. RZF, 1962, 10V109.

Exp.; E - e: T - e: T_d: d^1.

2.1.24. Feofilov, P. P., Kaplyanskiy, A. A., Latent optical anisotropy of cubic crystals which contain local centers and methods for its investigation. [Russ.]. - Usp. Fiz. Nauk, 1962, 76, N2, 201-238. - CA, 1962, 56, 15009h.

Rev.

2.1.25. Liehr, A. D., Forbidden transitions in organic and inorganic systems. - In: Adv. Chem. Phys. New York, etc., 1962, Vol. 5, p. 241-260.

Rev.; 1.5.

1963

2.1.26. Bedon, H. D., Horner, S. M., Tyree, S. Y., Jr., A molecular orbital treatment of the spectrum of TiF_6^{3-}. - 1963. - 21 p. - US Dept. Com., Office Tech. Serv., AD, 417633. - CA, 1964, 60, 15304a.

Theor.

2.1.27. Dunn, T. M., Appraisal of experiment and theory in the spectra of complexes. - Pure & Appl. Chem., 1963, 6, N1, 1-21. - CA, 1963, 59, 115e.

Rev.

2.1.28. Joergensen, C. K., Absorption spectra of osmium(III), osmium(IV), and platinum(VI) mixed halide and hexaiodide complexes. - Acta Chem. Scand., 1963, 17, N4, 1043-1048. - CA, 1963, 59, 7072a.

Exp.; d^1; d^4.

2.1.29. Kristofel', N. N., Zavt, G. C., The Condon approximation and optical properties of impurity centers. [Russ.]. - Fiz. Tverd. Tela, 1963, 5, N5, 1279-1285. - CA, 1963, 59, 2290c.

Theor.

2.1.30. Runciman, W. A., Syme, R. W. G., The optical absorption of divalent chromium in $CrCl_2 \cdot 4H_2O$ and $CrSO_4 \cdot 7H_2O$. - Philos. Mag., 1963, 8, N88, 605-613. - CA, 1963, 58, 13294d. RZF, 1963, 11D398.

Exp.; d^4.

1964

2.1.31. Bedon, H. D., Horner, S. M., Tyree, S. Y., Jr., Molecular orbital treatment of the spectrum of the hexafluorotitanium(III) anion. - Inorg. Chem., 1964, 3, N5, 647-652. - CA, 1964, 60, 14003b.

Theor.; T; O_h; d^1.

2.1.32. Dijkgraff, C., Electronic transitions in α-, β-, and γ-titanium trichloride. - Nature, 1964, 201, N4924, 1121-1122. - CA, 1964, 60, 12790e.

Exp.; E; T; O_h; d^1.

2.1.33. Fackler, J. P., Jr., Chawla, I. D., Spectra of manganese(III) complexes. I. Aquo-manganese(III) ion and hydroxide, fluoride, and chloride complexes. - Inorg. Chem., 1964, 3, N8, 1130-1134.

Exp.; E - e; O_h; d^1.

2.1.34. Fackler, J. P., Jr., Holah, D. G., Chawla, I. D., Dynamic Jahn-Teller effects in d^4 systems — aqueous manganese(III) and other complexes of Mn(III) and Cr(III). - In: Proc. 8th Int. Conf. Coord. Chem. Vienna; New York, 1964, p. 75-77. - CA, 1967, 66, 119417.

Exp.; d^4.

2.1.35. Ferguson, J., Guggenheim, H. J., Wood, D. L., Electronic absorption spectrum of Ni^{2+} in cubic perovskite fluorides. I. - J. Chem. Phys., 1964, 40, N3, 822-830. - RZF, 1964, 8D192.

Exp.; E; T; O_h; d^8.

2.1.36. Fukuda Atsuo, Alkali halide phosphors containing impurity ions with s^2 configuration. - Sci. Light, 1964, 13, N2/3, 64-114. - RZF, 1966, 3D646.

Exp.; T; 1.5; 2.2.

2.1.37. Fukuda Atsuo, Inohara Koichi, Onaka Ryumyo, A, B, and C bands in KCl:In and KCl:Sn. - J. Phys. Soc. Jpn., 1964, 19, N8, 1274-1280. - CA, 1964, 61, 10197b.

Exp.; T - e.

2.1.38. Israeli, Y. J., Tentative interpretation of the ultraviolet spectra of xenon tetrafluoride. - Bull. Soc. Chim. France, 1964, N3, 649-650. - CA, 1964, 61, 1400f.

Exp.; E; D_{4h}.

2.1.39. Martin, D. S., Jr., Lenhardt, C. A., The vibronic absorption spectrum and dichroism of potassium tetrachlorplatinate(II). - Inorg. Chem., 1964, 3, N10, 1368-1373. - CA, 1964, 61, 11489f. RZF, 1965, 5D380.

Exp.; E.

2.1.40. Schläfer, H. L., Absorption spectra and magnetic moments of octahedral titanium(III) complexes. [Germ.]. - In: Theory and Struct. Complex Compounds: Papers Symp., Wroclaw, 1962. London; New York, 1964, p. 181-193. - CA, 1965, 63, 12522c.

Exp.; E; O_h; d^1.

1965

2.1.41. Bersuker, I. B., Zhigunova, I. A., Kovner, M. A., Nul'man, V. S., Analysis of electronic, vibrational, and electron-vibrational spectra of some rare-earth chelates. [Russ.]. - In: Tez. Dokl. 2 Soveshch. Primeneniyu Fiz. Metodov Issled. Kompleks. Soedin. Kishinev, 1965, p. 94.

Exp.

2.1.42. Colson, S. D., Bernstein, E. R., First and second triplet of solid benzene. - J. Chem. Phys., 1965, 43, N8, 2661-2669.

Exp.; E.

2.1.43. Fackler, J. P., Jr., Davis, T. S., Chawla, I. D., Hydrogen bonding to manganese(III) β-ketoenolates: influence on the low-energy electronic band. - Inorg. Chem., 1965, 4, N1, 130-132.

Exp.; E; T; O_h; d^4.

2.1.44. Fackler, J. P., Jr., Holah, D. G., Properties of chromium(II) complexes. I. Electronic spectra of the simple salt hydrates. - Inorg. Chem., 1965, 4, N7, 954-958.

Exp.; E; T; O_h; d^4.

2.1.45. Ferguson, J., Guggenheim, H. J., Kamimura, H., Tanabe, Y., Electronic structure of Ni^{2+} in MgF_2 and ZnF_2. - J. Chem. Phys., 1965, 45, N2, 775-786. - CA, 1965, 62, 6025e.

Exp.; T; d^8.

2.1.46. Holah, D. G., Fackler, J. P., Jr., Preparation and properties of Cr(II) complexes. II. Complexes with pyridine. - Inorg. Chem., 1965, 4, N8, 1112-1116. - CA, 1965, 63, 6589g.

Exp.; E; T; O_h; d^4.

2.1.47. Holah, D. G., Fackler, J. P., Jr., Preparation and properties of Cr(II) complexes. III. Complexes with dimethylsulfoxide. - Inorg. Chem., 1965, 4, N12, 1721-1725. - RZK, 1966, 19V103.

Exp.; O_h; d^4; 2.7; 4.6.

2.1.48. Katsuura Kanji, Remarks on excitation transfer between unlike molecules. - J. Chem. Phys., 1965, 43, N11, 4149-4157. - PA, 1966, 69, 7380.

Theor.; 1.5.

2.1.49. Koswig, H. D., Kunze, I., Optical absorption of silver halides doped with Cr. I. Cr^{2+}. [Germ.]. - Phys. Status Solidi, 1965, 9, N2, 451-461. - CA, 1965, 63, 7782e.

Exp.; E - e; O_h; d^3; 2.7.

2.1.50. Moran, P. R., Effects of dynamic lattice distortions on the structure of the F band in the cesium halides. - Phys. Rev. A, 1965, 137, N3, 1016-1027. - CA, 1965, 62, 5978h.

Theor.; T - (a_1 + e); Γ_8 - e; O_h.

2.1.51. O'Brien, M. C. M., The Jahn-Teller coupling of $3d^6$ ions in a cubic crystal. - Proc. Phys. Soc. London, 1965, 86, Pt. 4, N552, 847-856. - CA, 1965, 63, 14239c.

Theor.; E - e; T - e; O_h; d^6; 3.1.

2.1.52. Onaka, R., Fukuda, A., Mabuchi, T., A new type of absorption band in NaCl:Pb and KCl:Pb. - J. Phys. Soc. Jpn., 1965, 20, N3, 466. - CA, 1966, 64, 10598g.

Exp.; T_{1u}; O_h.

2.1.53. Perlin, Yu. E., Tsukerblat, B. S., On the theory of the multiphonon processes in paramagnetic local centers. [Russ.]. - In: Tez. Dokl. 12 Vsesoyuz. Soveshch. Fiz. Nizk. Temp. Kazan', 1965, p. 72.

Theor.; T - e; d^3.

2.1.54. Toyozawa Yutaka, Inoue Masaharu, Vibration-induced splitting of the A and C absorption bands of heavy metal ions in alkali-halides. - J. Phys. Soc. Jpn., 1965, 20, N7, 1289-1290.

Theor.; T_{1u} - t_{2g}.

2.1.55. Walsh, A. D., Electronic spectra of polyatomic molecules. - In: Ann. Rept. Progr. Chem., 1964. London, 1965, Vol. 61, p. 8-26. - CA, 1966, 64, 12053g.

1966

2.1.56. Bersuker, I. B., Vekhter, B. G., Tsukerblat, B. S., Features of the optical spectra of ionic crystals with degenerate impurity terms. [Russ.]. - In: Tez. Dokl. 15 Soveshch. Luminestsentsii. Tbilisi, 1966, p. 14.

Theor.; E - e; T - e; O_h.

2.1.57. Ferguson, J., Guggenheim, H. J., Electronic absorption spectrum of Ni(II) in cubic perovskite fluorides. II. Concentration and exchange effects. - J. Chem. Phys., 1966, 44, N3, 1095-1102.

Exp.; E; T; d^8; 1.5.

2.1.58. Holah, D. G., Fackler, J. P., Jr., Preparation and properties of Cr(II) complexes. IV. Complexes with acetonitrile and observations on tetrahedral Cr(II). - Inorg. Chem., 1966, 5, N3, 479-482. - CA, 1966, 64, 15351c.

Exp.; E; T; T_d; d^4; 4.6.

2.1.59. Julg, A., Generalization of the concept of configuration interaction: vibronic interaction. Application to the calculation of vertical transition energies. [Fr.]. - J. Chim. Phys. & Phys. - Chim. Biol., 1966, 63, N11-12, 1459-1466. - CA, 1967, 67, 37915. PA, 1967, 70, 13715.

Theor.

2.1.60. Mabuchi Teruhiko, Fukuda Atsuo, Onaka Ryumyo, Optical properties of gallium-doped alkali halides. - Sci. Light, 1966, 15, N1, 79-96. - CA, 1967, 66, 24135.

Exp.; T; O_h; 2.2.

2.1.61. Merer, A. J., Travis, D. N., The absorption spectrum of CNC. - Can. J. Phys., 1966, 44, N2, 353-372. - CA, 1966, 64, 9092a.

Exp.; 1.6.

2.1.62. Pryce, M. H. L., Interaction of lattice vibrations with electrons at point defects. - In: Phonons in Perfect Lattices and in Lattices with Point Imperfections. Edinburg; London, 1966, p. 403-448.

Rev.

2.1.63. Toyozawa Yutaka, Inoue Masaharu, Dynamical Jahn-Teller effect in alkali halide phosphors containing heavy metal ions. - J. Phys. Soc. Jpn., 1966, 21, N9, 1663-1679. - CA, 1967, 66, 6404.

Theor.; E - e; T - (a + t).

2.1.64. Tsukerblat, B. S., Optical bands in paramagnetic crystals with degenerate impurity terms. [Russ.]. - Zh. Eksp. & Teor. Fiz., 1966, 51, N3, 831-841. - CA, 1967, 66, 15179.

Theor.; T - (a_1 + e + t); O_h; d^3.

2.1.65. Vekhter, B. G., Influence of intrinsic asymmetry on optical spectrum of transition metal complexes. [Russ.]. - Opt. & Spektrosk., 1966, 20, N2, 258-264. - CA, 1966, 64, 16830a.

Theor.; E - e; T - e; T - t.

1967

2.1.66. Einführung in die Ligandenfeldtheorie. [Germ.]./Ed.: H. L. Schläfer, G. Glieman. - Frankfurt/Main:Akad. Verl., 1967. - 535 S. - CA, 1967, 67, 121917.

Introduction to Ligand Field Theory.

2.1.67. Fischer, F., Optical absorption of U_2-centers in alkali metal halides of NaCl-type. [Germ.]. - Z. Phys., 1967, 204, N4, 351-374. - CA, 1968, 68, 7534.

Exp.; T - (e + t).

2.1.68. Fukuda, A., Makishima, S., Vibration-induced fine structures of the A and C absorption of In^+ ion in CsBr crystal. - Phys. Lett. A, 1967, 24, N5, 267-269.

Exp.; E; T; O_h.

2.1.69. Joergensen, C. K., Invariance and distortions of octahedral chromophores. - In: Adv. Chem. Ser. Washington, 1967, N62, p. 161-177. - Bibliogr.: 127 ref. CA, 1967, 66, 108933.

Exp.

2.7.70. Jones, G. D., Jahn-Teller splitting in the optical-absorption spectra of divalent iron compounds. - Phys. Rev., 1967, 155, N2, 259-261. - CA, 1967, 66, 120420.

Exp.; E; O_h; d^6.

2.1.71. Koswig, H. D., Retter, U., Ulrici, W., Jahn-Teller effect on transition metal ions in silver halides. I. Titanium(III) ($3d^1$) and iron(II) ($3d^6$). [Germ.]. - Phys. Status Solidi, 1967, 24, N2, 605-614. - CA, 1968, 68, 44495.

Exp.; E; O_h; d^1; d^6.

2.1.72. Kroto, H. W., The ${}^1\Pi_u \rightarrow {}^1\Delta_g$ electronic spectrum of NCN. - Can. J. Phys., 1967, 45, N3, 1439-1450. - CA, 1967, 67, 37829.

Exp.; 1.6.

2.1.73. Schmitz-Dumont, O., Grim, D., Color and constitution of inorganic solids. XVI. Light absorption of divalent copper in binary and ternary fluorides. [Germ.]. - Z. Anorg. & Allg. Chem., 1967, 355, N5/6, 280-294. - CA, 1968, 68, 44507.

Exp.; E; O_h; d^9.

2.1.74. Toyozawa Yutaka, Vibration-induced structures in the absorption spectra of localized electrons in solids. - In: Dynamical Processes in Solid State Optics: Tokyo Summer Lect. Theor. Phys., 1966. Tokyo; New York, 1967, Pt. I, p. 90-115. - RZF, 1968, 2D312.

Theor.

2.1.75. Warren, R. W., Shape of the F-aggregate bands in KCl and KBr. II. Analysis. - Phys. Rev., 1967, 155, N3, 948-958. - CA, 1967, 67, 37899.

Exp.; T - t.

2.1.76. Wells, C. F., Davies, G., A spectrophotometric investigation of the aquomanganese(III) ion in perchlorate media. - J. Chem. Soc. Ser. A, 1967, N11, 1858-1861. - CA, 1968, 68, 17227.

Exp.; T; O_h; d^4; d^5.

1968

2.1.77. Cho Kikuo, Optical absorption line shapes due to transition from orbital singlet to triplet states of defect centers with cubic symmetry. - J. Phys. Soc. Jpn., 1968, 25, N5, 1372-1387. - CA, 1969, 70, 42509.

Theor.; $^2T_{1u}$ - (a + e + t).

2.1.78. Faye, G. H., Optical absorption spectra of iron in six-coordinate sites in chlorite, biotite, phlogopite, and vivianite: Pleochroism in the sheet silicates. - Can. Mineral., 1968, 9, N3, 403-425. - CA, 1968, 69, 23362.

Exp.; O_h; d^6.

2.1.79. Fetterman, H. R.,, M' and R' centers in lithium fluorride: Comparative study of orbital degeneracy and the dynamic Jahn-Teller effect in optical spectra. - 1968. - 158 p. - [Order N 69-7381]. Diss. Abstr. B, 1969, 29, N12, 4802. CA, 1969, 71, 130054. PA, 1970, 73, 15478.

2.1.80. Holton, W. C., Defects and their role in determining bulk properties of crystals. - 1968. - 19 p. - US Clearinghouse Fed. Sci., Tech. Inform. AD 669814. - CA, 1969, 70, 15722.

Exp.; d^1.

2.1.81. Honma, A., Energy parameters of indium(I) and thallium(I) ions in potassium chloride. - Sci. Light, 1968, 17, N1, 34-44. - CA, 1969, 70, 110135.

Theor.; T_{1u} - t_{2g}; O_h.

2.1.82. Kamimura Hiroshi, Yamaguchi, T., Theory of Jahn-Teller-induced optical transitions. - J. Phys. Soc. Jpn., 1968, 25, N4, 1138-1147. - CA, 1969, 70, 15637. PA, 1969, 72, 7928.

Theor.

2.1.83. Kristofel', N. N., About the absorption spectra of impurity centers with close energy levels "mixed" by vibrations. [Russ.]. - Opt. & Spektrosk., 1968, 24, N4, 565-572. - CA, 1968, 69, 23371.

Theor.; 1.5.

2.1.84. Lannoo, M., Jahn-Teller effect for a single vacancy in covalent solids: Application to optical absorption and paramagnetic resonance spectra. [Fr.]. - Ann. Phys. (France), 1968, 3, N6, 391-423. - CA, 1970, 72, 105558. PA, 1970, 73, 44726.

Theor.; E - e; T - (e + t_2); 3.1.

2.1.85. Perlin, Yu. E., Optical transitions to degenerate local levels. [Russ.]. - Fiz. Tverd. Tela, 1968, 10, N7, 1941-1949.

Theor.

2.1.86. Rosenfeld, Yu. B., Vekhter, B. G., Tsukerblat, B. S., Optical bands in electron-vibrational systems with degeneracy. - Phys. Status Solidi, 1968, 28, N2, K179-K182. - CA, 1968, 69, 101367.

Theor.; E - e; T - e; T - t; T - (e + t).

2.1.87. Rozenfel'd, Yu. B., Vekhter, B. G., Tsukerblat, B. S., Optical bands in electron-vibrational systems with electronic degeneracy. [Russ.]. - Zh. Eksp. & Teor. Fiz., 1968, 55, N12, 2252-2261.

Theor.

2.1.88. Volchenskova, I. I., Yatsimirskiy, K. B., The interpretation of the absorption spectra of hexaaquo- and hexaamine ions of cobalt(III). [Russ.]. - Teor. & Eksp. Khim., 1968, 4, N5, 653-661. - CA, 1969, 70, 72343.

Theor.; T_{1g}; O_h.

2.1.89. Wagner, M., The Jahn-Teller effect for the U-center local mode. - In: Localized Excitations in Solids: Proc. 1st Int. Conf., 1967. New York, 1968, p. 551-558. - CA, 1968, 69, 101404.

Theor.; T - t.

1969

2.1.90. Allen, G. C., Warren, K. D., Electronic spectra of the hexafluoronickelate(III), hexafluorocuprate(III), and hexafluoroargentate(III) anions. - Inorg. Chem., 1969, 8, N9, 1895-1901. - CA, 1969, 71, 75772.

Exp.; T; T - e; O_h; d^8; d^9.

2.1.91. Baldini, G., Bosacchi, B., Jahn-Teller effect in the Γ exciton of lithium bromide. - Phys. Rev. Lett., 1969, 22, N5, 190-192. - CA, 1969, 70, 62186. PA, 1969, 72, 19882.

Exp.; Γ_8.

2.1.92. Brokopf, H., Reinen, D., Schmitz-Dumont, O., Ligand field spectrum and the Jahn-Teller effect of ferrous ions in oxidic and fluoridic crystal lattices. [Germ.]. - Z. Phys. Chem. (Frankfurt/Main), 1969, 68, N3/6, 228-241. - CA, 1970, 73, 30259.

Exp.; E - e; d^6.

2.1.93. Dingle, R., Crystal spectra of hexaurea complexes. I. Jahn-Teller effect in hexaurea titanium(III) triodide. - J. Chem. Phys., 1969, 50, N1, 545-546. - CA, 1969, 70, 82697. PA, 1969, 72, 14057; 39848.

Exp.; E - e; O_h; d^1.

2.1.94. Freeman, T. E., Jones, G. D., Jahn-Teller splitting in the optical absorption spectra of divalent iron in cadmium chloride-type crystals. - Phys. Rev., 1969, 182, N2, 411-415. - CA, 1969, 71, 65675. PA, 1970, 73, 2322.

Exp.; E - e; T_d; d^6; 1.7.

2.1.95. Fukuda Atsuo, Jahn-Teller effect in color centers. [Jap.]. - Funko Kenkyu, J. Spectrosc. Soc. Jpn., 1969, 18, N1, 11-30. - CA, 1969, 71, 95982.

Rev.

2.1.96. Fukuda Atsuo, Structure of the C absorption band of thallium(I)-type centers in alkali halides due to the Jahn-Teller effect. - J. Phys. Soc. Jpn., 1969, 27, N1, 96-109. - CA, 1969, 71, 43399. PA, 1970, 73, 4972.

Exp.; T - (a + e + t).

2.1.97. Gebhardt, W., Jahn-Teller effect. [Germ.]. - In: Festkörperprobleme. Berlin, 1969, Vol. 9, p. 99-137. - Bibliogr.: 79 ref. CA, 1970, 73, 60669. PA, 1970, 73, 56472.

Rev.

2.1.98. Grasso, V., Saitta, G., Modulated absorption in thallium-doped KBr. - Phys. Rev. Lett., 1969, 22, N11, 522-523.

Exp.; T - t.

2.1.99. Honma, A., Asymmetry of the triplet structure of the C absorption band for Tl^+-type impurity in alkali halides. - Sci. Light, 1969, 18, N1, 33-38. - CA, 1970, 72, 16996.

Theor.; T_{1u} - t_{2g}; O_h.

2.1.100. King, G. W., Santry, D. P., Warren, C. H., Fluorosulfate radical: Structural theory and assignment of electronic absorption systems. - J. Mol. Spectrosc., 1969, 32, N1, 108-120. - CA, 1969, 71, 107231.

Exp.; E - e.

2.1.101. Kristofel', N. N., Influence of the vibrational interaction between local and band energy levels on the optical spectra of impurity centers. [Russ.]. - Opt. & Spektrosk., 1969, 26, N4, 547-553.

Theor.; 1.5.

2.1.102. Merer, A. J., Schoonveld, L., Electronic spectrum of ethylene. I. The 1744-Å: Rydberg transition. - Can. J. Phys., 1969, 47, N16, 1731-1743. - CA, 1969, 71, 55115.

Exp.

2.1.103. Perlin, Yu. E., Tsukerblat, B. S., Rosenfel'd, Yu. B., Kharchenko, L. S., Gamurar', V. Yu., The electron-vibrational coupling effect on optical spectra of rare-earth and transition metal ions in crystals. - In: Int. Symp. Theory Electronic Shells, Atoms, and Molecules: Summaries of Papers, Vilnius, 1969, p. 46.

Theor.

2.1.104. Theissing, H. H., Ewanizky, T. F., Caplan, P. J., Grosse, D. W., Y^{2+} as a color center in irradiated calcium fluoride. - J. Chem. Phys., 1969, 50, N6, 2657-2671. - CA, 1969, 70, 110243.

Exp.; E - e; T - e; O_h; d^1.

2.1.105. Tsuboi Taiju, Kato Riso, Vibration-assited absorption (B band) in alkali halide-thallium crystals. - J. Phys. Soc. Jpn., 1969, 27, N5, 1192-1196.

Exp.; T; O_h.

2.1.106. Tsukerblat, B. S., Brunshtein, S. Kh., The Jahn-Teller effect for a transition metal ion in a crystal. [Russ.]. - In: Issled. Khim, Koord. Soedin. Fiz.-Khim. Metodam Analiza, Kishinev, 1969, p. 60-63. - CA, 1971, 74, 117751. RZK, 1970, 13B43.

Theor.; T - e; O_h; d^3.

2.1.107. Tsukerblat, B. S., Perlin, Yu. E., Kharchenko, L. S., The calculation of the multiplet band shapes of impurity light absorption by the method of moments. [Russ.]. - In: Tr. 2 Seminara Spektrosk. Svoystvam Lumineforov, Aktivir. Redkimi Zemlyami, Moscow, 1969, p. 181.

Theor.; E - e; T - $(a_1 + e + t_2)$.

2.1.108. Vekhter, B. G., Tsukerblat, B. S., Bersuker, I. B., Ablov, A. V., Band shapes of the optical absorption and luminescence in transition metal complexes taking into account the Jahn-Teller effect. [Russ.]. - Zh. Strukt. Khim., 1969, 10, N4, 712-717. - CA, 1969, 71, 107292.

Theor.; E; T.

2.1.109. Zgierski, M. Z., Towards the theory of vibronic coupling in molecular crystals. - Phys. Lett. A, 1969, 30, N6, 354-355.

Theor.; 1.5.

1970

2.1.110. Andriesh, I. S., Tsukerblat, B. S., Perling, Yu. E., Kharchenko, L. S., The forbidden transitions to degenerate impurity energy levels. [Russ.]. - In: Tez. Dokl. 3 Simp. Spektrosk. Kristallov, Aktivir. Ionami Perekhodnykh i Redkozem. Elementov Leningrad, 1970, p. 87.

Theor.; E - e; T - (a + e + t).

2.1.111. Englman, R., Caner, M., Toaff, S., The optical bandshape of a vibronically coupled orbital triplet. - J. Phys. Soc. Jpn., 1970, 29, N2, 306-310. - RZF, 1971, 1D546.

Theor.; T - t.

2.1.112. Ferguson, J., Spectroscopy of 3d complexes. - In: Progr. Inorg. Chem., New York; London, 1970, Vol. 12, p. 159-293. - Bibliogr.: 384 ref. CA, 1971, 74, 47390.

Rev.

2.1.113. Fukuda Atsuo, Jahn-Teller effect on the structure of the emission produced by excitation in the A band of potassium iodide: thallium type phosphor: Two kinds of minima on the Γ_4 - ($^3T_{1u}$) adiabatic potential energy surface. - Phys. Rev. B, 1970, 1, N10, 4161-4178. - CA, 1970, 73, 71791. PA, 1970, 73, 56621. RZF, 1970, 12D744.

Exp.; T - e; 1.2.

2.1.114. Fukuda Atsuo, Six-dimensional configuration coordinate scheme and KI:Tl type phosphors. - In: Proc. Int. Conf. Luminescence, Newark, 1969. Amsterdam, 1970, p. 376-384. CA, 1971, 74, 17747.

Exp.; 2.2.

2.1.115. Giorgianni, U., Grasso, V., Saitta, G., Absorption and emission of thallium-doped KBr in an external electric field. - Nuovo Cimento B, 1970, 68, N1, 100-108.

Exp.; T - t; 2.2.

2.1.116. Hagston, W. E., Possible new approach to Hartree model for defects in terms of the dynamical Jahn-Teller effect with applications to diamond. - J. Phys. C, 1970, 3, N4, 791-805. - CA, 1970, 73, 40332. PA, 1970, 73, 47220.

Theor.; E; T; T_d; 3.1.

2.1.117. Inoue, M., Sati, R., Wang, S., Electronic structure and dynamic Jahn-Teller effect on R' centers in alkali halides. - Can. J. Phys., 1970, 48, N14, 1694-1707. - CA, 1970, 73, 48654. PA, 1970, 73, 53549. RZF, 1971, 1D571.

Theor.; E - e.

2.1.118. Khizhnyakov, V. V., On the theory of the dynamic Jahn-Teller effect. [Russ.]. - Izv. Akad. Nauk Est.SSR, Fiz. Mat., 1970, 19, N2, 244-247. - CA, 1970, 73, 91335. PA, 1970, 73, 56471. RZF, 1970, 9D472.

Theor.; T - (a + t).

2.1.119. Kleemann, W., Optical transitions of Ag^- centers in alkali halides. - Z. Phys., 1970, 234, N4, 362-378. - CA, 1970, 73, 20089.

Exp.

2.1.120. Kristofel', N. N., Application of the microtheory of small-radius impurity centers to alkali halide crystal phosphors with mercury-like activators. [Russ.]. - In: Tr. Inst. Fiz. i Astron. Akad. Nauk Est.SSR, 1970, N38, p. 125-143. - Bibliogr.: 89 ref. CA, 1972, 76, 52142.

Rev.; T - e.

2.1.121. Lemos, A. M., Stauber, M. C., Marion, J. F., Structure of the A, B, and C absorption bands in KCl:Tl. - Phys. Rev. B, 1970, 2, N10, 4161-4168.

Theor.; T - (e + t); O_h; 1.5.

2.1.122. Lohr, L. L., Jr., Theoretical band shapes for vibronically induced electronic transitions. - J. Am. Chem. Soc., 1970, 92, N8, 2210-2217.

Theor.; 1.5.

2.1.123. Mathe, F., Evidence of neighbor cation influence and correlation effect from ligand field calculations. - Rev. Roum. Chim., 1970, 15, N2, 283-290. - CA, 1970, 73, 29020.

Theor.; d^4; d^5; d^9.

2.1.124. Petelin, A. N., On the Renner effect in an acetylene molecule. [Russ.]. - Opt. & Spektrosk., 1970, 29, N6, 1153-1155. - CA, 1971, 74, 104848. RZF, 1971, 4D173.

Theor.; 1.6.

2.1.125. Shaha, R. K., Bose, A., Choudhury, M., Polarized absorption spectrum of Fe^{2+} doped in cesium:cadmium trichloride at 77°C. - Indian J. Phys., 1970, 44, N1, 59-60. - CA, 1971, 75, 82109.

Exp.; E - e; C_{3v}; d^6.

2.1.126. Skorobogatova, I. V., Zvyagin, A. I., Mechanism of allowed transitions of Co^{2+} ions in $ZnWO_4$. [Russ.]. - Fiz. Tverd. Tela, 1970, 12, N2, 3651-3652. - RZF, 1971, 5D404.

Exp.; T; d^7; 1.5.

2.1.127. Sugano Satoru, Tanabe Yukito, Kamimura Hiroshi, Multiplets of Transition-Metal Ions in Crystals. - New York; London: Academic Press, 1970. - XI, 331 p. - (Pure and Appl. Phys.; Vol. 33.)

Ch. IX. Interaction between Electron and Nuclear Vibration, p. 213-248.

2.1.128. Vala, M., Jr., Mongan, P., McCarthy, P., The electronic spectra of tetrahaloferrate(III) complexes. - In: Abstr. 25th Symp. Mol. Structure & Spectrosc. Ohio, 1970, p. 80. - PA, 1971, 74, 13445.

Exp.

2.1.129. Wagner, M., A method of moments for the optimal Jahn-Teller problem. - Z. Phys., 1970, 230, N5, 460-480. - PA, 1970, 73, 25992.

Theor.; E - e; T - e.

1971

2.1.130. Bose, S. K., Electronic spectra of trigonal complexes of cobalt. - Indian J. Chem., 1971, 9, N5, 493-495. - CA, 1971, 75, 42643. PA, 1971, 74, 72319.

Exp.

2.1.131. Brand, J. C. D., Goodman, G. L., Weinstock, B., Optical absorption of iridium hexafluoride. - J. Mol. Spectrosc., 1971, 37, N3, 464-485. - CA, 1971, 74, 105064.

Exp.; Γ_8.

2.1.132. Cavallone, F., Pollini, I., Spinolo, G., Optical properties of α-titanium trichloride at the magnetic transition temperature. - Phys. Status Solid b, 1971, 45, N2, 405-410. - CA, 1971, 75, 55789. PA, 1971, 74, 53233.

Exp.; E - e; C_{3v}; d^1.

2.1.133. Devillers, C., Ramsay, D. A., The $\tilde{A}^3\pi i$ - $\tilde{X}^3\Sigma^-$ band system of the CCO molecule. - Can. J. Phys., 1971, 49, N22, 2839-2858. - CA, 1972, 76, 39684.

Exp.; 1.6.

2.1.134. Duran, J., Billardon, M., Electron-lattice interactions in KBr:Pb^{2+}. - Phys. Status Solid b, 1971, 44, N2, K73-K75. - PA, 1971, 74, 41444.

Exp.; 2.4.

2.1.135. Escribe, C., Hughes, A. E., Vibronic properties of the F^+ absorption band in calcium oxide. - J. Phys. C, 1971, 4, N16, 2537-2549. - CA, 1972, 76, 39693. PA, 1972, 75, 7068.

Exp.; T - (a + e + t); 2.3.

2.1.136. Ferguson, J., Guggenheim, H. J., Krausz, E. R., Optical absorption by copper-manganese pairs in potassium trifluorozincate. - J. Phys. C, 1971, 4, N13, 1866-1873. - CA, 1971, 75, 135437. PA, 1971, 74, 76965.

Exp.; E - e; d^5; d^9.

2.1.137. Kanda Teinosuke, Ebisu Takeo, Optical absorption spectra of cesium atoms in rare-gas matrices. - J. Phys. Soc. Jpn., 1971, 31, N3, 957-958. - PA, 1972, 75, 10392. RZF, 1972, 3D394.

Exp.; T - t.

2.1.138. Khlopin, V. P., Tsukerblat, B. S., Rozenfel'd, Yu. B., Bersuker, I. B., Optical band shapes of singlet-doublet transitions in Renner-type molecules. [Russ.]. - In: Tez. Dokl. 17 Vsesoyuz. S'ezda Spektrosk.: Mol. Spektrosk. Minsk, 1971, p. 237.

Theor.; E - e; 1.6.

2.1.139. Krausz, E. R., Optical absorption by Cu-Mn pairs in $KZnF_3$. - In: Abstr. 8th Austral. Spectrosc. Conf. Clayton, 1971, Ip. - PA, 1972, 75, 10388.

Exp.; E - e; d^5; d^9.

2.1.140. Medvedev, E. S., Opticheskiye Perekhody v Lokal'nykh Tsentrakh Kristallov: Avtoref. Dis. ... Kand. Fiz. - Mat. Nauk. [Russ.]. - Moscow, 1971. - 12 p. - Mosk. Fiz.-Tekhn. Inst.

Optical Transitions in Local Centers in Crystals.

Theor.; E - e; 2.3; 2.4.

2.1.141. Meyer, B., Bajema, L., Gouterman, M., Spectrum of matrix-isolated carbon disulfide. - J. Phys. Chem., 1971, 75, N14, 2204-2208. - CA, 1971, 75, 69173.

Exp.; 1.6.

2.1.142. Parke, S., Webb, R. S., Optical properties of Sn^{2+} and Sb^{3+} in calcium metaphosphate glass. - J. Phys. D, 1971, 4, N6, 825-828. - CA, 1971, 75, 27755. PA, 1971, 74, 45701.

Exp.; T.

2.1.143. Serikov, A. A., Vibronic spectra of molecular crystals with participation of non-totally symmetric vibrations. - Phys. Status Solid b, 1971, 44, N2, 733-746.

Theor.; 1.5.

2.1.144. Small, G. J., Herzberg-Teller vibronic coupling and the Duschinsky effect. - J. Chem. Phys., 1971, 54, N8, 3300-3306.

Theor.; E; 1.5; 1.7; 2.2.

2.1.145. Stufkens, D. J., Schenk, A., Dynamical Jahn-Teller effect in the excited state of $TeCl_6^{2-}$: Temperature dependence of the splitting of the C absorption band. - Recl. Trav. Chim. Pays-Bas, 1971, 90, N2, 190-192. - CA, 1971, 74, 117739.

Exp.; T; O_h.

2.1.146. Wieghardt, K., Siebert, H., Hexafluoromanganate(III). [Germ.]. - Z. Anorg. Allg. Chem., 1971, 381, N1, 12-20. - CA, 1971, 74, 117761.

Exp.; E; O_h; d^4; 2.7.

2.1.147. Witkowski, A., Zgierski, M. Z., Theory of vibronic spectra in molecular crystals with four molecules in the unit cell: perylene. - Phys. Status Solid b, 1971, 46, N1, 429-441. - CA, 1971, 75, 68428.

Theor.

1972

2.1.148. Bernstein, E. R., Reilly, J. P., Absorption spectrum of the 2000-Å system of borazine in the gas phase. - J. Chem. Phys., 1972, 57, N9, 3960-3969. - CA, 1972, 77, 158333.

Exp.; D_{3h}.

2.1.149. Bersuker, I. B., Perlin, Yu. E., Polinger, V. Z., Khlopin, V. P., Rosenfeld, Yu. B., Tsukerblat, B. S., Vekhter, B. G., Optical electronic band shapes in coordination systems with Jahn-Teller and Renner type vibronic coupling. - In: Proc. 14th Int. Conf. Coord. Chem.: Summaries of Papers. Toronto, 1972, p. 757-759.

Theor.; E; T.

2.1.150. Bertini, I., Gatteschi, D., Single crystal polarized spectra of the $Cu(en)_3SO_4$. - Inorg. & Nucl. Chem. Lett., 1972, 8, N2, 207-210.

Exp.; E - e; D_3; d^9.

2.1.151. Boksha, O. N., Varina, T. M., Popova, A. A., Smirnova, E. F., Conditions of synthesis and optical spectra of crystals containing transition elements. II. Corundum with titanium. [Russ.]. - Kristallografiya, 1972, 17, N6, 1246-1248. - CA, 1973, 78, 64504. RZF, 1973, 4D627.

Exp.; E - e; C_{3v}; d^1.

2.1.152. Brus, L. E., Transition dipole moments in extreme Renner effect molecules with application to the visible $A_1 \leftrightarrow B_1$ bands in CH_2, NH_2, BH_2. - J. Chem. Phys., 1972, 57, N8, 3167-3175. - CA, 1972, 77, 145671.

Theor.; 1.6.

2.1.153. Chen Ming-yi, McClure, D. S., Solomon, E. I., Jahn-Teller effect in the $^4T_{1g}$ (I) state of Mn^{++} in $RbMnF_3$. I. - Phys. Rev. B, 1972, 6, N5, 1690-1696. - CA, 1972, 77, 95052.

Theor.; T - (e + t); d^5.

2.1.154. Chlopin, V. P., Tsukerblat, B. S., Rosenfeld, Yu. B., Bersuker, I. B., Optical electronic band shapes and magneto-optical effects in Renner type systems. - In: Abstr. Int. Conf. Mol. Spectrosc. Wroclaw, 1972, p. 29.

Theor.; 1.6.

2.1.155, Chlopin, V. P., Tsukerblat, B. S., Rosenfeld, Yu. B., Bersuker, I. B., Optical band shape in Renner-type systems with strong vibronic coupling. - Phys. Status Solid b, 1972, 52, N2, K73-K75. - CA, 1972, 77, 158045. PA,1972, 75, 65248. RZF, 1973, 1E389.

Theor.; E; 1.6.

2.1.156. Dawson, J. W., Gray, H. B., Hix, J. E., Jr., Preer, J. R., Venanzi, L. M., Solution and solid state electronic spectra of low-spin trigonal bipyramidal complexes containing nickel(II), palladium(II), and platinum(II): Temperature dependence of the lowest energy ligand-field band. - J. Am. Chem. Soc., 1972, 94, N9, 2979-2987. - CA, 1972, 76, 160389. PA, 1972, 75, 53031.

Exp.; E - e; C_{3v}; D_{3h}.

2.1.157. Hansen, K. H., Schenk, H. J., Polarized crystal spectra of some hexaurea complexes at low temperatures. I. Absorption. [Germ.]. - Theor. Chim. Acta, 1972, 24, N2/3, 207-215. - CA, 1972, 76, 119387. PA, 1972, 75, 51289.

Exp.; E - e; d^1; d^3.

2.1.158. Khlopin, V. P., Tsukerblat, B. S., Rozenfel'd, Yu. B., Bersuker, I. B., Optical absorption and luminescence band shapes in systems with the Renner effect. [Russ.]. - Fiz. Tverd. Tela, 1972, 14, N4, 1060-1068. - CA, 1972, 77, 40898. RZF, 1972, 8D839.

Theor.; 1.6; 2.2.

2.1.159. Khlopin, V. P., Tsukerblat, B. S., Rosenfeld, Yu. B., Bersuker, I. B., The optical band shape of Renner-type system. - Phys. Lett. A, 1972, 38, N6, 437-438. - RZF, 1972, 8D176.

Theor.; E; 1.6.

2.1.160. Loorits, V., Hizhnyakov, V., Optical Jahn-Teller effect: strong electron-phonon interaction. - In: Phys. Impurity Cent. Cryst.: Proc. Int. Semin. "Selected Probl. Theory Impurity Cent. Cryst.," 1970. Tallin, 1972, p. 453-462. - CA, 1973, 78, 166313. RZF, 1973, 1E365.

Theor.

2.1.161. Luschik, N. E., Zazubovich, S. G., Electronic excitations of mercury-like centers in alkali halides. - In: Phys. Impurity Cent. Cryst.: Proc. Int. Semin. "Selected Probl. Theory Impurity Cent. Cryst.," 1970. Tallin, 1972, p. 483-504. - Bibliogr.: 81 ref. CA, 1973, 78, 166407.

Exp.; T.

2.1.162. McPherson, G. L., Kistenmacher, T. J., Folkers, J. B., Stucky, G. D., Effect of exchange coupling on the spectra of transition metal ions: The ligand field spectrum and crystal structure of $CsCrCl_3$. - J. Chem. Phys., 1972, 57, N9, 3771-3780. - RZF, 1973, 2D526.

Exp.; E - e; O_h; D_4; 4.6.

2.1.163. Menzel, E. R., Gruber, J. B., Ryan, J. L., Absorption spectra of inequivalent sites of tetravalent Np-237 in $[(C_2H_5)_4N]_2$-$NpCl_6$. - J. Chem. Phys., 1972, 57, N10, 4287-4290. - CA, 1973, 78, 9688. PA, 1973, 76, 2557.

Exp.; Γ_8.

2.1.164. Nicklaus, E., Fischer, F., F centers of two types in barium fluoride: chloride crystals. - Phys. Status Solidi b, 1972, 52, N2, 453-460. - CA, 1972, 77, 157282.

Exp.; E - e.

2.1.165. Nygren, B., Vallin, J. T., Slack, G. A., Direct observation of the Jahn-Teller splitting in $ZnSe:Cr^{2+}$. - Solid State Commun., 1972, 11, N1, 35-38. - CA, 1972, 77, 94997. PA, 1972, 75, 65270. RZF, 1972, 12E844.

Exp.; T - e.

2.1.166. Sati, R., Wang, S., Inoue, M., Line shape of C bands in alkali halide phosphors by the method of moments. - Can. J. Phys., 1972, 50, N12, 1370-1378. - CA, 1972, 77, 68036. RZF, 1972, 12E834.

Theor.; T - t.

2.1.167. Solomon, E. I., MacClure, D. S., Jahn-Teller effect in the $^4T_{1g}(I)$ state of Mn^{++} in $RbMnF_3$. II. - Phys. Rev. B, 1972, 6, N5, 1697-1708. - CA, 1972, 77, 95051. PA, 1972, 75, 65268. RZF, 1973, 3E1054.

Theor.; T_d; d^5.

2.1.168. Ulrici, W., Jahn-Teller effect of transition metal ions in silver halides. III. Interpreation. - Phys. Status Solidi b, 1972, 51, N1, 129-138. - CA, 1972, 77, 11732. PA, 1972, 75, 40657. RZF, 1972, 10E839.

Theor.; E - e.

2.1.169. Vekhter, B. G., Perlin, Yu. E., Polinger, V. Z., Rosenfeld, Yu. B., Tsukerblat, B. S., Optical bands of Jahn-Teller systems with orbital doublet. I. General expression. Semiclassical approximation. - Cryst. Lattice Defects, 1972, 3, N2, 61-68. - CA, 1972, 77, 107001. PA, 1972, 75, 65197. RZF, 1972, 12E487.

Theor.; E.

2.1.170. Vekhter, B. G., Perlin, Yu. E., Polinger, V. Z., Rosenfeld, Yu. B., Tsukerblat, B. S., Optical bands of Jahn-Teller systems with orbital doublet. II. Path integration method. Applicability conditions. - Cryst. Lattice Defects, 1972, 3, N2, 69-76. - CA, 1972, 77, 107000. PA,1972, 75, 65198.

Theor.; E - e.

2.1.171. Vekhter, B. G., Tsukerblat, B. S., Rosenfeld, Yu. B., Optical manifestation of Jahn-Teller effect in square-planar complexes. - Theor. Chim. Acta, 1972, 27, N1, 49-54. - CA, 1972, 77, 132801.

Theor.; E - (b_1 + b_2).

2.1.172. Vekhter, B. G., Perlin, Yu. E., Polinger, V. Z., Rozenfeld, Yu. B., Tsukerblat, B. S., Optical transitions in systems with electronic degeneracy. [Russ.]. - In: Phys. Impurity Cent. Cryst.: Proc. Int. Semin. "Selected Probl. Theory Impurity Cent. Cryst.," 1970. Tallin, 1972, p. 463-478. - RZK, 1973, 1E366.

Theor.; E - e.

1973

2.1.173. Ammeter, J. H., Schlosnagle, D. C., Spectroscopic investigations of Van der Waals complexes of aluminum and gallium atoms with noble gases. - Chimia, 1973, 27, N7, 372-375. - CA, 1973, 79, 109947.

Exp.; 3.1.

2.1.174. Ammeter, J. H., Schlosnagle, D. C., Electronic quenching of aluminum and gallium atoms isolated in rare gas matrixes. - J. Chem. Phys., 1973, 59, N9, 4784-4820. - CA, 1974, 80, 65226. PA, 1974, 77, 25136.

Exp.; T - e; T - t; O_h.

2.1.175. Asano Sumitada, Tomishima Yasuo, Dynamical Jahn-Teller splitting of the B absorption band in alkali halide phosphors activated with thallium(1+)-type ions. - Rep. Res. Lab. Surface Sci. Okayama Univ., 1973, 4, N1, 1-22. - CA, 1974, 80, 112985. PA, 1974, 77, 53483.

Theor.; E; T.

2.1.176. Bolman, P. S. H., Brown, J. M., Renner-Teller effect and vibronically induced bands in the electronic spectrum of isocyanato. - Chem. Phys. Lett., 1973, 21, N2, 213-216. - CA, 1973, 79, 141122.

Exp.; 1.6.

2.1.177. Brom, J. M., Jr., Weltner, W., Jr., Absorption spectrum of BS_2 at 4°K. - J. Mol. Spectrosc., 1973, 45, N1, 82-98. - CA, 1973, 78, 50003.

Exp.; 1.6.

2.1.178. Brunshtein, S. Kh., Tsukerblat, B. S., Multiphonon optical bands of excited ruby. [Russ.]. - In: Tez. Dokl. 4 Vsesoyuz. Soveshch. Spektrosk. Kristallov Aktivir. Ionami Redkozem. Perekhodnykh Elementov. Sveredlovsk, 1973, p. 6.

Theor.; d^3.

2.1.179. Canters, G. W., van Egmond, J., Schaafsma, T. J., van der Waals, J. H., Jansen, G., Optical and Zeeman studies of metal porphins in Shpolski-hosts. - In: Abstr. 11th Eur. Congr. Mol. Spectrosc. Tallin, 1973, 186. - PA, 1973, 76, 72737.

Exp.; D_{4h}.

2.1.180. Cho, K., Jahn-Teller effect in the parity-forbidden transition $^2P_{1/2} \rightarrow {}^2P_{3/2}$ of Tl° in KCl. - J. Phys. (France), 1973, 34, Colloq. N9, 127-130. - CA, 1974, 80, 150494.

Theor.; T - (e + t).

2.1.181. Cho, Kikuo, Line shape analysis of the optical transition $^2P_{1/2} \rightarrow {}^2P_{3/2}$ of Tl° in alkali halides. - Solid State Commun., 1973, 13, N4, 439-441. - CA, 1973, 79, 109904. PA, 1973, 76, 64698. RZF, 1974, 1E438.

Exp.; $(P_{1/2} + P_{3/2}) - (e_g + t_{2g})$.

2.1.182. Clack, D. W., Williams, W. T., Site symmetry and electronic spectra of copper(II) in cubic lattices. - J. Inorg. & Nucl. Chem., 1973, 35, N10, 3535-3539. - CA, 1974, 80, 19787.

Exp.; E - e; O_h; d^9.

2.1.183. Comes, F. J., Haensel, R., Nielsen, U., Schwarz, W. H. E., Spectra of the xenon fluorides XeF_2 and XeF_4 in the far u.v. region. - J. Chem. Phys., 1973, 58, N2, 516-529. - CA, 1973, 78, 77659. PA, 1973, 76, 17423.

Exp.; E - b; D_{4h}; d^9.

2.1.184. Englman, R., Steggles, P., A calculation of the lineshape for the singlet to doublet transition in the E × e system [Jahn-Teller effect]. - J. Phys. C, 1973, 6, N9, 1673-1676. - PA, 1973, 76, 43465.

Theor.; E - e.

2.1.185. Fukuda Atsuo, Relaxed excited states of thallium(+) ion-type centers. [Jap.]. - Nippon Butsuri Gakkaishi, Proc. Phys. Soc. Japn, 1973, 28, N8, 695-698. - CA, 1974, 80, 65081.

Exp.

2.1.186. Gutlyanski, E. D., Khartsiev, V. E., On the optical spectral of polyexcitonic and biexcitonic molecules in crystals. - In: Abstr. 11th Eur. Congr. Mol. Spectrosc. Tallin, 1973, 100. - PA, 1973, 76, 68583.

Theor.

2.1.187. Habitz, P., Schwarz, W. H. E., Vibronic and spin-orbit splitting in spectra of systems exhibiting the Jahn-Teller effect. - Theor. Chim. Acta, 1973, 28, N3, 267-282. - CA, 1973, 78, 117245. PA, 1973, 76, 67763. RZK, 1973, 6D161.

Theor.; E - (a + b); E - e.

2.1.188. Joergensen, W. L., Borden, W. T., Chemical consequences of orbital interactions in hydrocarbons containing unsaturatively bridged small rings. - J. Am. Chem. Soc., 1973, 95, N20, 6649-6654. - PA, 1974, 77, 19772.

Theor.; 1.5.

2.1.189. Masunaga Shoji, Matsuyama Eiichi, Structures of absorption bands in alkali halide crystals containing thallium(1+)-like ions under high pressure. - J. Phys. Soc. Jpn., 1973, 34, N5, 1234-1239. - CA, 1973, 78, 166576.

Exp.

2.1.190. Perlin, Yu. E., Kharchenko, L. S., Electron-vibration interaction in the quasimolecular F-center theory. - In: Abstr. 11th Eur. Congr. Mol. Spectrosc. Tallin, 1973, 63. - PA, 1973, 76, 69039.

Theor.

2.1.191. Petelenz, P., Zgierski, M. Z., Vibronic interpretation of the silicon phthalocyanine dimer absorption spectrum. - Mol. Phys., 1973, 25, N1, 237-239.

Theor.; 1.5.

2.1.192. Polinger, V. Z., The quadratic Jahn-Teller effect in the optical bands of the A $\rightarrow$ T transition. [Russ.]. - In: Tez. Dokl. 4 Vsesoyuz. Sovesch. Spectrosk. Kristallov, Aktivir. Ionami Redkozem. Perekhodnykh Elementov. Sverdlovsk, 1973, p. 3.

Theor.; T.

2.1.193. Polinger, V. Z., Rozenfel'd, Yu. B., Vitiu, E. V., The application of the method of path integrals to degenerate and pseudodegenerate systems. [Russ.]. - In: Materialy Dokl. IX Nauch.-Tekhn. Konf. Kishinev. Politekhn. Inst., 1973, p. 157.

Theor.

2.1.194. Rabalais, J. W., Karlsson, L., Werme, L. O., Bergmark, T., Siegbahn, K., Analysis of vibrational structure and Jahn-Teller effects in the electron spectrum of ammonia. - J. Chem. Phys., 1973, 58, N8, 3370-3372. - CA, 1973, 79, 25399. PA, 1973, 76, 38230. RZF, 1973, 10D318.

Exp.; E - e.

2.1.195. Radhakrishna, S., Pande, K. P., Lead centers in cesium halides. - Phys. Rev. B, 1973, 7, N1, 424-431. RZF, 1973, 7D468.

Exp.; T - t; O_h.

2.1.196. Rozenfel'd, Yu. B., Vekhter, B. G., Tsukerblat, B. S., The band shape of the impurity absorption and luminescence in systems with twofold degenerate electronic E level. [Russ.]. - In: Spektrosk. Kristallov: Materialy 3 Simp. Spektrosk. Kristallov, Aktivir. Inonami Redkozem. Perekhodnykh Metallov, 1970. Leningrad, 1973, p. 131-134.

Theor.; E - e; O_h; C_{3v}.

2.1.197. Sakoda Shoichiro, Toyozawa Yutaka, Theory of the Jahn-Teller effect on the optical spectra of degenerate excitons. - 1973. - 29 p. - Techn. Rept. ISSP, A, N568. - RZF, 1973, 10E764.

Theor.; E; T; C_{3i}; O_h.

2.1.198. Sakoda Shoichiro, Toyozawa Yutaka, Theory of the Jahn-Teller effect on the optical spectra of degenerate excitons. - J. Phys. Soc. Jpn., 1973, 35, N1, 172-179. - CA, 1973, 79, 59411. PA, 1973, 76, 50152. RZF, 1973, 12E815.

Theor.

2.1.199. Sastry, S. B. S., Viswanathan, V., Ramasastry, C., Lead centers in alkali halides: NaCl, KCl, and KBr. - J. Phys. Soc. Jpn., 1973, 35, N2, 508-513. - RZF, 1974, 1E1405.

Exp.; E; T.

2.1.200. Sastry, S. B. S., Viswanathan, V., Ramasastry, C., Structure of A and B bands of the lead(II) ion in sodium chloride crystals. - Phys. Status Solidi b, 1973, 55, N1, K21-K23. - CA, 1973, 78, 77541.

Exp.; E; T.

2.1.201. Shaha, R., Mukhopadhyay, A. K., Choudhury, M., Polarized absorption spectrum and magnetic behavior of iron(2+)-doped calcium trichlorocadmate. - In: Proc. 17th Nucl. Phys. & Solid State Phys. Symp., 1972. Bombay, 1973, Vol. C, p. 179-181. - CA, 1974, 80, 76197.

Exp.; E; O_h; D_6.

2.1.202. Shaha, R., Bera, S. C., Choudhury, M., Optical and ESR studies of copper(2+)-doped calcium trichlorocadmate. - In: Proc. 17th Nucl. Phys. & Solid State Phys. Symp., 1972. Bombay, 1973, Vol. C, p. 183-185. - CA, 1974, 80, 76198.

Exp.; E; C_{3v}; d^9; 3.1.

2.1.203. Tsuboi, T., Nakai, Y., Oyama, K., Jacobs, P. W. M., Vibration-induced absorption (B band) of s^2-configuration ions in alkali-halide crystals. - Phys. Rev. B, 1973, 8, N4, 1698-1707.

Exp.; T - (e + t); O_h; 1.5.

2.1.204. Tsukerblat, B. S., Brunshtein, S. Kh., The optical spectrum of the excited ruby. [Russ.]. - In: Spektrosk. Kristallov: Materialy 3 Simp. Spektrosk. Kristallov, Aktivir. Ionami Redkozem. Perekhodnykh Metallov, 1970. Leningrad, 1973, p. 128-131.

Theor.; E - e; T - (a + e + t).

2.1.205. Volkov, S. V., Tumanova, N. Kh., Buryak, N. I., Spectroscopic and polarographic study of chromium(II) complexing in molten potassium thiocyanate. [Russ.]. - Zh. Neorg. Khim., 1973, 18, N6, 1536-1439. - CA, 1973, 79, 108700.

Exp.; E - e; d^4.

2.1.206. Volkov, S. V., Buryak, N. I., Yatsimirskiy, K. B., Spectroscopical investigation of copper(II) complexes in melting salts. [Russ.]. - Teor. & Eksp. Khim., 1973, 9, N1, 8186. - RZK, 1973, 15B183.

Exp.; T_d; d^9.

2.1.207. Zgierski, M. Z., Fluorescence-absorption energy gap in the stable anthracene dimer spectra. - J. Chem. Phys., 1973, 59, N3, 1052-1053.

Theor.; 1.5.

2.1.208. Zgierski, M. Z., Herzberg-Teller interaction and spectra of dimers: Stable anthracene dimer. - J. Chem. Phys., 1973, 59, N6, 3319-3322.

Theor.; 1.5.

1974

2.1.209. Davies, G., Vibronic spectra in diamond. - J. Phys. C, 1974, 7, N20, 3797-3809. - CA, 1975, 82, 9607.

Exp.; E - e.

2.1.210. Goltzene, A., Mayer, B., Schwab, C., Cho, K., The chalcogenide centers in cuprous halides: The dynamic Jahn-Teller effect on a P pole in Td symmetry. - In: Extended Abstr. Int. Conf. Color Centers Ionic Crystals, Sendai, 1974, 178, 2 p. - PA, 1975, 78, 25615.

Exp.; T; T_d.

2.1.211. Honma, A., Ooaku, S., Mabuchi, T., Optical absorption line shape for the B band of Tl^{+}-like ions in alkali halides. - J. Phys. Soc. Jpn., 1974, 36, N6, 1708.

Theor.; T - (e + t).

2.1.212. Jacobs, P. W. M., Oyama, K., Jahn-Teller effect in the A- absorption band of alkali halide phosphors. - In: Extended Abstr. Int. Conf. Color Centers Ionic Crystals. Sendai, 1974, 179, 2 p. - PA, 1975, 78, 26159.

Exp.; T - (a + e + t); O_h.

2.1.213. Jacobs, P. W. M., Thorsley, S. A., Electronic spectra of potassium bromide crystals containing thallium. I. A-band absorption. - Cryst. Lattice Defects, 1974, 5, N1, 51-64. - CA, 1974, 81, 70467.

Exp.; T - t.

2.1.214. Kristofel', N. N., Teoriya Primesnykh Tsentrov Malykh Radiusov v Ionnykh Kristallakh. [Russ.]. - Moscow: Nauka, 1974. - 336 p.

Theory of Small-Radius Impurity Centers in Ionic Crystals.

2.1.215. Matsushima, A., Fukuda, A., Calculation of the B-band shape in the In^{+} center in alkali halides. - Phys. Status Solidi b, 1974, 66, N2, 663-667. - PA, 1975, 78, 18964.

Theor.

2.1.216. Mulazzi, E., Terzi, N., Optical response function of a dynamic Jahn-Teller system. - Phys. Rev. B, 1974, 10, N8, 3552-3559. - PA, 1975, 78, 22322. RZF, 1975, 5E70.

Theor.; E - e; T - t; E - (a + e); T - (a + e + t).

2.1.217. Muramatsu, S., Sakamoto, N., Temperature dependence of the band shape for transitions to Jahn-Teller coupled states. - J. Phys. Soc. Jpn., 1974, 36, N3, 839-842. - PA, 1974, 77, 33386.

Exp.

2.1.218. Nasu, K., Kojima, T., Theory of optical band shape for $A_{1g} \rightarrow T_{1u}$ transition I. - Prog. Theor. Phys., 1974, 51, N1, 26-42.

Theor.; T - (a + e + t); 1.7.

2.1.219. Natsume Yuhei, Vibration-Induced Transitions in Optical Spectra of Localized Electrons: Submitted in partial fulfillment of the requirements for the degree of doctor of phylosopy. - Tokyo: Univ. of Tokyo, 1974. - 127 p.

Theor.; (E_u + B_{1g}) - e_g; E - e; 1.3; 1.5.

2.1.220. Perlin, Yu. E., Tsukerblat, B. S., Effekty Elektronno-Kolebatel'nogo Vzaimodeystviya v Opticheskikh Spektrakh Primesnykh Paramagnitnykh Ionov. [Russ.]. - Kishinev: Shtiintsa, 1974. - 368 p.

Effects of Electron-Vibrational Interaction in the Optical Spectra of Impurity Paramagnetic Ions.

2.1.221. Perlin, Yu. E., Polinger, V. Z., Tsukerblat, B. S., Manifestations of the dynamic Jahn-Teller effect in the optical transition to the accidentally degenerate ^4A + ^4E mulitplet in d^5 ions. [Russ.]. - In: Fiz. Mat. Metody Koord. Khim.: Tez. Dokl. 5 Vsesoyuz. Soveshch. Kishinev, 1974, p. 74.

Theor.; (^4A + ^4E) - e.

2.1.222. Solomon, E. I., McClure, D. S., Comparison of the Jahn-Teller effect in four triply degenerate states of Mn^{++} in $RbMnF_3$. - Phys. Rev. B, 1974, 9, 11, 4690-4718. - CA, 1974, 81, 43404. PA, 1974, 77, 69987. RZF, 1974, 11E966.

Exp.; T - (e + t); d^5.

2.1.223. Tsuboi Taiju, Oyama Keiko, Jacobs, P. W. M., Fine structure of the B band in alkali halide:Tin(2+) crystals. - J. Phys. C, 1974, 7, N1, 221-229. - CA, 1974, 80, 89180. PA, 1974, 77, 21029.

Exp.; E; T.

2.1.224. Van Egmond, J., Van der Waals, J. H., Vibronic interaction in lower electronic states of benzene: III. Two active vibrational modes in pseudocylindrical approximation. - Mol. Phys., 1974, 28, N2, 457-467. - CA, 1975, 82, 85498.

Theor.; (B_{1u} + E_{1u}) - e_g; 1.5; 1.7.

2.1.225. Volkov, S. V., Buryak, N. I., Electronic absorption spectra and structure of iron(II) complexes in sulfate-containing melts. [Russ.]. - Teor. & Eksp. Khim., 1974, 10, N4, 523-528. - CA, 1974, 81, 161608.

Exp.; E - e; d^6.

2.1.226. Witkowski, A., Zgierski, M. Z., Vibronic spectra of cyclic diacetylene dimers. - Acta Phys. Pol. A, 1974, 46, N4, 445-450. - CA, 1975, 82, 57011.

Theor.; 1.5.

1975

2.1.227. Abdelaziz Daoud, Perret, R., Characterization of some bis(ammonioethyl)ammonium halide complexes. [Fr.]. - C.R. Hebd. Sean. Acad. Sci. Ser. C, 1975, 280, N22, 1377-1379. - CA, 1975, 83, 107511.

Exp.; E; O_h; d^9.

2.1.228. Bhattacharyya, B. D., Jahn-Teller effect in the ligand field theory of trigonal bipyramidal vanadium(4+) complexes. - Phys. Status Solidi b, 1975, 71, N2, 427-433. - CA, 1975, 83, 199596. PA, 1976, 79, 5761.

Theor.; E - e; D_{3h}; d^1; 3.1; 3.6.

2.1.229. Bolman, P. S. H., Brown, J. M., Carrington, A., Kopp, I., Ramsay, D. A., Reinvestigation of the $\tilde{A}^2\Sigma+$ - $\tilde{X}^2\Pi i$ band system of isocyanate free radical. - Proc. R. Soc. London Ser. A, 1975, 343, N1632, 17-44. - CA, 1975, 82, 131458.

Exp.; 1.6.

2.1.230. Bron, W. E., Dynamical charge overlap, electron-lattice coupling strength, and lattice dynamics of ionic crystals. - Phys. Rev. B, 1975, 11, N10, 3951-3959. - PA, 1975, 78, 72780. RZF, 1975, 11E62.

Exp.; f.

2.1.231. Brunstein, S. H., Tsukerblat, B. S., Calculation of the optical absorption band width of excited ruby. [Fr.]. - Phys. Status Solidi b, 1975, 70, N1, 385-400. - CA, 1975, 83, 87614. PA 1975, 78, 73169.

Theor.; E; D_{3d}.

2.1.232. Causley, G. C., Russel, B. R., Vacuum ultraviolet absorption spectra of the bromomethanes. - J. Chem. Phys., 1975, 62, N3, 848-857. - CA, 1975, 82, 138739.

Exp.; E.

2.1.233. Duxbury, G., Electronic spectra of triatomic molecules and the Renner-Teller effect. - Mol. Spectrosc., 1975, 3, 497-573. - CA, 1975, 83, 199552.

Rev.; 1.6.

2.1.234. Estle, T. L., The role of the Jahn-Teller effect in the optical spectra of ions in solids. - In: Opt. Prop. Ions Solids. New York; London, 1975, p. 419-448. (NATO Adv. Study Inst. Ser., Ser. B, 1974, 8.) - CA, 1977, 86, 162862. RZF, 1976, 11D535.

Theor.

2.1.235. Higginson, B. R., McAuliffe, C. A., Venanzi, L. M., Static and dynamic distortions in five-coordinate complexes of cobalt(II) and rhodium(I). - Helv. Chim. Acta, 1975, 58, N5, 1261-1271. - CA, 1975, 83, 90139. RZF, 1975, 12D289.

Exp.; E; C_{3v}; C_{2v}; d^7.

2.1.236. Ippolitova, G. K., Omel'yanovskiy, E. M., Optical absorption due to intracenter transitions in the iron(2+) ion in gallium arsenide. [Russ.]. - Fiz. & Tekhn. Poluprovodn., 1975, 9, N2, 236-241. - CA, 1975, 83, 17995. PA, 1976, 79, 14543. RZF, 1975, 6D494.

Exp.; E; T; d^6.

2.1.237. Jacobs, P. W. M., Oyama Keiko, Optical absorption of s^2 configuration ions in alkali halide crystals. I. Lineshape of the A band in In^+-doped crystals. - J. Phys. C, 1975, 8, N6, 851-864. - CA, 1975, 82, 177672. PA, 1975, 78, 47231. RZF, 1975, 10D542.

Exp.; T - t; O_h.

2.1.238. Jacobs, P. W. M., Oyama Keiko, Optical absorption of s^2 configuratin ions in alkali halide crystals. II. Lineshape of the C band in $KBr{:}In^+$. - J. Phys. C, 1975, 8, N6, 865-871. - CA, 1975, 82, 177673. PA, 1975, 78, 47232. RZF, 1975, 10D543.

Exp.; T - t; O_h.

2.1.239. Kushkuley, L. M., Perlin, Yu. E., Tsukerblat, B. S., On the role of odd crystal vibrations in the F-band formation. [Russ.]. - In: Tez. Dokl. 8 Soveshch. Teor. Poluprovodn. Kiev, 1975, p. 108.

Theor.; 1.5.

2.1.240. Muramatsu, S., Nasu, K., Optical lineshape for A_1 - E transition: Comparison of IOA with numerical method. - Phys. Status Solidi b, 1975, 68, N2, 761-766.

Theor.; E - e.

2.1.241. Nasu, K., Theory of optical band shape for A_{1g} - T_{1u} transition. II. - Z. Naturforsch. a, 1975, 30, N8, 1060-1070. - PA, 1975, 78, 73142.

Theor.; T - (a + e + t); O_h; 1.7.

2.1.242. Oyama Keiko, Jacobs, P. W. M., Optical absorption of s^2 configuration ions in alkali halide crystals. III. The A band in Sn^+-doped crystals. - J. Phys. & Chem. Solids, 1975, 36, N12, 1375-1382. - CA, 1976, 84, 51691. RZF, 1976, 4D488.

Exp.; T - (a + t).

2.1.243. Oyama Keiko, Jacobs, P. W. M., Optical absorption of s^2-configuration ions in alkali halide crystals. IV. Line shape of the C band in Sn^{2+}-doped crystals. - J. Phys. & Chem. Solids, 1975, 36, N12, 1383-1388. - RZF, 1976, 4D489.

Exp.; T - (e + t); O_h.

2.1.244. Rosenfeld, A., Boyn, R., Ruszczynski, G., Jahn-Teller effect on the optical absorption spectra of $CdS:Ti^{2+}$ and $CdSe:Ti^{2+}$. - Phys. Status Solidi b, 1975, 70, N2, 601-610. - CA, 1975, 83, 139266. PA, 1975, 78, 77633.

Exp.; T - t; O_h ; d^2.

2.1.245. Roterman, M., Witkowski, A., Zgierski, M. Z., Theoretical analysis of the absorption spectra of N,N'-paracyclophanes. - Acta. Phys. Pol. A, 1975, 47, N3, 385-389. - CA, 1975, 83, 50297.

Theor.; 1.5.

2.1.246. Sastry, S. B. S., Balasubramanyam, K., Jahn-Teller splitting of lead center bands in rubidium iodide crystals at 77 K. - In: Proc. Nucl. Phys. & Solid State Phys. Symp. Calcutta, 1975, Vol. 18C, p. 190-193. - CA, 1977, 87, 93042.

Exp.; T - (a_1 + t_2).

2.1.247. Stephens, P. J., Vibrational-electronic interactions. - In: Electronic States in Inorg. Compounds: New Exp. Techn. Dordrecht, 1975, p. 95-112. - PA, 1976, 79, 14518.

Theor.

2.1.248. Stoneham, A. M., Theory of Defects in Solids: Electronic Structure of Defects in Insulators and Semiconductors. - Oxford: Clarendon Press, 1975. - XIX, 955 p.

Theor.

2.1.249. Tsuboi Taiju, Oyama, K., Jacobs, P. W. M., Lineshape of the A band of s^2 configuration ions in alkali halide crystals. - Can. J. Phys., 1975, 53, N2, 192-199. - CA, 1975, 82, 162361. PA, 1975, 78, 33876.

Exp.; T - t.

2.1.250. Tsuboi Taiju, Chaney, R. E., Jacobs, P. W. M., Optical absorption spectra of the A band in lead(2+) ion-doped potassium bromide crystals. - Can. J. Phys., 1975, 53, N2, 200-201. CA, 1975, 82, 162362. PA, 1975, 78, 33877.

Exp.; T - t.

2.1.251. Tsukerblat, B. S., Elektronno-Kolebatel'nye i Magnitnye Vzaimodeystviya v Primesnykh Kristallakh: Avtoref. Diss. ... Dokt. Fiz.-Mat. Nauk. [Russ.]. - Tartu, 1975, - 44 p. - Tart. Univ.

Electron-Vibrational and Magnetic Interactions in Impurity Crystals.

Theor.; E - e; T - t; 1.5; 1.6; 2.2; 2.4

1976

2.1.252. Bernstein, E. R., Meredith, G. R., Interaction in inorganic molecule crystals: Electronic spectra of rhenium hexafluoride pure and mixed crystals. - J. Chem. Phys., 1976, 64, N1, 375-403. - CA, 1976, 84, 82023. PA, 1976, 79, 40895.

Exp.; Γ_8; O_h; d^1.

2.1.253. Giorgianni, U., Mondio, G., Saitta, G., Vermiglio, G., Role of the Jahn-Teller effect in the excited states of thallium-doped potassium bromide type crystals. - Phys. Status Solidi b, 1976, 74, N1, 317-322. - CA, 1976, 84, 171613. PA, 1976, 79, 45263.

Theor.; T; O_h.

2.1.254. Harrison, T. G., Patterson, H. H., Hsu, M. T., Optical spectra of the tetrabromopalladate(II) ion doped in cesium hexabromozirconate(IV) at 2 K. - Inorg. Chem., 1976, 15, N12, 3018-3024. - CA, 1976, 85, 184232.

Exp.; E - e.

2.1.255. Matsushima Akira, Fukuda Atsuo, Structure of the Jahn-Teller induced B absorption band of Tl^+-type centers in alkali halides. - Phys. Rev. B, 1976, 14, N8, 3664-3671. - CA, 1977, 86, 10298. PA, 1977, 80, 12410.

Theor.; T - (a + e + t); O_h.

2.1.256. Schirmer, O. F., Optical absorption of small polarons bound in octahedral symmetry: V-type centers in alkaline earth oxides. - Z. Phys. B, 1976, 24, N3, 235-244. - CA, 1976, 85, 101598.

Exp.; 1.5.

2.1.257. Sigmund, E., Wagner, M., An operator method for optical line shape calculations in non-adiabatic electron-phonon systems. - Phys. Status Solidi b, 1976, 76, N1, 325-336. - CA, 1976, 85, 101407. PA, 1976, 79, 73006.

Theor.; E - e.

2.1.258. Sigmund, E., Moments and absorption line shape of an antiresonant electron-phonon system in the strong coupling limit. - Phys. Rev. B, 1976, 14, N10, 4702-4703. - CA, 1977, 86, 48653.

Theor.

2.1.259. Stedman, G. E., Interference between optical transitions broadened by phonon interaction. - J. Phys. C, 1976, 9, N3, 535-551. - PA, 1976, 79, 31514.

Theor.

2.1.260. Sviridov, D. T., Sviridova, R. K., Smirnov, Yu. E., Opticheskiye Spektry Ionov Perekhodnykh Metallov v Kristallakh, [Russ.]. - Moscow: Nauka, 1976. - 267 p. - Bibliogr.: 868 ref.

Optical Spectra of Transition Metal Ions in Crystals.

Ch. 5. Electron-Vibrational Interaction.

Ch. 7. Optical Spectra of d^n Ions in Crystals.

2.1.261. Tsuboi Taiju, Jacobs, P. W. M., Line shape of the A band in potassium bromide:lead(2+) ion crystals. - Kyoto Sangyo Daigaku Ronshu, 1976, 5, N3, 48-57. - CA, 1978, 87, 75966.

Exp.

2.1.262. Yeh Chin-yah, Richardson, F. S., Near ultraviolet optical activity of chiral pyridine derivatives. - J. Chem. Soc. Faraday Trans. II, 1976, 72, N2, 331-350. - CA, 1976, 84, 120667.

Theor.; 1.5.

1977

2.1.263. Alonso, P. J., Alcala, R., On the optical absorption spectrum of cobalt(2+) ion in calcium fluoride. - Phys. Status Solidi b, 1977, 81, N1, 333-339. - CA, 1977, 87, 31407. PA, 1977, 80, 55806.

Exp.; T; d^7.

2.1.264. Baker, J. M., Buisson, R., Vial, J. C., Orbit-lattice interaction in Γ_8 quartets. II. A rigorous test for various models of the interaction for Er^{3+}:MgO and $Dy^{3+}:CaF_2$. - J. Phys. C, 1977, 10, N21, 4407-4417. - CA, 1978, 88, 128620.

Exp.; Γ_8.

2.1.265. Dushin, R. B., Mitchenko, N. S., Predtechenskiy, Yu. B., Shcherba, L. D., Manifestation of the electron-vibrational interaction in the spectra of impurity atoms in solid inert gases. [Russ.]. - In: Teor. Spektrosk. Moscow, 1977, p. 183-186. - CA, 1979, 90, 94660.

Exp.; T; O_h.

2.1.266. Ehlhardt, W. J., Lohr, L. L., Jr., A theoretical study of the optical absorption band shape for xenon hexafluoride. - J. Chem. Phys., 1977, 67, N5, 1935-1937. - CA, 1977, 87, 175054.

Theor.; $(^1T_{1u} + {}^3T_{1u}) - t_{1u}$; 1.5.

2.1.267. Fisher, G., Jacobson, S., Naaman, R., On anomalies in the absorption spectrum of benzene vapor and the mechanism of electronic relaxation. - Chem. Phys. Lett., 1977, 49, N3, 427-430.

Exp.; D_{6h}; 1.5.

2.1.268. McCarthy, P. J., Berman, R. D., The single crystal polarized optical spectra of hexaaquoiron(II) perchlorate. - J. Coord. Chem., 1977, 6, N3, 129-133. - CA, 1977, 86, 148202. PA, 1977, 80, 48188.

Exp.; D_{3d}.

2.1.269. Osadko, I. S., Electron-vibrational transitions in the presence of several minima. - Phys. Status Solidi b, 1977, 82, N2, K107-K110.

Theor.; 1.2; 2.2.

2.1.270. Pritchard, D. E., Ahmad-Bitar, R., Lapatovich, W. P., Laser spectroscopy of bound NaNe and related atomic physics. - In: Laser Spectrosc. 3: Proc. 3rd Int. Conf. Jackson Lake Lodge, Wyo.,

US, Berlin, etc., 1977, p. 355-364. (Springer Ser. Opt. Sci.) - CA, 1978, 88, 67682. PA, 1978, 81, 86871. RZF, 1979, 9D417.

Exp.; E - e.

2.1.271. Rueff, M., Sigmund, E., E - e Jahn-Teller system: optical response studies. - Phys. Status Solidi b, 1977, 80, N1, 215-223. - CA, 1977, 86, 162902. PA, 1977, 80, 35898.

Theor.; E - e.

2.1.272. Rueff, M., Untersuchung theoretischer Modelle trigonaler Jahn-Teller Zentren, Optische Absorption und Phonenstreung: Diss. [Germ.]. - Stuttgart, 1977. - 105 p.

The Theoretical Model for the Trigonal Jahn-Teller Centers: Optical Absorption and Phonon Scattering.

Theor.; E - e; C_{3v}; D_{3h}, D_{3d}; 1.7.

2.1.273. Ulrici, W., $CdF_2:Cr^{2+}$ - an example for an orthorhombic Jahn-Teller distortion. - Phys. Status Solidi b, 1977, 84, N2, K155-K157. - CA, 1978, 88, 56721. PA, 1978, 81, 19007.

Exp.; T - (e + t_2); O_h; d^8; 3.1.

2.1.274. Wilson, T. M., Wood, R. F., Electronic structure of the F center in strontium oxide. - Phys. Rev. B, 1977, 16, N10, 4594-4598. - CA, 1978, 88, 96936. PA, 1978, 81, 30887.

Theor.; T - (a + e + t_2).

2.1.275. Zolovkina, I. S., About the nonelementary structure of the A absorption bands of thallium(1+) type centers in alkali halide crystals. [Russ.]. - Izv. Akad. Nauk Latv. SSR. Ser. Fiz. Tekh., 1977, N3, 64-66. - CA, 1977, 87, 191409.

Exp.; T.

1978

2.1.276. Boyrivent, A., Duval, E., $^4T_{2g}$ excited level of Cr^{3+} ion in cubic site of the MgO crystal. - J. Phys. C, 1978, 11, N6, L227-L230.

Exp.; T; d^3.

2.1.277. Brawer, S., Dynamical effects on nearly degenerate electronic levels in solids. - J. Chem. Phys., 1978, 68, N8, 3352-3359.

Theor.; 1.5.

2.1.278. Duran, J., The Jahn-Teller effect in ionic crystals: recent developments. - Semicond. & Insul., 1978, 3, N4, 329-350. - CA, 1978, 89, 82177. PA, 1978, 81, 64015.

Rev.

2.1.279. Fujiwara Kitao, Umezawa Yoshio, Fujiwara Shizuo, Fuwa Keiichiro, Shima Nobuyuki, Kamimura Hiroshi, Transient spectra due to mercury atoms in aqueous solutions. - Nature (GB), 1978, 276, N 5683, 47-48. - CA, 1979, 90, 143692.

Exp.; E - e.

2.1.280. Gordon, K. R., Warren, K. D., Spectroscopic and magnetic studies of the 3d metallocenes. - Inorg. Chem., 1978, 17, N4, 987-994. - CA, 1978, 88, 135802.

Exp.; E; d^4; d^7; 3.3.

2.1.281. Kalnin, A. E., Trinkeler, M. F., Spectroscopy of mixed alkali halide crystals activated by thallium. - Phys. Status Solidi b, 1978, 88, N2, 445-455. - CA, 1978, 89, 120240. PA, 1978, 81, 80385.

Exp.; T.

2.1.282. Mulazzi, E., Optical response function from Jahn-Teller systems: the F band in cesium halides. - Phys. Rev. B, 1978, 18, N12, 7109-7121. - CA, 1979, 90, 112480. PA, 1979, 82, 43005.

Theor.; T - (a + e + t).

2.1.283. Nagel, D., Sonntag, B., 3s-Exitation of Na atoms trapped in Xe matrices. - Ber. Bunsenges. Phys. Chem., 1978, 82, N1, 38-40. - CA, 1978, 88, 112766. PA, 1978, 81, 46291.

Exp.; T.

2.1.284. Piepho, S. B., Krausz, E. R., Schatz, P. N., Vibronic coupling model for calculation of mixed valence absorption profiles. - J. Am. Chem. Soc., 1978, 100, N10, 2996-3005.

Theor.

2.1.285. Sastry, S. B. S., Balasubramanyam, K., Optical and electrical properties of $RbI:Pb^{2+}$ crystals. - J. Phys. C, 1978, 11, N20, 4213-4221. - PA, 1979, 82, 15493.

Exp.; T - $(a + t_2)$.

2.1.286. Sharan, V. B., Srivastawa, D. N. S., Sen, S. C., Structure of the A band in thallium-doped ammonium chloride. - In: Proc. Nucl. Phys. & Solid State Phys. Symp., 1978, Vol. 21C, p. 348. - CA, 1980, 92, 155452.

Exp.

1979

2.1.287. Arkhangel'skaya, V. A., Lushchik, N. E., Reiterov, V. M., Soovik, Kh. A., Trofimova, L. M., Absorption and luminescence of Pb^{2+} ions in alkaline earth fluoride crystals. [Russ.]. - Opt. & Spektrosk., 1979, 47, N4, 708-715. - CA, 1980, 92, 31320.

Exp.

2.1.288. Bakhtin, A. I., Crystal chemistry of iron in a chloritoid according to optical spectroscopic data. [Russ.]. - Geokhimiya, 1979, N2, 305-308. - CA, 1979, 90, 144443.

Exp.; E - e; O_h; d^6.

2.1.289. Brickmann, J., Vibronic coupling via the nonttotally symmetric C-H bending mode in pyrazine. - Ber. Bunsenges. Phys. Chem., 1979, 83, N1, 70-75. - CA, 1979, 90, 112492.

Exp.; D_{6h}; 1.5.

2.1.290. Davies, G., Transitions between E and T electronic states each coupled to E modes. - Solid State Commun., 1979, 32, N9, 745-747. - CA, 1980, 92, 85198. PA, 1980, 83, 20677.

Theor.; (T + E) - e; T - e.

2.1.291. Farge, Y., Fontana, M. P., Electronic and Vibrational Properties of Point Defects in Ionic Crystals. - Amsterdam, etc.: North-Holland, 1979, - 264 p. - (Ser. Defects in Crystalline Solids; N11.)

Ch. 3. Effect of Electron-Vibrational Interaction in Optical Transitions, p. 105-145.

Ch. 4. The Electronic States of Defects in Ionic Solids, p. 147-222.

E - e; T - t; 2.2; 2.8.

2.1.292. Henneker, W. H., Siebrand, W., Zgierski, M. Z., Non-Condon effects in the $\tilde{X} \rightarrow \tilde{A}$ electronic absorption band system of sulfur dioxide. - Chem. Phys. Lett., 1979, 68, N1, 5-8. - RZF, 1980, 4D414.

Theor.; 1.5.

2.1.293. Karkach, S. P., Osherov, V. I., Optical transitions to degenerate electronic states. - Mol. Phys., 1979, 38, N6, 1955-1971. - CA, 1980, 92, 188413.

Theor.; E - e.

2.1.294. Kupka, H., Ensslin, W., Wernicke, R., Schmidtke, H. H., Intensity distribution in the progressions of vibronic transitions of metal-ion complexes. - Mol. Phys., 1979, 37, N6, 1693-1701. - CA, 1980, 92, 31256. PA, 1979, 82, 78735.

Exp.; T.

2.1.295. Muramatsu, S., Sakamoto, N., Calculation of the A band shape in the $KCl:Au^-$ on D-mode model. - Phys. Status Solidi b, 1979, 91, N2, K87-K90. - CA, 1979, 90, 142303.

Theor.; T - (e + t).

2.1.296. Webb, J. D., Bernstein, E. R., A ligand field theory analysis of the spectra of the t_{2g}^3 levels of IrF_6. - Mol. Phys., 1979, 37, N5, 1509-1519. - CA, 1979, 91, 219559. PA, 1979, 82, 72248.

Exp.; T; Γ_8.

2.1.297. Webb, J. D., Spectroscopic studies of the Jahn-Teller interaction in transition metal hexafluorides: - Princeton Univ. - 1979. - 273 p. - [Order N 7928494.] Diss. Abstr. Int. B, 1980, 40, N7, 3194-3195. CA, 1980, 92, 138027.

Exp.; Γ_8 - (e + t); O_h; 2.8.

See also: 1.1.-30, 36, 62, 68; 1.2.-10; 1.3.-3, 18, 49, 70, 71, 73, 76, 77, 81, 85, 88, 99; 1.5.-5, 24, 26, 35, 48, 52; 1.6.-17; 1.7.-22, 37, 38, 40, 47; 2.2.-70, 79, 108, 137, 154; 2.3.-4, 44, 46, 48, 60, 68, 69, 86, 100, 108, 122; 2.4.-2, 12, 21, 31, 42, 52, 67, 70, 80, 86, 91, 92, 102, 103, 107, 110, 130, 136, 142, 143, 155, 157, 160, 162, 163, 164, 178, 180, 183, 186, 189, 195, 197, 208; 2.6.-1; 2.7.-5, 50, 82; 2.8.-87; 3.1.-6, 27, 93, 96, 101, 102, 117, 161, 199, 206; 3.2.-29, 116, 137; 3.6.-10, 20; 4.1.-82; 4.3.-1, 2, 3, 4, 5, 6, 10, 11, 12, 15, 19, 20, 21, 24, 26, 27, 33, 34, 35, 37, 39, 41, 43, 44, 49, 52, 57, 60, 61, 63, 65, 68, 70, 73, 74, 75; 4.4.-16, 34, 37, 115; 4.6.-38, 92, 181; 5.1.-29, 60; 5.2.-17; 19, 25, 42; 5.3.-1, 2, 3, 5, 6, 27.

2.2. Luminescence and Radiationless Transitions

The most pronounced nonadiabatic effects arise due to the contribution of the region of small nuclear displacements from the high-symmetry configuraton, and this creates great difficulties for the absorption band theory. From this point of view, the problem of luminescence seems to be simpler, since the emission arises from the point of the adiabatic potential minima ($Q \neq 0$). The evaluation of the equilibrium configurations dates back to Refs. [2.1.122; 1.1.63; 2.2.3]. Low-symmetry equilibrium configurations are manifested in the polarization of luminescence at low temperatures [2.2.-12, 17, 19, 21, 36, 33, 34, 87]. An increase of temperature leads to equal population of all the adiabatic potential minima, resulting in the depolarization of the emission [2.2.-15, 38, 49, 55, 114, 102]. The degree of polarization of emission depends on the incident light frequency [2.2.-38, 49]. Useful information can be obtained from investigation of the luminescence bands when uniaxial stress is applied [2.4.47].

The nonradiative relaxation theory for Jahn-Teller centers is developed in Refs. [2.2.-8, 14, 16, 30]. The theory was applied to study the kinetics of excited ruby luminescence [2.2.-44, 45, 46]. In the cited papers the spin-forbidden transitions, as well as the quadratic vibronic interactions, are discussed. The book [2.1.220] includes a review of the theory and its applications. The papers devoted to molecular systems involve complex problems of relaxation of the system with discrete vibronic levels. For a review of this subject, see Ref. [2.1.37].

1958

2.2.1. Klick, C. C., Compton, W. D., Polarized luminescent emission from KCl:Tl at low temperatures. - J. Phys. & Chem. Solids, 1958, 7, N1, 170-174. - CA, 1959, 53, 6784.

Exp.

1959

2.2.2. Kamimura Hiroshi, Sugano Satoru, Luminescence processes in the KCl:Tl phosphor in a multidimensional configuration space. - J. Phys. Soc. Jpn., 1959, 14, 1612-1621. - CA, 1960, 54, 23809d.

Theor.

1960

2.2.3. Kristofel', N. N., The Jahn-Teller effect in luminescence centers in crystals. [Russ.]. - In: Tr. Inst. Fiz. i Astron. Akad. Nauk Est.SSR, 1960, N12, p. 20-41. - CA, 1962, 56, 5505b.

Theor.; T - e.

2.2.4. Kristofel', N. N., On the interaction of the activator with local nontotally symmetric vibrations. [Russ.]. - Opt. & Spektrosk., 1960, 9, N5, 615-620. - CA, 1961, 55, 14069i.

Theor.; T - (e + t); O_h.

1961

2.2.5. Kristofel', N. N., On the interaction of the activator with nontotally symmetric local vibrations. [Russ.]. - Izv. Akad. Nauk SSSR, Ser. Fiz., 1961, 25, N4, 533-535.

Theor.; T - e; O_h.

2.2.6. Kristofel', N. N., On the influence of the structure type of the base crystal on the spectral properties of phosphorus. [Russ.]. - In: Tr. Inst. Fiz. i Astron. Akad. Nauk EstSSR, 1961, N17, p. 3-27.

Theor.; T - e + t; O_h.

1962

2.2.7. Fowler, W. B., Dexter, D. L., Relation between absorption and emission probabilities in luminescent centers in ionic solids. - Phys. Rev., 1962, 128, N5, 2154-2165. - CA, 1963, 58, 106d. RZF, 1963, 8D394.

Theor.

1963

2.2.8. Perlin, Yu. E., Kovarskiy, V. A., Tsukerblat, B. S., The theory of nonradiative transitions with spin inversion. [Russ.]. - In: Tez. Dokl. 3 Soveshch. Kvant. Khim. Kishinev, 1963, p. 52.

Theor.; T - e; d^3.

1964

2.2.9. Koda Takao, Shionoya Shigeo, Nature of the self-activated blue luminescence center in cubic ZnS:Cl single crystals. - Phys. Rev., 1964, 136, N2A, 541-555. - CA, 1964, 61, 12808h.

Exp.; E.

2.2.10. Perlin, Yu. E., Kovarskiy, V. A., Tsukerblat, B. S., On the theory of multiphonon nonradiative transitions with spin inversion. [Russ.]. - In: Tez. Dokl. 4 Soveshch Teorii Poluprovodnikov. Kishinev, 1964, p. 65.

Theor.; T - e; d^3.

2.2.11. Witkowski, A., Boileau, E., Resonance interactions between degenerate electronic states: intensity. [Fr.]. - C.R. Hebd. Seances Acad. Sci., 1964, 258, N8, 2331-2333. - CA, 1964, 60, 12679d.

Theor.

1965

2.2.12. Kristofel', N. N., Low-temperature polarization of luminescence of "isotropic" impurity centers and the Jahn-Teller effect. [Russ.]. - Opt. & Spektrosk., 1965, 18, N5, 798-802. - CA, 1965, 63, 6501c.

Theor.; T.

2.2.13. Nelson, D. F., Sturge, M. D., Relation between absorption and emission in the region of the R lines of ruby. - Phys. Rev., 1965, 137, N4A, 1117-1130. - RZF, 1965, 8D552.

Exp.; T; O_h; d^3.

2.2.14. Perlin, Yu. E., Pokatilov, E. P., Kabisov, K. S., Gifeysman, Sh. N., Tsukerblat, B. S., Rusanov, M. M., Investigations on the theory of local states in crystals. [Russ.]. - In: Uch. Zap. Kishinev. Univ., 1965, Vol. 80, p. 3-50. - RZF, 1966, 7E395.

Theor.

2.2.15. Tomura Massao, Nishimura Hitoshi, Polarized emission from Tl^+ ions in KCl single crystals. - J. Phys. Soc. Jpn., 1965, 20, N8, 1536-1537. - CA, 1966, 64, 1482c.

Exp.; T.

2.2.16. Tsukerblat, B. S., Perlin, Yu. E., On the theory of multiphonon radiationless transitions in paramagnetic local centers. [Russ.]. - Fiz. Tverd. Tela, 1965, 7, N11, 3278-3288. - CA, 1966, 64, 5906b.

Theor.; T; O_h; d^3.

1966

2.2.17. Nagli, L. J., On the NaI(Tl) crystal luminescence. [Russ.]. - Izv, Akad. Nauk LatvSSR. Fiz. & Tekh., 1966, N2, 50-56. - CA, 1966, 65, 6536g.

Exp.; T; O_h.

2.2.18. Perlin, Yu. E., Tsukerblat, B. S., On the theory of temperature quenching of luminescence of paramagnetic centers in ionic crystals. [Russ.]. - In: Tez. Dokl. 15 Soveshch. Luminestsentsii. Tbilisi, 1966, p. 61.

Theor.; T - e; O_h; d^3.

2.2.19. Tale, A., On the mechanism of intracenter fluorescence in KI-In crystals. [Russ.]. - Izv. Akad. Nauk Latv.SSR. Fiz. & Tekh., 1966, N3, 7-14. - CA, 1966, 65, 16210c.

Exp.; T - (e + t_2); O_h.

2.2.20. Tomura Masao, Nishimura Hitoshi, Kawashima Yukhiro, Emission of forbidden transition violated by phonons in KCl: Tl single crystals. - J. Phys. Soc. Jpn., 1966, 21, N10, 2081-2082.

Exp.; T.

2.2.21. Zazubovich, S. G., Polarized luminescence of alkali halide crystals activated by monovalent mercury-like ions. [Russ.]. - In: Tr. Inst. Fiz. i Astron. Akad. Nauk EstSSR, 1966, N34, p. 171. - CA, 1968, 69, 6878.

Exp.

1967

2.2.22. Fukuda, A., Makishima, S., Mabuchi, T., Onaka, R., Polarization of luminescence in KBr:Tl type crystals due to the Jahn-Teller effect. - J. Phys. & Chem. Solids, 1967, 28, N9, 1763-1780. - CA, 1967, 67, 95392. PA, 1967, 70, 33413.

Exp.; T_{1u} - (a + e + t).

2.2.23. Hizhnyakov, V., Tehver, I., Theory of resonant secondary radiation due to impurity centers in crystals. - Phys. Status Solid, 1967, 21, N2, 755-768. - RZF, 1967, 11E322.

Theor.; 2.7.

2.2.24. Kristofel', N. N., The calculations of the cation type exciton properties for the TlCl crystal. [Russ.]. - In: Tr. Inst. Fiz. i Astron. Akad. Nauk Est.SSR, 1967, N32, p. 69-74.

Theor.

2.2.25. Kristofel', N. N., The theory of impurity energy levels. [Russ.]. - In: Tr. Inst. Fiz. i Astron. Akad. Nauk Est.SSR, 1967, N32, p. 97-121.

Theor.

2.2.26. Kirstofel', N. N., On the origin of nonelementary features of the luminescent A-band centers in KCl-Tl type crystals. [Russ.]. - Opt. & Spektrosk., 1967, 22, N1, 74-79. - CA, 1967, 66, 60505.

Theor.; T - (e + t).

2.2.27. Nieman, G. C., Tinti, D. S., Geometry of the lowest triplet state of benzene. - J. Chem. Phys., 1967, 46, N4, 1432-1436.

Exp.; E.

2.2.28. Tale, A. K., Rising time of the photoscillations of potassium bromide-indium. [Russ.]. - Izf. Akad. Nauk. Latv.SSR, Fiz. & Tekh., 1967, N6, 39-42. - CA, 1968, 69, 63320.

Exp.; T - (e + t).

2.2.29. Tsukerblat, B. S., Perlin, Yu. E., Rozenfel'd, Yu. B., Non-radiative transitions in impurity ions with unfilled d-shells. [Russ.]. - In: Tez. Dokl. 2 Simp. Spektrosk. Kristallov, Aktivir. Redkozem. Elementami i Elementami Gruppy Zheleza. Khar'kov, 1967, p. 65.

Theor.

2.2.30. Tsukerblat, B. S., The non-radiative transitions in the unfilled d-shells of the impurity ions of the iron group. [Russ.]. - Izf. Akad. Nauk Mold.SSR, Ser. Biol. Khim. Nauk, 1967, N9, 8-17.

Theor.; T - (a + e); O_h; d^3.

1968

2.2.31. Ahrenkiel, R. K., Excited state structure of the I-center in AgCl. - Solid State Commun., 1968, 6, N10, 741-744. - PA, 1969, 72, 12057.

Exp.; T.

2.2.32. Arnold, G. W., Brice, D. K., Near band-edge luminescence in GaAs:Zn. - Appl. Phys. Lett., 1968, 13, N2, 51-53. - CA, 1968, 69, 63329.

Exp.; T_d.

2.2.33. Fukuda Atsuo, Makishima, S., Mabuchi Teruhiko, Onaka, R., Polarization spectra for luminescence of KBr:Tl type crystals. - In: Proc. Int. Conf. Luminescence, 1966. Budapest, 1968, Vol. 1, p. 713-718. - CA, 1969, 70, 42631.

Exp.; T; O_h.

2.2.34. Kamejima, T., Fukuda, A., Shionoya, S., Polarization of luminescence of α-center in potassium iodide. - Phys. Lett. A, 1968, 26, N11, 555-556. - CA, 1968, 69, 31824.

Exp.; T; O_h.

2.2.35. Kristofel', N. N., Some questions of the theory of luminescence centers in alkali halide crystal phosphors. - In: Proc. Int. Conf. Luminescence, 1966, Budapest, 1968, Vol. 1, p. 824-829. - CA, 1969, 70, 52101.

Theor.; T - e; T - t.

2.2.36. Tsukerblat, B. S., Ablov, A. V., Bersuker, I. B., Vekhter, B. G., Polarization of optical spectral bands of transition metal complexes taking into account the Jahn-Teller effect. [Russ.]. - In: Primenenie Noveyshikh Fiz. Metodov Issled. Koord. Soedin.: Tez. Dokl. 3 Vsesoyuz. Soveshch. Kishinev, 1968, p. 128.

Theor.

2.2.37. Uehara Yasuo, Structure of luminescence centers of zinc sulfide and alkaline earth orthophosphate phosphors activated with impurity ions of s^2 configuration. - In: Proc. Int. Conf. Luminescence, 1966. Budapest, 1968, Vol. 1, p. 1149-1157. - CA, 1969, 70, 42607.

Exp.; E - e.

2.2.38. Vekhter, B. G., Tsukerblat, B. S., Luminescence polarization and its dependence on the exciting light frequency for Jahn-Teller states of impurity ions. [Russ.]. - Fiz. Tverd. Tela, 1968, 10, N5, 1574-1576. - CA, 1968, 69, 23426. PA, 1968, 71, 41176.

Theor.; E - e; T - e; T - t; O_h.

1969

2.2.39. Dingle, R., Crystal spectra of hexaurea complexes. II. Absorption and emission in crystals containing the hexaureachromium(III) ion. - J. Chem. Phys., 1969, 50, N5, 1952-1955. - CA, 1969, 70,82690. PA, 1969, 72, 33551.

Exp.; T; O_h; d^3.

2.2.40. Gamurar', V. Ya., Perlin, Yu. E., Tsukerblat, B. S., The non-radiative energy transfer and optical line broadening for paramagnetic impurity ions. [Russ.]. - In: Tr. Vsesoyuz. Seminara Vopr. Prirody Ushireniya Spektr. Liniy Izluch. Kondens. Aktivnykh Sred OKG, 1968, Kiev, 1969, p. 42-49.

Theor.

2.2.41. Giorgianni, U., Grasso, V., Perillo, P., KBr:Tl emission in an electric field. - Phys. Rev. Lett., 1969, 23, N12, 640-641.

Exp.; T; O_h.

2.2.42. Jortner, J., Rice, S. A., Hoshstrasser, R. M., Radiationless transitions in photochemistry. - In: Adv. in Photochem. New York, 1969, Vol. 7, p. 149-309. - Bibliogr.: 230 ref.

Rev.

2.2.43. Perlin, Yu. E., Gamurar', V. Ya., Rozenfel'd, Yu. B., Tsukerblat, B. S., The non-radiative relaxation for rare earth and transition metal impurties. [Russ.]. - In: Tez. Dokl. 2 Vsesoyuz. Konf. Teor. Tverd. Tela, Moscow, 1969, p. 88.

Theor.

2.2.44. Perlin, Yu. E., Rozenfel'd, Yu. B., Tsukerblat, B. S., Temperature dependence of ruby luminescence. I. The $^4T_{2g}$ - $^4A_{2g}$ non-radiative transition. [Russ.]. - Ukr. Fiz. Zh., 1969, 14, N8, 1307-1316. - CA, 1969, 71, 107266. RZF, 1970, 1D409.

Theor.; T - (a + e); O_h; d^3.

2.2.45. Perlin, Yu. E., Rozenfel'd, Yu. B., Tsukerblat, B. S., Temperature dependence of ruby luminescence. II. Quantum yields and lifetimes. [Russ.]. - Ukr. Fiz. Zh., 1969, 14, N8, 1317-1323. - CA, 1970, 72, 17023. RZF, 1970, 1D410.

Theor.; T - (a + e); O_h; d^3.

2.2.46. Perlin, Yu. E., Rozenfel'd, Yu. B., Tsukerblat, B. S., Multiphonon radiationless transitions between the Cr^{3+} levels in ruby. [Russ.]. - In: Tr. Vsesoyzn. Seminara Vopr. Priorody Ushireniya Spektr. Liniy Izluch. Kondens. Aktivnykh Sred. OKG, 1968, Kiev, 1969, p. 35-41. - CA, 1970, 73, 103893. RZF, 1969, 11D679.

Theor.

2.2.47. Spiller, M., Gerhardt, V., Gebhardt, W., Investigation of the static and dynamic Jahn-Teller effect in KCl:Tl^+ by time-resolved emission spectroscopy. - Bull. Am. Phys. Soc., 1969, 14, N8, 872.

Exp.; T.

2.2.48. Tsukerblat, B. S., Perlin, Yu. E., Forbidden non-radiative transitions. [Russ.]. - In: Issled. Khim. Koord. Soedin. Fiz. Khim. Metodam Analyza. Kishinev, 1969, p. 195.

Theor.; E - e.

2.2.49. Tsukerblat, B. S., Vekhter, B. G., Luminescence polarization and its dependence on the incident light frequency for the Jahn-Teller states of impurity ions. [Russ.]. - In: Tez. Dokl. 17 Soveshch. Luminestsentsii. Irkutsk, 1969.

2.2.50. Uehara Yasuo, Electronic structure of luminescence centers of zinc sulfide phosphors activated with impurity ions of s^2 configuration. II. ZnS:Pb (zinc-blende) phosphors. - J. Chem. Phys., 1969, 51, N10, 4385-4400. - CA, 1970, 72, 26288.

Exp.; T - (a + e + t); T_d.

2.2.51. Zazubovich, S. G., On the nature of the low symmetry of the luminescent centers in alkali halide crystals activated by indium. [Russ.]. - Opt. & Spektrosk., 1969, 26, N2, 235-243. - CA, 1969, 70, 109916.

Exp.; T.

1970

2.2.52. Morgan, T. N., Theory of luminescent states in semiconductors: Vibronic interaction. - In: Proc. Int. Conf. Luminescence, Newark, 1969, Amsterdam, 1970, p. 420-434. - CA, 1970, 73, 39988.

Theor.; T - (e + t); 1.4; 3.1.

2.2.53. Spiller, N., Gerhardt, V., Gebhardt, W., Investigation of the static and dynamic Jahn-Teller effect in $KCl:Tl^+$ by time-resolved emission spectroscopy. - In: Proc. Int. Conf. Luminescence, Newark, 1969, Amsterdam, 1970, p. 651-655. - CA, 1971, 74, 17691.

Exp.; T.

2.2.54. Tsukerblat, B. S., Vekhter, B. G., Bersuker, I. B., Ablov, A. V., Polarization of bands in the optical spectra of transition metal complexes taking into account the Jahn-Teller effect. [Russ.]. - Zh. Strukt. Khim., 1970, 11, N1, 102-107. - CA, 1970, 73, 9115. RZK, 1970, 12B661.

Theor.

2.2.55. Zazubovich, S. G., The Jahn-Teller effect and polarized luminescence in indium-activated alkali halide crystals.

[Russ.]. - In: Tr. Inst. Fiz. i Astron. Akad. Nauk Est.SSR, 1970, N38, p. 144-162. - CA, 1972, 77, 107271. PA, 1971, 74, 80460. RZF, 1971, 11D931.

Exp.

2.2.56. Zazubovich, S. G., Jahn-Teller effect and the partially covalent bond as causes of the low symmetry of Ga^+, In^+, and Tl^+ centers in cesium chloride-type alkali halides. - Phys. Status Solidi b, 1970, 38, N1, 119-129. - CA, 1970, 72, 105595. PA, 1970, 73, 75984.

Exp.; T.

1971

2.2.57. Chan, I. Y., Van Dorp, W. G., Schaafsma, T. J., Van der Waals, J. H., Lowest triplet state of zinc porphyrin. II. Investigation of its dynamics by microwave-induced delayed phosphorescence. - Mol. Phys., 1971, 22, N5, 753-760. - CA, 1972, 76, 78893.

Exp.; E - $(b_1 + b_2)$; 2.4.

2.2.58. Choyke, W. J., Patrick, L., Photoluminescence of radiation defects in cubic SiC: localized modes and Jahn-Teller effect. - Phys. Rev. B, 1971, 4, N6, 1843-1847. - CA, 1971, 75, 103146. PA, 1971, 74, 66115. RZF, 1972, 3D1024.

Exp.

2.2.59. Fukuda Atsuo, Effect of a vacancy on the Jahn-Teller-distorted Γ_4 excited state in KI:Sn^{2+} as observed in polarized luminescence. - Phys. Rev. Lett., 1971, 26, N6, 314-318. - CA, 1971, 74, 93199. PA, 1971, 74, 28729.

Exp.; T - (e + t).

2.2.60. Goykhman, A. Ya., Electron structure of excitons in alkali halide crystals and noble atom crystals. [Russ.]. - Vestn. Mosk. Univ. Fiz. i Astron., 1971, 12, N1, 104-106. - CA, 1971, 75, 13121. PA, 1971, 74, 36904. RZF, 1971, 8E604.

Theor.; T; O_h.

2.2.61. Kamejima Taibun, Shionoya Shigeo, Fukuda Atsuo, Luminescence of KI containing ns^2 configuration ions under excitation by exciton absorption. - J. Phys. Soc. Jpn., 1971, 30, N4, 1124-1131. - CA, 1971, 74, 148749.

Exp.; T; O_h.

2.2.62. Li, Y. H., Lim, E. C., Pseudo-Jahn-Teller effect and luminescence of nitrogen heterocyclic compounds. - Chem. Phys. Lett., 1971, 9, N4, 279-284. - CA, 1971, 75, 69240. PA, 1971, 74, 44388.

Exp.; 1.5.

2.2.63. McDiarmid, R., Jahn-Teller effect in the $^2E^5_{2g}$ - $^2G^3_{2g}$ transition of rhenium hexafluoride. - J. Mol. Spectrosc., 1971, 38, N3, 495-502. - CA, 1971, 75, 55847. PA, 1972, 75, 3060. RZF, 1972, 2D460.

Exp.; Γ_8 - t_2.

2.2.64. Perrins, N. C., Simons, J. P., Smith, A. L., Molecular distortions in the triplet state of benzenepolyhaloalkane complexes. - Trans. Faraday Soc., 1971, 67, N12, 3415-3424. - CA, 1972, 76, 8492.

Exp.; E - b; D_{6h}; 1.5.

2.2.65. Van Egmond, J., Burland, D. M., Van der Waals, J. H., Zeeman effect in the benzene $^3B_{1u}$ state: Evidence for a dynamic pseudo-Jahn-Teller distortion. - Chem. Phys. Lett., 1971, 12, N1, 206-210. - CA, 1972, 76, 92499. PA, 1972, 75, 20109.

Exp.; (E + B) - e; D_{6h}; 1.5.

1972

2.2.66. Belyaeva, A. A., Predtechenskii, Yu. B., Scherba, L. D., Luminescence of alkali metal impurities in crystalline noble gases. - In: Abstr. Int. Conf. Lumin., Leningrad. Chernogolovka, 1972, p. 255-256.

Exp.; T; O_h.

2.2.67. Duxbury, G., Horani, M., Rostas, J., Rotational analysis of the electronic emission spectrum of the H_2S^+ ion radical. - Proc. R. Soc. London. Ser. A, 1972, 331, N1584, 109-137. - CA, 1972, 77, 170863.

Exp.; 1.6.

2.2.68. El-Sayed, M. A., Moomaw, W. R., Chodak, J. B., Sensitivity of pseudo-Jahn-Teller distortion of benzene to host crystal structure. - J. Chem. Phys., 1972, 57, N9, 4061-4063. - CA, 1972, 77, 170846. PA, 1972, 75, 80952. RZF, 1973, 3E952.

Exp.; D_{6h}; 1.5.

2.2.69. Flint, C. D., Greenough, P., $^2E_g \rightarrow {}^4A_{2g}$ transition of the hexa-ammino-chromium(III) ion in cubic environments: Intensity mechanism and Jahn-Teller effect. - J. Chem. Soc. Faraday Trans. II, 1972, 68, N6, 897-904. - CA, 1972, 77, 41019. RZF, 1972, 12D470.

Exp.; E.

2.2.70. Henderson, B., Chen, Y., Silbey, W. A., Temperature dependence of luminescence of F^+ and F centers in CaO. - Phys. Rev. B, 1972, 6, N10, 4060-4068. - PA, 1972, 75, 84537.

Exp.; T - (e + t_2).

2.2.71. Ishido Masahiro, Shimada Toshikazu, Ishiguro Masakazu, Emission light in thallous-doped potassium halide phosphors. - In: Mem. Inst. Sci. Ind. Res. Osaka Univ., 1972, Vol. 29, p. 21-22. - CA, 1972, 77, 26790.

Exp.; E.

2.2.72. Kamejima Taibun, Shionoya Shigeo, Fukuda Atsuo, Investigation of luminescence from bound exciton in KBr:I by using polarized light. - J. Phys. Soc. Jpn., 1972, 32, N3, 729-735. - CA, 1972, 76, 105780. PA, 1972, 75, 26236.

Exp.; O_h.

2.2.73. Khizhnyakov, V., Tehver, I., Theory of polarized luminescence of impurity centers. - In: Phys. Impurity Cent. Cryst.: Proc. Int. Semin. "Selected Probl. Theory Impurity Cent. Cryst.," 1970. Tallin, 1972, p. 607-626. - CA, 1973, 78, 166258.

Theor.; E - e; E - b; 1.5.

2.2.74. Koyanagi Motohiko, Zwarich, R. J., Goodman, L., Phosphorescence spectrum of pseudo-Jahn-Teller distortion. - J. Chem. Phys., 1972, 56, N6, 3044-3060. - CA, 1972, 76, 105895. PA, 1972, 75, 29966.

Exp.; 1.5.

2.2.75. Krausbauer, L., Wehner, R. K., Czaja, W., Null field splitting of the B-line in nitrogen-endowed GaP. [Germ.]. - Helv. Phys. Acta, 1972, 45, N1, 61-62, - PA, 1972, 75, 61692.

Exp.; Γ_8 - (e + t).

2.2.76. Morgan, T. N., Lorenz, M. R., Onton, A., Rotational levels of shallow acceptor states: the undulation spectra of N in GaP. - Phys. Rev. Lett., 1972, 28, N14, 906-909.

Exp.; Γ_8 - (e + t); O_h; 1.7.

2.2.77. Perlin, Yu. E., Tsukerblat, B. S., Perepelitsa, E. I., Multiphonon radiationless relaxation in impurity-phonon systems with static Jahn-Teller effect. [Russ.]. - Zh. Eksp. & Teor. Fiz., 1972, 62, N6, 2265-2278. - CA, 1972, 77, 81618. PA, 1973, 76, 41151. RZF, 1972, 9E392.

Theor.; T; d^3.

2.2.78. Ranfagni, A., Pazzi, G. P., Fabeni, P., Viliani, G., Fontana, M. P., Origin of the 4750-Å emission in KCl:Tl phosphors. - Phys. Rev. Lett., 1972, 28, N16, 1035-1037.

Exp.; T - e.

2.2.79. Tkachuk, A. M., Perepelitsa, E. I., Kharchenko, L. S., Tsukerblat, B. S., Electron-phonon interaction in spectra of $LaAlO_3$-Cr^+ luminescence. - In: Abstr. Int. Conf. Lumin., Leningrad, Chernogolovka, 1972, p. 158-159.

Theor.; E - e; T - (a + e - t).

2.2.80. Vink, A. T., Van Gorkom, G. G. P., Characteristic infrared luminescence in gallium phosphide due to manganese. - J. Lumin., 1972, 5, N5, 379-384. - CA, 1972, 77, 170893. PA, 1972, 75, 81424.

Exp.; T; d^5.

1973

2.2.81. Fukuda Atsuo, Cho Kikuo, Paus, H. J., Relaxed excited states of KI-Tl^+-type phosphors. - In: Lumin. Cryst., Mol. Solutions: Proc. Int. Conf., 1972. New York, 1973, p. 478-488. - CA, 1973, 79, 59470.

Exp.; T; 1.2.

2.2.82. Fukuda Atsuo, Cho Kikuo, Paus, H. J., Relaxed excited states of the KI-Tl^+ type phosphors. [Russ.]. - Izv. Akad. Nauk SSSR. Ser. Fiz., 1973, 37, N3, 577-584. - CA, 1973, 78, 153302. PA, 1973, 76, 36917.

Theor.; T - e.

2.2.83. Gerhardt, V., Gebhardt, W., The Jahn-Teller effect in the 3000-Å emission of KCl:Tl^+: An investigation by time-resolved emission spectroscopy. - Phys. Status Solidi b, 1973, 59, N1, 187-198. - CA, 1973, 79, 120116. PA, 1973, 76, 69061. RZF, 1974, 1D759.

Exp.; T - e.

2.2.84. Haaland, D. M., Nieman, G. C., Toluene: Phosphorescence spectrum and distortion in the triplet state. - J. Chem. Phys., 1973, 59, N8, 4435-4457. - CA, 1974, 80, 54248. PA, 1974, 77, 21058.

Exp.; T; D_{6h}.

2.2.85. Kazanskiy, S. A., Natadze, A. L., Ryskin, A. I., Khil'ko, G. I., Luminescence of $A_{II}B_{VI}$ crystals activated by transition metal ions. [Russ.]. - Izv. Akad. Nauk SSSR. Ser. Fiz., 1973, 37, N3, 670-676. - RZF, 1973, 9D560.

Exp.; T; T_d.

2.2.86. Khizhnyakov, V. V., Tehver, I., Polarization of resonant secondary glow luminescence centers taking into account the Jahn-Teller effect. [Russ.]. - Izv. Akad. Nauk SSSR. Ser. Fiz., 1973, 37, N3, 585-590. - CA, 1973, 79, 71757. PA, 1973, 76, 36918.

Theor.; T - (a_1 + e + t_2).

2.2.87. Kristofel', N. N., Khizhnyakov, V. V., Temperature depolarization of luminescence of KCl-Sn^{2+} type centers. [Russ.]. - Opt. & Spektrosk., 1973, 34, N6, 1236-1237. - CA, 1973, 79, 85229. PA, 1973, 77, 10480.

Theor.; T - e.

2.2.88. Perlin, Yu. E., Effects of electron-phonon interaction on the luminescence spectra of impurity ions of transition and rare earth elements in crystals. [Russ.]. - Izv. Akad. Nauk SSSR. Ser. Fiz., 1973, 37, N3, 532-539. - CA, 1973, 78, 153256. RZF, 1973, 7D483.

Theor.; T - e; 2.3.

2.2.89. Perlin, Yu. E., Effects of electron-phonon interaction on luminescence spectra of rare earth and transition metal impurity ions in crystals. - J. Lumin., 1973, 8, N2, 183-192. - CA, 1974, 80, 54137.

Theor.

2.2.90. Rozenfel'd, Yu. B., Vekhter, B. G., Tsukerblat, B. S., The band shape of impurity absorption and luminescence in systems with a doubly degenerate electronic E-term. [Russ.]. - In: Spektrosk. Kristallov: Materialy 3 Simp. Spektrosk. Kristallov, Aktivir. Ionami Perekhodnykh Redkozem. Metallov, 1970. Leningrad, 1973, p. 131-134.

Theor.; E - e.

2.2.91. Shimakura Noriyuki, Fujimura Yuichi, Nakajima Takeshi, Calculation of the nonradiative transition rare for the $^1B_{1u} \rightarrow {}^1B_{2u}$ transition in benzene. - Bull. Chem. Soc. Jpn., 1973, 46, N5, 1297-1300. - CA, 1973, 79, 31126.

Theor.; (B + E) - e; D_{6h}; 1.5.

1974

2.2.92. Clerjaud, B., Gelineau, A., Jahn-Teller effect in $^2T_2$ state of copper(2+) ion in zinc sulfide. - Phys. Rev. B, 1974, 9, N7, 2832-2837. - CA, 1974, 81, 18707. PA, 1974, 77, 57548. RZF, 1975, 7E191.

Exp.; T - e; T_d; d^9.

2.2.93. Duben, A. J., Goodman, L., Koganagi Motohiko, Interstate interaction in aromatic aldehydes and ketones. - In: Excited States. vol. 1, New York; London, 1974, p. 295-329. - RZF, 1976, 8D184.

Rev.

2.2.94. Gächter, B. F., Köningstein, J. A., Zero-phonon transitions and interacting Jahn-Teller phonon energies from the fluorescence spectrum of α-Al_2O_3:Ti^{3+}. - J. Chem. Phys., 1974, 60, N5, 2003-2006. - CA, 1974, 80, 138815. PA, 1974, 77, 49429. RZF, 1974, 8D645.

Exp.; T; d^1.

2.2.95. Grebe, G., Schulz, H. J., Luminescence of Cr^{2+} centers and related optical interactions involving crystal field levels of chromium ions in zinc sulfide. - Z. Naturforsch. a, 1974, 29, N12, 1803-1819. - PA, 1975, 78, 19001.

Exp.; T; d^4.

2.2.96. Hurrell, J. P., Kam, Z., Cohen, A., Analysis of the sideband fluorescence of Eu^{2+}:CaF_2. - Solid State Commun., 1974, 14, N11, 1111-1113. - PA, 1974, 77, 57621.

Theor.; E.

2.2.97. Kanamaru, N., Long, M. E., Lim, E. C., Dual phosphorescence from rigid glass solutions of phenyl alkyl ketones. - Chem. Phys. Lett., 1974, 26, N1, 1-9. - PA, 1974, 77, 48064.

Exp.; 1.5.

2.2.98. Kanezaki Eiji, Nishi Nobuyuki, Kinoshita Minoru, Niimori Kazumi, Vibronic interactions in the phosphorescent triplet state of p-chloroaniline as studied by the method of microwave-induced delayed phosphorescence. - Chem. Phys. Lett., 1974, 29, N4, 529-533. - CA, 1975, 82, 117928. PA, 1975, 78, 22334.

Exp.; 1.5.

2.2.99. Moreau, N., Boccara, A. C., Badoz, J., Jahn-Teller effect and relaxation processes in the $^1T_{2g}$ state of nickel(2+) in magnesium oxide. - Phys. Rev. B, 1974, 10, N1, 64-70. - CA, 1974, 81, 70497. PA, 1975, 78, 7347.

Exp.; T - e; d^8.

2.2.100. Stikeleather, J. A., The nature of the lowest state of cinnoline. - Chem. Phys. Lett., 1974, 24, N2, 253-256. - PA, 1974, 77, 25497.

Exp.; T.

2.2.101. Treadaway, M. J., Powell, R. C., Luminescence of calcium tungstate crystals. - J. Chem. Phys., 1974, 61, N10, 4003-4011.

Theor.; T; T_d.

2.2.102. Zazubovich, S. G., Polarized luminescence of Sn^{2+} centers in cubic alkali halide crystals. [Russ.]. - Opt. & Spektrosk., 1974, 37, N4, 711-716. - PA, 1975, 78, 51020. RZF, 1975, 2D832.

Exp.; T - t.

1975

2.2.103. Bacci, M., Fabeni, P., Fontana, M. P., Pazzi, G. P., Ranfagni, A., Viliani, G., Optical measurements on alkali-halide phosphors at low temperature. - Alta Frequenza, 1975, 44, N5, 267-270.

Exp.; T.

2.2.104. Black, A. M., Flint, C. D., Jahn-Teller effect of the $\Gamma_8(^2T_{2g}, t_{2g}^3)$ state of hexabromorhenium(IV). - J. Chem. Soc. Faraday Trans. II, 1975, 71, N11, 1871-1876. - CA, 1975, 83, 210895. PA, 1976, 79, 34818.

Exp.; Γ_8 - t_2.

2.2.105. Kristofel', N. N., Temperature quenching of the intensity of impurity optical bands. [Russ.]. - Opt. & Spektrosk., 1975, 38, N4, 822-823. - CA, 1975, 83, 50233.

2.2.106. Looritz, V., Khizhnyakov, V. V., The Jahn-Teller effect in electronic states with a nonzero spin. [Russ.]. - In: Kvant. Khim.: Tez. Dokl. 6 Vsesoyuz. Soveshch. Kishinev, 1975, L - 158. - CA, 1976, 84, 81888. RZK, 1975, 22B20.

Theor.; s; p.

2.2.107. Trinkler, M. F., Zolovkina, I. S., On the nature of the longwave A band luminescence in KI:Tl crystals. [Russ.]. - Izv. Akad. Nauk Latv.SSR. Ser. Fiz. i Tekh. Nauk, 1975, N4, 117-119. - PA, 1976, 79, 26922.

Theor.; T - e; T - t.

1976

2.2.108. Bacci, M., Bhattacharyya, B. D., Ranfagni, A., Viliani, G., Jahn-Teller structured emission in O_h symmetry. - Phys. Lett. A, 1976, 55, N8, 489-490. - CA, 1976, 84, 113650. PA, 1976, 79, 36334.

Theor.; T; O_h.

2.2.109. Benci, S., Fontana, M. P., Manfredi, M., Temperature dependence of the decay modes of the 420-nm emission in $KI:Tl^+$. - Solid State Commun., 1976, 18, N11/12, 1423-1426.

Exp.

2.2.110. Campion, A., El-Sayed, M. A., The mechanism of the $S_1 \rightarrow T_1$ nonradiative process in duraldehyde. - J. Phys. Chem., 1976, 80, N20, 2201-2206. - CA, 1976, 85, 159061. RZK, 1977, 5B1321.

Exp.; 1.5.

2.2.111. Chodos, S. L., Black, A. M., Flint, C. A., Vibronic spectra and lattice dynamics of Cs_2MnF_6 and $A_{1/2}M^{IV}F_6:MnF_6^-$. - J. Chem. Phys., 1976, 65, N11, 4816-4824. - CA, 1977, 86, 62942. PA, 1977, 80, 20904.

Exp.; T; O_h; d^3.

2.2.112. Despres, A., Migirdicyan, E., Blanco, L., Temperature effect on the phosphorescence spectra and lifetimes of 2,4-, 2,5-, and 3,4-dimethylbenzaldehydes dispersed in durene single crystals. - Chem. Phys., 1976, 14, N2, 229-240. - CA, 1976, 85, 38975.

Exp.; 1.5.

2.2.113. Fontana, M. P., Villiani, G., Bacci, M., Ranfagni, A., Effect of hyperfine structure interaction in the emission of KI-Tl. - Solid State Commun., 1976, 18, N11/12, 1615-1617.

Theor.; T; O_h.

2.2.114. Kaplyanskiy, A. A., Smolyanskiy, P. L., Polarized luminescence and structure of the radiative state of CaF_2-Yb^{2+}. [Russ.]. - Izv. Akad. Nauk SSSR. Ser. Fiz., 1976, 40, N9, 1962-1965. - CA, 1977, 86, 35937.

Exp.; O_h.

2.2.115. Kayanuma Yosuke, Vibronic problem for the relaxed exicted state of the F-center in alkali halides. II. Analysis of the experiments. - J. Phys. Soc. Jpn., 1976, 40, N2, 363-370.

Theor.; T - (e + t).

2.2.116. Khalil, O. S., Goodman, L., Hankin, S. H. W., Concerted matrix and pseudo-Jahn-Teller perturbations in the $^2\pi\pi^* \rightarrow S_0$ phosphorescence of p-chlorobenzaldehyde at 4.2 K. - Chem. Phys. Lett., 1976, 39, N2, 221-225. - CA, 1976, 85, 38889. PA, 1976, 79, 52548.

Exp.; 1.5.

2.2.117. Koidl, P., Jahn-Teller effect in the $^4T_1(I)$ and $^4T_2(I)$ states of tetrahedrally coordinated manganese(2+). - Phys. Status Solidi b, 1976, 74, N2, 477-484. - CA, 1976, 85, 11853. PA, 1976, 79, 53563.

Theor.; T - e; T_d; d^5.

2.2.118. Lin, S. H., Study of vibronic, spin-orbit, and vibronic-spin-orbit coupling of formaldehyde with applications to radiative and non-radiative processes. - Proc. R. Soc. London. Ser. A, 1976, 352, N1668, 57-71.

Theor.; 1.5.

2.2.119. Perepelitsa, E. I., Elektron-Kolebatel'nye Perekhody v Primesnykh Paramagnitnykh Ionakh: Avtoref. Dis. ... Kand. Fiz.-Mat. Nauk. [Russ.]. - Kishinev, 1976. - 18 p. - Kishinev. Univ.

Electron-Vibrational Transitions in Impurity Paramagnetic Ions.

Theor.; T - (a + e + t); O_h; d^3.

2.2.120. Radhakrishna, S., Polarized luminescence from lead centers in cesium halides. - J. Lumin., 1976, 12/13, 409-411. - PA, 1976, 79, 54218.

Exp.

2.2.121. Ranfagni, A., Pazzi, G. P., Fabeni, P., Bacci, M., Fontana, M. P., Viliani, G., Stark effect in the emission of thallium-doped alkali halides. - In: Proc. 12th Eur. Congr. Mol. Spectrosc. Amsterdam, 1976, p. 471-473. - CA, 1977, 86, 24037.

Exp.; 2.3; 2.4.

2.2.122. Rebane, K., Saari, P., Hot luminescence and relaxation processes in centers of luminescence. - J. Lumin., 1976, 12/13, 23-30. - RZF, 1976, 11D841.

Theor.

2.2.123. Scharmann, A., Schwabe, D., Weyland, D., Polarization of the luminescence of lead-doped zinc sulfide single crystals due to the Jahn-Teller effect. - J. Lumin., 1976, 12/13, 479-483. - CA, 1976, 85, 11885. PA, 1976, 79, 54229.

Exp.; T.

2.2.124. Schmidtke, H. H., Wernicke, R., Vibrational assignments from sharp line luminescence of a higher excited state of hexachlororhenate(IV) in cubic host crystals. - Chem. Phys. Lett., 1976, 40, N2, 339-341. - CA, 1976, 85, 84962.

Exp.; Γ_8; O_h.

2.2.125. Srivastava, J. K., Nagarajan, R., A new combined microwave-optical-perturbed angular correlation experiment. - In: Proc. Nucl. Phys. & Solid State Phys. Symp., 1976, Vol. 19C, p. 347-351. - CA, 1978, 89, 50995.

Exp.; 3.3.

2.2.126. Trinkler, M., Zolovkina, I. S., A-luminescence of alkali halide crystals activated by monovalent mercury-like ions. [Russ.]. - Izv. Akad. Nauk SSSR. Ser. Fiz., 1976, 40, N9, 1939-1943. - CA, 1977, 86, 162961. RZF, 1977, 1D802.

Exp.

2.2.127. Zazubovich, S. G., Soovik, T., Khizhnyakov, V. V., A polarized A-radiation of tin centers in alkali halide crystals and the model of their excited states. [Russ.]. - Izv. Akad. Nauk SSSR. Ser. Fiz., 1976, 40, N9, 1944-1948. - CA, 1977, 86, 162962. RZF, 1977, 1D804.

Exp.

1977

2.2.128. Belyi, M. U., Okhrimenko, B. A., Yablochkov, S. M., Luminescence of potassium borate glasses activated with lead(II) oxide. [Russ.]. - Ukr. Fiz. Zh., 1977, 22, N10, 1625-1629. - CA, 1978, 88, 43309. RZF, 1978, 6D1033.

Exp.; E - $(a + b_1 + b_2)$; C_{nv}.

2.2.129. Casalboni, M., Grassano, U. M., Scacco, A., Tanga, A., Emission spectra of thallium(+) ion-doped sodium bromide excited in the A band. - Phys. Status Solidi b, 1977, 84, N1, 23-26. - CA, 1978, 88, 29833.

Exp.

2.2.130. Fujimura, Y., Konno, H., Nakajima, T., Theory of nonradiative decays in the non-Condon scheme. - J. Chem. Phys., 1977, 66, N1, 199-208.

Theor.; 1.5.

2.2.131. Hoshina, T., Imanaga, S., Yokono, S., Charge transfer effects on the luminescent properties of Eu^{3+} in oxysulfides. - J. Lumin., 1977, 15, N4, 455-471. - CA, 1978, 88, 29847. PA, 1978, 81, 15214.

Exp.; f^6.

2.2.132. Lemoyne, D., Duran, J., Badoz, J., Jahn-Teller effect and relaxation processes in an ns^2 system: potassium chloride: gold(1-). - J. Phys. C, 1977, 10, N8, 1255-1265. - CA, 1977, 87, 124793. PA, 1977, 80, 48220.

Theor.; T.

2.2.133. Lukaszewicz, T., Niewinska, A., Zabludowska, K., Luminescence of lead in zinc sulfide-type compounds. [Pol.]. - In: Zesz. Nauk. Politech. Bialostockiey. [Ser.] Mat., Fiz. Chem., 1976, (Publ. 1977), Vol. 2, p. 57-70. - CA, 1979, 91, 11574.

Exp.; 2.4.

2.2.134. Nasu, K., Lattice relaxation in E × (b_1, b_2) Jahn-Teller system and polarization correlation between photoabsorption and luminescence. - Prog. Theor. Phys., 1977, 57, N2, 361-379. - PA, 1977, 80, 51708.

Theor.; E - $(b_1 + b_2)$; C_{4v}.

2.2.135. Ohkura Hiroshi, Imanaka Koichi, Kamada Osamu, Mori Yuzo, Iida Takeshi, Star effects in F center emission in KBr, KI, RhBr, and RbI crystals. - J. Phys. Soc. Jpn., 1977, 42, N6, 1942-1948. - RZF, 1977, 11D667.

Exp.; (P + S) - t.

2.2.136. Penner, A. P., Siebrand, W., Zgierski, M. Z., Nature of an optically prepared state undergoing radiationless decay. - Chem. Phys. Lett., 1977, 51, N2, 227-229.

Theor.; 1.5.

2.2.137. Ranfagni, A., Viliani, G., Cetica, M., Molesini, G., Nonradiative transitions in color centers by WKB approximation. - Phys. Rev. B, 1977, 16, N2, 890-894. - CA, 1977, 87, 124738. RZF, 1978, 1E1233.

Theor.; d^3.

2.2.138. Trinkler, M. F., Zolovkina, I. S., A-Luminescence of alkali halides activated by monovalent mercury-like ions. - Phys. Status Solidi b, 1977, 79, N1, 49-59. - CA, 1977, 86, 98353. PA, 1977, 80, 20914.

Exp.; T.

2.2.139. Zolovkina, I. S., Struktura i Priroda A-Polos Luminestsentsii Kristallofosforov KI-Tl-Tipa: Avtoref. Dis. ... Kand. Fiz.-Mat. Nauk. [Russ.]. - Riga, 1977. - 14 p. - Akad. Nauk SSSR. Inst. Fiz.

The Structure and Nature of the A-Band Luminescence of KI-Tl-Type Crystallophosphors.

Exp.; T - e; T - t; D_{3h}.

1978

2.2.140. Benci, S., Capelletti, R., Fermi, F., Manfredi, M., Damm, J. Z., Mugenski, E., Spectral and time decay analysis of photoluminescence excited in the A-band region in KCl:Pb. - Phys. Status Solidi b, 1978, 90, N2, 657-666. - CA, 1979, 90, 63838. PA, 1979, 82, 33823.

Exp.

2.2.141. Bernstein, E. R., Meredith, G. R., Webb, J. D., The Jahn-Teller effect in the lowest charge-transfer state of uranium hexafluoride. - J. Chem. Phys., 1978, 68, N9, 4066-4072. - CA, 1978, 89, 50888. PA, 1978, 81, 67992.

Exp.; T - t.

2.2.142. Bondybey, V. E., English, J. H., Miller, T. A., Laser fluorescence spectra of the hexafluorobenzene cation in solid argon. - J. Am. Chem. Soc., 1978, 100, N16, 5251-5252. - CA, 1978, 89, 107201.

Exp.; E - e; D_6h; 2.3.

2.2.143. Campion, A., El-Sayed, M. A., Spin labels and the mechanism of the $S_1 \to T_1$ nonradiative process in duraldehyde; possible manifestation of pseudo-Jahn-Teller forces on nonradiative processes. - NBS Spec. Publ. (U.S.), 1978, 526, 271-273. - CA, 1979, 90, 14349.

Exp.; T; 1.5.

2.2.144. Cossart-Magos, C., Cossart, D., Leach, S., Jahn-Teller effects in the emission spectra of 1,3,5-trifluorobenzene and hexafluorobenzene positive ions. - J. Chem. Phys., 1978, 69, N9, 4313-4314. - CA, 1979, 90, 14317. PA, 1979, 82, 10120.

Exp.; E - e; D_6h.

2.2.145. Hizhnyakov, V., Zazubovich, S., Vacancy-induced splitting of excited states and the structure of $Sn^{2+}V_C$ centers in alkali halides. - Phys. Status Solidi b, 1978, 86, N2, 733-739. - RZF, 1978, 9E65.

Exp.

2.2.146. Ignat'ev, I. V., Ovsyankin, V. V., The pseudo-Jahn-Teller effect in $CaF_2:Sm^{2+}$ and $SrF_2:Sm^{2+}$ crystals. [Russ.]. - Fiz. Tverd. Tela, 1978, 20, N8, 2484-2487. - CA, 1978, 89, 206752. PA, 1979, 82, 69476.

Exp.; 1.5.

2.2.147. Khan, S. M., Patterson, H. H., Engstrom, H., Multiple state luminescence for the d^4 hexachloroosmate(2) impurity ion in hexachloroplatinate and cesium hexachlorozirconate cubic crystals. - Mol. Phys., 1978, 35, N6, 1623-1636. - CA, 1979, 90, 14188.

Exp.; T; d^4.

2.2.148. Le Si Dang, Romestain, R., Simkin, D., Fukuda Atsuo, Luminescence decay times of $KBr:Ga^+$. - Phys. Rev. B, 1978, 18, N7, 2989-2994.

Exp.

2.2.149. Matsuzaki Kazuo, Azumi Tohru, Molecules distortion in the pyrazine triplet state: Analysis in terms of the MIDP experiments under polarized light. - J. Chem. Phys., 1978, 69, N9, 3907-3909. - CA, 1979, 90, 14314.

Exp.; T; 1.5.

2.2.150. Miller, T. A., Bondybey, V. E., Laser fluorescence excitation spectra of the 1,3,5-trifluorobenzene radical cation. - Chem. Phys. Lett., 1978, 58, N3, 454-456. - CA, 1979, 90, 31437. PA, 1979, 82, 3470.

Exp.; 2.3.

2.2.151. Nasu Keiichiro, Kayanuma Yosuke, Lattice relaxation and time-resolved polarization of emission from localized electron with level crossing. - J. Phys. Soc. Jpn., 1978, 45, N4, 1341-1348.

Theor.

2.2.152. Paulusz, A. G., The predictive use of the configurational coordinate model for luminescence centers. - J. Lumin., 1978, 17, N4, 375-384. - PA, 1979, 82, 19918.

Theor.

2.2.153. Penner, A. P., Siebrand, W., Zgierski, M. Z., Radiationless decay of vibronically coupled electronic states. - J. Chem. Phys., 1978, 69, N12, 5496-5508.

Theor.; 1.5.

2.2.154. Ranfagni, A., Viliani, G., Origin of the triple decay time in the low-energy emission of KI:Tl. - Solid State Commun., 1978, 25, N9, 731-733. - CA, 1978, 89, 14043.

Theor.; T - e.

2.2.155. Romestein, R., Le Si Dang, Merle D'Aubigne, Y., Fukuda Atsuo, Jahn-Teller effect in the $^{3}T_{1u}$ excited state of gallium(1+) in alkali halides. - Semicond. & Insul., 1978, 3, N2/3, 175-193. - CA, 1978, 88, 200530.

Exp.; T - (a + e + t_2); 3.1.

2.2.156. Strokach, N. S., Shigorin, D. N., On the problem of electron-vibrational interaction between near-lying excited electronic states in polyatomic molecules. [Russ.]. - Opt. & Spektrosk., 1978, 45, N3, 454-462. - CA, 1978, 89, 188460.

Exp.; (A + B) - b; 1.5; 2.3.

2.2.157. Trinkler, M. F., The static Jahn-Teller effect in activated alkali halide crystals. [Russ.]. - In: Tez. Dokl. 4 Vsesoyuz. Soveshch. Radiats. Fiz. i Khim. Ionnykh Kristallov, Riga. Salaspils, 1978, p. 459-460.

Theor.; T - (e + t); O_h.

2.2.158. Watts, R. J., Missimer, D., Environmentally hindered radiationless transitions between states of different orbital parentage in iridium(III) complexes: Application of rigid-matrix induced perturbations of the pseudo-Jahn-Teller potential to the rigidochromic effect in d^6 metal complexes. - J. Am. Chem. Soc., 1978, 100, N17, 5350-5357. - CA, 1978, 89, 154953.

Exp.; 1.5.

1979

2.2.159. Bondybey, V. E., Miller, T. A., English, J. H., Matrix laser fluorescence spectra of several fluorobenzene radical cations. - J. Am. Chem. Soc., 1979, 101, N5, 1248-1253. - CA, 1979, 90, 167501. PA, 1980, 83, 1009.

Exp.; E; D_{6h}; 2.3.

2.2.160. Bondybey, V. E., Miller, T. A., Laser-induced fluorescence from fluorobenzene radical cations in the gas phase. - J. Chem. Phys., 1979, 70, N1, 138-146. - CA, 1979, 90, 120569. PA, 1979, 82, 27022.

Exp.; 2.3.

2.2.161. Bondybey, V. E., Matrix studies of the $\tilde{B}^2A_2 \rightarrow \tilde{X}^2E''$ transition of the sym-trichlorotrifluorobenzene radical cation. - J. Chem. Phys., 1979, 71, N9, 3586-3591. - CA, 1980, 92, 13226. PA, 1980, 83, 14172.

Exp.; 2.3.

2.2.162. Champagnon, B., Duval, E., Emission spectrum of V^{4+}-α-Al_2O_3: Jahn-Teller effect in the fundamental $^2T_{2g}$ and excited ^2E_g states. - J. Phys. C, 1979, 12, N11, L425-L429. - CA, 1979, 91, 201447. PA, 1979, 82, 83007.

Exp.; T; C_{3v}; d^1; 2.3.

2.2.163. Cibert, J., Edel, P., Merle d'Aubigne, Y., Romestain, R., Relaxation in the $^3T_{1u}$ state of F centers in calcium oxide. - J. Phys. (France), 1979, 40, N12, 1149-1160. - CA, 1980, 92, 49810. PA, 1980, 83, 24443.

Exp.; T - (e + t).

2.2.164. Clark, M. G., Dean, P. J., Luminescence of divalent transition metals in gallium phosphide. - In: Phys. Semicond.: Invited and Contributed Papers 14th Int. Conf., Edinburgh, 1978. Birstol; London, 1979, p. 291-294. - CA, 1979, 90, 159551.

Exp.; d^6; d^7; d^8.

2.2.165. Cossart-Magos, C., Cossart, D., Leach, S., Jahn-Teller effects in substituted benzene cations. II. Gas-phase emission spectrum of 1,3,5-$C_6F_3D_3^+$ and comparison with 1,3,5-$C_6F_3H_3^+$. - Chem. Phys., 1979, 41, N3, 345-362. - CA, 1979, 91, 165797. PA, 1979, 82, 89680.

Exp.; E - e; 2.3.

2.2.166. Cossart-Magos, C., Cossart, D., Leach, S., Jahn-Teller effects in substituted benzene cations. III. Gas-phase emission spectrum of sym-$C_6Cl_3H_3^+$. - Chem. Phys., 1979, 41, N3, 363-372. - CA, 1980, 92, 5705. PA, 1979, 82, 89681.

Exp.; E - e; 2.3.

2.2.167. Cossart-Magos, C., Cossart, D., Leach, S., Emission spectra of seven fluorobenzene cations: Dynamical Jahn-Teller effect in sym-$C_6F_3H_3^+$ and $C_6F_6^+$. - Mol. Phys., 1979, 37, N3, 793-830. - CA, 1979, 91, 192269. PA, 1979, 82, 50034.

Exp.; 2.3.

2.2.168. Despres, A., Lijeune, V., Migirdicyan, E., Siebrand, W., Spectroscopic effects of vibronic coupling between nearly $^3n\pi^*$ and $3\pi\pi^*$ states of dimethylbenzaldehydes in a durene matrix. - Chem. Phys., 1979, 36, N1, 41-61. - CA, 1979, 90, 143743. RZF, 1979, 7D427.

Exp.; 1.5.

2.2.169. Le Si Dang, Merle d'Aubigne, Y., Romestain, R., Relaxed excited states of Tl^+-like centers in alkali halides; polarization effect of Ga^+ center. - J. Lumin., 1979, 18/19, N1, 331-335. - CA, 1979, 90, 129927. RZF, 1979, 6D981.

Exp.; T; O_h; 2.1.

2.2.170. Merle d'Aubigne, Y., Le Si Dang, Lifetimes of the triplet states of Tl^+-like ions in O_h symmetry-hyperfine effect. - Phys. Rev. Lett., 1979, 43, N14, 1023-1026. - CA, 1979, 91, 184416. RZF, 1980, 4D559.

Exp.; T; d^3; O_h; 2.1.

2.2.171. Miller, J. C., Allison, S. W., Andrews, L., Laser spectroscopy of matrix-isolated uranium(VI) fluoride at 12 K. - J. Chem. Phys., 1979, 70, N7, 3524-3530. - CA, 1979, 91, 11558.

Exp.

2.2.172. Miller, T. A., Bondybey, V. E., English, J. H., The vibrationally resolved electronic spectrum of the sym-trichlorobenzene radical cation in gas and condensed phases. - J. Chem. Phys., 1979, 70, N6, 2919-2925. - CA, 1979, 90, 212545. PA, 1979, 82, 68252.

Exp.; E - e; D_{3h}; 2.3.

2.2.173. Miller, T. A., Bondybey, V. E., Zegarski, B. R., Laser-induced fluorescence spectrum of gaseous 2,4-hexadiyne radical ion(1+) $C_6H_6^+$. - J. Chem. Phys., 1979, 70, N11, 4982-4985. - CA, 1979, 91, 65619. PA, 1979, 82, 72345.

Exp.; D_{6h}; 2.3.

2.2.174. Perlin, Yu. E., Kaminskii, A. A., Enakii, V. N., Vylegzhanin D. N. Nonradiative multiphon relaxation in $Y_3Al_5O_{12}$: Nd^{3+}. - Phys. Status Solidi b, 1979, 92, N2, 403-410. - CA, 1979, 90, 195049. PA, 1979, 82, 56853.

Exp.; T.

2.2.175. Ranfagni, A., Englman, R., Nearly classical approaches to the kinetics of luminescence. - J. Lumin., 1979, 18/19, N1, 353-356. - CA, 1979, 90, 129932.

Theor.; T.

2.2.176. Yu, P. W., Effect of oxygen in photoluminescence from chromium-doped semi-insulating gallium arsenide. - Solid State Commun., 1979, 32, N11, 1111-1114. - CA, 1980, 92, 67188. PA, 1980, 83, 29799.

Exp.; d^4.

See also: 1.6.-15, 16; 2.1.-62, 251; 2.3.-20, 21, 56, 113; 2.4.-182; 4.1.-50; 4.3.-48.

2.3. Vibronic Fine Structure of Spectra. Zero-Phonon Lines

When the Jahn-Teller stabilization energy is comparable with the energy of active vibrations and electronic splittings caused by spin-orbit interaction, low-symmetric fields, etc., the adiabatic approximation is not valid, even for the lowest vibronic levels. A correct interpretation requires numerical diagonalization of the vibronic Hamiltonian, including the above-mentioned perturbations. The transition metal ions in the A_2B_6 semiconductors are typical systems having moderate vibronic coupling (see [2.1.244; 2.3.-9, 14, 17, 18, 26, 39, 42, 59, 63, 67, 77, 79, 95, 120]).

Numerical calculations of vibronic energy levels for Cu^{2+} in ZnO, CdS, and ZnS were performed in Ref. [3.1.161], which take into account both the trigonal crystal field and anisotropic spin-orbit interactions.

If the vibronic coupling is weak enough, the intensity of the zero-phonon lines becomes most essential. Since they are due to the $n \to n$ transitions (and not only to the $0 \to 0$ transition), the splitting of the excited vibronic levels resulting from the weak Jahn-Teller effect essentially influences the zero-phonon line structure. The $A \to E$ and $A \to T - t_2$ transitions were considered in Ref. [2.3.84], where a single-mode cluster model was employed. The zero-phonon line shape shows a complicated temperature dependence. The theory [2.3.84] was confirmed by experimental data [2.3.87].

Provided the vibronic coupling is strong enough, the zero-phonon line can be observed in the long-wave tail of the broad absorption band. If the splitting of this line caused by external fields is known, one can estimate the reduction factors. Investigation of the Jahn-Teller effect in the $^4T_2(t_2^2e)$ state of Cr^{3+} in ruby [2.4.-4, 108] may serve as an example. In the cited papers the uniaxial stress experiment was performed, and the zero-phonon line splitting showed that tetragonal modes give the dominant contribution of the Jahn-Teller effect. A complex problem concerning the nature of the zero-phonon line in $MgO{:}V^{2+}$ has long been discussed [2.3.-4, 12, 85]. The pseudo-Jahn-Teller effect was investigated in Ref. [2.3.5]. The $Al_2O_3{:}V^{3+}$ zero-phonon line was considered in Refs. [2.3.-10, 46; 2.4.-98, 144].

Second-order reduction factors for $Al_2O_3{:}Cr^{3+},V^{3+}$ are given in Ref. [2.3.10]. Theoretical calculations of the zero-phonon line, taking into account the reduction factors for the excited vibronic states, were made in Ref. [2.3.76]. Provided both the linear and quadratic interactions with E and T_2 modes are essential, the adiabatic surface possesses orthorhombic minima. The zero-phonon line tunneling splitting for this case was considered in Ref. [2.3.80].

If the Jahn-Teller coupling is relatively weak, the reduction of the electronic energy level splitting also takes place, the reduction factor being close to unity. The Co^{2+} ion in semiconducting crystals can be mentioned as a typical example. Manifestations of the dynamic Jahn-Teller effect for transition metal ions in A_2B_6 crystals were investigated in Refs. [2.2.85; 2.3.67; 2.4.118]. The spin-oribt splitting is modified significantly by Jahn-Teller interactions, and the mirror law for the absorption and emission spectra is broken.

The vibronic structure of molecular-system spectra was investigated in Refs. [2.3.86; 2.2.-144, 165, 166, 167, 142, 159, 172].

1950

2.3.1. Craig, D. P., The role of E_g^+ vibrations in the 2600-Å benzene band system. - J. Chem. Soc., 1950, 59, 59-62. - CA, 1950, 44, 5212i.

Theor.; 1.5.

1960

2.3.2. Ford, R. A., Hill, O. F., Vibronic coupling in the $^4T_{2g}$ excited state of ruby. - Spectrochim. Acta, 1960, 16, N4, 493-496. - CA, 1960, 54, 17066g. RZK, 1961, 9B199.

Exp.; T.

1962

2.3.3. McClure, D. S., Optical spectra of transition-metal ions in corundum. - J. Chem. Phys., 1962, 36, N10, 2757-2779. - CA, 1962, 57, 10665b.

Exp.; E - e; C_{3v}; d^1.

2.3.4. Sturge, M. D., Jahn-Teller effect in the $^4T_{2g}$ excited state of V^{2+} in MgO. - Phys. Rev. A, 1965, 140, N3, 880-891. - CA, 1965, 63, 15745h. RZF, 1966, 4E428.

Exp.; T - (e + t_2); O_h; d^3.

1966

2.3.5. Bron, W. E., Wagner, M., Vibrational coupling of nearly degenerate electronic states. - Phys. Rev., 1966, 145, N2, 689-698. - RZF, 1966, 11D265.

Exp.; C_{2v}; 1.5.

2.3.6. Broser, I., Maier, H., Analysis of zero-phonon lines in absorption and emission of copper impurities in ZnS crystals. - J. Phys. Soc. Jpn., 1966, Suppl. 21, 254-258. - CA, 1967, 66, 50523.

Exp.; E - e; d^9.

2.3.7. Herzberg, G., Molecular Spectra and Molecular Structure. Vol. 3, Electronic Spectra and Electronic Structure of Polyatomic Molecules. - Toronto, etc.: Van Nostrand, 1966. - XVIII, 745 p.

2.3.8. Scott, W. C., Sturge, M. D., Dynamic Jahn-Teller effect in the $^3T_2$ excited term of $Al_2O_3:V^{3+}$. - Phys. Rev., 1966, 146, N1, 262-267. - CA, 1966, 65, 1603a. PA, 1966, 69, 24756. RZF, 1966, 11D274.

Exp.; T; d^2; 1.4.

2.3.9. Slack, G. A., Ham, F. S., Chrenko, R. M., Optical absorption of tetrahedral $Fe^{2+}(3d^6)$ in cubic ZnS, CdFe, and $MgAl_2O_4$. - Phys. Rev., 1966, 152, N1, 376-402. - CA, 1967, 66, 24048. RZF, 1967, 7D398.

Exp.; T - e; T_d; d^6.

2.3.10. Sturge, M. D., Second-order Jahn-Teller interaction in T_2 excited terms of V^{3+} and Cr^{3+} in Al_2O_3. - Bull. Am. Phys. Soc., 1966, 11, N7, 886.

Exp.; T - t; d^2; d^3.

1967

2.3.11. Anderson, R. S., Lattice-vibration effects in the spectra of ZnO:Ni and ZnO:Co. - Phys. Rev., 1967, 164, N2, 398-405. - CA, 1968, 68, 34449.

Exp.; T; T_d; d^8.

2.3.12. Ham, F. S., A model for the dynamical Jahn-Teller effect in the $^4T_{2g}$ excited state of V^{2+} in MgO. - In: Opt. Prop. Ions Cryst., Baltimore, 1966, New York, etc., 1967, p. 357-374. - CA, 1968, 69, 55984. RZF, 1969, 9D498.

Theor.; T - (e + t); O_h; d^3.

2.3.13. Marfunin, A. S., Platonov, A. N., Fedorov, W. E., The optical spectra of Fe^{2+} in sphalerite. [Russ.]. - Fiz. Tverd. Tela, 1967, 9, N12, 3616-3618.

Exp.; d^6.

2.3.14. Slack, G. A., Roberts, S., Ham, F. S., Far-infrared optical absorption of F^{2+} in ZnS. - Phys. Rev., 1967, 155, N2, 170-177. - RZF, 1967, 12D351.

Exp.; T; T_d; d^6.

2.3.15. Sturge, M. D., Crozier, M. H., Is there a dynamic Jahn-Teller effect in the ^2D-E state of Cr^{3+} in CaF_2? - J. Chem. Phys., 1967, 46, N11, 4551-4552. - CA, 1967, 67, 48698.

Theor.

2.3.16. Zazubovich, S. G., Lushchik, N. E., Luminescence anisotropy and luminescence center microstructure in activated ionic cyrstals. [Russ.]. - Izv. Akad. Nauk SSSR. Ser. Fiz., 1967, 31, N5, 798-802. - CA, 1968, 68, 63838. RZF, 1967, 10D714.

Exp.

1968

2.3.17. Brand, J. C. D., Goodman, G. L., Vibronically induced Jahn-Teller progressions in the electronic spectrum of iridium hexafluoride. - Can. J. Phys., 1968, 46, N15, 1721-1724. - CA, 1968, 69, 63179.

Exp.; Γ_8; O_h; 1.5.

2.3.18. Fetterman, H. R., Fitchen, D. B., Jahn-Teller vibronic structure of the R' center in lithium fluoride. - Solid State Commun., 1968, 6, N7, 501-503. - CA, 1968, 69, 55906.

Exp.; E - e.

2.3.19. Garcia, Sucre, M., Geny, F., Lefebvre, R., Intensity patterns in absorption and fluorescence spectra of molecular dimers: A comparison between the vibronic and the Born-Oppenheimer approach. - J. Chem. Phys., 1968, 49, N1, 458-464.

Theor.; 1.5.

2.3.20. Ralph, J. E., Townsend, M. G., Near-infrared fluorescence and absorption spectra of Co^{2+} and Ni^{2+} in MgO. - J. Chem. Phys., 1968, 48, N1, 149-154.

Exp.; T - (e + t); O_h; d^8.

1969

2.3.21. Ferguson, J., Wood, D. L., Van Uitert, L. G., Crystal field spectra of $d^{3,7}$ ions. V. Tetrahedral Co^{2+} in $ZnAl_2O_4$ spinel. - J. Chem. Phys., 1969, 51, N7, 2904-2910. - CA, 1969, 71, 130429.

Exp.; T; T_d; d^7.

2.3.22. Gradyushko, A. T., Mashenkov, V. A., Solov'ev, K. N., Studies of metalloporphin complexes by the method of quasi-linear spectra. [Russ.]. - Biofizika, 1969, 14, N5, 827-835. - CA, 1970, 72, 7568. RZF, 1970, 1D707.

Exp.; D_{4h}

2.3.23. King, G. W., Santry, D. P., Warren, C. H., Fluorosulfate radical: Structural theory and assignment of electronic absorption systems. - J. Mol. Spectrosc., 1969, 32, N1, 108-120. - CA, 1969, 71, 107231.

Exp.; E - e; 2.2; 2.7.

2.3.24. King, G. W., Warren, C. H., Fluorosulfate radical. Vibratwonal analysis of the 5160-Å absorption system. - J. Mol. Spectrosc., 1969, 32, N1, 121-137. - CA, 1969, 71, 107183.

Exp.; E.

2.3.25. Wagner, M., Exact solution of the optical Jahn-Teller problem in trigonal systems. - Phys. Lett. A, 1969, 29, N8, 472-473. - PA, 1970, 73, 6851.

Theor.; E - e.

2.3.26. Wallin, J. T., Slack, G. A., Roberts, S., Hughes, A. E., Near- and far-infrared absorption in chromium-doped zinc selenide. - Solid State Commun., 1969, 7, N17, 1211-1214. - CA, 1969, 71, 107233.

Exp.; E - e; d^4.

2.3.27. Young, J. P., Absorption spectra of several 3d transition metal ions in molten fluoride solution. - Inorg. Chem., 1969, 8, N4, 825-827.

Exp.; T; O_h; d^4; d^6.

2.3.28. Zazubovich, S. G., Polarized luminescence of alkali-halogen crystals activated with mercury-like ions. [Russ.]. - In: Tr. Inst. Fiz. Astron. Akad. Nauk Est.SSR, 1969, N36, p. 109-153. - Bibliogr.: 103 ref. CA, 1970, 74, 117558. RZF, 1969, 12D771.

Exp.; T - (e + t); D_{3d}; D_{4h}.

1970

2.3.29. Bersuker, I. B., Vekhter, B. G., Inversion splitting of the zero-phonon lines in impurity absorption spectra taking into account the low-symmetric crystal field effects. [Russ.]. - In:

Spektrosk. Kristallov: Materialy 2 Simp. Spektrosk. Kristallov, Soderzhashchikh Redkozem. Elementy i Elementy Gruppy Zheleza, Khar'kov, 1967. Moscow, 1970, p. 102-105.

Theor.

2.3.30. Burland, D. M., Castro, G., Robinson, G. W., Experimental observation of singlet-triplet absorption in pure crystalline benzene. - J. Chem. Phys., 1970, 52, N8, 4100-4108. - CA, 1970, 72, 126913. PA, 1970, 73, 41229.

Exp.; (E + B) - e; D_{6h}; 1.5; 2.2.

2.3.31. Stufkens, D. J., Dynamic Jahn-Teller effect in the excited states of $SeCl_6^{2-}$, $SeBr_6^{2-}$, $TeCl_6^{2-}$, and $TeBr_6^{2-}$: Interpretation of electronic absorption and Raman spectra. - Recl. Trav. Chim. Pays-Bas, 1970, 89, N11, 1185-1201. - CA, 1971, 74, 47671. RZF, 1971, 6D485.

Theor.; T; O_h; 2.8.

2.3.32. Sturge, M. D., Dynamic Jahn-Teller effect in the $^4T_2$ excited states of $d^{3,7}$ ions in cubic crystals. I. V^{2+} in manganese potassium fluoride. - Phys. Rev. B, 1970, 1, N3, 1005-1012. - CA, 1970, 73, 20041. PA, 1970, 73, 41077.

Exp.; T - e; T_d; d^3.

2.3.33. Vallin, J. T., Dynamic Jahn-Teller effect in the orbital ^5E state of Fe^{2+} in cadmium telluride. - Phys. Rev. B, 1970, 2, N7, 2390-2397. - CA, 1971, 74, 47414.

Theor.; E; O_h; d^6.

2.3.34. Wallin, J. T., Slack, G. A., Roberts, S., Hughes, A. E., Infrared absorption in some II-VI compounds doped with Cr. - Phys. Rev. B, 1970, 2, N11, 4313-4334. - PA, 1971, 74, 33967.

Exp.; d^4.

1971

2.3.35. Boyn, R., Ruszczynski, G., Jahn-Teller structure in the $^3A_2(^3F) \rightarrow {}^3T_1(^3F)$ absorption band of Ti^{2+}-activated cadmium sulfide. - Phys. Status Solidi b, 1971, 48, N2, 643-655. - CA, 1972, 76, 52411; PA, 1972, 75, 10371.

Exp.; T - t; d^2.

2.3.36. Brand, J. C. D., Goodman, G. L., Weinstock, B., Optical absorption of iridium hexafluoride. - J. Mol. Spectrosc., 1971, 37, N3, 464-485. - CA, 1971, 74, 105064.

Exp.; T - t.

2.3.37. Brand, J. C. D., Goodman, G. L., Weinstock, B., Near-infrared band system of rhenium hexafluoride. - J. Mol. Spectrosc., 1971, 38, N3, 449-463. - CA, 1971, 75, 55948.

Exp.; T - t; O_h.

2.3.38. Flint, C. D., Second-order Jahn-Teller effect in the ^2E_g state of d^3 ions. - Chem. Phys. Lett., 1971, 11, N1, 27-28. - CA, 1972, 76, 19677. PA, 1971, 74, 83687.

Exp.; E; O_h; d^3.

2.3.39. Manson, N. B., One-phonon vibronic transition $^3T_{2g}$-$^3A_{2g}$ of Ni^{2+} in MgO. - Phys. Rev. B, 1971, 4, N8, 2645-2656.

Exp.; T; O_h; d^8.

2.3.40. Sigmund, E., Wagner, M., Dynamical resonances in trigonal systems (optical Jahn-Teller problem). - In: Abstr. 11th Eur. Congr. Mol. Spectrosc. Tallin, 1971, 59. - PA, 1973, 76, 68555.

Theor.; E - e; C_{3v}.

2.3.41. Sturge, M. D., Merritt, F. R., Johnson, L. F., Guggenheim, H. J., Ziel, J. P., van der, Optical and microwave studies of divalent vanadium in octahedral fluoride coordination. - J. Chem. Phys., 1971, 54, N1, 405-413. - CA, 1971, 74, 26463. PA, 1971, 74, 16609. RZF, 1971, 7D476.

Exp.; d^3; 2.4; 3.1.

2.3.42. Ulrici, W., Jahn-Teller effect on excited triplet states in eight-fold coordination: Fe^2 and Co^{2+} in CdF_2. - Phys. Status Solid bi, 1971, 44, N1, K29-K32. - CA, 1971, 74, 148700. PA, 1971, 74, 37459. RZF, 1971, 9D392.

Exp.; E - e; O_h; d^6; d^7.

2.3.43. Vekhter, B. G., Kovarskiy, V. A., Rozenfel'd, Yu. B., Tsukerblat, B. S., On the possibility of ultrasound control of the zero-phonon line parameters. [Russ.]. - Pis'ma, Zh. Eksp. & Teor. Fiz., 1971, 13, N7, 365-368.

Theor.

2.3.44. Wagner, M., Quasi-exact solution of the optical Jahn-Teller problem. - Z. Phys., 1971, 244, N3, 275-288. - PA, 1971, 74, 49413.

Theor.; E - e; T - t.

1972

2.3.45. Bernstein, E. R., Reilly, J. P., Absorption spectrum of the 2000-Å system of borazine in the gas phase. - J. Chem. Phys., 1972, 57, N9, 3960-3969. - CA, 1972, 77, 158333.

Exp.; D_{3h}.

2.3.46. Champagnon, B., Etude sous contrainte de la transition $^3A_2 \rightarrow {}^3T_2$ de l'ion V^{3+} dans le corindon α-Al_2O_3: effet Jahn-Teller: Thèse doct. phys. Univ. Claude-Bernard. [Fr.]. - Lyon, 1972. - 85 p. - RZF, 1975, 3E55D.

Stress Investigation of the $^3A_2 \rightarrow {}^3T_2$ Transition in the V^{2+} Ion Doped in α-Al_2O_3: The Jahn-Teller Effect.

Exp.; T - e; C_{3v}; d^2.

2.3.47. Farge, Y., Fontana, M. P., Dynamics of F_2^+ centers in LiF: One-phonon sideband and Raman scattering. - Solid State Commun., 1972, 10, N4, 333-336.

Exp.; 2.8.

2.3.48. Fischer, G., Small, G. J., Jahn-Teller distortion of s-triazine in its lowest excited singlet state. - J. Chem. Phys., 1972, 56, N12, 5934-5944. - CA, 1972, 77, 26711. PA, 1972, 75, 46631. RZF, 1972, 11D460.

Exp.; E - e; D_{3h}; 1.7.

2.3.49. Gehring, G. A., Magnetic side bands induced by zero point motion. - Phys. Lett. A, 1972, 41, N3, 203-204. - PA, 1972, 75, 81376.

Theor.; E; 2.2.

2.3.50. Hare, A. P., Davies, G., Collins, A. T., The temperature dependence of vibronic spectra in irradiated silicon. - J. Phys. C, 1972, 5, N11, 1265-1276. - RZF, 1972, 12D726.

Exp.; 2.2.

2.3.51. Levin, I. W., Abramowitz, S., Muller, A., Jahn-Teller vibrations of ReF_6. - J. Mol. Spectrosc., 1972, 41, N2, 415-419. - CA, 1972, 76, 105679. PA, 1972, 75, 31825. RZF, 1972, 8D206.

Exp.; T - (e + t); 2.2.

2.3.52. McPherson, G. L., Kistenmacher, T. J., Folkers, J. B., Stucky, G. D., Effect of exchange coupling on the spectra of transition metal ions: Ligand field spectrum and crystal structure of cesium chromium trichloride. - J. Chem. Phys., 1972, 57, N9, 3771-3780. - CA, 1972, 77, 158322. PA, 1972, 75, 8136.

Exp.; E - e; T - e.

2.3.53. Maier, H., Dynamical Jahn-Teller interaction in cubic $ZnS-Cu^{2+}$. - In: Abstr. Int. Conf. Lumin., Leningrad. Chernogolovka, 1972, p. 108-109.

Exp.; T - e; T_d; d^9; 1.4.

2.3.54. Milstein, J. B., Ackerman, J., Holt, S. L., MacGarvey, B. R., The electronic structures of Cr(V) and Mn(V) in phosphate and vanadate hosts. - Inorg. Chem., 1972, 11, N6, 1178-1184. - RZF, 1972, 11D451.

Exp.; T; T_d; d^2.

2.3.55. Miyakawa, T., Vibronic structure in the luminescent spectra of ions in crystals. - In: Abstr. Int. Conf. Lumin., Leningrad. Chernogolovka, 1972, p. 157.

Theor.; E - e.

2.3.56. Parmenter, C. S., The structure of resonance fluorescence from some single vibronic levels in the S_1 state of benzene vapor and its use in observing vibratinal relaxation. - In: Abstr. Int. Conf. Lumin., Leningrad. Chernogolovka, 1972, p. 213.

Exp.; (A + B) - b; D_{6h}; 1.5; 2.2.

2.3.57. Petelin, A. N., Influence of the Renner effect on the intensities of electron-vibrational transitons. [Russ.]. - Opt. & Spektrosk., 1972, 33, N4, 631-635. - CA, 1973, 78, 50142. RZF, 1973, 2D392.

Theor.; 1.6.

2.3.58. Polinger, V. Z., Rozenfel'd, Yu. B., Tsukerblat, B. S., Vekhter, B. G., The vibronic reduction factors for Jahn-Teller systems. [Russ.]. - Fiz. Tverd. Tela, 1972, 14, N8, 2493-2495. - CA, 1973, 78, 64911. PA, 1973, 76, 36539. RZF, 1972, 11E401.

Theor.; E - b; d^4; 1.5; 3.1.

2.3.59. Schenk, H. J., Schwarz, W. H. E., Spin-orbit interaction and Jahn-Teller effect: Analysis of the $^4A_2 \rightarrow {}^2E$-vibrational progression of $Cr(ur)_6I_3$. [Germ.]. - Theor. Chim. Acta, 1972, 24, N2-3, 225-233. - CA, 1972, 76, 119497.

Exp.; E - e; d^3.

2.3.60. Stedman, G. E., Phonon-induced shift of spectral lines. II. Jahn-Teller splittings. - J. Phys. C, 1972, 5, N18, 2665-2668. - CA, 1972, 77, 145870. PA, 1972, 75, 75245. RZF, 1973, 3E490.

Theor.

2.3.61. Vasil'chenko, E. A., Zazubovich, S. G., Lushchik, N. E., Polarized C-emission of a tin-doped potassium chloride crystal and refinement of the energy diagram of Sn^{2+} centers. [Russ.]. - Opt. & Spektrosk., 1972, 32, N4, 749-755. - CA, 1972, 77, 26825.

Exp.; E; T; C_{4v}; O_h.

1973

2.3.62. Bersuker, I. B., Polinger, V. Z., The inversion splitting of the zero-phonon line of the A → T transition. [Russ.]. - In: Tez. Dokl. 4 Vsesoyuz. Soveshch. Spektrosk. Kristallov., Aktivir. Ionami Redkozem. Perekhodnykh Elementov. Sverdlovsk, 1973, p. 5.

Theor.; T - (e + t).

2.3.63. Boyn, R., Ruszczynski, G., Rosenfeld, A., Jahn-Teller structure in absorption spectrum of $CdS:Tl^{2+}$ and $CdSe:Ti^{2+}$. - In: Abstr. 11th Eur. Congr. Mol. Spectrosc. Tallin, 1973, 54.

Exp.; T - t; d^2.

2.3.64. Brillante, A., Taliani, C., Zauli, C., The 2000-Å absorption system of the benzene single crystal in polarized light. - Mol. Phys., 1973, 25, N6, 1263-1271.

Exp.; (B + E) - (e + b_1 + b_2); D_{6h}; 1.5.

2.3.65. Helmholz, L., Russo, M. E., Spectra of manganese(IV) hexafluoride ion (MnF_6^{2-}) in environments of O_h and D_{3d} symmetry. - J. Chem. Phys., 1973, 59, N10, 5455-5470. - PA, 1973, 77, 25137.

Exp.; D_{3d}; O_h.

2.3.66. Rosenfeld, Yu. B., Polinger, V. Z., Tsukerblat, B. S., Vekhter, B. G., Zero-phonon lines of singlet-multiplet transitions in Jahn-Teller systems. - In: Abstr. 11th Eur. Congr. Mol. Spectrosc. Tallin, 1973, 47. - PA, 1973, 76, 68554.

Theor.; E - e; T - t.

2.3.67. Ryskin, A. I., Natadze, A. L., Kazanskiy, S. A., The dynamic Jahn-Teller effect in the structure of quartet terms of the tetrahedrally coordinated cobalt(2+) ion in zinc sulfide-cobalt systems. [Russ.]. - Zh. Eksp. & Teor. Fiz., 1973, 64, N3, 910-919. - CA, 1973, 78, 153230. PA, 1975, 78, 7306. RZF, 1973, 7E817.

Theor.; T - (e + t); d^7.

2.3.68. Sigmund, E., Wagner, M., Dynamical resonance phenomena of the optical Jahn-Teller problem in trigonal systems. - Phys. Status Solidi b, 1973, 57, N2, 635-646. - PA, 1973, 76, 46800. RZF, 1973, 11E871.

Theor.; E - e.

2.3.69. Wagner, M., Dynamical resonance phenomena in nonadiabatic vibronic systems. - J. Phys. [France], 1973, 34, Colloq. N9, 133-136.

Theor.; E - e; T - e.

2.3.70. Wittekoek, S., Van Stapele, R. P., Wijma, A. W. J., Optical-absorption spectrum of tetrahedral Fe^{2+} in $CdIn_2S_4$: influence of a weak Jahn-Teller coupling. - Phys. Rev. B, 1973, 7, N4, 1667-1677. - CA, 1973, 78, 90545. PA, 1973, 76, 21455. RZF, 1973, 9E743.

Exp.; T; T_d; d^6.

1974

2.3.71. Baranowski, J. M., Noras, J. M., Allen, J. W., Optical absorption dips caused by vibronic antiresonances in nickel-doped zinc selenide and titanium-doped zinc selenide. - J. Phys. C, 1974, 7, N24, 4529-4536. - CA, 1975, 82, 66222.

Exp.; T; d^8.

2.3.72. Baranowski, J. M., Noras, J. M., Allen, J. W., Vibronic antiresonance and autoionization in the optical spectra of titanium-(d^2) impurities in semiconductors. - In: Proc. 12th Int. Conf. Phys. Semicond. Stuttgart, 1974, p. 416-420. - CA, 1975, 83, 35156.

Exp.; T; T_d; d^2.

2.3.73. Gehlhoff, W., Ulrici, W., Trigonal nickel(2+) ion center in cadmium fluoride. - Phys. Status Solidi b, 1974, 65, N2, K93-K96. - CA, 1975, 82, 49620. PA, 1975, 78, 1507.

Exp.; T - t; O_h; d^8; 3.1.

2.3.74. Hizhnyakov, V., Zazubovich, S., Soovik, T., Kinetics and temperature dependences of polarized emission of anisotropic tin centers in alkali halides. - Phys. Status Solidi b, 1974, 66, N2, 727-738.

Exp.; T - e; O_h.

2.3.75. Kaufmann, U. G., Koidl, P., Jahn-Teller effect in the $^3T_1(P)$ absorption band of Ni^{2+} in ZnS and ZnO. - J. Phys. C, 1974, 7, N4, 791-806. - CA, 1974, 80, 114369. PA, 1974, 77, 28872. RZF, 1974, 8D656.

Exp.; T - (e + t); d^8.

2.3.76. Polinger, V. Z., Rosenfeld, Yu. B., Vekhter, B. G., Tsukerblat, B. S., Static field splitting of zero-phonon lines in Jahn-Teller systems. - Phys. Status Solidi b, 1974, 64, N2, 765-769. - PA, 1974, 77, 75797. RZF, 1975, 2D79.

Theor.; E - b; 1.4; 1.5.

2.3.77. Sauer, U., Scherz, U., Maier, H., The dynamic Jahn-Teller effect of cubic ZnS:Cu^{2+}: attempt at coupling to T_2 modes. - Phys. Status Solidi b, 1974, 62, N2, K71-K74. - CA, 1974, 80, 150604. PA, 1974, 77, 41408. RZF, 1974, 10E801.

Theor.; T - t; d^9.

2.3.78. Stolov, A. L., Yakovleva, Zh. S., On the origin of the band fine structure of Co^{2+} impurities in MeF_2 crystals. [Russ.]. - Opt. & Spektrosk., 1974, 36, N6, 1234-1235. - CA, 1974, 81, 83967. RZF, 1974, 10D673.

Exp.; T - t; O_h; d^7; 2.7.

2.3.79. Ulrici, W., Jahn-Teller effect on excited orbital triplet states of 3d" ions in eightfold cubic coordination of Co^{2+} and Fe^{2+} in CdF_2. - Phys. Status Solidi b, 1974, 62, N2, 431-441. CA, 1974, 80, 138941. PA, 1974, 77, 37643. RZF, 1974, 9D542.

Theor.; T - t; d^6; d^7.

1975

2.3.80. Bersuker, I. B., Polinger, V. Z., The tunneling splitting of the zero-phonon A → T line. [Russ.]. - In: Spektrosk. Kristallov., Moscow, 1975, p. 254-257.

Theor.; T - (e + t).

2.3.81. Kalnin', A. E., Trinkler, M. F. I., Intraimpurity luminescence of KBr•KI-Tl crystal. [Russ.]. - Izv. Akad. Nauk Latv.SSR, Ser. Fiz. Tekh. Nauk, 1975, N5, 58-64.

Exp.; T.

2.3.82. Lowther, J. E., Theoretical models for the GR1 doublet in diamond. - J. Phys. C, 1975, 8, N20, 3448-3456. - CA, 1975, 83, 211015. PA, 1976, 79, 9716.

Theor.

2.3.83. Manson, N. B., Vibronic transitions induced by T_{1u} and T_{2u} vibrations. - J. Phys. C, 1975, 8, N21, L483-L486.

Theor.; T - t; O_h; 1.5.

2.3.84. Tsukerblat, B. S., Rozenfel'd, Yu. B., Polinger, V. Z., Vekhter, B. G., Zero-phonon lines of Jahn-Teller systems in the case of weak electron-phonon coupling. [Russ.]. - Zh. Eksp. & Teor. Fiz., 1975, 68, N3, 1117-1126. - PA, 1975, 78, 46574. RZF, 1975, 7E67.

Theor.; E - e; T - t.

1976

2.3.85. Bacci, M., Quadratic Jahn-Teller effect in octahedral clusters: An interpretation of the excitation spectrum of vanadium-(2+) ion-doped magnesium oxide. - Phys. Lett. A, 1976, 57, N5, 475-476. - CA, 1976, 85, 133306. PA, 1976, 79, 80351.

Theor.; T - (e + t); O_h; d^3.

2.3.86. Cossart-Magos, C., Leach, S., Two-mode vibronic interaction between neighboring 1^2A_2 and 2^2B_2 excited electronic states of the benzyl radical. - J. Chem. Phys., 1976, 64, N10, 4006-4019.

Exp.; 1.5.; 1.7.

2.3.87. De Martini, F., Giuliani, G., Mataloni, P., Simoni, F., Nonlinear optical excitation and photoconductivity of the vibronic states of F-centers. - Opt. Commun., 1976, N1, 190. - PA, 1976, 79, 76558.

Exp.

2.3.88. Gnatenko, Yu. P., Rozhko, A. Kh., Structure of a zero-phonon line of a Jahn-Teller system with weak vibronic coupling to pseudo-local vibrations. [Russ.]. - Pis'ma, Zh. Eksp. & Teor. Fiz., 1976, 24, N3, 125-128. - CA, 1976, 85, 16870. PA, 1976, 79, 91456.

Exp.; T; d^8.

2.3.89. Kaplyanskiy, A. A., Medvedev, V. N., Smolyanskiy, P. L., Spectra, kinetics, and polarization of the luminescence of CaF_2-Yb^{2+} crystals. [Russ.]. - Opt. & Spektrosk., 1976, 41, N6, 1043-1050. - PA, 1977, 80, 69892. RZF, 1977, 4D805.

Exp.; 2.2.

2.3.90. Koidl, P., Blazey, K. W., Optical absorption of manganese-doped magnesium oxide. - J. Phys. C, 1976, 9, N6, L167-L170. - CA, 1976, 85, 26880. PA, 1976, 79, 45830.

Exp.; T - e; T - t; O_h; d^4; d^5.

2.3.91. Lowther, J. E., Transitions between E and T states each coupled to e vibrational modes: The GR1 band in diamond. - Solid State Commun., 1976, 20, N9, 933-935. - CA, 1977, 86, 23720. PA, 1977, 80, 12430.

Theor.; E - e; T - e.

2.3.92. Maier, H., Jahn-Teller effect, stacking faults, and the formation of polytypes in ZnS. - Phys. Status Solidi b, 1976, 78, N1, K65-K68. - CA, 1977, 86, 24573. PA, 1977, 80, 4660.

Exp.; E; T; d^9.

2.3.93. Manson, N. B., Wong, K. Y., Vibronic splitting in the $^3T_{1g}a$ crystal-field level of CaO:Ni^{2+}. - J. Phys. C, 1976, 9, N4, 611-626. - RZK, 1976, 17B586.

Exp.; T - (e + t_2); O_h; d^8.

2.3.94. Mulazzi, E., Maradudin, A. A., Dynamical Jahn-Teller effects on the zero-phonon line. - J. Phys. [France], 1976, 37, Colloq., N7, 114-116. - CA, 1977, 86, 180020.

Theor.; T; d^3.

2.3.95. Natadze, A. L., Ryskin, A. I., Manifestations of the electron-vibrational interaction in the absorption and luminescence spectra of $A^{II}B^{VI}$ crystals activated by transition metal ions. [Russ.]. - Izv. Akad. Nauk SSSR, Ser. Fiz., 1976, 140, N9, 1846-1850. - RZF, 1977, 1D520.

Exp.

2.3.96. Parrott, R., Naud, C., Gendron, F., Structure of a $^4T_2$ level of Mn in tetrahedral symmetry, dynamical Jahn-Teller effect and selective intensity transfer. - Phys. Rev. B, 1976, 13, N9, 3748-3763. - CA, 1976, 85, 38938. PA, 1976, 79, 65986. RZK, 1976, 23B624.

Exp.; T - e; T_d; d^5.

1977

2.3.97. Bernstein, E. R., Webb, J. D., On the Jahn-Teller effect in iridium hexafluoride: the Γ_{8g} triply degenerate t_{2g} state at 6800-Å. - 1977. - 59 p. - US NTIS, AD Rep., AD-A046563. - CA, 1978, 88, 160913.

Exp.; Γ_8 - (e + t); D_{4h}.

2.3.98. Blazey, K. W., Optical absorption of iron-doped magnesium oxide. - J. Phys. & Chem. Solids, 1977, 38, N6, 671-675. - CA, 1977, 87, 174964.

Exp.; d^6.

2.3.99. Gardavsky, K., Barvik, I., Zvara, M., Jahn-Teller effect on the $^3T_1(^3F)$ state and resonance of the $^3T_1(^3p)$ state with the conduction band in titanium(2+)-doped cadmium telluride. - Phys. Status Solidi b, 1977, 84, N2, 691-698. - CA, 1978, 88, 56714. PA, 1978, 81, 27271.

Exp.; T; d^2.

2.3.100. Giorgianni, U., Mondio, G., Perillo, P., Saitta, G., Vermiglio, G., Thermally induced absorption - intensity transfer between Tl^+ bands in alkali halides. - Phys. Rev. B, 1977, 15, N12, 5983-5990.

Exp.; T - (a + e + t); E - e; O_h; 1.5.

2.3.101. Hjortsberg, A., Nygren, B., Vallin, J. T., Indirect measurements of the Jahn-Teller quenched spin-orbit splitting in the $^5T_{2g}$ state of $KMgF_3:Fe^{2+}$. - Solid State Commun., 1977, 22, N10, 619-621. - CA, 1977, 87, 108803. PA, 1977, 80, 63442.

Exp.; E; d^6; 1.4.

2.3.102. Itah, J., Katz, B., Scharf, B., Direct observations of the splitting of the vibrationless level in a Jahn-Teller active state in deuterobenzenes of D_{2h} symmetry. - Chem. Phys. Lett., 1977, 48, N1, 111-114. - CA, 1977, 87, 92890. PA, 1977, 80, 58080.

Exp.; D_{2h}.

2.3.103. Itah, J., Katz, B., Scharf, B., The effect of vibrational perturbation on the vibronic structure in Jahn-Teller active states. - Chem. Phys. Lett., 1977, 52, N1, 92-97. - CA, 1978, 88, 29827. RZF, 1978, 7D169.

Exp.; E - e; (A + B) - (a + b); D_{2h}; D_{6h}.

2.3.104. Lukashin, A. V., Frank-Kamenetskii, M. D., Comparison of different theoretical descriptions of vibronic spectra of molecular aggregates. - Chem. Phys. Lett., 1977, 45, N1, 36-40.

Theor.; 1.5.

2.3.105. Manson, N. B., Shah, G. A., Coupling of T_{1u} and T_{2u} vibrations to the $^2E_g \rightarrow {}^4A_{2g}$ transition in $MgO:V^{2+}$ and $MgO:Cr^{3+}$. - J. Phys. C, 1977, 10, N11, 1991-2003. - CA, 1977, 87, 191392.

Exp.

2.3.106. Natadze, A. L., Elektronno-Kolebatel'noye Vzaimodeystviye v Kristallakh Sulfida Tsinka, Aktivirovannykh Ionamy Perekhodnykh Metallov: Avtoref. Dis. ... Kand. Fiz.-Mat. Nauk. [Russ.]. - Leningrad, 1977. - 22 p. - Opt. Inst. im. S. I. Vavilova.

Electron-Vibrational Interactions in Zinc Sulfide Crystals Activated by Transition Metal Ions.

Exp.; T - (e + t); d^6; d^7; d^8; d^9.

2.3.107. Ortega, J. M., Spectra distribution of the triplet-to-singlet luminescence of the F_2 center in potassium chloride, bromide, and iodide. - Commun. Phys., 1977, 2, N3, 45-51. - CA, 1977, 87, 174970.

Exp.; T.

2.3.108. Ranfagni, A., Viliani, G., Spin-orbit allowed tunneling splitting in the $T \times E_g$ Jahn-Teller problem: Splitting of the zero-phonon line. - Phys. Status Solidi b, 1977, 84, N1, 393-399. - CA, 1978, 88, 29676. PA, 1978, 81, 19006.

Theor.; T - e; O_h; d^3.

2.3.109. Rozhko, A. Kh., Opticheskiye Proyavleniya Primesey Gruppy Zheleza v Kristallakh CdS: Avtoref. Dis. ... Kand. Fiz.-Mat. Nauk. - Kiev, 1977. - 22 p. - Akad. Nauk UkrSSR, Inst. Fiz.

Optical Manifestigations of Iron Group Impurities in CdS Crystals.

Exp.; T - e; T_d; d^7; d^8.

2.3.110. Sigmund, E., A practical method to derive canonical transformations in a closed form and its application to an anti-resonant electron-phonon system. - Z. Naturforsch. a, 1977, 32, N2, 113-118.

Theor.; E - e; T - t.

2.3.11. Stoneham, A. M., The low-lying levels of the GR1 center in diamon. - Solid State Commun., 1977, 21, N4, 339-341. - CA, 1977, 86, 130347. PA, 1977, 80, 20269.

Exp.; E - e.

1978

2.3.112. Brickmann, J., Vibronic coupling in molecular excited states: Model approach to emission and absorption spectra. - Mol. Phys., 1978, 35, N1, 155-176. - CA, 1978, 89, 67770.

Theor.; 1.5.

2.3.113. Guedel, H. U., Shellgrove, T. R., Jahn-Teller effect in the $^4T_{2g}$ state of chromium(III) in dicesium sodium indium(III) hexachloride. - Inorg. Chem., 1978, 17, N6, 1617-1620. - CA, 1978, 88, 200167.

Exp.; T - e; O_h; d^3; 1.4.

2.3.114. Guest, A., Howard-Lock, H. E., Lock, C. J. L., Dynamical Jahn-Teller effects in the near ultraviolet vapor phase absorption spectra of perrhenyl chloride, ReO_3Cl, and pertechnyl chloride, TcO_3Cl. - J. Mol. Spectrosc., 1978, 72, N1, 143-157. - CA, 1978, 89, 97445. PA, 1978, 81, 86881.

Exp.; E - e; C_{3v}; 2.2.

2.3.115. Le Si Dang, D'Aubigne, Y. M., Romestain, R., Inhomogeneous broadening of the zero-phonon line and EPR lines of F centers in CaO. - Semicond. & Insul., 1978, 3, N2/3, 151-161.

Exp.; T - (e + t); O_h; 3.1; 3.2.

2.3.116. Reut, E. G., Study of the characteristics of wide-band luminescence of europium and ytterbium ions in crystals with fluorite structure. [Russ.]. - Opt. & Spektrosk., 1978, 45, N3, 518-524. - CA, 1978, 89, 206773.

Exp.

2.3.117. Sasaki Naoyuki, Fine structure of charge transferred excited levels in copper gallium disulfide:iron. II. [Jap.]. - Nippon Shika Daigaku Kiyo, 1978, 7, 99-106. - CA, 1978, 89, 223273.

Theor.; T - e.

2.3.118. Takaoka Yoichi, Suzuki Naoshi, Moitzuki Kazuko, Jahn-Teller effect on optical absorption spectra of MnO. - J. Phys. Soc. Jpn., 1978, 44, N4, 1168-1176. - CA, 1978, 88, 200325. PA, 1978, 81, 52484.

Exp.; T; d^5.

2.3.119. Trinkler, M. F., Trinkler, L. E., Kalnin', A. E., Intercenter luminescence of KCl-Pb in the 8-80 K temperature region. [Russ.]. - Izv. Akad. Nauk Latv.SSR, Ser. Fiz. Tekh. Nauk, 1978, N3, 12-17. - CA, 1978, 89, 82523.

Exp.; 2.2.

2.3.120. Uba, S. M., Baranowski, J. M., Depression of vibronic levels and transition from the dynamic to static Jahn-Teller effect in the $^4T_1$ multiplet: the case of cobalt(2+) in zinc selenide. - Phys. Rev. B, 1978, 17, N1, 69-84. - CA, 1978, 88, 143784. PA, 1978, 81, 43337.

Exp.; T - (e + t_2); T - e; d^7.

1979

2.3.121. Davies, G., Dynamic Jahn-Teller distortions at trigonal optical centers in diamond. - J. Phys. C, 1979, 12, N13, 2551-2566. - CA, 1980, 92, 13098. PA, 1979, 82, 87069.

Exp.; E; C_{3v}; D_{3d}.

2.3.122. Kaminska, M., Baranowski, J. M., Uba, S. M., Vallin, J. T., Absorption and luminescence of $Cr^{2+}(d^4)$ in II-VI compounds. - J. Phys. C, 1979, 12, N11, 2197-2214. - CA, 1979, 91, 219550. PA, 1979, 82, 83000.

Exp.; E; T; d^4; 2.2; 2.7.

2.3.123. Kaufmann, U., Koschel, W. H., Schneider, J., Weber, J., Optical and EPR study of the nickel two-electron-trap state in GaP. - Phys. Rev. B, 1979, 19, N7, 3343-3352. - CA, 1979, 91, 65766. PA, 1979, 82, 73806.

Exp.; E; T; d^9; 3.1.

2.3.124. Lowther, J. E., Fine structure of the GR1 and UV band in diamond. - Solid State Commun., 1979, 32, N9, 755-756. - CA, 1980, 92, 67205. PA, 1980, 83, 20678.

Exp.; E - e.

2.3.125. Manson, N. B., Hasan, Z., Flint, C. D., Jahn-Teller effect in the $^4T_{2g}$ state of Mn^{4+} in Cs_2SiF_6. - J. Phys. C, 1979, 12, N24, 5483-5488. - CA, 1980, 92, 188516. PA, 1980, 83, 29367.

Exp.; T; d^3; 1.4.

2.3.126. Montagna, M., Pilla, O., Viliani, G., Dynamical Jahn-Teller effect on spin-orbit multiplets: intensity quenching. - J. Phys. C, 1979, 12, N17, L699-L704. - CA, 1980, 92, 15020. PA, 1979, 82, 95374.

Exp.; d^3; 1.4.

2.3.127. Naud, C., Porte, C., Gendron, F., Parrot, R., Orbit-lattice interaction and Jahn-Teller effect in the ^4E levels of Mn^{2+} in nearly tetrahedral clusters. - Phys. Rev. B, 1979, 20, N8, 3333-3344. - CA, 1980, 92, 67070. PA, 1980, 83, 24860.

Theor.; E - e; T_d; d^5; 1.4.

2.3.128. Trinkler, M. F., Trinkler, L. E., A-luminescence of KBr-Pb in the 8-80 K temperature region. [Russ.]. - Opt. & Spektrosk., 1979, 46, N1, 91-97. - CA, 1979, 90, 143817.

Exp.; T - (e + t); 2.2.

See also: 1.3.-40, 49, 97, 105; 2.1.-87, 135, 140, 251; 2.2.-88, 89, 101, 162; 2.4.-3, 21, 22, 48, 60, 97, 133, 135, 141, 144, 147, 149, 150, 169, 187, 190, 203; 2.7.-78; 4.3.-53.

2.4. Effects of External Fields and Stresses. Polarization Dichroism

The possibility of investigating Jahn-Teller centers by means of polarization dichroism was pointed out for the first time in Ref. [2.4.2]. The most general matrix formulation of the method of moments and the group-theoretic approach to dichroism effects were suggested in Refs. [2.4.-7, 110, 130, 131, 136, 142]. The change of the third moment under uniaxial stress turned out to be proportional to the change of the first moment. The appropriate proportionality coefficients are linear combinations of different Jahn-Teller coupling parameters. By extracting the observable changes of moments from experimental data, one can determine the Jahn-Teller coupling parameters. The data of piezodichroism and magnetic circular dichroism allow one to determine also the orbital g-factor of excited electronic states and the spin-orbit coupling parameters [2.4.-104, 105]. The theory of moments combined with polarization spectroscopy seems to be the only possible method of estimating the above-mentioned parameters for the excited electronic states in the case of a strong Jahn-Teller effect. The theory was applied to transition metal ions [2.4.-67, 86, 98, 129], mercury-like ions [2.4.-25, 127, 133, 180], and F centers [2.4.-2, 6, 8, 46, 57, 48, 52]. Many papers are devoted to the electrooptics of Jahn-Teller impurities without inversion centers (see [2.4.166] and references therin).

More recently the polarization dichroism theory was generalized in several directions. The equations of the method of moments have been derived for the case of multiplet-multiplet transitions, when the Jahn-Teller effect is active in the excited multiplet but the external and/or internal fields split both the ground and excited multiplets [2.4.-142, 143, 186, 204]. The theory of moments for spin-forbidden transition is presented in Ref. [2.4.155], where the essential features of magnetic circular dichroism are discussed. The generalization of the polarization dichroism theory for vibrationally induced transitions is given in Ref. [2.4.197]. The theoretical technique of latent anisotropy investigation for cubic crystals with anisotropic Jahn-Teller centers is suggested in Refs. [2.4.-163, 183]. The change of the third moment of the broad optical bands is found to be proportional to the change of only the Stark part of the first moment, whereas the pseudo-Stark splitting of the electronic energy levels does not affect the third moment. The theory of piezo- and electroabsorption (including the quadratic Stark effect) was developed in Ref. [2.4.183]. In Ref. [2.4.178] the polarization dichroism theory for Jahn-Teller Frenkel-type exitions is given. It is shown that the method of moments allows one to estimate not only the Jahn-Teller coupling parameters, but also the parameters of the exitonic zone.

1964

2.4.1. Pryce, M. H. L., Agnetta, G., Garofano, T., Palma-Vitorelli, M. B., Palma, M. U., Low-temperature optical absorption of nickel fluorosilicate crystals. - Philos. Mag., 1964, 10, N105, 477-496. - CA, 1964, 61, 12802g.

Exp.; d^8.

1965

2.4.2. Henry, C. H., Schnatterly, S. E., Slichter, C. P., Effect of applied fields on the optical properties of color centers. - Phys. Rev. A, 1965, 137, N2, 583-602.

Theor.

2.4.3. Hughes, A. E., Runciman, W. A., Stress spectra of color centers in LiF. - Proc. Phys. Soc. London, 1965, 86, N551, Pt. 3, 615-628. - CA, 1965, 63, 10859e.

Exp.; E; 2.3.

2.4.4. Kaplyanskii, A. A., Przhevuskii, A. K., The strain-induced splitting of the U-band and Jahn-Teller effect in excited ^4T-state of Cr^{3+} ions in ruby. - Phys. Status Solidi, 1965, 11, N2, 629-634. - RZF, 1966, 6D406.

Exp.; T; d^3.

2.4.5. Kaplyanskiy, A. A., Przhevuskiy, A. K., Stress splitting and rise of spectral lines and structure of excited levels of Eu^{2+} in crystals of alkali halide fluorides. [Russ.]. - Opt. & Spektrosk., 1965, 19, N4, 597-610.

Exp.; E; d^1.

2.4.6. Silsbee, R. H., R center in KCl: stress effects in optical absorption. - Phys. Rev. A, 1965, 138, N1, 180-197. - CA, 1965, 62, 11260f. RZF, 1965, 9D257.

Theor.; E; 1.7.

2.4.7. Weigang, O. E., Jr., Vibrational structuring in optical activity. I. Vibronic coupling in the circular dichroism of dimers. - J. Chem. Phys., 1965, 43, N1, 71-72. - CA, 1965, 63, 2531h. PA, 1965, 68, 22355.

Theor.; 1.5.

1966

2.4.8. Duval, P., Gareyte, J., Merle d'Aubigne, Y., Zeeman effect of R centers in KCl. [Fr.]. - Phys. Lett., 1966, 22, N1, 67-69. - CA, 1966, 65, 12966b.

Exp.; E - e; C_{3v}.

2.4.9. Fulton, T. A., Fitchen, D. B., Excited states of the F center in CsF. - Bull. Am. Phys. Soc., 1966, 11, N4, 245.

Exp.; Γ_8.

2.4.10. Kaplyanskiy, A. A., Przhevuskiy, A. K., Piezoelectric spectroscopic investigations of the level scheme and transitions in Sm^{2+} ions in crystals of alkali halide fluorides. [Russ.]. - Opt. & Spektrosk., 1966, 20, N6, 1045-1057.

Exp.; d^1.

1967

2.4.11. Dubenskiy, K. K., Ryskin, A. I., Khil'ko, G. I., The splitting of the 4274 cm^{-1} absorption line of the Ni^{++} ion in ZnS single crystal and the Jahn-Teller effect. [Russ.]. - Fiz. Tverd. Tela, 1967, 9, N7, 1974-1982. - CA, 1967, 67, 103692.

Exp.; T; d^8.

2.4.12. Lanzi, F., Linear Stark effect of the R-center. - Phys. Lett. A, 1967, 25, N8, 596-598. - PA, 1968, 71, 1992.

Theor.; E - e.

2.4.13. Merle d'Aubigne, Y., Duval, P., Circular dichroism of R centers in KCl. [Fr.]. - J. Phys. [France], 1967, 28, Colloq. N4, 112-114. - CA, 1968, 68, 100075.

Exp.

2.4.14. Onaka Ryumyo, Mabuchi Teruhiko, Yoshikawa Akira, Zeeman effect of the A absorption bands in Tl- and Pb-doped alkali halides. - J. Phys. Soc. Jpn., 1967, 23, N5, 1036-1037. - CA, 1968, 68, 44510.

Exp.; T - (e + t); O_h.

1968

2.4.15. Burke, W., Magnetic circular dichroism of the R center in KCl. - Phys. Rev., 1968, 172, N3, 886-899.

Exp.; E; D_{3h}.

2.4.16. Davis, J. A., Fitchen, D. B., Magnetic circular dichrosim of the R' center in LiF. - Solid State Commun., 1968, 6, N7, 506-509.

Exp.; E - e.

2.4.17. Fulton, T. A., Fitchen, D. B., F-center in cesium fluoride: optical absorption properties. - 1968. - 63 p. - US At. Energy Comm. NYO-3464-14. - CA, 1969, 70, 72449.

Exp.; T.

2.4.18. Ludlow, I. K., Spin-orbit coupling and stress spectra color centers in magnesium oxide. - J. Phys. C, 1968, 1, N5, 1194-1204.

Exp.; E - e.

2.4.19. Merle d'Aubigne, Y., Duval, P., Circular dichroism of R centers in KCl. [Fr.]. - J. Phys. [France], 1968, 29, N10, 896-910.

Exp.; E - e; 1.4.

2.4.20. Shankland, T. J., Pressure shift of absorption bands in $MgO:Fe^{2+}$ and the dynamic Jahn-Teller effect. - J. Phys. & Chem. Solids, 1968, 29, N10, 1907-1909. - CA, 1969, 70, 7799. PA, 1969, 72, 5927.

Exp.; E - e; d^6.

2.4.21. Shepherd, I. W., Magnetic circular dichroism of the R_2 band in potassium chloride and potassium fluoride. - Phys. Rev., 1968, 165, N3, 985-993. - CA, 1968, 68, 53459.

Exp.; E - e; C_{3v}; 2.3.

2.4.22. Stephens, P. J., Low-Pariseau, M., Zeeman study of the Jahn-Teller effect in the $^3T_{2g}$ state of $Al_2O_3:V^{3+}$. - Phys. Rev., 1968, 171, N2, 322-335. - CA, 1968, 69, 39946. PA, 1968, 71, 43965.

Exp.; T - e; O_h; d^2; 1.4; 2.3.

1969

2.4.23. Bimberg, D., Dultz, W., Gebhardt, W., Stress-induced dichroism in Tl^+-doped potassium halides. - Phys. Status Solidi, 199969, 31, N2, 661-671. - CA, 1969, 70, 62647.

Exp.; E - e; O_h.

2.4.24. Chen Ming-yu, McClure, D. S., Uniaxial stress effect in the $^4T_1(G)$ state of $RbMnF_3$. - Bull. Am. Phys. Soc., 1969, 14, N1, 79.

Exp.; T; 1.4.

2.4.25. Cho Kikuo, Theory of stress and magnetic circular dichroism in alkali halide phosphors. - J. Phys. Soc. Jpn., 1969, 27, N3, 646-658. - CA, 1969, 71, 96449. PA, 1970, 73, 4973.

Theor.; T - (e + t); O_h.

2.4.26. Davis, J. A., Fitchen, D. B., Magnetic circular dichroism of the F_3^+ center in NaF. - Solid State Commun., 1969, 7, N18, 1363-1365. - CA, 1969, 71, 117911. PA, 1970, 73, 7943.

Exp.; E - e.

2.4.27. Fontana, M. P., Davis, J. A., Magnetic circular polarization of the Tl luminescence in $KI:Tl^{+}$. - Phys. Rev. Lett., 1969, 23, N17, 974-977.

Exp.; T - e; O_h; 2.2.

2.4.28. Fontana, M. P., Fitchen, D. B., Magnetic circular polarization of F-center emission. - Phys. Rev. Lett., 1969, 23, N26, 1497-1499.

Exp.; T; 1.4; 3.1.

2.4.29. Fulton, T. A., Fitchen, D. B., F-center in cesium fluoride: optical absorption properties. - Phys. Rev., 1969, 179, N3, 846-859.

Exp.; T - (a + e + t).

2.4.30. Gillard, R. D., Mitchell, P. R., Vibronic fine structure of formally forbidden transitions in circular dichroism. - Trans. Faraday Soc., 1969, 65, N10, 2611-2620.

Exp.; D_{6h}.

2.4.31. Perlin, Yu. E., Kharchenko, L. S., Tsukerblat, B. S., The application of the method of moments to the calculation of the multiplet bands of the impurity absorption in ruby. [Russ.]. - Fiz. Tverd. Tela, 1969, 11, N4, 1065-1067.

Theor.; T - (a + e + t); O_h.

2.4.32. Wit, M., de., Zeeman effect in the absorption spectrum of copper-doped zinc sulfide. - Phys. Rev., 1969, 177, N2, 441-447.

Exp.; T - e; T_d; d^9.

1970

2.4.33. Bersuker, I. B., Ablov, A. V., Vekhter, B. G., Tsukerblat, B. S., Polarizations in the optical spectra of transition metal complexes taking into account the Jahn-Teller effect. [Russ.]. - Zh. Strukt. Khim., 1970, 11, N1, 102-107.

Theor.; T - (e + t).

2.4.34. Bogan, L. D., Fitcher, D. B., Stark effects in F-center emission. - Phys. Rev. B, 1970, 1, N10, 4122-4135.

Exp.

2.4.35. Broser, I., Scherz, U., Wohlecke, M., Zeeman effect of Cu^{2+} centers in II-VI compounds. - In: Proc. Int. Conf. Lumin., 1969, Amsterdam, 1970, p. 39-50. - CA, 1971, 74, 17755.

Exp.; d^9; 2.3; 2.7; 3.1.

2.4.36. Cho Kikuo, External-field-induced dichroism in the optical spectra of localized excitation. [Jap.]. - Kotai Butsuri, Solid State Phys., 1970, 5, N3, 131-140. - CA, 1971, 74, 92834.

Exp.

2.4.37. Davis, J. A., R' center in lithium fluoride: an optical study of Jahn-Teller distorted system. - 1970. - 135 p. - [Order N70-12-623.]. Diss. Abstr. Int. B, 1970, 31, N1, 351. CA, 1971, 74, 117883.

Exp.; E - e; C_{3v}.

2.4.38. Hughes, A. E., A solution of the calcium oxide F^+ center problem using a two-mode Jahn-Teller effect. - J. Phys. C, 1970, 3, N3, 627-637. - CA, 1970, 73, 19987. PA, 1970, 73, 50610.

Exp.; T - (e + t).

2.4.39. Kobayashi Hiroshi, Shimizu Masayuki, Kaizu Youkoh, Magnetic circular dichroism of $[Fe(CN_6)]^{3-}$. - Bull. Chem. Soc. Jpn., 1970, 43, N8, 2321-2325. - CA, 1970, 73, 103685.

Exp.; d^5; 1.4.

2.4.40. Onton, A., Morgan, T. N., Effect of uniaxial stress on excitons bound to bismuth in gallium phosphide. - Phys- Rev. B, 1970, 1, N6, 2592-2604. - CA, 1970, 73, 20155.

Exp.

2.4.41. Piepho, S. B., Lester, T. E., McCaffery, A. J., Dickinson, J. R., Electronic absorption and magnetic circular dichroism spectra of $IrBr^{2-}$ in several host crystals. - Mol. Phys., 1970, 19, N6, 781-802. - CA, 1971, 74, 47434.

Exp.; T - t; O_h.

2.4.42. Sati, R., Inoue, M., Wang, S., Magnetic circular dichroism and moments for $A_2 \rightarrow E$ transition in C_{3v} defect centers. - Can. J. Phys., 1970, 48, N22, 2769-2779. - PA, 1971, 74, 5817. RZF, 1971, 6D491.

Theor.; E - e.

2.4.43. Stiles, L. F., Jr., Fontana, M. P., Fitchen, D. B., Effect of electric field and temperature on the radiative lifetime of the F center. - Phys. Rev. B, 1970, 2, N6, 2077-2089. - CA, 1970, 73, 135560.

Exp.

1971

2.4.44. Chan, C. S., Corelli, J. C., Piezospectroscopic study (78 K) of radiation-produced absorption bands in Si. - In: Int. Conf. Radiat. Eff. Semicond. Albany, 1970. London, 1971, p. 197-201. - PA, 1972, 75, 48193.

Exp.

2.4.45. Chen, C. S., Corelli, J. C., Piezospectroscopic study (78 K) of radiation-produced absorption bands in silicon. - Radiat. Eff., 1971, 9, N1/2, 75-79. - CA, 1971, 75, 42606.

Exp.; 2.7.

2.4.46. Harmer, A. L., Hayes, W., O'Brien, M. C. M., Stark effect in the F^+ center in CaO. - J. Phys. C, 1971, 4, N1, L108-L111.

Exp.; T - (e + t_2).

2.4.47. Medvedev, E. S., The luminescence of local centers in crystals in the presence of axial strain. [Russ.]. - Pis'ma, Zh. Teor. & Eksp. Fiz., 1971, 13, N1, 59-61. - PA, 1971, 74, 41632. RZF, 1971, 6D753.

Exp.; T - e; 2.2.

2.4.48. Merle d'Aubigne, Y., Roussel, A., Magnetooptic study of the Jahn-Teller effect in the excited level of the F^+ centers in CaO. - Phys. Rev. B, 1971, 3, N4, 1421-1427. - CA, 1971, 74, 105123. PA, 1971, 74, 28645. RZF, 1971, 8D509.

Exp.; 1.4; 2.3.

2.4.49. Sturge, M. D., Selection rules for vibronic d → d transitions of impurity ions in cubic $KMgF_3$. - Solid State Commun., 1971, 9, N12, 899-902.

Exp.; T; d^3; d^8.

1972

2.4.50. Bird, B. D., Osborne, G. A., Stephens, P. J., Magnetic circular dichroism of impurities in solids: vibrationally induced d → d transitions in MgO:Ni. - Phys. Rev. B, 1972, 5, N5, 1800-1812.

Exp.; T - t.

2.4.51. Canters, G. W., Van Egmond, J., Shaafsma, T. J., Van der Waals, J. H., Optical and Zeeman studies of the first excited singlet state of zinc porphine in a single crystal of n-octane: Evidence for Jahn-Teller instability. - Mol. Phys., 1972, 24, N6, 1203-1215. - CA, 1973, 78, 49980. PA, 1973, 76, 8599.

Exp.; E; 3.1.

2.4.52. Duran, J., Merle d'Aubigne, Y., Romestain, R., Jan-Teller coupling in the excited level of the F^+ center in CaO: uniaxial stress effects. - J. Phys. C, 1972, 5, N16, 2225-2245. - CA, 1972, 77, 107093. PA, 1972, 75, 68635. RZF, 1973, 1D418.

Exp.; T - (e + t); 1.4.

2.4.53. Duval, E., Louat, R., Lacroix, R., Stress study of the Jahn-Teller effect in the $^4T_2$ state of Cr^{3+} in α-Al_2O_3. - Phys. Status Solidi b, 1972, 50, N2, 627-640. - CA, 1972, 76, 160227. PA, 1972, 75, 36787. RZF, 1972, 10E826.

Exp.; T - e; d^3.

2.4.54. Fukuda Atsuo, Jahn-Teller effect and Tl^+ type centers in alkali halides. - In: Phys. Impurity Cent. Cryst.: Proc. Int. Semin. "Selected Probl. Theory Impurity Cent. Cryst.," 1970. Tallin, 1972, p. 505-527.

Exp.; T.

2.4.55. Fukuda Atsuo, Yuster, P., Unusually large change in radiative lifetime of the A-band emission in $KI:In^+$ and $KI:Sn^{2+}$ induced by a magnetic field. - Phys. Rev. Lett., 1972, 28, N16, 1032-1034. - CA, 1972, 76, 16492. RZF, 1972, 8D852.

Exp.; T - e; O_h; 2.2.

2.4.56. Grasso, V., Perillo, P., Vermiglio, G., Magneto-optical measurements of the C band in $KBr:Tl^+$. - Solid State Commun., 1972, 11, N4, 563-565. - CA, 1972, 77, 120150. PA, 1972, 75, 6806. RZF, 1973, 1D428.

Exp.; T.

2.4.57. Hughes, A. E., Pells, G. P., and Somber, E., Symmetry assignments of one-phonon sidebands in the emission spectrum of the F^+ center in calcium oxide. - J. Phys. C, 1972, 5, N6, 709-715. - CA, 1972, 76, 133727. PA, 1972, 75, 29911.

Exp.; T - (e + t).

2.4.58. Iida, T., Kurata, K., Muramatsu, S., Dynamical Jahn-Teller effect in relaxed excited state of F-center. - J. Phys. & Chem. Solids, 1972, 33, N6, 1225-1233. - CA, 1972, 77, 41051. PA, 1972, 75, 436747. RZF, 1972, 10E825.

Theor.; T - e; O_h; 1.4.

2.4.59. Kanno Kenichi, Dichroism due to Jahn-Teller distortion in silver chloride-doped cadmium chloride. [Jap.]. - Nagoya Kagyo Daigaku Gakuho, Bull. Nagoya Inst. Technol., 1972, 24, 85-95. - CA, 1974, 80, 114528. PA, 1975, 78, 13238.

Exp.; E; d^9.

2.4.60. Kazanskiy, S. A., Ryskin, A. I., Electronic structure of an ion-hole complex at a nickel ion in a zinc sulfide crystal. [Russ.]. - Fiz. Tverd. Tela, 1972, 14, N6, 1818-1820. - CA, 1972, 77, 80850.

Exp.; d^8; d^9; 2.3.

2.4.61. Vala, M., Jr., Mongan, P., McCarthy, P. J., Spectra of tetrahedral complexes of transition metals: Jahn-Teller effect in the tetrabromoferrate(III) ion. - J. Chem. Soc., Dalton Trans., 1972, N17, 1870-1875. - CA, 1972, 77, 95012. RZF, 1973, 3D597.

Exp.; T; d^5.

2.4.62. Van Gorkom, G. G. P., Theoretical optical Zeeman patterns for Mn^{2+} in an octahedral or tetrahedral crystal field. - Solid State Commun., 1972, 11, N9, 1253-1256. - CA, 1973, 78, 21840. PA, 1973, 76, 2011.

Exp.; T; O_h; d^5.

2.4.63. Vekhter, B. G., Rozenfel'd, Yu. B., Tsukerblat, B. S., Nonlinear optical and acoustic phenomena in the Jahn-Teller systems. [Russ.]. - In: Tez. Dokl. 7 Vsesoyuz. Soveshch. Kvant. Akustike Tverd. Tela, Khar'kov, 1972, p. 145.

Theor.; 3.6.

2.4.64. Washimiya Shuko, Oscillatory behavior of g values and magnetic circular dichroism in a Jahn-Teller coupled system. - Phys. Rev. Lett., 1972, 28, N9, 556-559. - CA, 1972, 76, 105693. PA, 1972, 75, 26154. RZF, 1972, 7D206.

Theor.; E; 1.4; 3.1.

2.4.65. Wöhlecke, M., The Zeeman effect of the infrared luminescence of Cu^{2+} in ZnS. - In: Abstr. Int. Conf. Lumin., Leningrad. Chernogolovka, 1972, p. 108.

Theor.; T - e; T_d; d^9.

1973

2.4.66. Binet, G., Margerie, J., D'Aubigne, Y. M., Magnetic circular dichroism of M and R centers in potassium chloride. II. Technique of moments in the R center problem. - J. Phys. C, 1973, 6, N17, 2745-2756. - CA, 1973, 79, 120007. PA, 1973, 76, 69027.

Theor.; E.

2.4.67. Donecker, J., Kluge, J., Magnetic circular dichroism of the vibrationally induced d → d transition $^4A_{2g}(F) \rightarrow {}^4T_{1g}(p)$ on Co^{2+} ions in CdF_2. - Phys. Status Solidi b, 1973, 59, N1, 163-173. - CA, 1973, 79, 120016. PA, 1973, 76, 64631. RZF, 1974, 1D496.

Exp.; T; d^7.

2.4.68. Fukuda Atsuo, Influence of magnetic field on the triplet-term relaxation-excitation states of thallium-ion-type fluorescence. [Jap.]. - Bussei, Solid State Phys. & Chem., 1973, 14, N10, 607-620. - CA, 1974, 80, 76174.

Exp.; T; 2.3.

2.4.69. Fukuda Atsuo, Negative magnetic circular polarization of the A-band emission in $KI:Sn^{2+}$. - Solid State Commun., 1973, 12, N10, 1039-1043. - CA, 1973, 79, 59283.

Exp.; T; O_h.

2.4.70. Grasso, V., Perillo, P., Vermiglio, G., Measurements of magnetic circular dichroism in the C-band of KBr:Tl. - Nuovo Chimento B, 1973, 13, N1, 42-48. - CA, 1973, 78, 77474. PA, 1973, 76, 18430. RZF, 1973, 6D899.

Exp.; 3.1.

2.4.71. Kaplyanskiy, A. A., Smolyanskiy, P. L., Polarized luminescence of CaF_2-Yb^{2+} ion crystals. [Russ.]. - Opt. & Spektrosk., 1973, 34, N3, 624-625. - CA, 1973, 78, 141997. RZF, 1973, 7D766.

Exp.; O_h; 2.2.

2.4.72. Lemos, A. M., Krolik, C., Stress-induced dichroism and energy shifts in the A, B, and C absorption bands in KCl:Tl. - Phys. Rev. B, 1973, 7, N4, 1608-1616. - CA, 1973, 78, 90581. PA, 1973, 76, 21454.

Exp.

2.4.73. Modine, F. A., Magnetooptical study of a charge-transfer band in vanadium-doped magnesium oxide. - Phys. Rev. B, 1973, 8, N2, 854-863. - CA, 1973, 79, 59451.

Exp.; T; 2.2.

2.4.74. Muramatsu Shinji, Application of vibronic model to stress-induced polarization of F-center emission. - Phys. Status Solidi b, 1973, 56, N2, 631-636. - RZF, 1973, 9E739.

Theor.; (A + T) - e; 2.3.

2.4.75. Perlin, Yu. E., Tsukerblat, B. S., Optical dichroism effects in systems with dynamical Jahn-Teller coupling. [Russ.]. - In: Tez. Dokl. 4 Vsesoyuz. Soveshch. Spektrosk. Kristallov, Aktivir. Ionami Redkozem. Perekhodnykh Elementov. Sverdlovsk, 1973, p. 3.

Theor.

2.4.76. Perlin, Yu. E., Kharchenko, L. S., The effect of electron-phon interaction on magneto-optical spectra of color centers. - In: Magn. Resonance & Related Phenomena: Proc. 17th Congr. AMPERE, Turki, 1972, Amsterdam; London, 1973, p. 77-78. - PA, 1973, 76, 2580.

Theor.

2.4.77. Perrin, M. H., Vibronic coupling. VII. Magnetic circular dichroism of trimers, aromatic hydrocarbons, and porphyrins. - J. Chem. Phys., 1973, 59, N4, 2090-2104.

Theor.; D_{4h}; 1.5.

2.4.78. Regis, M., Farge, Y., Fontana, M., Study of the optical transitions in nickel(2+)-doped magnesium oxide by magnetic circular dichroism. [Fr.]. - Phys. Status Solidi b, 1973, 57, N1, 307-314. - CA, 1973, 79, 11513. PA, 1973, 76, 41515.

Exp.; T; d^8.

2.4.79. Ryskin, A. I., Spektroskopiya Sulfida Tsinka i Drugikh Soedineniy $A^{II}B^{VI}$, Aktivirovannykh Ionami Perekhodnykh Metallov:

Avtoref. Diss. ... Dokt. Fiz.-Mat. Nauk. [Russ.]. - Leningrad, 1973. - 24 p. - Gos. Opt. Inst. im. S. I. Vavilova.

Spectroscopy of Zinc Sulfide and Other $A^{II}B^{VI}$ Compounds Activated by the Transition Metal Ions.

Exp.; E - e; T - (e + t); O_h.

<u>1974</u>

2.4.80. Brunshtein, S. Kh., Studies of optical bands of a coordinated chromium(3+) ion by the method of moments. [Russ.]. - In: Fiz. Mat. Metody Koord. Khimii: Tez. Dokl. 5 Vsesoyuz. Sovesch. Kishinev, 1974, c. 73. - CA, 1975, <u>82</u>, 147775.

Theor.; E; T; d^3.

2.4.81. Caliga, D., Richardson, F. S., Chiroptical spectra of molecules with nearly degenerate electronic states. - Mol. Phys., 1974, <u>28</u>, N5, 1145-1166.

Theor.; 1.5.

2.4.82. Edel, P., D'Aubigne, Y., Merle, Louat, R., Jahn-Teller coupling in the excited levels of the F-centers in calcium oxide: Effect of uniaxial stress. [Fr.]. - J. Phys. & Chem. Solids, 1974, <u>35</u>, N1, 67-70. - CA, 1974, <u>80</u>, 126344. PA, 1974, <u>77</u>, 25157. RZF, 1974, 7D462.

Exp.; T - e.

2.4.83. Fergusson, J., Guedel, H. U., Krausz, E. R., Guggenheim, H. J., High-resolution magnetic circular dichroism (MCD) and phosphorescence excitation spectroscopy of manganese(2+) in potassium magnesium trifluoride and potassium zinc trifluoride. - Mol. Phys., 1974, <u>28</u>, N4, 879-891. - CA, 1975, <u>82</u>, 117925. PA, 1974, <u>77</u>, 73171.

Exp.; T; O_h; d^5; 2.2.

2.4.84. Harada Masafumi, Tsujikawa Ikuji, Study of the initial tetragonal distortion of the $[CoX_4]^{2-}$ complex ion (X = chlorine or bromine) in cesium cobalt pentachloride and cesium cobalt pentabromide cyrstals by the effect of uniaxial stress on optical absorption lines. I. Experimental results. - J. Phys. Soc. Jpn., 1974, <u>37</u>, N5, 1353-1358. - CA, 1975, <u>82</u>, 9630. PA, 1975, <u>78</u>, 15988.

Exp.; E - e; T_d; 2.3.

2.4.85. Harada Masafumi, Tsujikawa Ikuji, Study of the initial tetragonal distortion of the $[CoX_4]^{2-}$ complex ion (X = chlorine or bromine) in cesium cobalt pentachloride and cesium cobalt pentabromide crystals by the effect of uniaxial stress on optical absorption lines. II. Analysis and estimation of the Jahn-Teller coupling constant. - J. Phys. Soc. Jpn., 1974, 37, N5, 1359-1366. - CA, 1975, 82, 23845. PA, 1975, 78, 15989.

Theor.; E - e; 1.5; 2.3.

2.4.86. Harding, M. J., Briat, B., Electronic absorption and magnetic circular dichroism spectra of the hexaaquochromium(III) ion. - Mol. Phys., 1974, 1974, 27, N5, 1153-1172. - CA, 1974, 81, 113000.

Exp.; T - t.

2.4.87. Kanno Ken-ichi, Nakai Yoshio, Dichroic optical absorption and thermal reorientation of Jahn-Teller distorted Ag^{2+} centers in $CdCl_2$. - In: Extended Abstr. Int. Conf. Color Centers in Ionic Crystals. Sendai, 1974, 174, 2 p. - PA, 1975, 78, 26156.

Exp.; E - e.

2.4.88. Linder, R. E., Bunnenberg, E., Seamans, L., Moscowitz, A., Calculation of vibronic magnetic rotational strengths in formaldehyde. - J. Chem. Phys., 1974, 60, N5, 1943-1951.

Theor.; 5.3.

2.4.89. Masunaga Shoji, Emura, S., Yamamoto, H., Fukuda Atsuo, Matsushima Akira, Hydrostatic pressure effect on the A-band emission in $KI:In^+$ and quadratic Jahn-Teller effect. - In: Extended Abstr. Int. Conf. Color Centers Ionic Crystals. Sendai, 1974, G132, 2 p. - PA, 1975, 78, 19007.

Exp.

2.4.90. Masunaga Shoji, Abe Takashi, Temperature dependence of the A band of thallium(+)-type centers in alkali crystals under high pressure. - J. Phys. Soc. Jpn., 1974, 37, N6, 1540-1544. - CA, 1975, 82, 49492. PA, 1975, 78, 13239.

Exp.

2.4.91. Matsushima Akira, Fukuda Atsuo, Calculation of the spectral bandshape and its change in magnetic fields for the transition $a^2{}_{eg} \rightarrow a_{eg}t_{eu}$ in Tl^+-type centers. - In: Extended Abstr. Int. Conf. Color Centers in Ionic Crystals. Sendai, 1974, 144, 2 p. - PA, 1975, 78, 26139.

Theor.; T.

2.4.92. Perlin, Yu. E., Tsukerblat, B. S., The method of moments in the theory of optical absorption. [Russ.]. - Usp. Fiz. Nauk, 1974, 113, N2, 343-345.

Theor.; T - (a + e + t); O_h; 2.1.

2.4.93. Robbins, D. J., Jahn-Teller instability and the magnetic circular dichroism of a tetragonal molecular system. - Theor. Chim. Acta, 1974, 33, N1, 51-62. - CA, 1974, 80, 150313. RZF, 1974, 8D258.

Theor.; E - b; C_{4v}.

2.4.94. Schairer, W., Schmidt, M., Strongly quenched deformation potentials of the manganese acceptor in gallium arsenide. - Phys. Rev. B, 1974, 10, N6, 2501-2505. - CA, 1975, 82, 9664. PA, 1975, 78, 18578.

Exp.; 1.5; 2.2; 2.3.

2.4.95. Vekhter, B. G., Rozenfel'd, Yu. B., Polinger, V. Z., Tsukerblat, B. S., Nonlinear optical and acoustic phenomena in Jahn-Teller systems. [Russ.]. - In: Tez. Dokl. 8 Vsesoyuz. Soveshch. Kvant. Akustike Akustoelektronike. Kazan', 1974, p. 109.

Theor.

2.4.96. Yeakel, W. C., Schatz, P. N., Ham effect in the $^2T_{2u}$ charge-transfer excited state of octahedral hexachloroiridate(IV). - J. Chem. Phys., 1974, 61, N2, 441-455. - CA, 1974, 81, 97368. PA, 1974, 77, 73127.

Exp.; T - t; O_h; 1.4.

2.4.97. Yeakel, W. C., Slater, J. L., Schatz, P. N., Confirmation of a Ham effect in the 22,900 cm^{-1} band of $Cs_2ZrCl_6:Ir^{4+}$. - J. Chem. Phys., 1974, 61, N11, 4868-4874. - CA, 1975, 82, 162168. PA, 1975, 78, 33879. RZF, 1975, 4D653.

Exp.; T; 1.4; 2.2; 2.3.

1975

2.4.98. Champagnon, B., Duval, E., Louat, R., Stress absorption spectra of vanadium(3+) in α-aluminum oxide: Jahn-Teller effect in the $^3T_2$ state. - Phys. Status Solidi b, 1975, 69, N2, 339-348. - CA, 1975, 83, 68526. PA, 1975, 78, 54475. RZF, 1975, 12D435.

Exp.; T - (a + e + t); C_{3v}; d^2.

2.4.99. Cox, P. A., Robbins, D. J., Day, P., Magnetic circular dichroism study of the Jahn-Teller effect in the $^1A_1 \to {}^1T_2$ change transfer band of MnO_4 in KIO4. - Mol. Phys., 1975, 30, N2, 405-411. - CA, 1975, 83, 199608. PA, 1975, 78, 77628. RZF, 1976, 4D502. RZK, 1976, 4B230.

Exp.; E - b.

2.4.100. Duval, E., Louat, R., Champagnon, B., Lacroix, R., Weber, J., Uniaxial stress effect on the $^1A_1 \to {}^1T_1$ transition of Co^{3+} in α-Al_2O_3. - J. Phys. [France], 1975, 36, N6, 559-569. - CA, 1975, 83, 68553. PA, 1975, 78, 58740.

Exp.; T; d^6.

2.4.101. Gatteschi, D., Krausz, E., Single crystal MCD [magnetic circular dichroism] of five-coordinate cobalt(II) and nickel-(II) complexes. - Chem. Phys. Lett., 1975, 34, N2, 348-351. - CA, 1975, 83, 154902. PA, 1975, 78, 81855.

Exp.

2.4.102. Giorgianni, U., Mondio, G., Perillo, P., Saitta, G., Vermiglio, G., Analysis of the electron-lattice interaction in KBr doped with thallium. [Fr.]. - Physica B + C, 1975, 79, N2, 175-191. - CA, 1975, 83, 105572. PA, 1975, 78, 77634. RZF, 1975, 12E1400.

Exp.; O_h.

2.4.103. Honma Ako, Theory of optical absorption and magnetic circular dichroism due to forbidden transitions of defect centers in solids. - Sci. Light, 1975, 24, N1, 33-56. - CA, 1975, 83, 105598.

Theor.; (A + E) - e; D_{4h}; 1.5.

2.4.104. Honma Akio, Theory of magnetic circular dichroism for the A band of thallium(+) ion-like centers in alkali halides. - Sci. Light, 1975, 24, N1, 57-61. - CA, 1975, 83, 105599.

Theor.

2.4.105. Honma Akio, Theory of stress dichroism in the optical absorption due to forbidden transitions of defect centers in solids. - Sci. Light, 1975, 24, N2, 75-96. - CA, 1976, 85, 11750. PA, 1976, 79, 58413.

Theor.

2.4.106. Kan'no Ken-ichi, Nakai Yoshio, Dichroic optical absorptions and thermal reorientation of Jahn-Teller distorted Ag^{2+}

centers in $CdCl_2$. - J. Phys. Soc. Jpn., 1975, 38, N5, 1420-1429. - CA, 1975, 83, 35226. PA, 1975, 78, 54874. RZF, 1975, 11D523.

Exp.; E - e; O_h; d^9.

2.4.107. Lemoyne, D., Duran, J., Yuste, M., Billardon, M., Excited states and electron-lattice coupling of low-symmetry F centers: the tetragonal F centers in BaClF. - J. Phys. C, 1975, 8, N10, 1455-1471. - CA, 1975, 83, 87570. PA, 1975, 78, 50541.

Exp.

2.4.108. Louat, R., Lacroix, R., Duval, E., Champagnon, B., Uniaxial stress effect on the $^4A_2 \rightarrow {}^4T_2$ zero-phonon transition of chromium(3+) in α-aluminum oxide. - Phys. Status Solidi b, 1975, 69, N1, 33-43. - CA, 1975, 83, 36078. PA, 1975, 78, 73182.

Exp.; E - $(e_E + e_T)$; C_{3v}; d_3.

2.4.109. Perlin, Yu. E., Kharchenko, L. S., Tsukerblat, B. S., Khoang Tkhi, Ngok Kam, Magnetic circular dichroism of Jahn-Teller electron-vibrational bands. [Russ.]. - In: Tez. Dokl. 8 Soveshch. Teorii Poluprovodnikov. Kiev, 1975, p. 134.

Theor.

2.4.110. Perlin, Yu. E., Tsukerblat, B. S., Piezoelectric spectroscopy of broad optical bands in systems with a dynamic Jahn-Teller effect. [Russ.]. - In: Spektrosk. Kristallov. Moscow, 1975, p. 61-72. - CA, 1976, 85, 11716.

Theor.; T.

2.4.111. Ranfagni, A., Pazzi, G. P., Fabeni, P., Bacci, M., Fontana, M. P., Viliani, G., Structured thallium-doped potassium chloride emission detected by electric field: Dynamical Jahn-Teller effect interpretation. - Phys. Rev. Lett., 1975, 35, N11, 752-754. - CA, 1975, 83, 155159. PA, 1975, 78, 86298.

Exp.; T - t.

2.4.112. Richardson, F. S., Hilmes, G., Jenkins, J. J., Influence of vibronic interactions on the chiroptical spectra of dissymmetric pseudotetragonal metal complexes. - Theor. Chim. Acta, 1975, 39, N1, 75-91. - CA, 1975, 83, 170213.

Theor.; (B + B) - e; D_{4h}; D_2; C_2; d^8; d^9.

2.4.113. Richardson, F. S., Hilmes, G., Theory of natural optical activity in crystalline copper(2+)-doped zinc selenate hexahydrate. - Mol. Phys., 1975, 30, N1, 237-255. - CA, 1975, 83, 154883. PA, 1975, 78, 69158.

Exp.; E - e; O_h; d^9; 1.5.

2.4.114. Richardson, F. S., Caliga, D., Hilmes, G., Jenkins, J. J., Vibronic effects in the chiroptical spectra of dissymetric trigonal systems. - Mol. Phys., 1975, 30, N1, 257-280. - CA, 1975, 83, 154884. PA, 1975, 78, 69159.

Theor.; (A + E) - e.

2.4.115. Schatz, P. H., MCD spectra of charge-transfer transitions: octahedral Ir^{4+}. - In: Electronic States Inorg. Compounds: New Exp. Techn. Dordrecht, 1975, p. 223-240. - PA, 1976, 79, 14533.

Exp.; 1.4.

2.4.116. Scheps, R., Lombardi, J. R., Field-induced perturbations in the ultraviolet spectrum of s-triazine. - Chem. Phys., 1975, 10, N2-3, 445-452. - CA, 1975, 83, 199868. PA, 1976, 79, 4834.

Exp.; E.

2.4.117. Shatwell, R. A., Gale, R., McCaffery, A. J., Sichel, K., Studies of the emitting states of some metalloporphyrins by magnetically induced circular emission (MCE). J. Am. Chem. Soc., 1975, 97, N24, 7015-7023. - CA, 1975, 83, 210903.

Exp.; E - $(b_1 + b_2)$; D_{4h}; d^8; d^9; 2.2.

2.4.118. Vasil'ev, A. V., Natadze, A. L., Ryskin, A. I., Piezospectroscopic effect on local and resonance vibrations linked with the degenerated electron state of an impurity center in a crystal. [Russ.]. - Pis'ma Zh. Eksp. & Teor. Fiz., 1975, 22, N1, 43-47. - CA, 1975, 83, 170271.

Exp.; T - e; T_d.

2.4.119. Vasil'ev, A. V., Malkin, B. Z., Natadze, A. L., Ryskin, A. I., Piezospectroscopic effect in an iron-doped zinc sulfide piezoelectric crystal. [Russ.]. - Fiz. Tverd. Tela, 1975, 17, N11, 3167-3173. - CA, 1976, 84, 36935.

Exp.; T; T_d; d^6.

1976

2.4.120. Craig, D. P., Stiles, P. J., Circular dichroism of D_{2d} molecules in chiral environments promoted by Jahn-Teller coupling. - Chem. Phys. Lett., 1976, 41, N2, 225-227. - CA, 1976, 85, 113961. PA, 1976, 79, 71619.

Theor.; D_{2d}; S_4.

2.4.121. Drotning, W. D., Drickamer, H. G., High-pressure optical studies of doped alkali halides. II. Jahn-Teller effects. - Phys. Rev. B, 1976, 13, N10, 4576-4585. - CA, 1976, 85, 54096. PA, 1976, 79, 69388.

Exp.; T - (e + t).

2.4.122. Fukuda Atsuo, Matsushima Akira, Masunaga Shoji, Relaxed excited states determined by Jahn-Teller effect in thallium(+)-type centers in alkali halides. - J. Lumin., 1976, 12/13, 139-149. - CA, 1976, 84, 187043. PA, 1976, 79, 54191.

Exp.; T - (e + t).

2.4.123. Fukuda Atsuo, Magnetic field effect on the triplet relaxed excited states responsible for the A_T and A_X emission bands of gallium(+) and indium(+) centers in alkali halides. - J. Phys. Soc. Jpn., 1976, 40, N3, 776-783. - CA, 1976, 84, 171682.

Exp.; T; O_h.

2.4.124. Fukuda Atsuo, Paus, H. J., Matsushima Akira, Magnetic circular dichroism of the A and B bands in $KI:Ga^+$, $KI:In^+$, and $KI:Sn^{2+}$. - Z. Phys. B, 1976, 25, N3, 211-218. - CA, 1977, 86, 10184.

Exp.; T; O_h.

2.4.125. Harada Masafumi, Tsujikawa Ikuji, The pseudo-Jahn-Teller effect in the excited state $^2E(^2D)$ of tetragonal crystals Cs_3CoCl_5 and Cs_3CoBr_5 studied by the effect of uniaxial stress and the Zeeman effect on absorption spectra. - J. Phys. Soc. Jpn., 1976, 40, N2, 513-523. - CA, 1976, 84, 113760. PA, 1976, 79, 31562. RZK, 1976, 15B632.

Exp.; E; 1.5.

2.4.126. Harada Masafumi, Tsujikawa Ikuji, The optical anisotropies and the mechanism of vibronic transitions in spectra $^4A_2 \rightarrow {}^2T_2(^2D)$ and $^2E(^2D)$ of Cs_3CoBr_5 and Cs_3CoCl_5 studied by the effect of uniaxial stress and the Zeeman effect. - J. Phys. Soc. Jpn., 1976, 41, N4, 1264-1272.

Exp.; T; T; 3.1.

2.4.127. Honma Akio, Ooaku Satoshi, Theory of optical absorption and magnetic circular dichroism for the vibrational-induced band (B-band) of Tl^+-like centers in alkali halides. - J. Phys. Soc. Jpn., 1976, 41, N1, 152-161. - CA, 1976, 85, 101471.

Theor.; (T + E) - (e + t); O_h.

2.4.128. Kielman-Van Luijt, E. C. M., Dekkers, H. P. J. M., Canters, C. W., MCD of transition to Jahn-Teller unstable states in the metalloporphins of Mg, Zn, Cu, Pd, and Pt. - Mol. Phys., 1976, 32, N4, 899-919. - PA, 1977, 80, 3391.

Exp.

2.4.129. Krylov, V. A., Ulrici, W., Effect of external electric field on the $^4A_{2g} \rightarrow {}^4T_{1g}(P)$ absorption band of the cobalt(2+) ion in calcium fluoride and cadmium fluoride with Jahn-Teller splitting. [Russ.]. - Fiz. Tverd. Tela, 1976, 18, N1, 275-277. - CA, 1976, 84, 113772. PA, 1976, 79, 87803.

Exp.; T - t; O_h.

2.4.130. Kushkuley, L. M., Perlin, Yu. E., Tsukerblat, B. S., Engel'gardt, G. R., Effects of dichroism in the absorption spectra of Jahn-Teller centers. [Russ.]. - Zh. Eksp. & Teor. Fiz., 1976, 70, N6, 2226-2235. - CA, 1976, 85, 101452. PA, 1976, 79, 84143. RZF, 1976, 10D375.

Theor.; T - (e + t).

2.4.131. Kushkuley, L. M., Perlin, Yu. E., Kharchenko, L. S., Tsukerblat, B. S., Piezospectroscopy of the wide absorption bands of impurity ions and color centers. [Russ.]. - In: Tez. Dokl. 5 Vsesoyuz. Symp. Spektrosk. Kristallov, Aktivir. Redkimi Zemlyami i Elementami Gruppy Zheleza. Kazan', 1976, p. 179.

Theor.; T - e; T - (e + t).

2.4.132. Lemoyne, D., Duran, J., Badoz, J., Vibronic model for an ns^2 system: $KCl:Au^-$. - J. Phys. [France], 1976, 37, Colloq. N7, 145-148. - CA, 1977, 86, 180023.

Exp.; T - (e + t).

2.4.133. Lemoyne, D., Duran, J., Billardon, M., Le Si Dang, Vibronic model for an ns^2 system: $KCl:Au^-$. - Phys. Rev. B, 1976, 14, N2, 747-753. - CA, 1976, 85, 133059. PA, 1976, 79, 84141.

Exp.; T - (e + t).

2.4.134. Muramatsu Shinji, Magnetic circular dichroism of the R_1 band in KCl. - J. Phys. C, 1976, 9, N21, 4057-4061. - PA, 1977, 80, 8551.

Theor.

2.4.135. Parrot, R., Curie, D., Piezospectroscopic study of $^4T_2$ levels of d^5 ions in tetrahedral symmetry: Jahn-Teller interactions and anomalous dipole strengths. - J. Lumin., 1976, 12/13, 811-817. - CA, 1976, 85, 11910. PA, 1976, 79, 58347.

Exp.; T; T_d; d^5; 2.3.

2.4.136. Perlin, Yu. E., Kharchenko, L. S., Tsukerblat, B. S., The Jahn-Teller effect in the optical bands of activated crystals. [Russ.]. - Izv. Akad. Nauk SSSR Ser. Fiz., 1976, 40, N9, 1770-1777. CA, 1977, 86, 48681. PA, 1978, 81, 6706. RZF, 1977, 1D515.

Theor.

2.4.137. Ruszczynski, G., Boyn, R., Jahn-Teller coupling parameters from stress-dichroism spectra for titanium(2+)-doped cadmium sulfide and cobalt(2+)-doped cadmium sulfide. - Phys. Status Solidi b, 1976, 76, N1, 427-436. - CA, 1976, 85, 101467. PA, 1976, 79, 73142.

Exp.; T - (e + t); d^2; d^7.

2.4.138. Stephens, P. J., Magnetic circular dichroism. - In: Adv. Chem. Phys. New York, etc., 1976, Vol. 35, p. 197-264. - CA, 1976, 85, 101340.

Rev.

2.4.139. Zapasskiy, V. S., Feofilov, P. P., Magnetooptical investigation of radiative states of divalent ytterbium in fluoride-type crystals. [Russ.]. - Opt. & Spektrosk., 1976, 41, N6, 1051-1055. - CA, 1976, 84, 171750.

Exp.; O_h.

1977

2.4.140. Bieg, K. W., Woracek, D. L., Drickamer, H. G., High-pressure luminescence studies of alkali-halide phosphors doped with lead(2+) and tin(2+). - J. Appl. Phys., 1977, 48, N2, 639-646. - CA, 1977, 86, 113247.

Exp.

2.4.141. Boyrivent, A., Duval, E., Louat, R., 0-Phonon lines associated with the broad band emission of $MgO:Cr^{3+}$. [Fr.]. - J. Phys. [France], 1977, 38, N4, L107-L110. - CA, 1977, 86, 130524.

Exp.; T; O_h; d^7; 2.3.

2.4.142. Brunshtein, S. Kh., Tsukerblat, B. S., The polarization dichroism of multiplet-multiplet optical transitions. [Russ.]. - In: Fiz. Mat. Metody Koord. Khim.: Tez. Dokl. 6 Vsesoyuz. Soveshch. Kishinev, 1977, p. 44.

Theor.; E; T.

2.4.143. Brunshtein, S. Kh., Elektoronno-Kolebatel'nye Polosy Jahn-Tellerovskikh Kompleksov s Nezapolnennymi di-Obolochkamy: Avtoref. Dis. ... Kand. Fiz.-Mat. Nauk. [Russ.]. - Kishinev, 1977. - 16 p. - Kishinev. Univ.

The Electron-Vibrational Bands of Jahn-Teller Complexes with Open d-Shells.

Theor.

2.4.144. Champagnon, B., Duval, E., Emission spectrum of V^{3+}-α-Al_2O_3. - J. Phys. [France], 1977, 38, N14, L299-L301. - CA, 1977, 87, 124900.

Exp.; T; O_h; d^6; 2.3.

2.4.145. Czarniecki, S., Witkowski, A., Vibronic structure of the circular dichroism and absorption spectra of bianthryls. - Chem. Phys. Lett., 1977, 47, N1, 28-31. - CA, 1977, 86, 179766.

Exp.; 1.5.

2.4.146. Czarniecki, S., Witkowski, A., Zgierski, M. Z., Theoretical analysis of UV, CD, and ORD spectra of two 1,1'-bianthryl derivatives. - Acta Phys. Pol. A, 1977, 51, N3, 451-462. - CA, 1977, 87, 67464.

Exp.

2.4.147. Davis, G., Nazare, M. H., Pseudo-Jahn-Teller effects in H3 vibronic spectrum in diamond. - In: Int. Conf. Radiat. Eff. Semicond., Dubrovnik, 1976. London, 1977, p. 332-338. - CA, 1977, 87, 92916. PA, 1977, 80, 90182.

Exp.; 1.5; 2.3.

2.4.148. Douglas, I. N., Magnetic circular dichroism of Co^{2+} ions in CdF_2 and CaF_2. - Phys. Status Solidi b, 1977, 79, N2, K99-K101. - PA, 1977, 80, 28603.

Exp.; T; d^7.

2.4.149. Fournier, D., Boccara, A. C., Rivoal, J. C., Magnetooptical study of $^4T_1$ state of Mn^{2+} in ZnS. - J. Phys. C, 1977, 10, N1, 113-121. - CA, 1977, 87, 13650. PA, 1977, 80, 20833.

Exp.; T - e; d^5; 2.3.

2.4.150. Hjortsberg, A., Nygren, B., Vallin, J., Ham, F. S., Jahn-Teller effect in the ^5E_g state of iron(2+) in magnesium oxide: observation of tunneling splitting and test of the vibronic model for an E-state Jahn-Teller ion. - Phys. Rev. Lett., 1977, 39, N19, 1233-1236. - CA, 1977, 87, 208996. PA, 1978, 81, 19517.

Exp.; E - e; d^6; 1.4; 2.2; 2.3.

2.4.151. Imanaka Koichi, Iida Takeshi, Ohkura Hiroshi, Vibronic theory of the Stark effects on the relaxed excited state of the F center. - J. Phys. Soc. Jpn., 1977, 43, N2, 519-528. - RZF, 1978, 1D498.

Theor.; T - t.

2.4.152. Kaplyanskiy, A. A., Medvedev, V. N., Smolyanskiy, P. L., Strain polarization of luminescence and the structure of the radiative state of CaF_2-Yb^{2+} crystals. [Russ.]. - Opt. & Spektrosk., 1977, 42, N1, 136-143. - CA, 1977, 86, 130305.

Exp.; 2.3.

2.4.153. Kazanskiy, S. A., The application of the induced dichroism for the investigation of the vibronic spectra of Eu^{2+} and Sm^{2+} in cubic crystals CaF_2. [Russ.]. - Opt. & Spektrosk., 1977, 42, N2, 318-326.

Exp.; T - (a + e + t); 2.2.

2.4.154. Krylov, V. A., Ulrici, W., Influence of an electric field on the Jahn-Teller split absorption band $^4A_{2g} \rightarrow {}^4T_{1g}(P)$ of cobalt(2+) in cadmium fluoride and calcium fluoride. - Phys. Status Solidi b, 1977, 84, N1, 215-225. - CA, 1978, 88, 29835.

Exp.; T; d^7.

2.4.155. Kushkuley, L. M., Perlin, Yu. E., Tsukerblat, B. S., Piezo- and magnetooptics of spin-forbidden transitions in Jahn-Teller impurity centers. [Russ.]. - Fiz. Tverd. Tela, 1977, 19, N7, 2178-2188. - CA, 1977, 87, 108891. RZF, 1977, 11E300.

Theor.; E - e; T - (a + e + t).

2.4.156. Kushkuley, L. M., Perlin, Yu. E., Tsukerblat, B. S., Polarization spectroscopy of forbidden transitions in optical impurity absorption. [Russ.]. - In: Tez. Dokl. 18 Vsesoyuz. S"ezda Spektrosk., Gorkiy, 1977, p. 168-171.

Theor.; E - e; T - (a + e + t).

2.4.157. Kushkuley, L. M., Effekty Polyarizatsionnogo Dikhroizma v Elektronno-Kolebatel'nykh Spektrakh Aktivirovannykh Kristallov: Avtoref. Dis. ... Kand. Fiz.- Mat. Nauk. - Kishinev, 1977. - 20 p. - Kishinev. Univ.

Polarization Dichroism Effects in Electron-Vibrational Spectra of Activated Crystals.

Theor.; E - e; T - (a + e + t); 1.5.

2.4.158. Masunaga Shoji, Goto Nobuyuki, Matsushima Akira, Fukuda Atsuo, Hydrostatic pressure effects on the triplet relaxed excited states of KI:Tl^+-type phosphors through the quadratic Jahn-Teller interaction. - J. Phys. Soc. Jpn., 1977, 43, N6, 2013-2020. - CA, 1978, 88, 56671. PA, 1978, 81, 23510. RZF, 1978, 6D1009.

Exp.; T - (e + t).

2.4.159. Parrot, R., Naud, C., Dynamic Jahn-Teller effect on the fine structure lines of transition metal ions. - In: Spectrosopie des elements lourds dan les solids. Paris, 1977, p. 109-117. (Colloq. Int. CNRS, N255.) - RZF, 1978, 2D442.

Exp.; T; T_d; d^3 - d^7.

2.4.160. Perlin, Yu. E., Tsukerblat, B. S., Dod, T. Singh, Moments of electroabsorption of Jahn-Teller impurity centers. - Phys. Status Solidi b, 1977, 80, N2, 703-707. - CA, 1977, 86, 180087. PA, 1977, 80, 48205.

Theor.

2.4.161. Shekhtman, V. L., Quadratic Srark effect for crystal impurity centers. III. The Jahn-Teller effect in crystals. [Russ.]. - Opt. & Spektrosk., 1977, 43, N3, 458-465. - CA, 1978, 88, 43301. PA, 1978, 81, 68536.

Theor.; T.

2.4.163. Singh, Dod T., Perlin, Yu. E., Tsukerblat, B. S., Electro-, piezo-, and magnetooptics of low-symmetry complexes in cubic crystals. [Russ.]. - In: Fiz. Mat. Metody Koord. Khim.: Tez. Dokl. 6 Vsesoyuz. Soveshch. Kishinev, 1977, p. 43-44.

Theor.

2.4.163. Singh, Dod T., Perlin, Yu. E., Tsukerblat, B. S., Electrooptic absorption of anisotropic Jahn-Teller centers in cubic crystals. [Russ.]. - Fiz. Tverd. Tela, 1977, 19, N6, 1569-1579. - PA, 1978, 81, 27188.

Theor.

2.4.164. Singh, Dod T., Momenty Jahn-Tellerovskikh Elektronno-Kolebatel'nykh Spectrov: Avtoref. Dis. ... Kand. Fiz.-Mat. Nauk. [Russ.]. - Kishinev, 1977. - 18 p. - Kishinev. Univ.

The Moments of the Jahn-Teller Electron-Vibrational Spectra.

Theor.; T - t; E - e; d^2.

2.4.165. Tsujikawa Ikuji, Harada Masafumi, Effect of uniaxial stress on absorption spectra of tetragonal crystals of cesium pentachlorocobaltate(II) and cesium pentabromocobaltate(II). - Rev. Roum. Chim., 1977, 22, N9-10, 1305-1313. - CA, 1978, 88, 29904.

Exp.; E.

2.4.166. Tsukerblat, B. S., Singh, Dod T., Perlin, Yu. E., The polarization dichroism of the electro-optical impurity absorption in wurtzite type crystals. [Russ.]. - In: Fiz. Protsessy v Poluprovodnikakh. Kishinev, 1977, p. 3-15.

Theor.; T - (a + e + t); C_{3v}.

2.4.167. Vasil'ev, A. V., Ryskin, A. I., Piezoelectric spectroscopic effect on electron-vibrational lines of the $^5E \to {}^5T_2$ band of the Fe^{2+} ion in a ZnS-Fe crystal. [Russ.]. - Opt. & Spektrosk., 1977, 42, N6, 1116-1120. - CA, 1977, 87, 92909. RZF, 1977, 10D604.

Exp.; T; O_h; d^6; 2.3.

2.4.168. Wong, K. Y., Manson, N. B., Osborne, G. A., Magnetic circular dichroism of transition metal ions in solids. II. $K_2NaGaF_6{:}Cr^{3+}$. - J. Phys. & Chem. Solids, 1977, 38, N9, 1017-1022. - CA, 1978, 88, 29699.

Exp.; T - t; d^3.

1978

2.4.169. Boyrivent, A., Duval, E., Broad-band luminescence in $MgO{:}Cr^{3+}$: Relaxed level on emission processes. - J. Phys. C, 1978, 11, N2, 439-448.

Exp.; T; O_h; d^7; 2.3.

2.4.170. Canters, G. W., Van der Waals, J. H., High-resolution Zeeman spectroscopy of metalloporphyrins. - In: The Porphyrins. New York, 1978, Vol. 3, p. 531-582.

Rev.; E - $(b_1 + b_2)$; D_{4h}; d^0; d^{10}.

2.4.171. Czarniecki, S., Witkowski, A., Optical activity of molecular dimers: Theoretical analysis of vibronic circular dichroism in model biaryls. - Acta Phys. Pol. A, 1978, 53, N2, 235-242. - CA, 1978, 88, 143615.

Theor.; 1.5.

2.4.172. Drickamer, H. G., Klick, D. I., High-pressure studies of luminescent phenomena. - J. Less-Common Met., 1978, 62, 381-396. - CA, 1979, 90, 129818. PA, 1979, 82, 43017. RZF, 1979, 6D979.

Exp.; 2.2.

2.4.173. Fournier, D., Boccara, A. C., Rivoal, J. C., Excited-state magnetic circular dichroism from the $^4T_1$ state of Mn^{++} ions. - Semicond. & Insul., 1978, 3, N2-3, 105-112. - CA, 1978, 88, 179550. PA, 1978, 81, 43907.

Exp.; T; T_d; d^5.

2.4.174. Hennel, A. M., Uba, S. M., Piezoabsorption measurements of GaAs:Co^{2+}: Tunnel splitting and Jahn-Teller effect in the $^4T_1(P)$ state. - J. Phys. C, 1978, 11, N22, 4565-4581. - CA, 1979, 90, 159347.

Exp.; T - (e + t); d^7.

2.4.175. Jansen, G., Noort, M., Canters, G. W., Van der Waals, J. H., Jahn-Teller instability of the first excited singlet state of zinc porphine: a study by Zeeman spectroscopy in some n-alkane host crystals at 4.2 K. - Mol. Phys., 1978, 35, N1, 283-294. - CA, 1978, 89, 67772. PA, 1978, 81, 27256.

Exp.; D_{4h}.

2.4.176. Klick, D., Drickamer, H. G., High-pressure studies of Jahn-Teller split luminescence in alkali halides doped with indium(+) and thallium(+). - Phys. Rev. B, 1978, 17, N3, 952-963. - CA, 1978, 88, 179709.

Exp.

2.4.177. Klokishner, S. I., Effekty Elektron-Fononnogo Vzaimodeystviya v Spektrakh Primesnykh Tsentrov i Eksitonov Malogo Radiusa: Avtoref. Dis. ... Kand. Fiz.-Mat. Nauk. [Russ.]. - Kishinev, 1978. - 16 p. - Kishinev. Univ.

Effects of Electron-Phonon Interaction on the Spectra of Impurity Centers and Small Radius Exitons.

Theor.

2.4.178. Klokishner, S. I., Perlin, Yu. E., Tsukerblat, B. S., The polarization dichroism of optical spectra of Jahn-Teller exitons. [Russ.]. - Fiz. Tverd. Tela, 1978, 20, N11, 3201-3210. - PA, 1979, 82, 69135. RZF, 1979, 2E1079.

Theor.; T - (a + e + t); O_h.

2.4.179. Koning, R. E., Znadvoort, H., Zandstra, P. J., MCD of the triphenylene anion radical: Influence of ion-pairing and Jahn-Teller effect. - Chem. Phys., 1978, 28, N3, 343-355. - CA, 1978, 88, 179539.

Exp.

2.4.180. Le Si Dang, Romestain, R., Simkin, D., Systematic study of magnetic circular dichroism of ns^2 ions in KBr and KI. - Phys. Rev. B, 1978, 18, N11, 6313-6323. - CA, 1979, 90, 129677.

Exp.; T - (a_1 + e + t_2); O_h.

2.4.181. Lowther, J. E., Jahn-Teller coupling at ND1 and GR1 centers in diamond. - J. Phys. C, 1978, 11, N9, L373-L380. - CA, 1978, 89, 171951.

Theor.; T - t; 1.4.

2.4.182. Parrot, R., Naud, C., Porte, C., Fournier, D., Boccara, A. C., Rivoal, J. C., Jahn-Teller effect in the fluorescent level of manganese(2+) in zinc selenide and zinc sulfide. - Phys. Rev. B, 1978, 17, N3, 1057-1066. - CA, 1978, 88, 179710.

Exp.; T - e; T_d; d^5.

2.4.183. Perlin, Yu. E., Tsukerblat, B. S., Singh, Dod T., Anisotropic Jahn-Teller impurity centers in cubic crystals. [Russ.]. - In: Spektrosk. Kristallov. Leningrad, 1978, p. 11-27. - RZF, 1979, 4D467.

Theor.; E; C_{3v}; C_{4v}; D_3; S_6; T_d; O_h.

2.4.184. Sharonov, Yu. A., Mineyer, A. P., Sharonova, N. A., Figlovskii, V. A., The differences in vibronic interactions in the active site of nitrosyl derivatives of ferrous hemoproteins as revealed by magnetic circular dichroism. - In: Pap. 11th IUPAC Int. Symp. Chem. Nat. Prod. Sofia, 1978, Vol. 2, p. 286-289. - CA, 1979, 91, 205895.

Exp.

2.4.185. Skowronski, M., Baranowski, J. M., Ludwicki, L. J., Optical properties of a zinc manganese telluride ($Zn_{0.9}Mn_{0.1}Te$) solid solution. [Pol.]. - Pr. Inst. Fiz. PAN, 1978, 75, 127-132. - CA, 1979, 90, 14151.

Exp.; T; 2.2.

2.4.186. Tsukerblat, B. S., Brunshtein, S. Kh., Effects of polarization dichroism in multiphonon optical bands of multiplet-multiplet transitions. [Russ.]. - Fiz. Tverd. Tela, 1978, 20, N4, 1220-1222. - CA, 1978, 89, 33569. PA, 1979, 82, 15125. RZF, 1978, 9D778.

Theor.; T - (e + t).

2.4.187. Viliani, G., Montagna, M., PIlla, O., Fontana, A., Bacci, M., Ranfagni, A., Magnetic circular dichroism in the split zero-phonon line of $MgO:V^{2+}$. - J. Phys. C, 1978, 11, N10, L439-L443.

Exp.; T; d^3; 2.3.

2.4.188. Wilson, R. B., Solomon, E. I., Spectroscopic studies of the photoactive $^4T_{2g}$ excited state of hexaamminechromium(III). - Inorg. Chem., 1978, 17, N7, 1729-1736. - CA, 1978, 89, 67780.

Exp.; T - e; O_h; 1.4.

1979

2.4.189. Bermudez, V. M., McClure, D. S., Spectroscopic studies of the two-dimensional magnetic insulators chromium trichloride and chromium tribromide. I. - J. Phys. & Chem. Solids, 1979, 40, N2, 129-147. - CA, 1979, 91, 99307.

Exp.; T; O_h; d^3; 1.5.

2.4.190. Boyrivent, A., Duval, E., Montagna, M., Viliani, G., Pilla, O., New experimental results for the interpreation of the $^4A_{2g} \rightarrow {}^4T_{2g}$ spectra of Cr^{3+} and V^{2+} in MgO. - J. Phys. C, 1979, 12, N20, L803-L807. - CA, 1980, 92, 118800. PA, 1980, 83, 11441.

Exp.; T - (e + t); O_h; d^3; 2.3.

2.4.191. Craig, D. P., Stiles, P. J., Palmieri, P., Zauli, C., Electronic circular dichroism of twisted [4,4-spirononatetraene]. - J. Chem. Soc. Faraday Trans. II, 1979, 75, N1, 97-104. - CA, 1979, 91, 73996.

Exp.; E - e.

2.4.192. Drickamer, H. G., High-pressure luminescence studies. - 1979. - 7 p. - Report COO-1198-1270, CONF-790709-15. - Energy Res. Abstr., 1979, 4, N22, 57782. CA, 1980, 92, 101756.

Exp.; 2.2.

2.4.193. Fabeni, P., Mugnai, D., Pazzi, G. P., Ranfagni, A., Traniello, R., Simkin, D. J., Negative magnetic circular polarization in the emission of Jahn-Teller systems. - Phys. Rev. B, 1979, 20, N8, 3315-3318. - CA, 1980, 92, 49670. PA, 1980, 83, 20694.

Exp.; T - e; O_h.

2.4.194. Fuke Kiyokazu, Schepp, O., Absorption and magnetic circular dichroism spectra of allene. - Chem. Phys., 1979, 38, N2, 211-216. - CA, 1979, 91, 29899.

Exp.; E.

2.4.195. Kamishina, Y., Sivasankar, V. S., Jacobs, P. W. M., Method of Gaussian quadrature in the calculation of optical absorption and magnetic circular dichroism spectra of s^2 ions in alkali halide crystals: application to $KBr:In^+$. - Can. J. Phys., 1979, 57, N10, 1614-1623. - RZF, 1980, 3D465.

Exp.; $(T_{1u} + T_{2u} + E_u) - (e_g + t_{2g})$; O_h.

2.4.196. Kielman-Van Luijt, E. C. M., Canters, G. W., Linear polarization studies of the high-resolution absorption spectrum of palladium porphin. - J. Mol. Spectrosc., 1979, 78, N3, 469-485. - CA, 1980, 92, 197364.

Exp.

2.4.197. Kushkuley, L. M., Tsukerblat, B. S., Polarization dichroism of optical bands of Jahn-Teller centers: Forbidden transitions induced by lattice vibrations. [Russ.]. - Fiz. Tverd. Tela, 1979, 21, N8, 2254-2263. - CA, 1979, 91, 148695. PA, 1980, 83, 55062.

Theor.; E; T.

2.4.198. Malakhovskiy, A. V., Edel'man, I. S., Zabluda, V. N., The Jahn-Teller effect in the $^6A_{1g} \rightarrow {}^4E_g(^4G)$ transition in iron borate ($FeBO_3$). [Russ.]. - Fiz. Tverd. Tela, 1979, 21, N7, 2164-2166. - CA, 1979, 91, 131635. PA, 1980, 83, 54662. RZF, 1979, 10E1606.

Exp.; E.

2.4.199. Mineyev, A. P., Sharonov, Yu. A., Sharonova, N. A., Figlovsky, V. A., Orbit-orbit interaction in the NO-Fe(II) hemoproteins studied by low-temperature magnetic circular dichroism: dynamic Jahn-Teller effect detected by MCD. - Int. J. Quantum Chem., 1979, 16, N4, 883-889. - PA, 1980, 83, 30645.

Exp.; E; D_{4h}; d^6.

2.4.200. Mineyev, A. P., Nizkotemperaturnyy Magnitnyy Krugovoy Dikhroism Gemsoderzhashchikh Belkov: Avtoref. Dis. ... Kand. Fiz.-Mat. Nauk. [Russ.]. - Moscow, 1979. - 22 p. - Mosc. Gos. Univ.

Low Temperature Magnetic Circular Dichroism of Hemoproteins.

Exp.

2.4.201. Mowery, R. L., Miller, J. C., Krausz, E. R., Schatz, P. N., Jacobs, S. M., Andrews, L., Magnetic circular dichroism of the $^1S \rightarrow {}^1P$ transition of magnesium atoms in noble gas matrixies. - J. Chem. Phys., 1979, 70, N8, 3920-3926. - CA, 1979, 91, 11495.

Exp.; T - t.

2.4.202. Rivoal, J. C., Briat, B., Vala, M., Magnetic circular and linear dichroism of cubic or tetragonally distorted high-spin d^5 systems. II. Study of tetrahalomanganates. - Mol. Phys., 1979, 38, N6, 1829-1854. - CA, 1980, 92, 173522. PA, 1980, 83, 29725.

Exp.; 2.3.

2.4.203. Ryskin, A. I., Vasil'ev, A. V., Natadze, A. L., Stress-induced alignment of Jahn-Teller center Cr^{2+} in ZnSe crystals: Optical study. - Phys. Status Solidi b, 1979, 93, N2, 459-467. - CA, 1979, 91, 81163. PA, 1979, 82, 78651.

Exp.; d^4.

2.4.204. Simkin, D., Martin, J. P., Le Si Dang, Kamishina, Y., Determination of excited state symmetry and g-value from the magnetic field dependence of the radiative decay time: $KI:In^+$ and $KI:Sn^{2+}$. - Chem. Phys. Lett., 1979, 65, N3, 569-573.

Exp.; T - e; 2.2; 3.1.

2.4.205. Tsuboi Taiju, Magnetic field effects on the A band in thallium(+1) ion-doped potassium chloride crystals. - J. Chem. Phys., 1979, 71, N12, 5376-5377. - CA, 1980, 92, 67010.

Exp.; T - (a + e + t).

2.4.206. Vasil'ev, A. V., Natadze, A. L., Ryskin, A. I., The dynamical aspect of the static Jahn-Teller effect in crystals of GaP-Cr and GaAs-Cr. [Russ.]. - In: Fizika Soedin. A^3B^5: Vsesoyuz. Konf. Leningrad, 1979, p. 63-66. - RZF, 1980, 8E1442.

Exp.; T - (e + t); O_h; d^4.

2.4.207. Vasil'ev, A. V., Ryskin, A. I., Piezodichroism induced by reorientation of Jahn-Teller centers in ZnSe-Cr crystals. [Russ.]. - Fiz. Tverd. Tela, 1979, 21, N4, 1054-1059. - CA, 1979, 91, 29782. PA, 1980, 83, 11363.

Exp.; d^4.

2.4.208. Yamaguchi Hiroyuki Akimoto Yoshinori, Ikeda Teruki, Yoneda Fumio, Magnetic circular dichroism spectra and photoelectron spectra of tricycloquinazoline. - J. Chem. Soc. Faraday Trans. II, 1979, 75, N11, 1506-1514.

Exp.; E - e; 1.4; 1.7; 2.5; 2.7; 2.8.

2.4.209. Yokoyama Yuko, Koshizuka Naoki, Okuda Takashi, Tsushima Tachiro, Magneto-optical effects in the cadmium dichromium-(III) sulfide-iron(II) dichromium(III) sulfide magnetic semiconductor series. [Jap.]. - Denshi Gijutsu Sogo Kenkyusho Iho, Bull. Electrotech. Lab., 1979, 40, N11, 817-827. - CA, 1977, 86, 149730.

Exp.; T; T_d; d^6.

2.4.210. Zgierski, M. Z., Pawlikowski, M., Jahn-Teller pseudo-Jahn-Teller coupling, and circular dichroism spectra of (E + A)* e systems. - J. Chem. Phys., 1979, 70, N7, 3444-3452. - CA, 1979, 90, 212486. PA, 1979, 82, 68247.

Theor.; (A + E) - e; 1.5.

See also: 2.1.-100, 134, 189, 251; 2.2.-65, 115, 121, 132, 135; 2.3.-41, 53, 89, 93, 125; 2.7.-8, 29, 33, 40; 4.3.-19, 21, 24, 25, 30, 35, 47, 49, 55, 71.

2.5. Photoelectron and X-Ray Spectroscopy

Photo- and X-ray spectra demonstrate the Jahn-Teller vibronic structure due to degenerate ground and/or excited states. The important influence on the band shape is inherent in the continuous spectra (see Ref. [2.5.3]). Photoelectronic spectra of some hydrocarbons and their analogues were examined in Refs. [2.5.-1, 4, 5, 6, 7, 16, 17]. The results concerning transition metal complexes are given in Ref. [2.5.2]. The effect of spin-oribt coupling reducing the Jahn-Teller effect is noted in Ref. [2.5.8]. The Renner type interaction in H_2S^+ was considered in Ref. [2.5.10] on the basis of optical and photoelectronic measurements. The experimental work [2.5.18] shows that the Jahn-Teller splitting of the A → E - e transition in some aromatic systems is important, similar effects being observed in sulfur trioxide. The Jahn-Teller vibronic interaction is manifest in the X-ray elastic scattering [2.5-22, 29].

In Ref. [2.5.24] the photoelectronic spectra of ethane (^2E ground state) was considered. Photodissociation spectra induced by laser sources are obtained in Ref. [2.5.27] (the E - e problem is discussed also in Refs. [2.5.-32, 33, 34, 37, 46]).

Photoelectronic spectra of binuclear mixed-valence complexes are considered in Ref. [2.5.28], where the results are discussed using the concept of the pseudo-Jahn-Teller effect.

1968

2.5.1. Baker, A. D., Baker, C., Brundle, C. R., Turner, D. W., The electronic structures of methane, ethane, ethylene, and formaldehyde studied by high-resolution molecular photoelectron spectroscopy. - Int. J. Mass Spectrom. & Ion Phys., 1968, 1, N4/5, 285-301. - CA, 1969, 70, 31766. RZF, 1974, 2D357.

Exp.

2.5.2. Davis, T. S., Fackler, J. P., Weeks, M. J., Spectra of manganese(III) complexes: Origin of the low-energy band. - Inorg. Chem., 1968, 7, N10, 1994-2002. - CA, 1968, 69, 101237.

Exp.; E; O_h; d^4.

2.5.3. Price, W. C., Wilkinson, G. R., Developments in molecular photoelectron spectroscopy. I. Studies by far infrared and Raman spectroscopy of intermolecular vibrations in solids and liquids. - 1968. - 89 p. - Clearinghouse Fed. Sci. Tech. Infrom., AD 689854. - CA, 1970, 72, 26826.

Exp.; 2.7; 2.8.

1969

2.5.4. Baker, C., Turner, D. W., Photoelectron spectra of allene and ketene: Jahn-Teller distortion in the ionization of allene. - J. Chem. Soc. Ser. D, 1969, N9, 480-481. - CA, 1969, 71, 16106.

Exp.

1970

2.5.5. Asbrink, L., Lindholm, E., Edqvist, O., Jahn-Teller effect in the vibrational structure of the photoelectron spectrum of benzene. - Chem. Phys. Lett., 1970, 5, N9, 609-612. - CA, 1970, 73, 55308. PA, 1970, 73, 53062.

Exp.; D_{6h}.

2.5.6. Brundle, C. R., Robin, M. B., Basch, H., Electronic energies and electronic structures of the fluoromethanes. - J. Chem. Phys., 1970, 53, N6, 2196-2213. - CA, 1970, 73, 93037.

Exp.

2.5.7. Pullen, B. P., Carlson, T. A., Moddeman, W. E., Schweitzer, G. K., Bull, W. E., Grimm, F., Photoelectron spectra of methane, silane, germane, methyl fluoride, difluoromethane, and trifluoromethane. - J. Chem. Phys., 1970, 53, N2, 768-782. - CA, 1970, 73, 60789.

Exp.

2.5.8. Ragle, J. L., Stenhouse, I. A., Frost, D. C., McDowell, C. A., Valence shell ionization potentials of halomethanes by photoelectron spectroscopy. I. Methyl chloride, methyl bromide, methyl iodide: Vibrational frequencies and vibronic interaction in CH_3Br^+ and CH_3Cl^+. - J. Chem. Phys., 1970, 53, N1, 178-184. - CA, 1970, 73, 48624.

Exp.

1971

2.5.9. Carlson, T. A., Anderson, C. P., Angular distribution of the photoelectron spectrum for benzene. - Chem. Phys. Lett., 1971, 10, N5, 561-564. - CA, 1972, 76, 19896.

Exp.; E; D_{6h}.

2.5.10. Dixon, R. N., Duxbury, G., Horani, M., Rostas, J., H_2S^+ radical ion comparison of photoelectron and optical spectroscopy. - Mol. Phys., 1971, 22, N6, 977-992. - CA, 1972, 76, 105972.

Exp.; 1.6; 2.1.

2.5.11. Rabalais, J. W., Bergmark, T., Werme, L. O., Karlsson, L., Siegbahn, K., The Jahn-Teller effect in the electron spectrum of methane. - Phys. Scr., 1971, 3, N1, 13-18. - CA, 1971, 75, 68983. PA, 1971, 74, 35830. RZF, 1972, 1D386.

Exp.; T; T_d.

2.5.12. Rowland, C. G., Jahn-Teller coupling and the photoelectron spectrum of cyclopropane. - Chem. Phys. Lett., 1971, 9, N2, 169-173. - CA, 1971, 75, 42648. PA, 1971, 74, 35807.

Exp.; 5.4.

1972

2.5.13. Bergmark, K. T., Rabalais, J. W., Werme, L. O., Karlsson, L., Siegbahn, K., High-resolution electron spectra of methane, thiophene, 2-bromothiophene, and 3-bromothiophene. - In: Proc. Int. Conf. Electron Spectrosc. Amsterdam, 1972, p. 413-423. - PA, 1972, 75, 35340.

Exp.; T.

2.5.14. Brundle, C. R., Kuebler, N. A., Robin, M. B., Basch, H., Ionization potentials of the tetraphosphorus molecule. - Inorg. Chem., 1972, 11, N1, 20-25. - CA, 1972, 76, 50329.

Exp.

2.5.15. Evans, S., Joachim, P. J., Orchard, A. F., A study of the orbital electronic structure of the P_4 molecule by photoelectron spectroscopy. - Int. J. Mass Spectrom. & Ion Phys., 1972, 9, N1, 41-49. - PA, 1972, 75, 46574.

Exp.

2.5.16. Potts, A. W., Price, W. C., Streets, D. G., Williams, T. A., Photoelectron spectra of benzene and fluorobenzenes. - Discuss. Faraday Soc., 1972, N54, 168-181. - CA, 1975, 84, 88984.

Exp.; E.

1973

2.5.17. Brogli, F., Crandall, J. K., Heilbronner, E., Kloster-Jensen, E., Sojka, S. A., The photoelectron spectra of methyl-substituted allenes and of tetramethylbis(allenyl). - J. Electron Spectrosc. & Relat. Phenom., 1973, 2, N5, 455-465. - CA, 1974, 80, 42678. PA, 1974, 77, 36571.

Exp.; 1.5.

2.5.18. Buezzli, J. C., Frost, D. C., Weiler, L., Photoelectron spectrum of triquinacene. - Tetrahedron Lett., 1973, N14, 1159-1162. - CA, 1973, 79, 17671.

Exp.; E - e; D_{3h}.

2.5.19. Foster, S., Felps, S., Cusachs, L. C., McGlynn, S. P., Photoelectron spectra of osmium and ruthenium tetroxides. - J. Am. Chem. Soc., 1973, 95, N17, 5521-5524. - CA, 1973, 79, 110093. PA, 1974, 77, 6118.

Exp.; T.

2.5.20. Heilbronner, E., Gleiter, R., Hoshi Toshihiko, DeMeijere, A., Interaction of Walsh orbitals in diademane and related hydrocarbons. - Helv. Chim. Acta, 1973, 56, N5, 1594-1604. - CA, 1974, 80, 2910.

Theor.; 1.5.

2.5.21. Nishikawa Shigeaki, Watanabe Tsutomu, Superexcited states of methane. - Chem. Phys. Lett., 1973, 23, N3, 590-594. - CA, 1974, 80, 32320. PA, 1974, 77, 5994.

Theor.; T; T_d.

1974

2.5.22. Kazantseva, L. A., Ogurtsov, I. Ya., Elastic scattering of X-rays by systems with degenerate electronic terms. [Russ.]. - Izv. Akad. Nauk Mold.SSR, Ser. Biol. Khim. Nauk, 1974, N4, 67-71.

Theor.; E - e.

2.5.23. Mines, G. W., Thomas, R. K., The photoelectron spectrum of sulfur trioxide: Jahn-Teller distortion in SO_3^+. - Proc. R. Soc. London, Ser. A, 1974, 336, N1606, 355-364. - CA, 1974, 80, 54339. PA, 1974, 77, 19899. RZF, 1974, 6D382.

Exp.; E; D_{3h}.

2.5.24. Rabalais, J. W., Katrib Ali, Electron states of ethane(1+) ($C_2H_6^+$). - Mol. Phys., 1974, 27, N4, 923-931. - CA, 1974, 81, 62713.

Exp.; E - e.

2.5.25. Thomas, R. K., Thompson, H., The photoelectron spectra of allene, deuterioallenes, and tetrafluoroallene. - Proc. R. Soc. London, Ser. A, 1974, 339, N1616, 29-36. - CA, 1974, 81, 119429. PA, 1974, 77, 56295.

Exp.; 2.7; 2.8.

1975

2.5.26. Botz, F. K., Glick, R. E., Methane temporary negative ion resonances. - Chem. Phys. Lett., 1975, 33, N2, 279-283. - CA, 1975, 83, 105937. PA, 1975, 78, 65099.

Exp.; T_d.

2.5.27. Freiser, B. S., Beauchamp, J. L., Laser photodissociation of benzene radical cations: Evidence for a two-photon process involving a long-lived intermediate. - Chem. Phys. Lett., 1975, 35, N1, 35-40. - CA, 1975, 83, 177823.

Exp.; D_{6h}.

2.5.28. Hush, N. S., Inequivalent XPS binding energies in symmetrical delocalized mixed-valence complexes. - Chem. Phys., 1975, 10, N2/3, 361-366. - RZF, 1976, 3E206.

Theor.; 1.5.

2.5.29. Kazantseva, L. A., Ogurtsov, I. Ya., The elastic scattering of X-rays by systems with degenerate electronic terms. [Russ.]. - In: Kvant. Khim.: Tez. Dokl. 6 Vsesoyuz. Soveshch. Kishinev, 1975, B-36.

Theor.; E - e; T_d; O_h.

2.5.30. Schmidt, W., Photoelectron spectrum of adamantane: Reply to comments. - J. Electron Spectrosc. & Relat. Phenom., 1975, 6, N2, 163-166. - CA,1975, 82, 162623.

Exp.

2.5.31. Schwarz, W. H. E., How to determine vertical ionization potentials from photoelectron spectra. - J. Electron Spectrosc. & Relat. Phenom., 1975, 6, N5, 377-390. - CA, 1975, 83, 15787. RZF, 1975, 11D307.

Theor.

1976

2.5.32. Dewar, M. J. S., Fonken, G. J., Jones, T. B., Photoelectron spectra of molecules. Part VII. Cyclopropyallallenes. - J. Chem. Soc. Perkin Trans. II, 1976, N6, 764-767. - CA, 1976, 85, 76958.

Exp.; E - e.

2.5.33. Price, W. C., Potts, A. W., Williams, T. A., The orbital interpretation of the photoelectron spectrum of benzene, 1,3,5-trifluorobenzene, and hexafluorobenzene. - Chem. Phys. Lett., 1976, 37, N1, 17-19. - CA, 1976, 84, 67571. PA, 1976, 79, 21378.

Exp.; E - e; D_{6h}.

1977

2.5.34. Bieri, G., Burger, F., Heilbronner, E., Maier, J. P., Valence ionization energies of hydrocarbons. - Helv. Chim. Acta, 1977, 60, N7, 2213-2233. - CA, 1978, 88, 37039.

Exp.; E - e; D_{3h}.

2.5.35. Cederbaum, L. S., Domcke, W., Köppel, H., Van Niessen, W., Strong vibronic coupling effects in ionization spectra: the "mystery band" of butatriene. - Chem. Phys., 1977, 26, N2, 169-177. - CA, 1978, 88, 61723.

Theor.; 1.5; 5.3.

2.5.36. Domcke, W., Cederbaum, L. S., Vibronic coupling and symmetry breaking in core electron ionization. - Chem. Phys., 1977, 25, N2, 189-196. - CA, 1978, 88, 13727. RZF, 1978, 7D167.

Theor.; 1.5.

2.5.37. Engelking, P. C., Lineberger, W. C., Laser photoelectron spectrometry of $C_5H_5^-$: A determination of the electron affinity and Jahn-Teller coupling in cyclopentadienyl. - J. Chem. Phys., 1977, 67, N4, 1412-1417. - CA, 1977, 87, 167025. PA, 1977, 80, 88499.

Exp.; E - e; D_{5h}.

2.5.38. Leng, F. J., Nyberg, G. L., The angular-distribution helium (He I)/neon (Ne I) photoelectron spectra of cyclopropane. - J. Electron Spectrosc. & Relat. Phenom., 1977, 11, N3, 293-299. - CA, 1977, 87, 76111.

Exp.

2.5.39. Richartz, A., Buenker, K. J., Bruna, P. J., Peyerimhoff, S. D., Stability and structure of the ethane cation: investigation of the photoelectron spectrum of ethane below 14 eV using ab initio methods. - Mol. Phys., 1977, 33, N5, 1345-1346. - CA, 1977, 87, 133609.

Theor.; D_{3d}; 5.4.

1978

2.5.40. Bally, T., Buser, U., Haselbach, E., Tetrakis(methylidene) cyclobutane-[4]-radialene: electronic states of the molecular ion. - Helv. Chim. Acta, 1978, 61, N1, 38-45. - CA, 1978, 88, 113032.

Exp.; E - (b_1 + b_2).

2.5.41. Cederbaum, L. S., Domcke, W., Jahn-Teller effect induced by non-degenerate vibrational modes in cumulenes. - Chem. Phys., 1978, 33, N3, 319-326. - CA, 1979, 90, 54078. PA, 1979, 82, 817.

Theor.

2.5.42. Kobayashi Tsunetoshi, A simple general tendency in photoelectron angular distributions of some monosubstituted benzenes. - Phys. Lett. A, 1978, 69, N2, 105-108. - CA, 1979, 90, 86204.

Exp.

2.5.43. Koenig, T., Chang, J. C., Helium(I) photoelectron spectrum of tropyl radical. - J. Am. Chem. Soc., 1978, 100, N7, 2240-2242. - CA, 1978, 89, 23311.

Exp.; E; D_{7h}.

2.5.44. Koppel H., Domcke, W., Cederbaum, L. S., von Niessen, W., Vibronic coupling effects in the photoelectron spectrum of ethylene. - J. Chem. Phys., 1978, 69, N9, 4252-4263. - CA, 1979, 90, 71318.

Theor.; 1.5.

2.5.45. Kovac, B., Klasinc, L., Photoelectron spectroscopy of adamantane and some adamantanones. - Croat. Chem. Acta, 1978, 51, N1, 55-74. - CA, 1978, 89, 120413.

Exp.

2.5.46. Robin, M. B., Kuebler, N. A., Electronic structure and spectra of small rings. VI. Multiphonon ionization spectra of the saturated three-membered rings. - J. Chem. Phys., 1978, 69, N2, 806-810. - CA, 1979, 90, 71284.

Exp.; E - e.

1979

2.5.47. Domcke, W., Vibrational state dependence of the photoelectron angular asymmetry parameter caused by vibronic coupling. - Phys. Scr., 1979, 19, N1, 11-15. - CA, 1979, 90, 159762.

Theor.

See also: 2.4.-208; 5.4.-38, 48.

2.6. Electron Scattering Spectroscopy

Electron scattering spectra contain information about electronic and vibronic states. Intensive investigations of Jahn-Teller systems started about 20 years ago. In Ref. [2.6.1] the presence of a strong dynamic Jahn-Teller effect was observed in the system CaF_2-YF_2 (^2E term). The electron scattering theory, taking into account the pseudo-Jahn-Teller effect in the excited state, was developed in Ref. [2.6.2] and applied to benzene in Ref. [2.6.3]. The E - e problem is discussed in Refs. [2.6.-14, 13, 20, 21, 22]. The Jahn-Teller effect in VCl_4 and xenon hexafluoride was examined in Refs. [2.6.4] and [2.6.5]. Hexafluorodies of other metals were investigated in Refs. [2.6.7]. The electronic scattering theory for the $\Sigma \rightarrow \Pi$ transition in linear Renner type systems was developed in Refs. [2.6.-8, 9]. Some simple hydrocarbons were investigated in Ref. [2.6.10]; the spectra of transition metal complexes are presented in Ref. [2.6.11].

1964

2.6.1. O'Connor, J. R., Chen, J. H., Energy levels of d^1 electrons in CaF_2: evidence of strong dynamic Jahn-Teller distortions. - Appl. Phys. Lett., 1964, 5, N5, 100-102. - CA, 1964, 61, 14004h.

Exp.; E; T_d; d^1; 2.1; 3.1.

2.6.2. Read, F. H., Electron impact spectroscopy. II. Vibronic interactions in excitation cross sections. - Proc. Phys. Soc. London, 1964, 83, Pt. 4, N534, 619-627. - CA, 1964, 60, 12820c.

Theor.; 1.5.

1965

2.6.3. Read, F. H., Whiterod, G. L., Electron impact spectroscopy. III. Calculated cross sections for inelastic scattering from benzene. - Proc. Phys. Soc. London, 1965, 85, Pt. 1, N543, 71-77. - CA, 1965, 62, 3548g.

Theor.; 1.5.

1966

2.6.4. Morino Yonezo, Uehara Hiromichi, Vibronic interactions in vanadium tetrachloride by gas-phase electron diffraction. - J. Chem. Phys., 1966, 45, N12, 4543-4551. - CA, 1967, 66, 32091. PA, 1967, 70, 6701.

Exp.

1968

2.6.5. Bartell, L. S., Gavin, R. M., Jr., Molecular structure of xenon hexafluoride. II. Internal motion and mean geometry deduced by electron diffraction. - J. Chem. Phys., 1968, 48, N6, 2466-2483. - CA, 1968, 68, 117798.

Exp.; T - t; O_h.

2.6.6. Ducros, P., Possible interpretation by the Jahn-Teller effect of permanent deformations detected by diffraction of slow electrons on the dense faces of silicon and germanium. [Fr.]. - Surface Sci., 1968, 10, N2, 295-298. - CA, 1968, 69, 30919. PA, 1968, 71, 26917.

Exp.

2.6.7. Kimura Masao, Schomaker, V., Smith, D. W., Weinstock, B., Electron-diffraction investigation of the hexafluorides of tungsten, osmium, iridium, uranium, neptunium, and plutonium. - J. Chem. Phys., 1968, 48, N9, 4001-4013. - CA, 1968, 69, 39487.

Exp.; T; O_h; d^0; d^2; d^3; f^0; f^1; f^2.

1971

2.6.8. Kiselev, A. A. Obyedkov, V. D., Yurova, I. Yu., The Renner effect in electron-molecular collisions. - In: Proc. 7th ICPEAC, 1971, p. 1062-1063.

Theor.; 1.6.

2.6.9. Kiselev, A. A., Petelin, A. N., Yurova, I. Yu., Possible observation of the Renner effect in molecules by electron scattering. [Russ.]. - Vest. Leningr. Univ., 1971, N4, Fiz. & Khim., N1, 114-116. - CA, 1971, 75, 56022. RZF, 1971, 9D164.

Theor.; 1.6.

1973

2.6.10. Harshbarger, W. R., Lassettre, E. N., On the electron impact spectra of CH_4 and CF_4. - J. Chem. Phys., 1973, 58, N4, 1505-1513. - CA, 1973, 78, 117379. PA, 1973, 76, 23543.

Exp.; T_d.

1976

2.6.11. Almenningen, A., Gard, E., Haaland, A., Brunvoll, J., Dynamic Jahn-Teller effect and average structure of dicyclopentadienylcobalt $(C_5H_5)_2CO$, studied by gas-phase electron diffraction. -

J. Organomet. Chem., 1976, 107, N2, 273-279. - CA, 1976, 84, 120650. RZK, 1976, 13B104.

Exp.

2.6.12. Kazantseva, L. A., Ogurtsov, I. Ya., Ishchenko, A. A., Taking into account the vibronic interaction in the determination of the structure of octahedral molecules by electronnographic methods. [Russ.]. - In: Tez. Dokl. Vsesoyz. Seminara Izucheniyu Stroenya Opredeleniyu Molek. Postoyannykh Prostykh Kompleks. Soedin. Gazovoy Faze. Moscow, 1976, p. 52-53.

Theor.

1977

2.6.13. Ogurtsov, I. Ya., Vibronic interaction in the inelastic scattering of electrons. [Russ.]. - In: Fiz. Mat. Metody Koord. Khim.: Tez. Dokl. 6 Vsesoyuz. Soveshch. Kishinev, 1977, p. 150.

Theor.; 1.5.

2.6.14. Ogurtsov, I. Ya., Kazantseva, L. A., Ishchenko, A. A., Vibronic effects in elastic scattering of electrons. - J. Mol. Struct., 1977, 41, N2, 243-251. - CA, 1978, 88, 12202.

Theor.; E - e.

1978

2.6.15. Challis, L. J., De Goer, A. M., Escribe, C., Fletcher, J. R., Hutchings, M. T., Jefferies, D. J., Sheard, F. W., Toombs, G. A., The observation of a critically coupled spin-phonon mode in chromium-doped magnesium oxide. - In: Proc. Int. Conf. Lattice Dynamics, 1977. Paris, 1978, p. 261-264. - CA, 1979, 90, 92670. PA, 1978, 81, 67779.

Exp.; d^4.

2.6.16. Spiridonov, V. P., Ishchenko, A. A., Zasorin, E. Z., Electron diffraction study of stereochemically flexible molecules. [Russ.]. - Usp. Khim., 1978, 47, N1, 101-126. - Bibliogr.: 201 Ref. CA, 1978, 88, 110634.

Rev.; T_d; O_h; 1.5; 1.6.

1979

2.6.17. Challis, L. H., Fletcher, J. R., Jefferies, D. J., Sheard, F. W., Toombs, G. A., De Goer, A. M., Hutchings, M. T., Investigation of critically coupled spin-phonon modes in chromium-doped MgO. - Phys. Rev. B, 1979, 19, N1, 296-299. - CA, 1979, 90, 178341.

Exp.; O_h; d^4.

2.6.18. Dixon, J. M., On the dynamic Jahn-Teller effect of Er^{3+} in Pd-Er. - Solid State Commun., 1979, 30, N10, 657-659. - CA, 1979, 91, 62914. PA, 1979, 82, 73478.

Theor.; T - t; 3.1.

2.6.19. Ivashkevich, L. S., Ishchenko, A. A., Spiridonov, V. P., Romanov, G. V., Vibronic interactions and molecular structure of gaseous vanadium tetrabromide: an electron diffraction study. - J. Mol. Struct., 1979, 51, N2, 217-227. - CA, 1979, 90, 127936. PA, 1979, 82, 31711.

Exp.; E.

2.6.20. Kazantseva, L. A., Ogurtsov, I. Ya., The influence of molecular electronic degeneracy on the scattering of X-rays and hot electrons. [Russ.]. - In: Sovrem. Sostoyanye Teorii Atomov i Molekul: Tez. Dokl. Konf. Teorii Atomov i Molekul. Vil'nius, 1979, p. 101.

Theor.; E - e; 2.5.

2.6.21. Ogurtsov, I. Ya., Kazantseva, L. A., Angular dependence of inelastic scattering of electrons by systems with the pseudo-Jahn-Teller effect. [Russ.]. - Zh. Strukt. Khim., 1979, 20, N3, 414-419. - CA, 1979, 91, 129213. PA, 1980, 83, 69246.

Theor.; (A + T) - t; O_h; 1.5.

2.6.22. Ogurtsov, I. Ya., Kazantseva, L. A., On the theory of scattering of fast electrons by Jahn-Teller molecules. - J. Mol. Struct., 1979, 55, N2, 301-307. - CA, 1979, 91, 149078. PA, 1979, 82, 89784.

Theor.

See also: 1.6.-13; 2.7.-57; 4.1.-65, 66, 118, 135; 4.4.-31, 33, 110, 113, 145, 148; 4.5.-22; 4.6.-42, 70, 71, 72, 73, 85, 93, 102, 119, 124, 125, 133, 134, 135, 139, 146, 152, 153, 157, 159, 162, 173, 176, 177, 178, 179, 180; 4.7.-6, 9, 15, 23, 24, 30, 33, 35, 36, 38; 5.2.-18, 26. 69.

2.7. Vibronic Infrared Spectra

The infrared spectra are observed in the spectral region of molecular or crystal lattice frequencies, where the Jahn-Teller effect is essential. In Ref. [2.7.2] the forbidden transitions induced by the Jahn-Teller effect were considered. In the case of strong vibronic coupling, the absorption is accompanied by tunneling processes. In Ref. [2.7.3] the E - e problem is discussed taking into account quadratic interaction, and the infrared spectra of acetylacetonates are considered. The complex structure of infrared bands, arising due to the Jahn-Teller effect, was detected in some octahedral hexafluorides [2.7.-4, 5, 6, 7, 11, 12, 15, 20, 35, 83].

The rotational and vibrational spectra of symmetric top molecules (X_3 type molecule, linear E - e problem) were considered in Rev. [2.7.8]. Inversion (tunneling) splitting of infrared absorption and Raman scattering spectra of octahedral transition metal complexes were predicted in the theoretical work [2.7.9].

The effects of spin-orbit coupling are discussed in Ref. [2.7.10] (linear vibronic E - e coupling, $\Gamma_8 - (e + t_2)$ problem), both the intensities and frequencies of the infrared absorption spectra being determined (see also Ref. [2.7.14]). The spectra of dimeric systems are considered in the well-known work [2.7.16]. The case of a strong Jahn-Teller effect has been investigated in Ref. [2.7.18] within the framework of the inversion (tunneling) splitting theory. In Ref. [2.7.23] the semiclassic theory is applied to the KCl:Tl^{0} system. The transition metal ions having degenerate ground states in the crystal field give rise to a complicated infrared spectra, which is due to the transitions between the vibronic states [2.7.-28, 29, 31, 32, 33]. The case of moderate vibronic coupling is of special interest and occurs in the A_2B_6 type crystals [2.7.-36, 40, 41, 46, 53, 60, 61, 67, 78, 85] and in other crystals and transition metal complexes [2.7.-55, 56, 57, 58, 59, 62, 63, 64, 68, 71, 72, 73]. Hydrogen bond spectroscopy [2.7.65] gives rise to additional problems.

1954

2.7.1. Gaunt, J., The vibrational spectrum and molecular structure of rhenium hexafluoride. - Trans. Faraday Soc., 1954, 50, N2, 209-213. - CA, 1954, 48, 13425h.

Exp.

1958

2.7.2. Thorson, W. R., Vibronically induced absorption of forbidden infrared transitions. - J. Chem. Phys., 1958, 29, N1, 938-944. - CA, 1959, 53, 3868f.

Theor.

1959

2.7.3. Forman, A., Orgel, L. E., Jahn-Teller effect in manganic acetylacetonate. - Mol. Phys., 1959, 2, N4, 362-366. - CA, 1961, 55, 25465e. RZF, 1960, N8, 21569.

Exp.; E - e; D_3; d^4.

2.7.4. Moffitt, W., Goodman, G. L., Fred, M., Weinstock, B., The color of transition metal hexafluorides. - Mol. Phys., 1959, N2, 109-122. - RZF, 1960, N3, 7212.

Exp.; T - (e + t); Γ_8 - e; O_h; 2.1.

2.7.5. Weinstock, B., Claassen, H. H., Jahn-Teller effect in the vibrational spectra of hexafluorides. - J. Chem. Phys., 1959, 31, N1, 262-263. - CA, 1960, 54, 2003f. PA, 1959, 62, 10763.

Exp.; O_h; 2.1.

1960

2.7.6. Eisenstein, J. C., Spectrum of ReF_6. - J. Chem. Phys., 1960, 33, N5, 1530-1531. - CA, 1961, 55, 14058b. RZF, 1961, 5B85.

Exp.; O_h.

2.7.7. Weinstock, B. H., Claassen, H., John, G. M., Vibrational spectra of OsF_6 and PtF_6. - J. Chem. Phys., 1960, 32, N1, 181-185. - CA, 1960, 54, 11698b.

Exp.; O_h; 2.8.

1961

2.7.8. Child, M. S., Longuet-Higgins, H. C., Studies of the Jahn-Teller effect. III. The rotational and vibrational spectra of symmetric-top molecules in electronically degenerate states. - Philos. Trans. R. Soc. London A, 1961, 254, N1041, 259-294. - CA, 1963, 58, 12086h. RZF, 1962, 6B61.

Theor.; E - e; D_{3h}; 2.8.

1962

2.7.9. Bersuker, I. B., Vekhter, B. G., The band splitting of infrared absorpiton and Raman spectra in octahedral transition metal complexes resulting from intrinsic asymmetry. [Russ.]. - Izv. Akad. Nauk Mold.SSR, 1962, N10, 11-17. - RZF, 1963, 7D76. CA, 1964, 60, 3618c.

Theor.; 2.8.

2.7.10. Child, M. S., Studies of the Jahn-Teller effect. IV. The vibrational spectra of spin-degenerate molecules. - Philos. Trans. R. Soc. London A, 1962, 255, N1050, 31-53. - CA, 1965, 62, 7253a. RZF, 1963, 4D119.

Theor.; E - e; Γ_8 - (e + t); 1.5; 2.8.

2.7.11. Claassen, H. H., Selig, H., Malm, J. G., Vibrational spectra of MoF_6 and TcF_6. - J. Chem. Phys., 1962, 36, N11, 2888-2889. - CA, 1962, 57, 11983i. RZF, 1963, 1D144.

Exp.; O_h; 2.8.

2.7.12. Claassen, H. H., Malm, J. G., Selig, H., Vibrational spectra of ReF_6. - J. Chem. Phys., 1962, 36, N11, 2890-2892. - CA, 1962, 57, 11986a.

Exp.; O_h; Γ_8; 2.8.

2.7.13. Dingle, R., The Jahn-Teller effect in manganic acetylacetonate. - J. Mol. Spectrosc., 1962, 9, N5, 426-427. - CA, 1963, 58, 4036e.

Exp.; E - e; O_h.

1963

2.7.14. Child, M. S., Anomalous spectroscopic properties accompanying a weak dynamical Jahn-Teller effect. - J. Mol. Spectrosc., 1963, 10, N5, 357-365. - CA, 1963, 59, 5930f. RZF, 1964, 3D51.

Theor.; E - $(b_1 + b_2)$; E - e; T - (e + t); Γ_8 - (e + t).

2.7.15. Weinstock, B., Claassen, H. H., Chernik, C. L., Jahn-Teller effect in the E_g vibrational mode of hexafluoride molecules: The infrared spectra of RuF_6 and RhF_6. - J. Chem. Phys., 1963, 38, N7, 1470-1475. - CA, 1963, 58, 10877b.

Exp.; E; T; Γ_8; O_h; d^1; d^2; d^3.

1964

2.7.16. Fulton, R. L., Gouterman, M., Vibronic coupling. II. Spectra of dimers. - J. Chem. Phys., 1964, 41, N8, 2280-2286. - CA, 1964, 61, 12805d. PA, 1965, 68, 2933.

Theor.; 1.5.

1965

2.7.17. Bader, R. F. W., Huang Kun Po, Jahn-Teller effect in the vibrational spectra of pentachlorides. - J. Chem. Phys., 1965, 43, N10, 3760-3761. - CA, 1966, 64, 7540d.

Exp.; E - e; 2.8.

2.7.18. Bersuker, I. B., Vekhter, B. G., The microwave "inversion" spectrum of the transition metal complexes. [Russ.]. - In: Tr. Komis. Spektrosk. Akad. Nauk SSSR. Vyp. I: Mater. 15 Soveshch. Spektrosk., Minsk, 1963. Moscow, 1965, Vol. 3, p. 520-528. - RZF, 1965, 11D101.

Theor.; E; T; T_d; O_h; $d^1 - d^9$.

2.7.19. Meret-Bailly, J., Symmetries of the proper functions and the selection rules in symmetric top molecules: application to spherical top molecules. [Fr.]. - Cahier Phys., 1965, 19, 253-265. - PA, 1966, 69, 17542.

Theor.; 2.8.

2.7.20. Weinstock, B., Goodman, G. L., Vibrational properties of hexagluoride molecules. - In: Adv. Chem. Phys., New York, etc., 1965, Vol. 9, p. 169-316. - CA, 1966, 64, 7388g. RZF, 1966, 8D248.

Theor.; 1.2; 2.8.

1966

2.7.21. Haas, H., Sheline, R. K., On the Jahn-Teller effect in vanadium hexacarbonyl. - J. Am. Chem. Soc., 1966, 88, N14, 3219-3220. - CA, 1966, 65, 8200e.

Exp.; O_h.

2.7.22. Müller, R., Günthard, H. H., Spectroscopic study of the reduction of nickel and cobalt ions in sapphire. - J. Chem. Phys., 1966, 44, N1, 365-373. - CA, 1966, 64, 6076g. RZF, 1966, 8D383.

Exp.; E; d^7.

1967

2.7.23. Knox, R. S., Semiclassical theory of the near-infrared spectrum of $KCl:Tl^0$. - Phys. Rev., 1967, 154, N3, 799-801. - CA, 1967, 66, 89838.

Exp.

2.7.24. Nelson, E. D., Wong, J. Y., Schawlow, A. L., Far infrared spectra of $Al_2O_3:Cr^{3+}$ and $Al_2O_3:Ti^{3+}$. - Phys. Rev., 1967, 156, N2, 298-308. - CA, 1967, 67, 58926. RZF, 1967, 12D385.

Exp.; d^1; d^3.

2.7.25. Yeranos, W. A., On the theory of vibronic interactions. - Z. Naturforsch. a, 1967, 22, N2, 183-187. - CA, 1968, 68, 73605. PA, 1967, 70, 25741.

Theor.

1968

2.7.26. Bersuker, I. B., On the possibility of a dynamically wide range Maser effect on inversion (tunneling) levels. - Phys. Lett. A, 1968, 28, N5, 357.

Theor.; E; 3.3.

2.7.27. Joyce, R. R., Richards, P. L., Far-infrared spectra of alumina doped with titanium, vanadium, and chromium. - 1968. 24 p. - US At. Energy Comm., UCRL-18229. - CA, 1969, 70, 24327.

Exp.; C_{3v}; d^1.

2.7.28. Joyce, R. R., Richards, P. L., Far-infrared spectra of $Al_2O_3:Ti^{3+}$ and $Al_2O_3:V^{4+}$. - Bull. Am. Phys. Soc., 1968, 13, N8, 435.

Exp.; E - e; d^1.

2.7.29. Macfarlane, R. M., Wong, J. Y., Sturge, M. D., Dynamic Jahn-Teller effect in octahedrally coordinated d^1 impurity systems. - Phys. Rev., 1968, 166, N2, 250-258. - CA, 1968, 68, 91453. PA, 1968, 71, 16990. RZF, 1968, 7E576.

Theor.; T - e; O_h; d^1; 3.1.

2.7.30. Vekhter, B. G., Electric dipole transitions between inversion states in complexes having no center of symmetry. [Russ.]. - Izv. Akad. Nauk Mold.SSR, Ser. Biol. Khim. Nauk, 1968, N6, 81-85.

Theor.; T_d.

2.7.30a. Zvyagin, A. I., Skorobogatova, I. V., Kut'ko, V. I., Infrared absorption spectrum of the cobaltous ion in a crystalline field of octahedral symmetry. [Russ.]. - Ukr. Fiz. Zh. 1968, 13, N1, 83-86.

1969

2.7.31. Brabers, V. A. M., Infrared spectra of cubic and tetragonal manganese ferrites. - Phys. Status Solidi, 1969, 33, N2, 563-572. - CA, 1969, 71, 26209. PA, 1969, 72, 39841.

Exp.; T_d.

2.7.32. Ham, F. S., Schwarz, W. M., O'Brien, M. C. M., Jahn-Teller effects in the far-infrared, EPR and Mössbauer spectra of $MgO:Fe^{2+}$. - Phys. Rev., 1969, 185, N2, 548-567. - CA, 1970, 72, 16948. PA, 1970, 73, 21476.

Exp.; T - (e + t); O_h; d^6; 1.4; 3.1; 3.2; 3.4.

2.7.33. Joyce, R. R., Richards, P. L., Far-infrared spectra of Al_2O_3 doped with Ti, V, and Cr. - Phys. Rev., 1969, 179, N2, 375-380. - PA, 1969, 72, 31100.

Exp.

2.7.34. King, G. W., Warren, C. H., Fluorosulfate radical: Rotational analysis of the 5160-Å system and molecular geometry. - J. Mol. Spectrosc., 1969, 32, N1, 138-150. - CA, 1969, 71, 107181.

Exp.

2.7.35. Kiseljov, A. A., Jahn-Teller effect in hexafluorides. - J. Phys. B, 1969, 2, N2, 270-273. - CA, 1969, 70, 72328. PA, 1969, 72, 34647. RZF, 1969, 8D296.

Theor.; O_h.

2.7.36. Vallin, J. T., Slack, G. A., Roberts, S., Hughes, A. E., Near and far infrared absorption in Cr-doped ZnSe. - Solid State Commun., 1969, 7, N17, 1211-1214.

Exp.; d^4.

1970

2.7.37. Burdett, J. K., Interpretation of the vibrational behavior of methyl and substituted methyl radicals perturbed by alkali halide molecules. - J. Mol. Spectrosc., 1970, 36, N3, 365-375. - CA, 1971, 74, 36476. RZF, 1971, 6D217.

Theor.; C_{3v}; 1.5.

2.7.38. Hancock, R. D., Thornton, D. A., Effect of d-orbital splitting on the metal-ligand frequencies of first transition period metal(II) complexes of acetylacetone. - J. S. Afr. Chem. Inst., 1970, 23, N2, 71-76. - CA, 1970, 73, 125281.

Exp.

2.7.39. Percy, G. C., Thornton, D. A., The application of infrared spectra to structural problems in copper(II) complexes of 2,2'-bipyridine. - Spectrosc. Lett., 1970, 3, N11/12, 323-328. - CA, 1971, 75, 12825. RZF, 1971, 10D355.

Exp.; E - e.

2.7.40. Vallin, J. T., Slack, G. A., Bradley, C. C., Far-infrared absorption of $ZnS:Fe^{2+}$ in strong magnetic fields. - Phys. Rev. B, 1970, 2, N11, 4406-4413. - CA, 1971, 74, 69711. RZF, 1971, 7D481.

Exp.; T_d; d^6.

2.7.41. Vallin, J. T., Slack, G. A., Roberts, S., Hughes, A. E., Infrared absorption in some II-VI compounds doped with Cr. - Phys. Rev. B, 1970, 2, N11, 4313-4334. - CA, 1971, 74, 69718.

Exp.; T - e; d^4.

1971

2.7.42. Bersuker, I. B., Ogurtsov, I. Ya., Shaparev, Yu. V., Vibrational and rotational spectra of symmetric "dipolarly unstable" polyatomic molecules. [Russ.]. - In: Molek. Spektrosk.: Tez. Dokl. 17 Vsesoyuz. S"ezda Spektrosk. Minsk, 1971, p. 29.

Theor.; (A + T) - t; T_d.

2.7.43. Bersuker, I. B., Ogurtsov, I. Ya., Shaparev, Yu. V., The dipole moments in symmetric "dipolarly unstable" molecules. [Russ.]. - In: Elektr. Svoystva Molekul: Tez. Dokl. Vsesoyz. Konf. Khar'kov, 1971, p. 6-7.

Theor.; T - t; D_{3h}; T_d.

2.7.44. Creighton, J. A., Vibrational studies of the Jahn-Teller effects. - In: Essays Struct. Chem. New York, 1971, p. 355-382. - Bibliogr. 70 ref. CA, 1974, 80, 126175.

Rev.

2.7.45. Fox, K., Theory of pure rotational transitions in the vibronic ground state of methane. - Phys. Rev. Lett., 1971, 27, N5, 233-236. - CA, 1971, 75, 81901.

Theor.; T_d.

2.7.46. Ham, F. S., Slack, G. A., Infrared absorption and luminescence spectra of Fe^{2+} in cubic zinc sulfide: Role of the Jahn-Teller coupling. - Phys. Rev. B, 1971, 4, N3, 777-798. - CA, 1971, 75, 82036. PA, 1971, 74, 53237.

Exp.; T - e; T_d; d^6; 2.2.

2.7.47. Hulett, L. G., Thornton, D. A., Infrared spectra of first transition series metal tropolonates. - Spectrochim. Acta A, 1971, 27, N1, 2089-2096. - CA, 1971, 75, 156543.

Exp.; d^4; d^9.

2.7.48. Shephard, G. S., Thornton, D. A., Infrared spectra of α-thenoyltrifluoroacetonates of metal ions of the first transition series. - Helv. Chim. Acta, 1971, 54, N7, 2212-2221. - CA, 1972, 76, 19799.

Exp.

1972

2.7.49. Bottger, G. L., Salwin, A. E., Vibrational spectra of alkali salts of hexahaloiridates. - Spectrochim. Acta A, 1972, 28, N5, 925-931. - CA, 1972, 76, 160283.

Exp.; 2.8.

2.7.50. Csaszar, J., Horvath, E., $[CuCl_5]^{3-}$ ion. - Acta Chim. Acad. Sci. Hung., 1972, 71, N2, 167-175. - CA, 1972, 76, 119291.

Exp.; E; T; T_d; D_{3h}; 2.1.

2.7.51. Fox, K., Theory of microwave transitions in the vibronic ground state of tetrahedal molecules. - Phys. Rev. A, 1972, 6, N3, 907-919. - CA, 1972, 77, 106996.

Theor.; T_d.

2.7.52. Ishii, M., Nakahira, M., Yamanaka, T., Infrared absorption spectra and cation distributions in $(Mn, Fe)_3O_4$. - Solid State Commun., 1972, 11, N1, 209-212. - CA, 1972, 1972, 77, 94879. PA, 1972, 75, 61648.

Exp.

2.7.53. Koswig, H. D., Retter, U., Ulrici, W., Jahn-Teller effect on transition metal ions in silver halides. II. $Cr^{2+}(3d^4)$. - Phys. Status Solidi b, 1972, 51, N1, 123-128. - CA, 1972, 77, 11737. PA, 1972, 75, 40656. RZF, 1972, 10E838.

Exp.; E - e; d^4.

2.7.54. Peytavin, S., Brun, G., Cot, L., Maurin, M., Vibrational study of sodium divalent metal double sulfate and selenate dihydrates, $Na_2M(SO_4)_2 \cdot 2H_2O$ (M = chromium, cadmium, manganese, copper), $Na_2M(SeO_4)_2 \cdot 2H_2O$ (M = cadmium, manganese). [Fr.]. - Spectrochim. Acta A, 1972, 28, N10, 1995-2003. - CA, 1973, 78, 9698. PA, 1972, 75, 83270.

Exp.; T_d; d^1; d^9.

2.7.55. Tomar, V. S., Ganguli, P., Bist, H. D., Manoharan, P. T., Jahn-Teller distortions and optical phonons in hexanitrocuprates. - In: Proc. Nucl. Phys. & Solid State Phys. Symp., Bombay, 1972, S.1., 1972, Vol. 14C, p. 541-544. - CA, 1973, 78, 166451. RZF, 1973, 5E911.

Exp.; E; O_h; d^9; 3.1.

1973

2.7.56. Akyuz, S., Dempster, A. B., Morehouse, R. L., Suzuki, S., Infrared and Raman spectroscopic study of some metal pyridine tetracyanonickelate complexes. - J. Mol. Struct., 1973, 17, N1, 105-125. - CA, 1973, 79, 85216. PA, 1973, 76, 63467.

Exp.; d^9.

2.7.57. Armstrong, J. R., Chadwick, B. M., Jones, D. W., Sarneski, J. E., Wilde, H. J., Yerkess, J., Single crystal vibrational spectroscopic and neutron diffraction studies of the hexacyano-metallates, $Cs_2LiM(CN)_6$ (M = Cr, Mn, Fe, and Co). - Inorg. & Nucl. Chem. Lett., 1973, 9, N10, 1025-1029. - CA, 1973, 79, 119415. PA, 1973, 76, 79294.

Exp.; 2.6.; 2.8.

2.7.58. Drissen-Fluer, A. H. M., Groeneveld, W. L., Complexes with ligands containing the sulfur-oxygen double bond group. X. Metal solvates with ligands 1,3-dithiane monosulfoxide and 2-phenyl-1,3-dithiane monosulfoxide. -Inorg. Chim. Acta, 1973, 7, N1, 139-143. - CA, 1973, 79, 26674.

Exp.; O_h; d^9; 3.1.

2.7.59. Hulett, L. G., Thornton, D. A., Effects of electron spin pairing and Jahn-Teller distortion on the infrared spectra of metal tropolonate complexes. - Spectrochim. Acta A, 1973, 29, N5, 757-763. - CA, 1973, 79, 36761. PA, 1973, 76, 36026.

Exp.; d^5; d^6.

2.7.60. Kaufmann, U., Koidl, P., Schirmer, O. F., Near infrared absorption of Ni^{2+} in ZnO and ZnS: dynamic Jahn-Teller effect in the $^3T_2$ state. - J. Phys. C, 1973, 6, N2, 310-322. - CA, 1973, 78, 90534. PA, 1973, 76, 21446. RZF, 1973, 6D487.

Exp.; T.

2.7.61. Koidl, P., Schirmer, O. F., Kaufmann, U., Near-infrared absorption of Co^{2+} in ZnS: weak Jahn-Teller coupling in the $^4T_2$ and $^4T_1$ states. - Phys. Rev. B, 1973, 8, N11, 4926-4934. - CA, 1974, 80, 76075. RZF, 1974, 7D477.

Exp.; T - e; T_d; d^7.

2.7.62. Lebedda, J. D., Palmer, R. A., The vibronic spectrum of bis(diethyldithiophosphato)nickel(II). II. Temperature dependence and fine structure analysis. - Inorg. Chem., 1973, 12, N1, 108-112. - CA, 1973, 78, 21910.

Exp.; d^8; 1.5; 2.8.

2.7.63. Ray, T., Regnard, J. R., Laurant, J. M., Ribeyron, A., Evidence of equal Jahn-Teller coupling to E_g and T_{2g} modes in $KMgF_3$: Fe^{2+} from far infrared studies. - Solid State Commun., 1973, 13, N12, 1959-1963. - CA, 1974, 80, 54035. PA, 1974, 77, 10461. RZF, 1974, 5D475.

Exp.; T - (e + t); O_h; d^6.

2.7.64. Stoelinga, J. H. M., Wyder, P., Challis, L. J., De Goer, A. M., Far-infrared studies of manganese(3+) in aluminum oxide. - J. Phys. C, 1973, 6, N24, L486-L489. - CA, 1974, 80, 65174. PA, 1974, 77, 17094.

Exp.; d^4.

2.7.65. Witkowski, A., Wojcik, M., Infrared spectra of hydrogen bond: a general theoretical model. - Chem. Phys., 1973, 1, N1, 916. - CA, 1973, 78, 166453.

Theor.; 1.5.

1974

2.7.66. Bersuker, I. B., Ogurtsov, I. Ya., Shaparev, Yu. V., Rotational spectra of symmetrical polyatomic systems with dipolar instability. [Russ.]. - Opt. & Spektrosk., 1974, 36, N2, 315-321. - CA, 1974, 81, 18639. RZF, 1974, 5D361.

Theor.; T_d.

2.7.67. Maier, H., Scherz, U., Vibronic spectra and the dynamic Jahn-Teller effect of cubic $ZnS:Cu^{2+}$. - Phys. Status Solidi b, 1974, 62, N1, 153-164. - CA, 1974, 80, 114445. PA, 1974, 77, 33416. RZF, 1974, 9D543.

Exp.; T - (e + t); d^9; 2.1.

2.7.68. Meyer, P., Regis, M., Farge, Y., Far infrared absorption of iron(2+) ions in magnesium oxide crystals. - Phys. Lett. A, 1974, 48, N1, 41-42. - CA, 1974, 81, 56019. PA, 1974, 77, 57583.

Exp.; T; d^6; 1.4; 3.1.

2.7.69. Milligan, D. E., Jacox, M. E., Vibrational spectrum and structure of CO_3. - Ber. Bunsenges. Phys. Chem., 1974, 78, N2, 200-201. - PA, 1974, 77, 48006.

Exp.; 2.8.

1975

2.7.70. Astheimer, L., Hauck, J., Schenk, H. J., Schwochau, K., Tetroxo anions of hexavalent technetium and rhenium. - J. Chem. Phys., 1975, 63, N5, 1988-1991. - CA, 1975, 83, 156940.

Exp.; T; T_d; d^1; 2.1; 3.6.

2.7.71. Aurbach, R. L., Richards, P. L., Far-infrared EPR of Al_2O_3:Mn^{3+} and γ-irradiated ruby. - Phys. Rev. B, 1975, 12, N7, 2588-2595. - CA, 1975, 83, 211000. PA, 1976, 79, 18619. RZF, 1976, 4D783.

Theor.; d^4; 1.4.

2.7.72. Aurbach, R. L., Far-infrared spectroscopy of solids. I. Impurity states in aluminum oxide. II. Electron-hole droplets in germanium. - 1975. - 229 p. - Report LBL-4102. - CA, 1976, 85, 101707. Nucl. Sci. Abstr., 1976, 33, N10, 23663.

Exp.; T; O_h; d^4.

2.7.73. Chau, J. Y. H., Hanprasopwattana Pakawadee, Infrared solvent shift studies. I. Acetylacetonates, of some trivalent transition metals. - Austral. J. Chem., 1975, 28, N8, 1689-1697. - CA, 1975, 83, 154951.

Exp.; E - e; d^3 - d^6.

2.7.74. Goltzene, A., Meyer, B., Schwab, C., Cho, K., Sulfur-(1-) and selenium(1-) ion centers in cuprous halides. - Phys. Status Solidi b, 1975, 69, N1, 237-247. - CA, 1975, 83, 50429.

Exp.; T - t; T_d; 3.1.

2.7.75. Hjortsberg, A., Nigren, B., Vallin, J. T., Far-infrared absorption of iron(2+)-doped magnesium oxide in magnetic fields. - Solid State Commun., 1975, 16, N1, 105-107. - CA, 1975, 82, 91831.

Exp.; T - (e + t); d^6.

1976

2.7.76. Aliev, M. R., Forbidden rotational transitions in molecules. [Russ.]. - Usp. Fiz. Nauk, 1976, 119, N3, 557-572. - Bibliogr.: 68 ref. CA, 1976, 85, 132954. PA, 1977, 80, 34612.

Rev.

2.7.77. Everstein, P. L. A., Zuur, A. P., Driessen, W. L., Complexes with ligands containing nitrile groups. Part XVIII. Hydrogen cyanide, a neutral nitrogen donor ligand to divalent metal ions. - Inorg. & Nucl. Chem. Lett., 1976, 12, N3, 277-284. - CA, 1976, 84, 144024.

Exp.; E - e; O_h; d^9.

2.7.78. Grebe, G., Roussos, G., Schulz, H. J., Infrared luminescence of chromium doped zinc selenide crystals. - J. Lumin., 1976, 12/13, 701-705. - CA, 1976, 85, 11899. PA, 1976, 79, 54254.

Exp.; T; T_d.

2.7.79. Matsumoto Yoshiaki, Short note on the infrared spectrum of antimony(V) chloride(1-) anion in the solids of cross-linked poly(N-vinylcarbazole). - Bull. Kyushu Inst. Technol., Math. Nat. Sci., 1976, 23, 31-38. - CA, 1976, 85, 94998.

Exp.; T - t.

1977

2.7.80. Dixon, R. N., Field, D., Hyperfine level-crossing spectroscopy of NH_2 with dye-laser excitation. - Mol. Phys., 1977, 34, N6, 1563-1576. - CA, 1978, 89, 33472. RZF, 1978, 8D407.

Exp.; 1.6.

2.7.81. Nishizawa Hitoshi, Mizogami Masayuki, Matsuoka Kiyoshi, Infrared spectra of synthetic vanadate garnet. [Jap.]. - Rep. Res. Lab. Hydrotherm. Chem. (Kochi, Jpn.), 1977, 2, 5-8. - CA, 1978, 89, 120041.

Exp.; T; T_d; d^9.

2.7. 82. Payne, S. H., Stedman, G. E., Ion-lattice coupling, spectral moments, and transition interference in Fe^{2+}:MgO. - J. Phys. C, 1977, 10, N9, 1549-1559. - CA, 1977, 87, 143418.

Theor.; T - (e + t); d^6; 3.1.

1978

2.7.83. Bernstein, E. R., Webb, J. D., On the Jahn-Teller effect in IrF_6: the Γ_{8g} $(t_{2g})^3$ state at 6800 Å. - Mol. Phys., 1978, 35, N6, 1585-1607. - CA, 1979, 90, 14187. PA, 1978, 81, 76032.

Exp.; Γ_8 - (e + t); O_h.

2.7.84. Bevis, M., Sievers, A. J., Harrison, J. P., Taylor, D. R., Thouless, D.J., Infrared absorption by elementary excitations of the one-dimensional XY system. - Phys. Rev. Lett., 1978, 41, N14, 987-990. - CA, 1978, 89, 188377. PA, 1978, 81, 95584.

Exp.

2.7.85. Kelley, C. S., Williams, F., Anomalous temperature dependence of the near-infrared absorption of ZnS:Cr crystals. - Solid State Commun., 1978, 28, N5, 389-391. - CA, 1979, 90, 112515. PA, 82, 28794.

Exp.; E; T.

See also: 1.2.-10; 1.3.-8, 12; 1.4.-2; 1.5.-39; 1.7.-4, 37, 38, 52; 2.1.-12, 41, 49, 146; 2.2.-80; 2.3.-14, 20, 23, 26, 34, 37, 78, 80, 122; 2.4.-35, 45, 65, 78, 208; 2.5.-3, 25; 2.8.-4, 15, 30; 3.1.-201; 3.2.-11; 3.4.-28; 3.6.-22; 4.1.-69; 4.2.-17; 4.3.-7, 9, 22, 28, 40, 43, 57, 62, 64, 74; 4.6.-38, 111, 126, 133, 191, 201; 5.1.-36; 5.2-18, 25, 61; 5.3.-10.

2.8. Vibronic Raman Spectra

The theory of Raman light scattering taking into account weak pseudo-Jahn-Teller interaction was developed in Ref. [2.8.2]. The possible depolarization of the Rayleigh scattering in Jahn-Teller systems was noted in Ref. [2.8.3]. The Raman scattering theory for tetrahedral and octahedral systems was developed in Refs. [2.8.-6, 7]. For a discussion of the E - e problem in the strong coupling limit, see Refs. [2.8.-14, 20]. Raman spectra of hexafluorides are presented in Refs. [2.8.4, 8, 10, 59, 64, 65, 73]. In Ref. [2.8.9] the Raman scattering spectra were used for restoration of the potential surface of the $KCl:Tl^+$ system (see also Refs. [2.8.-12, 15]). References [2.8.-21, 22, 23, 25, 26, 28, 58, 68, 80] are devoted to the spectra of transition metal complexes.

The Raman scattering method is useful for investigating relaxed and unrelaxed states of F centers [2.8.-30, 47, 48, 82].

A general theory of secondary radiation polarization is given in Ref. [2.2.86]. The group-theoretical approach to the light-scattering problem is discussed in detail in Refs. [2.8.-31, 32, 33, 46, 48, 49]. The important problem of vibronic level relaxation in the light scattering processes is considered in the theoretical work [2.8.52]. The theory of light scattering was developed in Refs. [2.8.-14, 86]. Raman spectra have also been used for the investigation of biologic molecules [2.8.-90, 78].

1960

2.8.1. Claassen, H. H., Weinstock, B., Search for a Jahn-Teller effect in IrF_6. - J. Chem. Phys., 1960, 33, N2, 436-437. - CA, 1961, 55, 4149d.

Exp.; Γ_8.

<u>1961</u>

2.8.2. Albrecht, A. C., On the theory of Raman intensities. - J. Chem. Phys., 1961, <u>34</u>, N5, 1476-1484.

Theor.; 1.5.

<u>1962</u>

2.8.3. Bersuker, I. B., On the possibility of a pure rotational Raman spectrum and depolarization of Rayleigh scattering in some hexafluorides and analogous systems. [Russ.]. - Opt. & Spektrosk., 1962, <u>12</u>, N4, 528-529. - RZF, 1962, 12V95.

Theor.; O_h.

<u>1964</u>

2.8.4. Woodward, L. A., Ware, M. J., Raman and infrared spectra of the hexachloro- and hexabromorhenate ions ($ReCl_6^{2-}$ and $ReBr_6^{2-}$), and the hexachloroosmate ion ($OsCl_6^{2-}$). - Spectrochim. Acta, 1964, <u>20</u>, N4, 711-720. - CA, 1964, <u>61</u>, 5106h.

Exp.; O_h.

<u>1968</u>

2.8.5. Barrett, J. J., Adams, N. I., Laser-excited rotation-vibration Raman scattering in ultra-small gas samples. - J. Opt. Soc. Am., 1968, <u>58</u>, N3, 311-319. - PA, 1968, <u>71</u>, 29683.

Exp.

2.8.6. Verlan, E. M., Static Jahn-Teller effect in the carbon tetrachloride molecule and its Raman spectrum. [Russ.]. - Opt. & Spektrosk., 1968, <u>24</u>, N3, 378-387. - CA, 1968, <u>69</u>, 6932. PA, 1969, <u>72</u>, 4950.

Theor.; T - t; T_d.

<u>1969</u>

2.8.7. Verlan, E. M., Influence of the Jahn-Teller effect on Raman spectra. [Russ.]. - Opt. & Spektrosk., 1969, <u>26</u>, N6, 899-907. - CA, 1969, <u>71</u>, 75973. PA, 1970, <u>73</u>, 49833.

Theor.; E - e; D_{3h}.

<u>1970</u>

2.8.8. Claassen, H. H., Goodman, G. L., Holloway, J.H., Selig, H., Raman spectra of MoF_6, TcF_6, ReF_6, UF_6, OsF_6, SeF_6, and TeF_6 in the vapor state. - J. Chem. Phys., 1970, <u>53</u>, N1, 341-347. - RZF, 1970, 12D321.

Exp.

2.8.9. Kravitz, L. C., Configuration coordinates of Tl^+ in KCl from Raman scattering. - Phys. Rev. Lett., 1970, 24, N16, 884-887.

Exp.; T - (e + t).

2.8.10. Selig, H., Claassen, H. H., Holleway, J. H., The Raman spectra of some hexafluorides and pentafluorides in the vapor state. - In: Abstr. 2nd Int. Conf. Raman Spectrosc., Oxford. Bradford, 1970, 1 p. - PA, 1971, 74, 39655.

Exp.; T - (e + t).

2.8.11. Stufkens, D. J., Dynamic Jahn-Teller effect in the excited states of $SeCl_6^{2-}$, $SeBr_6^{2-}$, $TeCl_6^{2-}$, and $TeBr_6^{2-}$: Interpretation of electronic absorption and Raman spectroscopy. - In: Abstr. 2nd Int. Conf. Raman Spectrosc., Oxford. Bradford, 1970, 1 p. - PA, 1971, 74, 41600.

Exp.; T - (e + t); 2.1; 2.3.

1971

2.8.12. Benedek, G., Terzi, N., The electron-phonon interaction induced by Tl^+ in K-halides in connection with the first-order Raman spectra. - In: Light Scattering in Solids: Proc. 2nd Int. Conf. Paris, 1971, p. 291-297. - RZF, 1973, 5E905.

Theor.; T.

2.8.13. Christie, J. H., Lockwood, D. J., Electronic Raman spectrum of Co^{2+} in $CoCl_2$. - Chem. Phys. Lett., 1971, 8, N1, 120-122. - RZF, 1971, 7D459.

Exp.; T; T_d; d^7.

2.8.14. Hizhnyakov, V., Tehver, I., Theory of the resonant Raman scattering by asymmetric vibrations of impurity centers. - In: Light Scattering in Solids: Proc. 2nd Int. Conf. Paris, 1971, p. 57-61. - CA, 1973, 78, 22019. PA, 1972, 75, 16513.

Theor.; E - e.

1972

2.8.15. Benedek, G., Terzi, N., An evaluation of Raman spectra, phonon relaxation time, and infrared and ultraviolet absorption induced by Tl^+ substituted into KI. - In: Phys. Impurity Cent. Cryst.: Proc. Int. Semin. "Selected Probl. Theory Impurity Cent. Cryst.," 1970. Tallin, 1972, p. 321-342. - RZF, 1973, 1D454.

Theor.; (^3T + ^1T) - (a + e + t); 2.1; 2.7; 3.7.

2.8.16. Bottger, G. L., Damsgard, C. V., The Raman spectra of crystalline K_2OsCl_2 and K_2OsBr_6. - Spectrochim. Acta A, 1972, 28, N8, 1631-1635. - PA, 1972, 75, 72369.

Exp.

2.8.17. Kalantar, A. H., Franzosa, E. S., Innes, K. K., Raman intensity as a function of exciting wavelength for a vibration known to mix electronic states. - Chem. Phys. Lett., 1972, 17, N3, 335-341. - RZK, 1973, 9B211.

Exp.; 1.5.

2.8.18. Khizhnyakov, V. V., Teoriya Rezonansnogo Vtorichnogo Svecheniya Primesnykh Tsentrov Kristallov: Avtoref. Dis. ... Dokt. Fiz.-Mat. Nauk. [Russ.]. - Tartu, 1972. - 22 p. - Tart. Univ.

The Theory of Resonance Secondary Radiation of Impurity Centers in Crystals.

Theor.; E - e; T - e; T - t.

1973

2.8.19. Bobovich, Ya. S., Aleksandrov, I. V., Maslov, V. G., Sidorov, A., On the selection rules for the intensity and polarization of the lines in the spectra of resonance Raman scattering. [Russ.]. - Pis'ma Zh. Eksp. & Teor. Fiz., 1973, 18, N3, 175.

Theor.

2.8.20. Khizhnyakov, V. V., Tehver, I., Theory of polarized resonant secondary radiation of impurity centers including the Jahn-Teller effect. - In: Lumin. Cryst., Mol. Solutions: Proc. Int. Conf., 1972. New York, 1973, p. 489-496. - CA, 1973, 79, 59644.

Theor.; E - e.

1974

2.8.21. Gächter, B. F., Köningstein, J. A., Observation of electronic Raman transitions from the transition metal ions Ti^{3+} and V^{4+} highly diluted in α-Al_2O_3. - Solid State Commun., 1974, 14, N4, 361-364.

Exp.; T - e; O_h; d^1; 1.4.

2.8.22. Galluzzi, F., Garozzo, M., Ricci, F. F., Resonance Raman scattering and vibronic coupling in aquo- and cyanocobalamin. - J. Raman Spectrosc., 1974, 2, N4, 351-362. - CA, 1975, 82, 9525.

Exp.; D_{4h}; 1.5; 2.1.

2.8.23. Guha, S., Chase, L. L., Raman scattering from vibronic levels of a Jahn-Teller distorted complex. - Phys. Rev. Lett., 1974, 32, N16, 869-872. - CA, 1974, 80, 150716. PA, 1974, 77, 41368. RZF, 1974, 9D551.

Exp.; E - e; O_h; d^9.

1975

2.8.24. Aleksandrov, I. V., Belyavskaya, N. M., Bobovich, Ya. S., Bortkevich, A. V., Raman scattering in systems with pronounced vibronic interaction. [Russ.]. - Zh. Eksp. & Teor. Fiz., 1975, 68, N4, 1274-1287. - RZF, 1975, 9D350.

Exp.

2.8.25. Chase, L. L., Hoo, C. H., Raman scattering from vibronic levels of nickel(3+) ion in aluminum oxide. - Phys. Rev. B, 1975, 12, N12, 5990-5994. - CA, 1976, 84, 67304. PA, 1976, 79, 36287.

Exp.; E; d^7.

2.8.26. Guha, S., Chase, L. L., Vibronic and impurity-induced Raman scattering from a Jahn-Teller distorted impurity. - Phys. Rev. B, 1975, 12, N5, 1658-1675. - CA, 1975, 83, 185839. PA, 1976, 79, 10334. RZK, 1976, 7B723.

Exp.; E - e; O_h; d^9.

2.8.27. Guha, S., Raman scattering study of the perturbation of crystal lattice phonons by a Jahn-Teller distorted impurity state. - 1975. - 160 p. - [Order N 75-23475]. Diss. Abstr. Int. B, 1975, 36, N5, 2328. CA, 1976, 84, 36976.

Exp.

2.8.28. Hamaguchi Hiroo, Harada Issei, Shimanouchi Takehiko, Anomalous polarization in the resonance Raman effect of octahedral hexachloroiridate(IV) ion. - Chem. Phys. Lett., 1975, 32, N1, 103-107. - CA, 1975, 83, 50349. PA, 1975, 78, 53790.

Exp.

2.8.29. Iijima, M., Udagawa, Y., Kaya, K., Ito, M., Raman spectra and Jahn-Teller effects of $C_4O_4{}^{2-}$ and $C_5O_5{}^{2-}$ ions. - Chem. Phys., 1975, 9, N1-2, 229-235. - PA, 1975, 78, 68533. RZF, 1975, 12D298.

Exp.; E; D_{4h}; D_{5h}.

2.8.30. Kondo Yasuhiro, Kanzaki Hiroshi, Infrared absorption spectrum of the relaxed excited state of the F center in potassium halides. - Phys. Rev. Lett., 1975, 34, N11, 664-667. - RZF, 1975, 8D532.

Exp.; 1.5.

2.8.31. Mulazzi, E., Terzi, N., Raman scattering selection rules for Jahn-Teller systems. - 1975. - 7 p. - US NTIS, AD Rep., AD-A021194. - CA, 1976, 85, 54085.

Theor.

2.8.32. Mulazzi, E., Breakdown of symmetry selection rules for defect induced Raman scattering. - 1975. - 8 p. - US NTIS, AD Rep., AD-A021195. - CA, 1976, 85, 54089.

Theor.

2.8.33. Mulazzi, E., Bishop, M. F., Dynamical lowering of site symmetry by Jahn-Teller effect in resonant Raman scattering. - 1975. - 10 p. - US NTIS, AD Rep., AD-A021197. - CA, 1976, 85, 54086.

Theor.

2.8.34. Small, G. J., Yeung, E. S., Nonadiabatic vibronic interactions and the Raman effect. - Chem. Phys., 1975, 9, N3, 379-383. - RZK, 1976, 2B224.

Theor.; D_{6h}.

2.8.35. Tsuboi Masamichi, Hirakawa Akiko, Y., Muraishi Shuichi, A possible correlation between Jahn-Teller coupling and a Raman scattering. - J. Mol. Spectrosc., 1975, 56, N1, 146-158. - CA, 1975, 83, 35119. PA, 1975, 78, 49707. RZK, 1975, 23B236. RZF, 1975, 11D169.

Theor.

2.8.36. Zgierski, M. Z., Non-adiabatic interaction and the resonance Roman scattering for a non-totally symmetric mode. - Chem. Phys. Lett., 1975, 36, N3, 390-392. - RZK, 1976, 8B272.

Theor.; 1.5.

1976

2.8.37. Bersuker, I. B., Ogurtsov, I. Ya., Shaparev, Yu. V., Rayleigh and rotational Raman scattering by spherical top molecular systems. [Russ.]. - In: Tez. Dokl. Simp. Molek. Spektrosk. Vysokogo Sverkhvysokogo Razresheniya. Novosibirsk, 1976, p. 15-17.

Theor.; E - e; T - t; (A + T) - t; T_d.

2.8.38. Chase, L. L., Glynn, T. J., Hayes, W., Rushworth, A. J., Ryan, J. E., Walsh, D., De Goer, A. M., Raman studies of some Jahn-Teller systems. - In: Proc. 5th Int. Conf. Raman Spectrosc. Freiburg; Breisgau, 1976, p. 660-661. - CA, 1977, 87, 208857.

Exp.; E; T; d^1; d^4; d^7.

2.8.39. Garozzo, M., Galluzzi, F., A comparison between different approaches to the vibronic theory of Raman intensities. - J. Chem. Phys., 1976, 64, N4, 1720-1723.

Theor.

2.8.40. Hamaguchi, H., Shimanouchi, T., Anomalous polarization in the resonance Raman effect of the octahedral hexabromoiridate(IV) ion. - Chem. Phys. Lett., 1976, 38, N2, 370-373. - PA, 1976, 79, 44337.

Exp.; O_h.

2.8.41. Hamaguchi, H., Shimanouchi, T., Anomalous polarization in the vibrational resonance Raman effect of octahedral complex ions with degenerate ground electronic states. - In: Proc. 5th Int. Conf. Raman Spectrosc. Freiburg; Breisgau, 1976, p. 294-295. - RZK, 1977, 4B221.

Exp.

2.8.42. Ito, M., Vibronic coupling as revealed by preresonance Raman effect. - In: Proc. 5th Int. Conf. Raman Spectrosc. Freiburg; Breisgau, 1976, p. 267-276. - RZK, 1977, 6B175.

Exp.

2.8.43. Kaya, K., Raman spectra and Jahn-Teller effect of oxocarbon ions. - In: Proc. 5th Int. Conf. Raman Spectrosc. Freiburg; Breisgau, 1976, p. 237-246. - CA, 1978, 88, 104223. RZK, 1977, 5B222.

Exp.; E.

2.8.44. Köningstein, J. A., Parameswaran, T., Origin of asymmetric scattering tensors of normal modes in electronically degenerate states. - Mol. Phys., 1976, 32, N5, 1311-1319. - CA, 1977, 86, 147924.

Theor.; E; T.

2.8.45. Lefrant, S., Buisson, J. P., Taurel, L., Experimental study of the Raman spectrum of cesium bromide containing F centers. [Fr.]. - C.R. Hebd. Seances Acad. Sci. Ser. B, 1976, 282, N17, 407-410. - CA, 1976, 85, 38991.

Exp.; O_h.

2.8.46. Mulazzi, E., Raman scattering in resonance with the optical transition of a Jahn-Teller system. - In: Light Scattering in Solids: Proc. 3rd Int. Conf., 1975. Paris, 1976, p. 567-570. - CA, 1976, 85, 133488. PA, 1976, 79, 69336.

Theor.; T - t; O_h.

2.8.47. Mulazzi, E., Bishop, M. F., Raman scattering in resonance with the absorption band of the F center in alkali halides: breakdown of the scattering selection rules due to the spin-orbit interaction. - J. Phys. [France]. 1976, 37, Colloq. N7, 109-113. - CA, 1977, 86, 180019.

Theor.; T - (e + t).

2.8.48. Mulazzi, E., Bishop, M. F., Dynamical lowering of site symmetry by Jahn-Teller effects in resonant Raman scattering. - Solid State Commun., 1976, 19, N1, 39-44. - CA, 1976, 85, 84978. PA, 1976, 79, 68816. RZF, 1976, 12D466.

Theor.

2.8.49. Mulazzi, E., Terzi, N., Raman scattering selection rules for Jahn-Teller systems. - Solid State Commun., 1976, 18, N6, 721-724. - PA, 1976, 79, 45805.

Theor.

2.8.50. Shelnutt, J. A., Yu. Nai-teng, Chang, R. C. C., Cheung, L. D., Felton, R. H., Effect of Jahn-Teller instability and excited state nuclear distortion on resonance Raman excitation profiles of metalloporphyrins: theory and experiment. - In: Proc. 5th Int. Conf. Raman Spectrosc. Freiburg; Breisgau, 1976, p. 336-337. - CA, 1977, 87, 175214.

Exp.; E; d^8.

2.8.51. Srinivasan, K., Jahn-Teller effect and Raman activity in molecules. - In: Proc. 5th Int. Conf. Raman Spectrosc. Freibourg; Breisgau, 1976, p. 314-315. - CA, 1977, 87, 175206; RZK, 1977, 5B185.

Theor.

2.8.52. Toyozawa Yutaka, Resonance and relaxation in light scattering. - J. Phys. Soc. Jpn., 1976, 41, N2, 400-441.

Theor.; 2.2.

2.8.53. Warshel, A., On the molecular origin of the anomalous polarization in resonance Raman polarization. - Chem. Phys. Lett., 1976, 43, N2, 273-278.

Theor.; $D_{4}h$; 1.5.

2.8.54. Young, A. P., Excitation in Jahn-Teller coupled systems. - In: Light Scattering in Solids: Proc. 3rd Int. Conf. 1975. Paris, 1976, p. 817-821. - CA, 1976, 85, 151080. PA, 1976, 79, 73096.

Theor.

1977

2.8.55. Aleksandrov, I. V., Bobovich, Ya. S., Study of vibronic interactions in anions of organic molecules with orbitally degenerate states by the method of Raman spectroscopy. [Russ.]. - In: Teor. Spektrosk. Moscow, 1977, p. 93-96.

Exp.; $D_{3}h$; $D_{6}h$.

2.8.56. Asher, S. A., Vickery, L. E., Schusters, T. M., Sauer, K., Resonance Raman spectra of methemoglobin derivatives: Selective enhancement of axial ligand vibrations and lack of an effect of inositol hexaphosphate. - Biochemistry, 1977, 16, N26, 5849-5859.

Exp.; C_{4v}; C_{2v}; d^6; 1.5.

2.8.57. Capozzi, V., Fontana, A., Fontana, M. P., Mariotto, C., Montagna, M., Viliani, G., Raman scattering in PbI_2. - Nuovo Cimento B, 1977, 39, N2, 556-560. - CA, 1977, 87, 92993.

Exp.; $C_{6}v$.

2.8.58. Chase, L. L., Hayes, W., Rushworth, A. J., Electronic and vibronic Raman scattering from the Mn^{3+} ion in Al_2O_3. - J. Phys. C, 1977, 10, N19, L575-L578. - CA, 1978, 88, 96835.

Exp.; E; d^4.

2.8.59. Clark, R. J. H., Turtle, P. C., Electronic Raman effect in the hexachloroiridate(IV) ion. - Chem. Phys. Lett., 1977, 51, N2, 265-268. - PA, 1978, 81, 5231.

Exp.; T; O_h.

2.8.60. Fontana, M. P., Viliani, G., Fontana, A., Montagna, M., Capozzi, V., Mariotto, G., Raman spectroscopy of thallium chloride in the melting region. - Phys. Lett. A, 1977, 64, N3, 319-321. - CA, 1978, 88, 96889.

Exp.

2.8.61. Henneker, W. H., Penner, A. P., Seibrand, W., Zgierski, M. Z., Resonance Raman scattering in molecules with double minimum potentials. - Chem. Phys. Lett., 1977, 48, N2, 197-200. - CA, 1977, 87, 60283. PA, 1977, 80, 58011.

Theor.; 1.5.

2.8.62. Hong Hwei-kwan, Theory of resonant Raman scattering in the strong vibronic coupling limit - a Green's function approach. - J. Chem. Phys., 1977, 67, N2, 801-812.

II. Overtones and hot bands. - Ibid., 813-823.

Theor.; 1.5.

2.8.63. Johnson, B. B., Nafie, L. A., Peticolas, W. L., Calculation of excitation profiles from the vibronic theory of Raman scattering. - Chem. Phys., 1977, 19, N3, 303-311. - RZK, 1977, 14B183.

Theor.; 1.5.

2.8.64. Meredith, G. R., Webb, J. D., Bernstein, E. R., On the Jahn-Teller effect in rhenium hexafluoride. - Mol. Phys., 1977, 34, N4, 995-1017. - CA, 1978, 88, 128430. PA, 1978, 81, 6675.

Exp.; O_h.

2.8.65. Meredith, G. R., Webb, J. D., Bernstein, E. R., On the Jahn-Teller effect in rhenium hexafluoride. - 1977. - 60 p. - US NTIS, AD Rep., AD-A039705. - CA, 1977, 87, 143648.

Exp.; O_h.

2.8.66. Muramatsu, S., Nasu, K., Takahashi, M., Kaya, K., Resonance Raman profile of the Jahn-Teller coupled vibrations in the corconate ion. - Chem. Phys. Lett., 1977, 50, N2, 284-288. - CA, 1977, 87, 175075. PA, 1977, 80, 83602.

Exp.

2.8.67. Nishimura Yoshifumi, Hirakawa Akiko, Y., Tsuboi Masamichi, A vibronic coupling in a degenerate electronic state via a nuclear momentum and an antisymmetric Raman scattering tensor. - J. Chem. Phys., 1977, 67, N3, 1009-1014. - CA, 1977, 87, 124742. RZK, 1978, 3B93.

Theor.; D_{4h}.

2.8.68. Parameswaran, T., Köningstein, J. A., Haley, L. V., Selection rules and Raman spectrum of a molecule with a degenerate ground state zero spin-orbit coupling: vanadium tetrachloride. - J. Mol. Spectrosc., 1977, 66, N3, 350-356. - CA, 1977, 87, 124937. PA, 1978, 81, 988. RZK, 1978, 4B190.

Exp.; T_d.

2.8.69. Pawlikowski, M., Zgierski, M. Z., Resonance Raman scattering and depolarization dispersion of a degenerate Jahn-Teller active mode. - Chem. Phys. Lett., 1977, 48, N2, 201-206. - CA, 1977, 87, 60284. PA, 1977, 80, 58012.

Theor.; E - e; C_{nv}; C_{nh}; D_h; D_{nd}; D_{nh} (n = 3, 4, 5).

2.8.70. Shelnutt, J. A., Cheung, L. D., Chang, R. C. C., Yu Nai-teng, Felton, R. H., Resonance Raman spectra of metalloporphyrins: Effects of Jahn-Teller instability and nuclear distortion on excitation profiles of Stokes fundamentals. - J. Chem. Phys., 1977, 66, N8, 3387-3398. - CA, 1977, 86, 180107. PA, 1977, 80, 58018.

Exp.; E - ($a_1 + b_1 + b_2$); C_{4v}; D_{4h}.

1978

2.8.71. Aleksandrov, I. V., Bobovich, Ya. S., Spectroscopy of Raman scattering by complex organic molecules in ionized states. [Russ.]. - In: Sovrem. Problemy Spektrosk. Kombinats. Rasseyaniya Sveta. Moscow, 1978, p. 279-296.

Rev.; E - e; D_{4h}.

2.8.72. Bernstein, E. R., Webb, J. D., Raman scattering of neat and mixed crystals of iridium fluoride (IrF_6): the Jahn-Teller interaction in the ground state. - 1978. - 31 p. - US NTIS, AD Rep., AD-A052754. - CA, 1978, 89, 120134.

Exp.; T - (e + t).

2.8.73. Bernstein, E. R., Webb, J. D., Absorption and electronic Raman scattering spectra of the Γ_{8g} state of iridium hexafluoride at 1.6 micrometers - a resolution of the Jahn-Teller problem. - 1978. - 35 p. - US NTIS, AD Rep., AD-A053553. - CA, 1978, 89, 155047.

2.8.74. Buisson, J. P., Lefrant, S., Ghomi, M., Taurel, L., Chapelle, J. P., Study of phonon-electron interactions of F centers in cesium chloride, cesium bromide, and cesium fluoride in first-order Raman scattering. - In: Proc. Int. Conf. Lattice Dynamics, 1977. Paris, 1978, p. 223-225. - CA, 1979, 90, 94723.

Exp.; T.

2.8.75. Chen Kuo-mei, Yeung, E. S., Rovibronic two-photon transitions of symmetric top molecules. - J. Chem. Phys., 1978, 69, N1, 43-52. - CA, 1978, 89, 119872.

Theor.; E; D_{3h}.

2.8.76. Cheung, L. D., Yu Nai-teng, Felton, R. H., Resonance Raman spectra and excitation profiles of Soret-excited metalloporphyrins. - Chem. Phys. Lett., 1978, 55, N3, 527-530. - CA, 1978, 89, 67911. PA, 1978, 81, 58698.

Exp.; E - $(b_1 + b_2)$; C_{4v}; D_{4h}.

2.8.77. Dallinger, R. F., Stein, P., Spiro, T. G., Resonance Raman spectroscopy of uranocene: Observation of an anomalously polarized electronic band and assignment of energy levels. - J. Am. Chem. Soc., 1978, 100, N25, 7865-7870. - CA, 1979, 90, 14243. PA, 1979, 82, 59451.

Exp.

2.8.78. Felton, R. H., Yu Nai-teng, Resonance Raman scattering from metalloporphyrins and hemoproteins. - In: The Porphyrins. New York, 1978, Vol. 3, p. 347-393.

Rev.; E - $(b_1 + b_2)$; (E + E) - $(b_1 + b_2)$; C_{4v}; D_{4h}.

2.8.79. Hamaguchi Hiro-O, Polarized resonance Raman spectra of hexahalide complexes of transition metals: The effect of electronic degeneracy on the polarization of vibrational Raman scattering. - J. Chem. Phys., 1978, 69, N2, 569-578. - Bibliogr.: 32 ref.

Exp.; E; O_h; d^5.

2.8.80. Johnstone, I. W., Lockwood, D. J., Mischler, G., Examples of a dynamic Jahn-Teller system: Raman scattering from electronic excitations and phonons in paramagnetic and antiferromagnetic ferrous chloride and bromide. - J. Phys. C, 1978, 11, N10, 2147-2164. - CA, 1978, 89, 206693. PA, 1978, 81, 56474.

Exp.; T - e; d^6.

2.8.81. Johnstone, I. W., Lockwood, D. J., Mischler, G., Fletcher, J. R., Bates, C. A., Temperature-dependent electron-phonon coupling in $FeCl_2$ observed by Raman scattering. - J. Phys. C, 1978, 11, N21, 4425-4438.

Exp.; T - e; d^6.

2.8.82. Mulazzi, E., Buisson, J. P., Study of the properties of the electron-phonon interaction in the first excited degenerate states Γ_6^- and Γ_8^- of the F centers in Cs halides. - In: Proc. Int. Conf. Lattice Dynamics, 1977. Paris, 1978, p. 220-222. - CA, 1979, 90, 94722.

Theor.; E - e.

2.8.83. Mulazzi, Terzi, N., The vibronic levels for Jahn-Teller impurities in polar crystals. - In: Proc. Int. Conf. Lattice Dynamics, 1977. Paris, 1978, p. 267-269. - CA, 1979, 90, 94725. PA, 1978, 81, 68520.

Theor.; T - (a + e + t); O_h.

2.8.84. Ogurtsov, I. Ya., Shaparev, Yu. V., Bersuker, I. B., The rotational Raman scattering of light by spherical top molecules in electron degenerate or pseudo degenerate states. [Russ.]. - Opt. & Spektrosk., 1978, 45, N4, 672-678. - CA, 1979, 90, 63810. RZF, 1979, 1D156.

Theor.; E - e; T - t; 1.5.

2.8.85. Pawlikowski, M., Zgierski, M. Z., Resonance Raman scattering from triply degenerate singlet electronic states. - J. Raman Spectrosc., 1978, 7, N2, 106-110. - CA, 1978, 89, 33641. PA, 1978, 81, 78714.

Theor.; T - e; T - t.

2.8.86. Rebane, K. K., Zavt, G. S., Rebane, L. A., The Raman scattering of light by impurity centers in crystals. [Russ.]. - In: Sovrem. Problemy Spektrosk. Kombinats. Rasseyaniya Sveta. Moscow, 1978, p. 103-149.

Rev.; E - e; O_h; d^9.

2.8.87. Shelnutt, J. A., O'Shea, D. C., Resonance Raman spectra of copper tetraphenylporphyrin: Effects of strong vibronic coupling on excitation profiles and the absorption spectrum. - J. Chem. Phys., 1978, 69, N12, 5361-5374.

Exp.; D_{4h}; d^9; 1.5; 2.1.

2.8.88. Takahashi Machiko, Kaya Koji, Ito Mitsuo, Resonance Raman scattering and the Jahn-Teller effect of oxocarbon ions. - Chem. Phys., 1978, 35, N3, 293-306. - CA, 1979, 90, 112409. PA, 1979, 82, 18190.

Exp.

2.8.89. Zgierski, M. Z., Theory of resonance Raman scattering from a nontotally symmetric mode that weakly couples three electronic states: Naphthalene. - Acta Phys. Pol. A, 1978, 53, N2, 243-251.

Theor.; 1.5.

2.8.90. Zgierski, M. Z., Pawlikowski, M., Theory of depolarization dispersion of inversely polarized modes in heme proteins. - Chem. Phys. Lett., 1978, 57, N3, 438-441.

Theor.; D_{4h}; 1.5.

1979

2.8.91 Bernstein, E. R., Webb, J. D., Absorption and electronic Raman spectra of the Γ_{8g} ($^2T_{1g}$) state of IrF_6 at 1.6 μ - a resolution of the Jahn-Teller problem. - Mol. Phys., 1979, 37, N1, 191-202. - CA, 1979, 91, 81096. PA, 1979, 82, 36648.

Exp.; T - (e + t); O_h.

2.8.92. Bernstein, E. R., Webb, J. D., Raman scattering of neat and mixed crystals of IrF_6: the Jahn-Teller interaction in the ground state. - Mol. Phys., 1979, 37, N1, 203-209. - CA, 1979, 91, 81097. PA, 1979, 82, 38540.

Exp.; T - (e + t); O_h.

2.8.93. Champion, P. M., Korenowski, G. M., Albrecht, A. C., On the vibronic theory of resonance Raman scattering. - Solid State Commun., 1979, 32, N1, 7-12.

Exp.; D_{4h}; D_6; 1.5.

2.8.94. Clark, R. J. H., Dines, T. J., The resonance Raman spectrum of titanium tetrabromide. - Chem. Phys. Lett., 1979, 64, N3, 499-502. - CA, 1979, 91, 131540. PA, 1979, 82, 85965.

Exp.; T.

2.8.95. Lanir, A., Yu Nai-teng, Felton, R. H., Conformational transitions and vibronic couplings in acid ferricytochrome c: a resonance Raman study. - Biochemistry, 1979, 18, N9, 1656-1660.

Exp.; D_{4h}; d^5; 1.5.

2.8.96. Mulazzi, E., Terzi, N., Vibrational Raman scattering induced by Jahn-Teller systems in polar crystals. - Phys. Rev. B, 1979, 19, N4, 2332-2342. - CA, 1979, 91, 29965. PA, 1979, 82, 69830.

Theor.; E - e; O_h.

2.8.97. Muramatsu Shinji, Nasu Keiichiro, Resonance Raman scattering in the $E \otimes e$ Jahn-Teller system. - J. Phys. Soc. Jpn., 1979, 46, N1, 189-197. - CA, 1979, 90, 94819. PA, 1979, 82, 28736.

Theor.; E - e.

2.8.98. Siebrand, W., Zgierski, M. Z., Theory of resonance Raman scattering by degenerate modes due to Renner-Teller interaction in porphyrins. - Chem. Phys. Lett., 1979, 63, N2, 226-229. - PA, 1979, 82, 59461.

Theor.; D_{4h}; 1.6.

2.8.99. Siebrand, W., Zgierski, M. Z., Resonance Raman spectroscopy. - a key to vibronic coupling. - In: Excited States. New York, 1979, Vol. 4, p. 1-136.

Theor.; 1.5.

2.8.100. Zgierski, M. Z., Pawlikowski, M., Resonance Raman scattering in the region of a forbidden nondegenerate electronic state by a doubly degenerate vibration. - Chem. Phys. Lett., 1979, 63, N2, 221-225. - CA, 1979, 91, 65574. PA, 1979, 82, 59460.

Theor.; 1.5.

2.8.101. Zgierski, M. Z., Pawlikowski, M., Theory of resonance Raman scattering by doubly degenerate modes. - J. Chem. Phys., 1979, 71, N5, 2025-2043. - PA, 1979, 82, 98108.

Theor.; E; 1.5; 1.6.

2.8.102. Zgierski, M. Z., Shelnutt, J. A., Pawlikowski, M., Interference between intra- and itermanifold coupling in resonance Raman spectra of metalloporphyrins. - Chem. Phys. Lett., 1979, 68, N2-3, 262-266. - CA, 1980, 92, 85213. PA, 1980, 83, 28020. RZF, 1980, 4D457.

Theor.; D_{4h}; 1.5.

See also: 1.2.-10; 1.3.-8, 12; 1.7.-4, 31, 37, 38, 52; 2.1.-22, 297; 2.2.-23; 2.3.-31, 47; 2.4.-208; 2.5.-3, 25; 2.7.-7, 8, 9, 10, 11, 12, 17, 19, 20, 49, 54, 56, 57, 62; 3.2.-137; 4.1.-42, 43, 50, 69; 4.3.-8, 13, 16, 17, 22, 23, 25, 31, 36, 38, 39, 45, 48, 51, 58, 68; 4.4.-104, 119a, 140; 4.7.-39; 5.2.-18.

3. MAGNETIC RESONANCE AND RELATED METHODS. MAGNETIC, ELECTRIC, AND IMPURITY THERMODYNAMIC PROPERTIES*

The widespread use of magnetic resonance methods for investigating vibronic effects is due to the high sensitivity of the magnetic center to the electronic state and dynamics of the surrounding atoms. The first attempt to use the Jahn-Teller theorem for elucidation of the origin of the magnetic features of some Ti, V, and Cr compounds was undertaken by Van Vleck [3.6.-1, 2]. During the decade after the work [3.2.-1] there were no studies of vibronic effects on magnetic properties.

Intensive research work in this area began after the publications on EPR [3.1.-1, 2, 3] in which the importance of the Jahn-Teller effect was revealed for the first time (see also the review article [3.1.-4]).

Being the main method of investigating vibronic effects in the fifties, the EPR method was supplemented in the next two decades by other resonance methods including the Mössbauer effect, nuclear magnetic, acoustic paramagnetic, and paraelectric resonances. Increasing attention has also been paid to relaxation phenomena. The main results are summarized in the book [3.1.-114] and the review articles [3.1.-120, 151, 253, 254; 3.5.-34, 42; 3.6.-15].

3.1. EPR. General. Angular Dependence, Fine and Hyperfine Structure of Spectra

Systematic investigation of the Jahn-Teller effect by the EPR method began with the work of Bleaney and Ingram [3.1.-2], in which the anisotropic EPR spectrum of the Cu^{2+} ion in $CuSiF_6 \cdot 6H_2O$ was attributed to the Jahn-Teller effect. At first only the so-called static Jahn-Teller resulting in an anisotropic spectrum due to the low-symmetry crystal field in the Jahn-Teller distorted configuration was studied [3.1.-32, 42, 64, 79, 84, 101, 106, 108, 109, 133,

*The introductory reviews for this section were written by I. Ya. Ogurtsov.

138, 155, 168, 189, 205, 237, etc.], the origin of localization of the nuclei in the adiabatic potential minimum being neglected. A more adequate description was obtained by taking into account the complex Jahn-Teller nuclear dynamics (dynamic Jahn-Teller effect) [3.1.-7, 8, 33, 39, 45, 52, 70, 132, 137, 231, 252].

The so-called static and dynamic Jahn-Teller effects are manifested in the EPR spectra as limiting cases dependent on the relation between relaxation rates, tunneling splitting [3.1.-18, 25, 26, 48, 49, 50], and random strain in crystals [1.4.-3; 3.1.-114, 151]. In the case where the averaged splitting produced by random strain is larger than the tunneling splitting, the static Jahn-Teller effect is manifested in the spectrum. In the opposite case, the dynamic Jahn-Teller effect spectrum occurs. Depending on the temperature, transitions from the static to the dynamic Jahn-Teller effect can also be observed (see Section 3.2.).

Also important in EPR spectra is the vibronic reduction of the orbital momentum [1.4.-1, 3, 19; 3.1.-151] studied in more detail in a series of works [3.1.-44, 65, 112, 115, 124, 146, 215, 230, 254, 266, etc.]. Some aspects of the reduction can be revealed by using external stress [3.1.-159, 221, 223, 228, 263, 270, 283]. The multimode aspects of the EPR problem are discussed in Refs. [3.1.-193, 253, 254].

1950

3.1.1. Abragam, A., Pryce, M. H. L., Theoretical interpretation of copper fluosilicate spectrum. - Proc. Phys. Soc. London, Ser. A, 1950, 63, 409-411. - CA, 1950, 44, 5703e.

Theor.; E - e; O_h; d^9; 3.2.

3.1.2. Bleaney, B., Ingram, J. E., Paramagnetic resonance in copper fluosilicate. - Proc. Phys. Soc. London, Ser. A, 1950, 63, 408-409. - CA, 1950, 44, 5661e.

Exp.; E - e; O_h; d^9.

1951

3.1.3. Abragam, A., Pryce, M. H. L., Theory of the nuclear hyperfine structure of paramagnetic resonance spectra in crystals. - Proc. R. Soc. London, Ser. A, 1951, 205, 135-153. - CA, 1952, 46, 339h.

Theor.; d^3; d^4; d^5; d^8; d^9; 3.3.

1953

3.1.4. Bleaney, B., Stevens, K. W. H., Paramagnetic resonance. - Rep. Prog. Phys., 1953, 16, N1, 108-159. - Bibliogr.: 103 ref.

Rev.; O_h; T_d; D_{3h}; C_{3v}; $d^1 - d^9$.

3.1.5. Owen, J., Stevens, K. W. H., Paramagnetic resonance and covalent bonding. - Nature, 1953, 171, N4358, 836. - CA, 1953, 47, 7843g.

Exp.; O_h; D_5.

1958

3.1.6. Low, W., Paramagnetic and optical spectra of divalent cobalt in cubic crystalline fields. - Phys. Rev., 1958, 109, N2, 256-265.

Exp.; O_h; d^7; 2.1.

3.1.7. Tucker, R. F., Paramagnetic resonance study of copper-doped silver chloride. - Phys. Rev., 1958, 112, N3, 725-731.

Exp.; E - e; O_h; d^9.

1959

3.1.18. Avvakumov, V. I., Dynamic nature of the Jahn-Teller effect and its influence on the paramagnetic resonance of copper(II) ions. [Russ.]. - Zh. Eksp. & Teor. Fiz., 1959, 37, N4, 1017-1025. - CA, 1961, 55, 5132a. RZF, 1960, 20496.

Theor.; E - e; O_h; d^9.

3.1.9. Ludwig, G. W., Woodbury, H. H., Carlson, R. O., Spin resonance of deep level impurities in germanium and silicon. - Phys. & Chem. Solids, 1959, 8, N4, 490-492. - CA, 1959, 53, 12850. PA, 1959, 62, 7186.

Exp.; T_d; 3.3.

3.1.10. Ludwig, G. W., Woodbury, H. H., Electron spin resonance in nickel-doped germanium. - Phys. Rev., 1959, 113, N4, 1014-1018. - CA, 1959, 53, 14708.

Exp.; T_d; 3.2.

3.1.11. Smith, W. V., Sorokin, P. P. Gelles, I. L., Lasser, G. J., Electron-spin resonance (ESR) of nitrogen donors in diamond. - Phys. Rev., 1959, 115, N6, 1546-1552. - CA, 1960, 54, 11707b.

Exp.; T_d; sp^3.

1960

3.1.12. Low, W., Weger, M., Paramagnetic resonance and optical spectra of bivalent iron in cubic fields. I. Theory. - Phys. Rev., 1960, 118, N5, 1119-1130. - CA, 1960, 54, 23819d.

Theor.; T; O_h; d^6.

3.1.13. Low, W., Weger, M., Paramagnetic resonance and optical spectra of bivalent iron in cubic fields. II. Experimental results. - Phys. Rev., 1960, 118, N5, 1130-1136. - CA, 1960, 54, 23819e.

Exp.; T; O_h; d^6.

3.1.14. Low, W., Paramagnetic Resonance in Solids. - New York; London: Academic Press, 1960. - VIII, 212 p. - (Solid State Phys.; Vol. 10, Suppl. 2.)

3.1.15. Ludwig, G. W., Woodbury, H. H., Electronic structure of transition metal ions in a tetrahedral lattice. - Phys. Rev. Lett., 1960, 5, N2, 98-100. - CA, 1961, 55, 9040h.

Exp.; T; T_d; d^7.

3.1.16. Townsend, M. G., Weissman, S. I., Possible symptom of the Jahn-Teller effect in the negative ions of coronene and triphenylene. - J. Chem. Phys., 1960, 32, N1, 309-310. - CA, 1960, 54, 12786h. RZF, 1960, 30289.

Exp.

3.1.17. Van Vleck, J. H., Note on the gyromagnetic ratio of Co^{++} and on the Jahn-Teller effect in Fe^{++}. - Physica, 1960, 26, N7, 544-552. - RZF, 1961, 3V328.

Theor.; T; O_h; d^6.

1962

3.1.18. Bersuker, I. B., Hindered motions of conformations of octahedral complexes and their connections with radiospectroscopy. [Russ.]. - In: Tez. Dokl. Soveshch. Primeniniyu Fiz. Metodov Issled. Komleks. Soedin. Kishinev, 1962, p. 15.

Theor.; E - e; O_h; d^9.

3.1.19. Bolton, J. R., Carrington, A., Forman, A., Orgel, L. E., The effects of near-degeneracy in the electron spin resonance spectra of aromatic negative ions. - Mol. Phys., 1962, 5, N1, 43-49. - RZF, 1962, 11V466.

Exp.; E - e; D_{6h}.

3.1.20. Geschwind, S., Remeika, J. P., Spin resonance of transition metal ions in corundum. - J. Appl. Phys., 1962, 33, N1, 370-377. - CA, 1963, 58, 2035b. RZF, 1962, 9V279.

Exp.; E - e; O_h; C_{3v}; d^7; d^9.

3.1.21. Ludwig, G. W., Woodbury, H. H., Electron spin resonance in semiconductors. - In: Solid State Phys.: Adv. Res. and Appl., New York; London, 1962, Vol. 13, p. 223-304. - CA, 1962, 57, 5444d.

Exp.; T; T_d.

3.1.22. Watkins, G. D., Feher, E., Effect of uniaxial stress on the EPR of transition element ions in MgO. - Bull. Am. Phys. Soc., 1962, 7, N1, 29.

Exp.; T; O_h; d^6.

3.1.23. Woodbury, H. H., Ludwig, G. W., Spin resonance of Pb and Pt in silicon. - Phys. Rev., 1962, 126, N2, 466-470. - RZF, 1962, 12V474.

Exp.; T; T_d.

1963

3.1.24. Anderson, P. W., Electron spin resonance spectroscopy in molecular solids. - In: Solid State Phys.: Adv. Res. and Appl. New York; London, 1963, Vol. 14, p. 99-280.

Rev.

3.1.25. Bersuker, I. B., Vekhter, B. G., The EPR spectrum and microwave spectrum of transwtion metal octahedral complexes of d^1 configuration taking into account inversion splitting. [Russ.]. - Fiz. Tverd. Tela, 1963, 5, N9, 2432-2440. - RZF, 1964, 4D471.

Theor.; E - e; O_h; C_{3v}; d^1.

3.1.26. Bersuker, I. B., Spin-inversion levels in magnetic fields and the EPR spectrum of octahedral complexes of Cu^{2+} ions. [Russ.]. - Zh. Eksp. & Teor. Fiz., 1963, 44, N4, 1239-1247. - RZF, 1963, 9D379.

Theor.; E - e; O_h; d^9; 3.2.

3.1.27. Delbecq, C. J., Hayes, W., O'Brien, M. C. M., Yuster, P. H., Paramagnetic resonance and optical absorption of trapped holes and electrons in irradiated KCl:Ag. - Proc. R. Soc., London, Ser. A, 1963, 271, N1345, 243-267. - RZF, 1963, 8D251.

Exp.; E - e; O_h; d^9; 2.1.

3.1.28. Englman, R., Horn, D., Anomalies of g-factor due to vibrational coupling. - In: Paramagnetic Resonance: Proc. 1st Int. Conf., Jerusalem, 1962. New York; London, 1963, Vol. 1, p. 329-346. - CA, 1964, 60, 3648e.

Theor.; E - e; O_h; d^9.

3.1.29. Estle, T. L., Walters, G. K., De Wit, M., Paramagnetic resonance of Cr in CdS. - In: Paramagnetic Resonance: Proc. 1st Int. Conf., Jerusalem, 1962. New York; London, 1963, Vol. 1, p. 144-154. - CA, 1964, 60, 3648d.

Exp.; T_d; d^4.

3.1.30. Hayes, W., Twidell, J. W., Paramagnetic resonance of divalent lanthanum in irradiated CaF_2. - Proc. Phys. Soc. London, 1963, 82, N526, 330-331. - CA, 1963, 59, 5965h.

Theor.; E - e; O_h.

3.1.31. Low, W., Suss, J. T., Jahn-Teller effect of Ni^+ and Cu^{++} in single crystals of calcium oxide. - Phys. Lett., 1963, 7, N5, 310-312. - CA, 1964, 60, 10088c.

Exp.; E - e; O_h; d^9.

3.1.32. Morigaki Kazuo, Electron spin resonance of Cr^{2+} in CdS. - J. Phys. Soc. Jpn., 1963, 18, N5, 733. - RZF, 1964, 2D360.

Exp.; T_d; d^4.

3.1.33. O'Brien, M. C. M., The ESR spectrum of a d^9 ion in an octahedral environment. - In: Paramagnetic Resonance: Proc. 1st Int. Conf., Jerusalem, 1962. New York; London, 1963, Vol. 1, p. 322-328. - CA, 1964, 60, 7596a.

Theor.; E - e; O_h; d^9.

3.1.34. Rubins, R. S., Low, W., Paramagnetic resonance spectra of some rare earth and iron-group impurities in strontium titanate. - In: Paramagnetic Resonance: Proc. 1st Int. Conf., Jerusalem, 1962. New York; London, 1963, Vol. 1, p. 59-67. - CA, 1964, 60, 36478.

Exp.; O_h; d^7; f^1.

3.1.35. Watkins, G. D., An electron paramagnetic resonance (EPR) study of the lattice vacancy in silicon. - J. Phys. Soc. Jpn., 1963, 18, Suppl. II, 22-27. - CA, 1963, 59, 5964a.

Exp.; T_d.

3.1.36. Wit, M., de, Estle, T. L., Paramagnetic resonance of Cu^{2+} in ZnO. - Bull. Am. Phys. Soc., 1963, 8, N1, 24.

Exp.; T; T_d; d^9.

1964

3.1.37. Bersuker, I. B., Budnikov, S. S., Vekhter, B. G., Chinik, B. I., The hyperfine structure of EPR spectra of copper complexes with inversion splitting. [Russ.]. - Fiz. Tverd. Tela, 1964, 6, N9, 2583-2589. - RZF, 1965, 3D504.

Theor.; E - e; O_h; d^9; 1.3.

3.1.38. Hayes, W., Wilkens, J., The Ni^+ ion in irradiated LiF and NaF. - Proc. R. Soc. London, Ser. A, 1964, 281, N1386, 340-365. - CA, 1964, 61, 8964e.

Exp.; E - e; O_h; d^9.

3.1.39. Krupka, D. C., Silsbee, R. H., Electron spin resonance (ESR) of the R center (in alkali halides). - Phys. Rev. Lett., 1964, 12, N8, 193-195. - CA, 1964, 60, 13966a.

Exp.; E - e; O_h.

3.1.40. Lacroix, R., Höchli, U., Müller, K. A., Strong field g-value calculation for d^7 ions in octahedral surroundings. - Helv. Phys. Acta, 1964, 37, N7/8, 627-629.

Theor.; E - e; O_h; C_{3v}; d^7.

3.1.41. Lawler, R. G., Bolton, J. R., Fraenkel, G. K., Brown, T. H., Orbital degeneracy and the electron spin resonance spectrum of the benzene-1-d negative ion. - J. Am. Chem. Soc., 1964, 86, N3, 520-521. - RZF, 1964, 10D345.

Exp.; E - e; D_{6h}.

3.1.42. Low, W., Suss, J. T., Evidence for tetragonal distortions from electron spin resonance (ESR) spectra in crystals of CaO. - Solid State Commun., 1964, 2, N1, 1-6. - CA, 1964, 61, 2624c.

Exp.; T; O_h; d^6.

3.1.43. Morigaki Kazuo, Electron spin resonance (ESR) of Cr^{2+} in cadmium sulfide single crystals. - J. Phys. Soc. Jpn., 1964, 19, N2, 187-197. - CA, 1964, 60, 7596h.

Exp.; T_d; d^4.

3.1.44. Morigaki Kazuo, Electron spin resonance of a photosensitive center in Cu-doped CdS. - J. Phys. Soc. Jpn., 1964, 19, N7, 1240. - RZF, 1965, 1D325.

Exp.; T; T_d; d^9.

3.1.45. O'Brien, M. C. M., The dynamic Jahn-Teller effect in octahedrally co-ordinated d^9 ions. - Proc. R. Soc. London, Ser. A, 1964, 281, N1386, 323-339. - CA, 1964, 61, 9051f. RZF, 1965, 8D88.

Theor.; E - e; O_h; d^9; 3.2.

3.1.46. Shuskus, A. J., Paramagnetic resonance of divalent iron in calcium oxide. - J. Chem. Phys., 1964, 40, N6, 1602-1604. - RZF, 1964, 9D314.

Exp.; T; O_h; d^6.

3.1.47. Silverstone, H. J., Wood, D. E., McConnell, H. M., Molecular orbital degeneracy in C_7H_7. - J. Chem. Phys., 1964, 41, N8, 2311-2323. - CA, 1964, 61, 12790c.

Exp.; E - e; D_{7h}.

1965

3.1.48. Bersuker, I. B., Intrinsic asymmetry and hindered motions in complex compounds. [Russ.]. - Teor. & Eksp. Khim., 1965, 1, N1, 5-10. - CA, 1965, 63, 6325g. PA, 1966, 69, 17584.

Theor.

3.1.49. Bersuker, I. B., Budnikov, S. S., EPR spectrum taking into account inversion splitting in transition metal complexes with an electronic ^{5}E-term. [Russ.]. - Izv. Akad. Nauk Mold.SSR, Ser. Khim., Biol. Nauk, 1965, N11, 14-25. - RZF, 1966, 12D434.

Theor.; E - e.

3.1.50. Bersuker, I. B., Budnikov, S. S., Vekhter, B. G., The influence of inversion splitting on the EPR spectra of complexes with spin S = 2 in crystals at low temperatures. [Russ.]. - In: Tez. Dokl. 12 Vsesoyuz. Soveshch. Fiz. Nizk. Temp. Kazan', 1965, p. 122.

Theor.

3.1.51. Carrington, A., Longuet-Higgins, H. C., Moss, R. E., Todd, P. F., Isotope effects in electron spin resonance: the negative ion of cyclo-octatetraene-ld-d. - Mol. Phys., 1965, 9, N2, 187-190. - RZF, 1966, 2D505.

Exp.; E; D_{6h}.

3.1.52. Coffman, R. E., On the existence of a third type of Jahn-Teller EPR spectrum in octahedrally coordinated Cu^{++}. - Phys. Lett., 1965, 19, N6, 475-476. - CA, 1966, 64, 12070h. RZF, 1966, 8D536.

Exp.; E - e; O_h; d^9; 3.2.

3.1.53. Groot, M. S., de, Hesselmann, I. A. M., Van der Waals, J. H., Electron resonance of phosphorescent mesitylene. - Mol. Phys., 1965, 10, N1, 91-93. - RZF, 1966, 6D643.

Exp.; E - e; C_{3v}.

3.1.54. Liebling, G. R., McConnell, H. M., Study of molecular orbital degeneracy in C_5H_5. - J. Chem. Phys., 1965, 42, N11, 3931-3934. - RZF, 1966, 4D661.

Exp.; D_{5h}.

3.1.55. Nussbaum, M., Voitländer, J., Electron paramagnetic resonance in organometallic sandwich compounds. II. - Z. Naturforsch. a, 1965, 20, N11, 1417-1424. - CA, 1966, 64, 9121b.

Theor.; D_{5h}; d^8.

3.1.56. Schirmer, O. F., Müller, K. A., Schneider, J., Electron spin resonance of donor centers in beryllium oxide. - Phys. Kondens. Mater., 1965, 3, N4, 323-334. - CA, 1967, 66, 6959. RZF, 1965, 11D393.

Exp.; O_h.

3.1.57. Watkins, G. D., Corbett, J. W., Defects in irradiated silicon: electron paramagnetic resonance of the divacancy. - Phys. Rev., 1965, 138, N2, A543-A555. - CA, 1965, 62, 12530g. RZF, 1965, 10D563.

Exp.; D_{3d}; 3.2.

1966

3.1.58. Coffman, R. E., Paramagnetic resonance of Cu^{++}:MgO at 1.2°K. - Phys. Lett., 1966, 21, N4, 381-383. - CA, 1966, 65, 11581h. RZF, 1966, 10D514.

Exp.; E - e; O_h; d^9.

3.1.59. Gelerinter, E., Silsbee, R. H., Electron spin resonance identification of an N_2^- defect in X-irradiated sodium azide. - J. Chem. Phys., 1966, 45, N5, 1703-1709. - CA, 1966, 65, 13050b.

Exp.; E - e.

3.1.60. Groot, M. S., de, Hesselmann, I. A. M., Van der Waals, J. H., Paramagnetic resonance in phosphorescent aromatic compounds. IV. Ions in orbitally degenerate states. - Mol. Phys., 1966, 10, N3, 241-251. - CA, 1966, 65, 6549g.

Exp.

3.1.61. Hobey, W. D., Spin densities in para-disubstituted benzene anions. - J. Chem. Phys., 1966, 45, N7, 2718-2720. - RZF, 1967, 3D116.

Exp.; D_{6h}.

3.1.62. Hudson, A., The effects of dynamic exchange on the electron resonance line shapes of octahedral copper complexes. - Mol. Phys., 1966, 10, N6, 575-581. - RZF, 1967, 3D451.

Theor.; E - e; O_h; d^9.

3.1.63. Jones, M. T., An electron spin resonance study of tetra- and pentacyanopyridine anion radicals. - J. Am. Chem. Soc., 1966, 88, N22, 5060-5062. - CA, 1967, 66, 42187.

Exp.

3.1.64. Krishnan, R., Influence of tetrahedral Ni^{2+} ions on the magnetic properties of YIG. - Phys. Status Solidi, 1966, 18, N1, K53-K56. - CA, 1967, 66, 60337.

Exp.; 3.2.

3.1.65. Krupka, D. C., Silsbee, R. H., R-center in KCl: electron-spin-resonance studies of the ground state. - Phys. Rev., 1966, 152, N2, 816-828. - CA, 1967, 66, 41681.

Theor.; C_{3v}.

3.1.66. Kuska, H. A., D'Itri, F. M., Popov, A. I., Electron spin resonance of pentamethylenetetrazole manganese(II) and copper(II) complexes. - Inorg. Chem., 1966, 5, N7, 1272-1277. - CA, 1966, 65, 4880b.

Exp.; T; E - e; T_d; O_h; d^9.

3.1.67. Marechal, Y., Vibronic coupling in ESR spectra of triplet states in molecules: temperature effects. - J. Chem. Phys., 1966, 44, N5, 1908-1912. - PA, 1966, 69, 14502.

Theor.; T - e; 3.2.

3.1.68. Müller, K. A., Berlinger, W., The existence of Slonczewski modes in absence of complete orbital degeneracy. - Helv. Phys. Acta, 1966, 39, N7, 590-591. - RZF, 1967, 7D530.

Exp.; E - e; O_h; C_4v; d^7.

3.1.69. Rao, L. S. R., Premaswarup, D., Estimation of g-factors for paramagnetic impurities in MgO. - Indian J. Pure & Appl. Phys., 1966, 4, N8, 296-298. - CA, 1966, 65, 16229g.

Theor.; E - e; O_h; d^9.

3.1.70. Shul'man, L. A., Zaritsky, I. M., Podzyarey, G. A., Reorientation of Jahn-Teller displacement in N impurity centers in diamond. [Russ.]. - Fiz. Tverd. Tela, 1966, 8, N8, 2307-2312. - CA, 1967, 66, 6969. RZF, 1967, 2D427.

Exp.; T_d.

3.1.71. Stevens, K. W. H., Persico, F., Jahn-Teller effects in the electron spin resonance spectra of ions with spin 1. - Nuovo Cimento B, 1966, 41, N1, 37-51. - CA, 1966, 64, 12069f. RZF, 1966, 10D471.

Theor.; T; O_h; d^6; d^8.

1967

3.1.72. Carter, M. K., Vincow, G., Electron spin resonance of the benzene positive-ion radical. - J. Chem. Phys., 1967, 47, N1, 292-302. - CA, 1967, 67, 59387.

Exp.; E - e; D_{6h}.

3.1.73. Hannon, D. M., Electron paramagnetic resonance of iron(III) and nickel(III) in potassium tantalate(V). - Phys. Rev., 1967, 164, N2, 366-371. - CA, 1968, 68, 34579.

Exp.; E - e; O_h; d^7.

3.1.74. Höchli, U. T., Jahn-Teller effect of a d^1 ion in eightfold cubic coordination. - Phys. Rev., 1967, 162, N2, 262-273. - CA, 1968, 68, 7969. RZF, 1968, 6D709.

Exp.; E - e; O_h; d^1.

3.1.75. Höchli, U. T., Estle, T. L., Paramagnetic resonance study of the dynamic Jahn-Teller effect in $CaF_2:Sc^{2+}$ and $SrF_2:Sc^{2+}$. - Phys. Rev. Lett., 1967, 118, N4, 128-130. - CA, 1967, 66, 100059. RZF, 1967, 6D515.

Exp.; T; O_h; d^1.

3.1.76. Müller, K. A., Jahn-Teller effects in magnetic resonance. - In: Magn. Resonance and Relaxation: Proc. Int. Conf., Ljubljiana, 1966. Amsterdam, 1967, 1967, p. 192-208. - Bibliogr.: 50 ref. CA, 1968, 69, 39924. PA, 1968, 71, 38910. RZF, 1969, 3D579.

Rev.; O_h.

3.1.77. Pratt, D. W., Magnetic resonance spectra of vanadium-(IV) chloride and other paramagnetic species. - 1967. - 266 p. - US At. Energy Comm., UCRL-17406. - CA, 1968, 69, 14737.

Exp.; E - e; T_d; d^1.

3.1.78. Rai, R., Savard, J. Y., Tousignant, B., Evidence of inversion splitting in chromium(III) ion doped in zinc selenide. - Phys. Lett. A, 1967, 25, N6, 443-444. - CA, 1968, 68, 100553.

Exp.; d^3.

3.1.79. Schneider, J., Dischler, B., Räuber, A., Jahn-Teller distortion of the V^{2+} ion in cubic ZnS. - Solid State Commun., 1967, 5, N8, 603-605. - CA, 1967, 67, 103828. PA, 1967, 70, 36812. RZF, 1968, 1D555.

Exp.; E - e; T_d; d^3.

3.1.80. Schneider, J., Electron spin resonance of defect centers in II-VI semiconductors. - In: 2-6 [Two-six] Semicond. Compounds: Int. Conf., Brown Univ. New York, 1967, p. 40-67. - CA, 1968, 69, 111464.

Rev.

3.1.81. Sierro, J., Paramagnetic resonance of the Ag^{2+} ion in irradiated alkali chlorides. - J. Phys. & Chem. Solids, 1967, 28, N3, 417-422. - CA, 1967, 66, 80705. RZF, 1967, 8D572.

Exp.; E - e; O_h; d^9.

3.1.82. Van Willigen, H., van Broekhoven, J. A. M., de Boer, E., An ESR study of the mono- and dinegative ions of triphenylene: Evidence for the Jahn-Teller instability of the triplet dianion. - Mol. Phys., 1967, 12, N6, 533-548. - CA, 1968, 68, 82992.

Exp.; E - e; C_{3v}.

3.1.83. Wit, M., de, Reinberg, A. R., Electron paramagnetic resonance of copper in beryllium oxide. - Phys. Rev., 1967, 163, N2, 261-265. - CA, 1968, 68, 64472.

Exp.; T; T_d; d^9.

3.1.84. Zaripov, M. M., Kropotov, V. S., Livanova, L. D., Stepanov, V. G., Electron paramagnetic resonance of vanadium and chromium in CaF_2. [Russ.]. - Fiz. Tverd. Tela, 1967, 9, N1, 209-214. - RZF, 1967, 6D518.

Exp.; O_h; d^3.

1968

3.1.85. Boettger, H., Jahn-Teller resolution of an ESR line. [Germ.]. - Phys. Status Solidi, 1968, 26, N2, 681-692. - CA, 1968, 68, 110122. PA, 1968, 71, 31841.

Theor.; 3.2.

3.1.86. Chase, L. L., Electron spin resonance of the excited $^2E(3d^3)$ level of Cr^{3+} and V^{2+} in MgO. - Phys. Rev., 1968, 168, N2, 341-348. - RZF, 1968, 9D528.

Exp.; E; O_h; d^3.

3.1.87. Coffman, R. E., Jahn-Teller effect in the EPR spectrum of Cu^{++}:MgO at 1.2°K. - J. Chem. Phys., 1968, 48, N2, 609-618. - CA, 1968, 68, 83002. RZF, 1968, 9D537.

Exp.; E - e; O_h; d^9.

3.1.88. Coffman, R. E., Lyle, D. L., Matisson, D. R., Small tunneling effect in the electron paramagnetic resonance spectrum of Cu^{2+}-CaO at 1.2°K. - J. Phys. Chem., 1968, 72, N4, 1392-1394. - CA, 1968, 68, 110176.

Exp.; E - e; O_h; d^9.

3.1.89. Johannesen, R. B., Candela, G. A., Tsang Tung, Jahn-Teller distortion: magnetic studies of vanadium tetrachloride. - J. Chem. Phys., 1968, 48, N12, 5544-5549. - CA, 1968, 69, 55689. PA, 1968, 71, 49619.

Exp.; E - e; T_d; d^1.

3.1.90. Kamimura, H., Mizuhashi, S., Magnetic anisotropy due to dynamical Jahn-Teller effect in $d\varepsilon^1$ and $d\varepsilon^5$ ions. - J. Appl. Phys., 1968, 39, N2, 684-686. - CA, 1968, 68, 73445. PA, 1968, 71, 36234.

Theor.; E - (b_1 + b_2); C_{4v}; d^1; d^5.

3.1.91. Krupka, D. C., Williams, F. I. B., Breen, D. P., Electron spin echo study of a Jahn-Teller system. - Bull. Am. Phys. Soc., 1968, 13, N8, 459.

Exp.; E - e; O_h; d^9.

3.1.92. Lin Wei Ching, McDowell, C. A., Ward, D. J., Electron paramagnetic resonance of X-ray-irradiated single crystals of potassium cobalticyanide, $K_3Co(CN)_6$. - J. Chem. Phys., 1968, 49, N7, 2883-2886. - CA, 1969, 70, 8017. PA, 1969, 72, 10278.

Exp.

3.1.93. Low, W., Rosental, A., ESR and optical spectrum of Ti^{3+} in CaF_2. - Phys. Lett. A, 1968, 26, N4, 143. - RZF, 1968, 5D559.

Exp.; E - e; O_h; d^1; 2.1.

3.1.94. Müller, K. A., Effective-spin Hamiltonian for "non-Kramers" doublets. - Phys. Rev., 1968, 171, N2, 350-354. - RZF, 1969, 2D752.

Theor.

3.1.95. Sharf, B., Jortner, J., Asymmetric vibronic effects on degenerate electronic states. - Chem. Phys. Lett., 1968, 2, N2, 68-70. - RZF, 1969, 1D464.

Theor.

3.1.96. Watkins, G. D., EPR and optical absorption studies in irradiated semiconductors. - In: Radiat. Eff. Semicond.: Proc. Santa Fe Conf., 1967. New York; London, 1968, p. 67-81. - PA, 1971, 74, 21950.

Rev.; 2.1.

3.1.97. Weeks, M. J., Fackler, J. P., Single-crystal electron paramagnetic resonance studies of copper diethyldithiocarbamate. - Inorg. Chem., 1968, 7, N12, 2548-2553. - CA, 1969, 70, 15905.

Exp.; D_{4h}; d^9; 5.4.

1969

3.1.98. Abragam, A., The Jahn-Teller effect in paramagnetic resonance. - Comments Solid State Phys., 1969, 2, N3, 69-75. - CA, 1970, 72, 105264. PA, 1970, 73, 12678.

Theor.; E; T.

3.1.99. Bates, C. A., Bentley, J. P., Lattice ion interactions of Ti^{3+} in corundum. I. Jahn-Teller effect and the theory of the effects of electric field and strain. - J. Phys. C, 1969, 2, N11, 1947-1963. - CA, 1970, 72, 7372. PA, 1970, 73, 24106.

Exp.; E - e; O_h; C_{3v}; d^1; 3.2.

3.1.100. Bhattacharyya, B. D., Role of Jahn-Teller effect in the ligand field theory of magnetic behavior of tetrahedral Fe^{2+} complexes. - In: Abstr. Symp. Nuclear Phys. & Solid State Phys. New Delhi, 1969, 1 p. - PA, 1970, 73, 41083.

Theor.; E - e; T_d; d^6.

3.1.100a. Chase, L. L., Identification of a Jahn-Teller tunneling level. - Phys. Rev. Lett., 1969, 23, N6, 275-277. - CA, 1969, 71, 65117. PA, 1969, 72, 45383.

Exp.; E - e; O_h.

3.1.101. Fedder, R. C., Paramagnetic resonance study of Ag^{2+} distortions of CaF_2. - Bull. Am. Phys. Soc., 1969, 14, N1, 62.

Exp.; E - e; O_h; d^9; 2.1.

3.1.102. Friebel, C., Propach, V., Reinen, D., The Jahn-Teller effect in diffuse reflection and ESR spectra of Cu^{2+} ions in solid oxides. - In: Vortragsberichte Symp. "Koordination-schemie Ubergangselemente." Jena, 1969, p. 76-87.

Exp.; E - e; O_h; d^9; 2.1.

3.1.103. Groot, M. S., de, Hesselmann, I. A. M., Van der Waals, J. H., Paramagnetic resonance in phosphorescent aromatic hydrocarbons. V. Benzene in perdeuterobenzene crystal. - Mol. Phys., 1969, 16, N1, 45-60. - CA, 1969, 70, 82875.

Exp.; $(E_u + B_{1u}) - e_g$; D_{6h}; 2.2.

3.1.104. Groot, M. S., de, Hesselmann, I. A. M., Van der Waals, J. H., Paramagnetic resonance in phosphorescent aromatic hydrocarbons. VI. Mesitylene in B-trimethylborazole. - Mol. Phys., 1969, 16, N1, 61-68. - CA, 1969, 70, 82876.

Exp.; $(E_u + B_{1u}) - e_g$; D_{6h}; D_{3h}; 2.2.

3.1.105. Hardeman, G. E. G., Gerritsen, G. B., Magnetic resonance in 6H silicon carbide. - In: **Silicon Carbide:** Proc. 2nd Int. Conf. New York, 1969, p. S261-S272. - CA, 1972, 76, 133888.

Exp.; E; D_{4h}; d^5; 3.4; 3.6.

3.1.106. Khaldre, Yu. Yu., EPR spectrum of Cu^{2+} ions in a NaCl-Cu crystal. [Russ.]. - In: Tr. Inst. Fiz. i Astron. Akad. Nauk Est.SSR, 1969, N35, 234-236. - CA, 1970, 73, 50595. RZK, 1979, 24B299.

Exp.; E - e; O_h; d^9.

3.1.107. MacFarlane, R. M., g-Values of the d^1 $^2T_2$ ground term in the presence of a dynamic Jahn-Teller effect. - Phys. Rev., 1969, 184, N2, 603. - CA, 1969, 71, 116679. PA, 1970, 73, 4904.

Theor.; T - e; O_h; d^1.

3.1.108. Mizuhashi Seiji, Anisotropy of g-values in low-spin ferrihemoglobin azide. - J. Phys. Soc. Jpn., 1969, 26, N2, 468-492. - CA, 1969, 70, 84364. PA, 1969, 72, 35600.

Theor.; E - $(b_1 + b_2)$; C_{4v}; d^5.

3.1.109. Pilbrow, J. R., Stevenson, R. W. H., ESR of copper(II) ions in lithium chloride. - Phys. Status Solidi, 1969, 34, N1, 293-300. - CA, 1969, 71, 55399. PA, 1969, 72, 43281.

Exp.; E - e; O_h; d^9.

3.1.110. Rai, R., Savard, J. Y., Tousignant, B., Jahn-Teller effect in the chromium(III) ion in a tetrahedral environment. - Can. J. Phys., 1969, 47, N11, 1147-1153. - CA, 1969, 70, 119934. PA, 1969, 72, 44259.

Theor.; T - e; T_d; d^3.

3.1.111. Von Hoene, D. C., Fedder, R. C., EPR spectra of copper in cadmium fluoride at 4.2°K. - Phys. Lett. A, 1969, 30, N1, 1-2. - CA, 1970, 72, 7815. PA, 1970, 73, 5053.

Exp.; E - e; O_h; d^9.

3.1.112. Watts, R. K., Electron paramagnetic resonance of Ni^+ and Ni^{3+} in ZnSe. - Phys. Rev., 1969, 188, N2, 568-571. - PA, 1970, 73, 41349.

Exp.; T_d; d^7; d^9.

3.1.113. Zdansky, K., Dynamic Jahn-Teller effect of $MgO:Cu^{++}$ at 4.2°K. - Phys. Rev., 1969, 177, N2, 490-493. - CA, 1969, 70, 92188. PA, 1969, 72, 24494.

Exp.; E - e; O_h; d^9.

1970

3.1.114. Abragam, A., Bleaney, B., Electron Paramagnetic Resonance of Transition Ions. - New York: Oxford Univ. Press, 1970. - 912 p. (Int. Ser. of Monogr. on Phys.). - CA, 1970, 73, 114908.

3.1.115. Bratashevskiy, Yu. A., Litvin, Yu. A., Samsonenko, N. D., Sobolev, E. V., Electron paramagnetic resonance in lattice defects in synthetic diamonds. [Russ.]. - Izv. Akad. Nauk SSSR. Neorg. Mater., 1970, 6, N2, 368-369. - CA, 1970, 72, 127137.

Exp.; T; T_d.

3.1.116. Fedder, R. C., Jahn-Teller distortions of Ag^{2+} ions in strontium difluoride and calcium difluoride by odd modes. - Phys. Rev. B, 1970, 2, N1, 32-39. - CA, 1971, 74, 70082. PA, 1970, 73, 76212.

Exp.; d^9.

3.1.117. Hagston, W. E., Manifestations of the dynamic Jahn-Teller effect in trigonally distorted defect centers. - Mol. Phys., 1970, 19, N5, 593-601. - CA, 1971, 74, 8219. PA, 1970, 73, 69055.

Theor.; T; T_d; 2.1.

3.1.118. Von Hoene, D. C., An investigation of the Jahn-Teller effect on some transition metal ions in the fluoride lattice: Univ. Cincinnati, Ohio, Thesis. - 130 p. [Order N69-6358.] PA, 1970, 73, 24105.

Exp.; E - e; O_h; d^9.

3.1.119. Watkins, G. D., Ham, F., Electron paramagnetic resonance studies of a system with orbital degeneracy: the lithium donor in silicon. - Phys. Rev. B, 1970, 1, N10, 4071-4098.

Exp.; $(E + T_2)$; T_d.

1971

3.1.120. Bates, C. A., Chandler, P. E., The Jahn-Teller interaction and $3d^9$ ions at tetrahedral sites. - J. Phys. C, 1971, 4, N16, 2713-2724. - CA, 1972, 76, 39702. PA, 1972, 75, 7013. RZF, 1972, 4E816.

Theor.; T; T_d; d^9.

3.1.121. Bersuker, I. B., Vekhter, B. G., Ogurtsov, I. Ya., Electronic degeneracy and quasi-degeneracy effects in coordination systems with strong electron-vibrational coupling. [Russ.]. - In: Theory of Electronic Shells of Atoms and Molecules: Rep. Int. Symp., 1969. Vilnius, 1971, p. 281-289.

Theor.

3.1.122. Bersuker, I. B., Vekhter, B. G., EPR in systems with inversion splitting. [Russ.]. - In: Paramagnitniy Rezonans. 1944-1969: Vsesoyuz. Yubileynaya Konf., Kazan', 1969. Moscow, 1971, p. 54-60.

Rev.

3.1.123. Bill, H., Jahn-Teller effect of the O^- ion in CaF_2. - Solid State Commun., 1971, 9, N8, 477-480. - PA, 1971, 74, 37465.

Exp.; O_h; 3.5.

3.1.124. Bratashevskiy, Yu. A., Bukhan'ko, F. N., Samsonenko, N. D., Shapiro, O. Z., Dynamic Jahn-Teller effect on defects in synthetic diamond powders. [Russ.]. - Fiz. Tverd. Tela, 1971, 13, N7, 2154-2156. - CA, 1971, 75, 103354. PA, 1971, 74, 77047. PZF, 1971, 12D705.

Exp.; E - e; T - (e + t); T_d.

3.1.125. Carrington, A., Fabris, A. R., Howard, B. J., Lucas, J. D., Electron resonance studies of the Renner effect. I. Gaseous NCO in its $^2\Pi_{3/2}$ (n = 1), $^2\Delta_{5/2}$ (n = 2), and $^2\Phi_{7/2}$ (n = 3) vibronic states. - Mol. Phys., 1971, 20, N6, 961-980.

Exp.; E - e; $C_{\infty v}$; 1.6.

3.1.126. Dixon, J. M., Modifications to the magnetic g-tensor of a Pt^+ ion in ruby due to the Jahn-Teller effect. - J. Phys. C, 1971, 4, N10, 1221-1230. - PA, 1971, 74, 62296. RZF, 1972, 2E953.

Theor.; E - e; C_{3v}.

3.1.127. Fletcher, J. R., The angular dependence of strain-broadened resonance lines. - J. Phys. C, 1971, 4, N9, L156-L158. - CA, 1971, 75, 69385.

Theor.; C_{3v}.

3.1.128. Harrowfield, B., Cesium titanium sulfate: a possible Jahn-Teller system [EPR measurement]. - In: Abstr. 8th Austral. Spectrosc. Conf. Clayton, 1971, K7. - PA, 1972, 75, 10288.

Exp.

3.1.129. Hiraki, A., EPR in electron-bombarded phosphorus-doped n-type germanium - an attempt to detect the presence of bombardment-induced interstitial phosphorus. - Radiat. Eff., 1971, 9, N1/2, 51-55. - CA, 1971, 75, 26860.

Exp.; T_d.

3.1.130. Kuwabara Goro, ESR investigation of $3d^9$ ions in alkali halides. - J. Phys. Soc. Jpn., 1971, 31, N4, 1074-1084. - CA, 1971, 75, 135568. PA, 1971, 74, 77046.

Exp.; E - e; O_h; d^9; 2.4.

3.1.131. Landi, A., Blanchard, C., Parrot, R., g-Factors of the first excited states of Mn^{2+} ion in zinc sulfide when a dynamical Jahn-Teller effect is present. - Phys. Lett. A, 1971, 36, N4, 267-268. - CA, 1972, 76, 8314. PA, 1971, 74, 73927.

Theor.; T; T_d; d^5.

3.1.132. Mehran, F., Morgan, T. N., Title, R. S., Blum, S. E., The effects of the dynamical Jahn-Teller interaction on the EPR of shallow accpetors in gallium phosphide. In: Abstr. 4th Int. Symp. Magnetic Resonance. Rehovot, 1971, 2 p. - PA, 1972, 75, 7119.

Exp.; T_d.

3.1.133. Miyanaga Takeshi, Kanno Kenichi, Naoe Shunichi, Matsumoto Hiroaki, Electron paramagnetic resonance study of Ag^{2+} ions in cadmium chloride. - J. Phys. Soc. Jpn., 1971, 30, N6, 1669-1675. - CA, 1971, 75, 42939. PA, 1971, 74, 49668.

Exp.; E - e; O_h; d^9.

3.1.134. Noack, M., Kokoszka, G. F., Gordon, G., Dynamic Jahn-Teller effects and magnetic anisotropies in aqueous solutions and water-ethanol glasses of copper(II) solvates and complexes with 2,2'-dipyridine. - J. Chem. Phys., 1971, 54, N3, 1342-1350. - CA, 1971, 74, 70068. PA, 1971, 74, 30993.

3.1.134a. Otka, A. J., El'chaninova, S. D., Zvyagin, A. I., The estimation of spin-phonon interaction parameters of Eu^{2+} in CaF_2 using uniaxial deformation influence on the EPR spectra. [Russ.]. - In: Fiz. kondensir. sostoyaniya. Kharkov, 1971, 11, 3-11.

Exp.; O_h.

3.1.135. Rumin, N., Walsh, D., The Jahn-Teller coupling of T_2 ions. - Phys. Canada, 1971, 27, N4, 49-50. - PA, 1971, 74, 65986.

Exp.; T - e; O_h; d^1.

3.1.136. Samoylovich, M. I., Bezrukov, G. N., Butuzov, V. P., Electron paramagnetic resonance of nickel in synthetic diamond. [Russ.]. - Pis'ma, Zh. Eksp. & Teor. Fiz., 1971, 14, N10, 551-553.- CA, 1972, 76, 92753.

E - e; C_{3v}; d^9.

3.1.137. Sturge, M. D., Guggenheim, H. J., Dynamic Jahn-Teller effect in the $^4T_2$ excited states of $d^{3,7}$ ions in cubic crystals. II. Co^{2+} in $KMgF_3$. - Phys. Rev. B, 1971, 4, N7, 2092-2099. - PA, 1971, 74, 69868. RZF, 1972, 4E817.

Exp.; T - (e + t_2); O_h; d^7; 2.1; 2.3; 2.4.

3.1.138. Wallin, J. T., Piper, W. W., Paramagnetic resonance of Fe^{2+} in potassium magnesium trifluoride. - Solid State Commun., 1971, 9, N11, 823-825. - CA, 1971, 75, 56499.

Exp.; T; O_h; d^6.

3.1.139. Wallin, J. T., Watkins, G. D., Jahn-Teller effect for Cr^{2+} in II-VI crystals. - Solid State Commun., 1971, 9, N13, 953-956. - CA, 1971, 75, 92914.

Exp.; T; T_d; d^4.

3.1.140. Watts, R. K., Holton, W. C., De Wit, M., Phosphorus and arsenic impurity centers in zinc selenide. I. Paramagnetic resonance. - Phys. Rev. B, 1971, 3, N2, 404-409. - CA, 1971, 74, 105482. PA, 1971, 74, 30576.

Exp.; T_d; s^2p^5.

3.1.141. Wertheim, G. K., Hausmann, A., Sander, W., The Electronic Structure of Point Defects as Determined by Mössbauer Spectroscopy and by Spin Resonance. - Amsterdam: North-Holland, 1971. - 222 p. - (Defects in Crystalline Solids; Vol. 4.) - CA, 1972, 76, 160695.

Rev.; 3.4.

1972

3.1.142. Al'tshuler, S. A., Kozyrev, B. M., Elektronniy Paramagnitniy Rezonans Soedineniy Elementov Promezhutochnykh Grupp. [Russ.]. - Izd. 2-e, pererab. - Moscow: Nauka, 1972. - 672 p.

Electron Paramagnetic Resonance of Compounds of the Intermediate Group Elements.

3.1.143. Ammeter, J. H., Swalen, J. D., Electronic structure and dynamic Jahn-Teller effect of cobaltocene from EPR and optical studies. - J. Chem. Phys., 1972, 57, N2, 678-698. - CA, 1972, 77, 68304. PA, 1972, 75, 56226. RZF, 1972, 11D613.

Exp.; E - e; D_{5h}; D_{5d}; 2.1.

3.1.144. Bates, C. A., Oglesby, M. J., Standley, K. J., The properties of Co^{2+} in zinc tungstate. I. The EPR spectrum and its interpretation. - J. Phys. C, 1972, 5, N20, 2949-2960. - PA, 1972, 75, 81457.

Exp.; d^7.

3.1.145. Bersuker, I. B., Vekhter, B. G., Rafalovich, M. L., The ESR spectra of systems with an electronically degenerate T-term: the influence of the electric field. [Russ.]. - Teor. & Eksp. Khim., 1972, 8, N6, 739-744. - PA, 1976, 79, 31371. RZK, 1973, 14B282.

Theor.; T - t.

3.1.146. Bhattacharyya, B. D., On lattice-ion interactions of tetrahedral-site Ni^{2+} ions. - In: Abstr. Symp. Nuclear Phys. & Solid State Phys. Bombay, 1972, 1 p. - PA, 1972, 75, 36791.

Theor.; T; T_d; d^8; 3.2.

3.1.147. Bill, H., Comment on Jahn-Teller distortion of Ag^{2+} ions in SrF_2 and CaF_2 by odd modes. - Phys. Rev. B, 1972, 6, N11, 4359. - CA, 1973, 78, 9939. PA, 1973, 76, 2356. RZF, 1973, 6D689.

Exp.; d^9.

3.1.148. Edel, P., Hennies, C., Merle d'Aubigne, Y., Romestain, R., Twarowski, Y., Optical detection of paramagnetic resonance in the excited state of F centers in CaO. - Phys. Rev. Lett., 1972, 28, N19, 1268-1271. - PA, 1972, 75, 44437.

Exp.; T; O_h; 2p.

3.1.149. Fackler, J. P., Jr., Levy, J. D., Smith, J. A., Electron paramagnetic resonance spectra of copper(II) and oxovanadium-(IV) complexes oriented in nematic glasses from liquid crystal solvent. - J. Am. Chem. Soc., 1972, 94, N7, 2436-2445. - CA, 1972, 76, 147081. RZK, 1972, 17B314.

Exp.; d^1; d^9.

3.1.150. Faughnan, B. W., Electron paramagnetic resonance spectrum of Mo^{5+} in strontium titanate: Example of the dynamic

Jahn-Teller effect. - Phys. Rev. B, 1972, 5, N12, 4925-4931. - CA, 1972, 77, 54541. PA, 1972, 75, 48295.

Exp.; O_h.

3.1.151. Ham, F. S., Jahn-Teller effect in electron paramagnetic resonance spectra. - In: Electron Paramagnetic Resonance. New York, 1972, p. 1-119. - Bibliogr.: 202 ref. CA, 1973, 78, 90363.

Rev.

3.1.152. Hausmann, A., Schreiber, P., Observation of the dynamic Jahn-Teller effect with Nb^{2+} ions in zinc oxide. - Solid State Commun., 1972, 10, N10, 957-959. - CA, 1972, 77, 26786. PA, 1972, 75, 48284. RZF, 1972, 10D629.

Exp.; E; T_d; d^3.

3.1.153. Herrington, J. R., Estle, T. L., Boatner, L. A., Observation of a quadrupole interaction for cubic imperfections exhibiting a dynamic Jahn-Teller effect. - Phys. Rev. B, 1972, 5, N7, 2500-2510. - CA, 1972, 76, 119606. RZF, 1972, 10D630.

Exp.; E - e; O_h; d^1.

3.1.154. Mehran, F., Morgan, T. N., Title, R. S., Blum, S. E., Effects of dynamical Jahn-Teller interaction on the EPR of shallow acceptors in gallium phosphide. - J. Magn. Resonance, 1972, 6, N4, 620-627. - CA, 1972, 77, 11970.

Exp.; Γ_8 - (e + t); T_d.

3.1.155. Miyanaga Takeshi, Silver(2+) ion center in cadmium chloride. [Jap.]. - Wakayama Daigaku Kyoikugabu Kiyo, Shizen Kagaku, 1972, 22, 5-12. - CA, 1973, 79, 110073.

Exp.; E - e; O_h; d^9.

3.1.156. Mollenauer, L. F., Baldacchini, G., Empirically determined wave function for the relaxed-excited state of the F center in KI [ESR and ENDOR]. - Phys. Rev. Lett., 1972, 29, N8, 465-468. - PA, 1972, 75, 68738.

Exp.; $(A_{1g} + T_{1u})$ - t_{1u}; O_h; 2s; 2p; 3.3.

3.1.157. Rai, R., Jahn-Teller theory of the Ti^{3+} ion in corundum. - Phys. Status Solidi b, 1972, 52, N2, 671-681. - CA, 1972, 77, 158479. PA, 1972, 75, 65331. RZF, 1973, 1E815.

Theor.; E - e; C_{3v}; d^1.

3.1.158. Shen, L., Nai-Sing, Dynamic Jahn-Teller effect in C_3 symmetry Ni^{3+} doped aluminum oxide. - 1972. - 236 p. - [Order N72-25008.]. Diss. Abstr. Int. B, 1972, 33, N4, 1740. CA, 1973, 70, 22300.

Exp.; E - e; C_{3v}; d^7.

3.1.159. Shen L., Nai-Sing, Estle, T. L., The effects of applied stress on the EPR spectra of a system exhibiting the dynamic Jahn-Teller effect. II. $Al_2O_3:Ni^{3+}$. - Bull. Am. Phys. Soc., 1972, 17, N3, 263.

Exp.; E - e; C_{3v}; d^7; 3.5.

3.1.160. Wilson, D. G., Lohr, L. L., Jr., Simple model of the dynamic Jahn-Teller effect in six-coordinated copper(II) complexes. - J. Chem. Phys., 1972, 57, N2, 702-709. - CA, 1972, 77, 68287. PA, 1972, 75, 53004. RZF, 1972, 12E885.

Theor.; E - e; O_h; d^9.

3.1.161. Yamaguchi Tsuyoshi, Kamimura Hiroshi, Dynamic Jahn-Teller interaction in optical and paramagnetic resonance spectra: Application to a Cu deep impurity in II-VI semiconductors. - J. Phys. Soc. Jpn., 1972, 33, N4, 953-966. - CA, 1972, 77, 158364. PA, 1972, 75, 80954. RZF, 1973, 3D786.

Exp.; E - e; T - e; O_h; T_d; d^9; 2.1.

1973

3.1.162. Amano Chikara, Fujiwara Shizuo, ESR of hot ions: Ni(I) complex ions produced in rigid solutions by γ-irradiation. - Bull. Chem. Soc. Jpn., 1973, 46, 1379-1383. - CA, 1973, 79, 35684. RZF, 1973, 11D1054.

Exp.; E - e; d^7; d^8; d^9.

3.1.163. Barksdale, A. O., Investigation by electron paramagnetic resonance of the Jahn-Teller effect for zinc sulfide: scandium(2+). - 1973. - 323 p. - [Order N 73-21532.]. Diss. Abstr. Int. B, 1973, 34, N3, 1228. CA, 1974, 80, 8819.

Exp.; E - e; T_d.

3.1.164. Barksdale, A. O., Estle, T. L., An electron paramagnetic resonance study of the dynamic Jahn-Teller effect for Sc^{2+} in ZnS. - Phys. Lett. A, 1973, 42, N6, 426-428. - CA, 1973, 78, 90706. PA, 1973, 76, 21302. RZF, 1973, 6D709.

Exp.; E - e; T_d.

3.1.165. Barksdale, A. O., Estle, T. L., Intermediate Jahn-Teller effect for ^2E states in cubic symmetry. - J. Chem. Phys., 1973, 59, N2, 962-963. - CA, 1973, 79, 85013. PA,1973, 76, 60987. RZF, 1974, 2E401.

Theor.; E - e; O_h; T_d.

3.1.166. Bates, C. A., Chandler, P. E., The Jahn-Teller interaction and $3d^9$ ions at tetrahedral sites. II. Tetragonal distortions. - J. Phys. C, 1973, 6, N11, 1975-1980. - CA, 1973, 79, 47261. PA, 1973, 76, 46799. RZF, 1973, 11E889.

Theor.; T - e; T_d; d^9.

3.1.167. Benoit a la Guillaume C., Lavallard, P., Magneto-emission of gallium antimonide: Free and bound excitons. - Phys. Status Solidi b, 1973, 59, N2, 545-549. - CA, 1973, 79, 131013.

Exp.

3.1.168. Bill, H., Milleret, C., Lacroix, R., EPR study of the Ag^{2+} ion in $SrCl_2$. - In: Magn. Resonance and Related Phenomena: Proc. 17th Cong. AMPERE, Turku, 1972. Amsterdam; London, 1973, p. 233-236.

Exp.; T - t; O_h; d^9.

3.1.169. Chandler, R. N., Bene, R. W., EPR study of the solid solutions $Ni_xFe_{1-x}S_2$, $Co_xFe_{1-x}S_2$, and $Co_xNi_yFe_{1-x-y}S_2$. - Phys. Rev. B, 1973, 8, N11, 4979-4988. - CA, 1974, 80, 76299. RZF, 1974, 8D862.

Exp.; d^7; d^8.

3.1.170. Chandler, R. N., Bene, R. W., EPR study of the pyrites $Co_xFe_{1-x}S_2$, $Ni_xFe_{1-x}S_2$, and $Co_xNi_yFe_{1-x-y}S_2$. - 1973. - 59 p. - US Nat. Techn. Inform. Serv. AD Rep., N759547. - CA, 1973, 79, 72029.

Exp.; d^7; d^8.

3.1.171. Fukui Minoru, Hayashi Yoshikazu, Yoshioka Hide, ESR study of growth and decay processes of self-trapped holes in silver chloride crystals doped with copper. - J. Phys. Soc. Jpn., 1973, 34, N5, 1226-1233. - CA, 1973, 78, 166674.

Exp.; E - e; O_h; d^9.

3.1.172. Goltzene, A., Schwab, C., Meyere, B., Nikitine, S., Absorption spectrum and electronic paramagnetic resonance of point defects associated to group VI impurities in CuCl. - In: Abstr. 11th Eur. Congr. Mol. Spectrosc. Tallin, 1973, 56. - PA, 1973, 76, 68844.

Exp.; 2.1.

3.1.173. Holuj, F., Wilson, R. G., Trigonal Jahn-Teller effects in the electron-spin resonance of Cu^{++} in $Ca(OD)_2$. - Phys. Rev. B, 1973, 7, N9, 4065-4072. - CA, 1973, 78, 166707. PA, 1973, 76, 39024. RZF, 1974, 4D748.

Exp.; E - e; C_{3v}; d^9.

3.1.174. Koch, R. C., Joesten, M. D., Venable, J. H., Jr., Single crystal electron paramagnetic resonance study of tris(octamethylpyrophosphoramide) copper(II) perchlorate. - J. Chem. Phys., 1973, 59, N12, 6312-6320. - CA, 1974, 80, 76296.

Exp.; E - e; D_3; d^9.

3.1.175. Kooistra, C., Van Dijk, J. M. F., Van Lier, P. M., Buck, H. M., Study of π-conjugation in Chichibabin's- and Schlenk's-like hydrocarbons with ESR. I. - Recl. Trav. Chim. Pays-Bas, 1973, 92, N9-10, 961-969. - CA, 1974, 81, 12618.

Exp.

3.1.176. Kozhuhar, A. Yu., Tsintsadze, G. A., Shapovalov, V. A., Seleznev, V. N., The Jahn-Teller effect in a lithium-gallium spinel. - Phys. Lett. A, 1973, 42, N5, 377-378. - CA, 1973, 78, 130270. PA, 1973, 76, 21300.

Exp.; E - e; O_h; d^9.

3.1.177. Misra, B.N., Sharma, S. D., Gupta, S. K., Evaluation of g values for compounds with a paramagnetic ion. - Z. Naturforsch. a, 1973, 28, N2, 246-248. - PA, 1974, 77, 41166.

Theor.; d^9.

3.1.178. Niewenhuijse, B., Reedijk, J., Jahn-Teller distortions in copper(II) complexes as determined from ESR powder spectra. - Chem. Phys. Lett., 1973, 22, N1, 201-203. - CA, 1974, 80, 32274. PA, 1974, 77, 1334. RZF, 1974, 4D761.

Exp.; E - e; O_h; d^9.

3.1.179. Schoenberg, A., EPR study in single crystals of magnesia enriched with oxygen-17. - 1973. - 92 p. - Report INIS-MF-1252. - CA, 1975, 83, 50427.

Exp.; E - e; O_h; d^7; d^9.

3.1.180. Ursu, I., Nistor, S. V., ESR studies of some irradiated paramagnetic centers in alkali halides. - In: Magn. Resonance and Related Phenomena: Proc. 17th Congr. AMPERE, Turku, 1972, Amsterdam; London, 1973, p. 166-176.

Exp.; $(A_{1g} + T_{1u})$ - t_{1u}; O_h; p^5.

3.1.181. Vincent, C., Walsh, D., Measurement of tunneling and strain parameters for octahedrally coordinated $3d^9$ ions. - Phys. Rev. B, 1973, 8, N11, 4935-4940. - CA, 1974, 80, 76300. PA, 1974, 77, 41245.

Exp.; E - e; C_{3v}; d^9.

1974

3.1.182. Ammeter, J. H., Brom, J. M., Jr., ESR spectra of cobaltocene in rare gas matrices. - Chem. Phys. Lett., 1974, 27, N3, 380-384. - PA, 1974, 77, 72226.

Exp.; E - e; O_h; d^4.

3.1.183. Bray, J. E., Magnetic resonance studies of tetravalent neptunium in an octahedral complex: Washington State Univ., Pullman. Thesis. - 1974. - 88 p. - [Order N 75-7625.]. - PA, 1976, 79, 10185.

Exp.; Γ_8; O_h.

3.1.184. Broser, I., Schulz, M., Electron paramagnetic resonance of $ZnS:Sc^{2+}$. - J. Phys. C, 1974, 7, N7, L147-L149.

Exp.; E - e; T_d.

3.1.185. Brower, K. L., EPR of a Jahn-Teller distorted (111) carbon interstitialcy in irradiated silicon. - Phys. Rev. B, 1974, 9, N6, 2607-2617. - CA, 1974, 81, 8069. PA, 1974, 77, 61035.

Exp.; C_{3v}.

3.1.186. Chandler, R. N., Bene, R. W., EPR study of FeS_2: Ni_ICo. - In: Magn. & Magn. Mater.: 19th AIP Annu. Conf., Boston, 1973. New York, 1974, Pt. 2, p. 534. - PA, 1975, 78, 1500.

Exp.; E - e; O_h; d^7.

3.1.187. De Siebenthal, J. M., Bill, H., A new O^- center in $SrCl_2$ crystals. - Phys. Status Solidi b, 1974, 65, N1, K35-K37. - CA, 1974, 81, 129464. PA, 1974, 77, 76159.

Exp.; p^5.

3.1.188. Edel, P., Le Si Dang, Merle d'Aubigne, Y., Jahn-Teller coupling in the excited level of F centers in CaO: dynamical aspects. - In: Extended Abstr. Int. Conf. Color Centers in Ionic Crystals. Sendai, 1974, F102, 1 p. - PA, 1975, 78, 22154.

Exp.; T - e; O_h; 2s2p.

3.1.189. Frey, W., Seidel, H., ESR investigations of tin(+) centers in potassium chloride. - Phys. Status Solidi b, 1974, 66, N2, K39-K41. - CA, 1975, 82, 49629.

Exp.; O_h.

3.1.190. Friebel, C., Reinen, D., Do tetragonally compressed coordination octahedrons with divalent copper exist? [Germ.]. - Z. Anorg. Allg. Chem., 1974, 407, N2, 193-200. - CA, 1974, 81, 113305.

Exp.; E - e; O_h; d^9; 4.6.

3.1.191. Graham, S. O., White, R. L., EPR study of the Jahn-Teller effect on copper(2+) ions in cesium cadmium chloride. - Phys. Rev. B, 1974, 10, N11, 4505-4509. - CA, 1975, 82, 66260. PA, 1975, 78, 29922. RZF, 1975, 6D682.

Exp.; E - e; d^9.

3.1.192. Hathaway, B. J., Hodgson, P. G., Power, P. C., Single-crystal electronic spin resonance spectra of three-chelate copper(II) complexes. - Inorg. Chem., 1974, 13, N8, 2009-2013. - CA, 1974, 81, 70651.

Exp.; E - e; O_h; d^9.

3.1.193. Jaussaud, P. C., Abou-Ghantous, M., Bates, C. A., Fletcher, J. R., Moore, W. S., Distinction between lattice and cluster models of the Jahn-Teller effect in an orbital doublet. - Phys. Rev. Lett., 1974, 33, N9, 530-533. - PA, 1974, 77, 73023.

Exp.; E - e; O_h; C_{3v}; d^7; 1.7.

3.1.194. Miyanaga Takeshi, EPR study of a Jahn-Teller system, $CdBr_2:Ag^{++}$. - In: Extended Abstr. Int. Conf. Color Centers in Ionic Crystals. Sendai, 1974, 195, 2 p. - PA, 1975, 78, 42529.

Exp.; E - e; O_h; d^9.

3.1.195. Park Yoon Chang, ESR studies of metal-ligand and metal-metal interactions in copper(II) complexes: Jahn-Teller and exchange effects. - 1974. - 226 p. - [Order N 74-23136.]. - Diss. Abstr. Int. B, 1975, 35, N9, 4411. - CA, 1975, 82, 177783.

Exp.; d^9; 1.8.

3.1.196. Shing, Y. H., Walsh, D., Quasi Γ_8 ground state of a Jahn-Teller ion. - Phys. Rev. Lett., 1974, 33, N18, 1067-1069. - CA, 1974, 81, 179566. PA, 1975, 78, 1284.

Theor.; Γ_8; d^1.

3.1.197. Shing, Y. H., Walsh D, Quasi Γ_8 ground state of titanium substituted alums. - J. Phys. C, 1974, 7, N18, L346-L348. - CA, 1974, 81, 161471. PA, 1974, 77, 79299.

Theor.; T - e; Γ_8 - e; O_h; d^1.

3.1.198. Tolparov, Yu. N., Bir, G. L., Sochava, L. S., Kovalev, N. N., A Jahn-Teller ion in an off-center position: The $SrO{\cdot}Cu^{2+}$ system. [Russ.]. - Fiz. Tverd. Tela, 1974, 16, N3, 895-905. - CA, 1974, 81, 8120. PA, 1975, 78, 10215.

Exp.; E - e; O_h; C_{3v}; d^9.

3.1.199. Volkova, L. A., Nizamutdinov, N. M., Dunin-Barkovskiy, R. L., Vinokurov, V. M., Samoylovich, M. I., Optical absorption spectra and electron paramagnetic resonance of Ni^{2+} and Cu^{2+} ions in zinc spinel ($ZnAl_2O_4$) single crystals. [Russ.]. - Minsk, 1974. - 11 p. -Deposited Doc., VINITI, 1974, N596-74 Dep. - CA, 1974, 81, 83947. RZF, 1974, 7D702Dep.

Exp.; T; T_d; d^8; d^9; 2.1.

3.1.200. Wallin, J. T., Watkins, G. D., EPR of chromium(2+) ion in II-VI lattices. - Phys. Rev. B, 1974, 9, N5, 2051-2072. - CA, 1974, 81, 8117. PA, 1974, 77, 49226.

Exp.; T - e; T_d; d^4.

3.1.201. Wohlecke, M., Zeeman effect of the Cu^{2+} centre in cubic ZnS. - J. Phys. C, 1974, 7, N14, 2557-2568. - CA, 1974, 81, 97330. RZF, 1975, 2D505.

Exp.; E - e; O_h; d^9; 2.7.

1975

3.1.202. Ammeter, J. H., Oswald, N., Bucher, R., Dynamic Jahn-Teller distortions and chemical bonding conditions in orbitally degenerate sandwich complexes. [Germ.]. - Helv. Chim. Acta, 1975, 58, N3, 681-682. - CA, 1975, 83, 27087.

Exp.; E - e; D_{5h}; d^5; d^7; 5.2.

3.1.203. Balestra, C., Bill, H., Investigation of the $4d^1$ ion Y^{2+} in CaF_2 crystals. - In: Magn. Resonance & Related Phenomena: Proc. 18th Congr. AMPERE, 1974. Amsterdam, 1975, Vol. 1, p. 155-156. - PA, 1975, 78, 81698.

Exp.; E - e; O_h; d^1.

3.1.204. Bertini, I., Gatteschi, D., Paoletti, P., Scozzafava, A., Jahn-Teller effect in the complex dipotassium lead(II) copper(II) hexanitrite: Comments. - Inorg. Chim. Acta, 1975, 13, N2, L5-L6. - CA, 1975, 83, 18551.

Exp.; E - e; O_h; d^9.

3.1.205. Bhattacharyya, B. D., Jahn-Teller effect in ESR study of chromium(2+) in II-VI semiconductors. - Phys. Status Solidi b, 1975, 71, N2, K181-K185. - CA, 1975, 83, 199970. PA, 1976, 79, 6087.

Theor.; T - e; T_d; d^4.

3.1.206. Cieplak, M. Z., Godlewski, M., Baranowski, J. M., Optical charge transfer spectra and EPR spectra of $Cr^{2+}(d^4)$ and $Cr^{1+}(d^5)$ in CdTe. - Phys. Status Solidi b, 1975, 70, N1, 323-331. - CA, 1975, 83, 88221. PA, 1975, 78, 77380.

Exp.; T; T_d; d^4; d^5; 2.1.

3.1.207. Clerjaud, B., Gelineau, A., Influence of the Jahn-Teller effect on the paramagnetic properties of the ground state of Zn/Cu^{2+}. - In: Magn. Resonance & Related Phenomena: Proc. 18th Congr. AMPERE, 1974. Amsterdam, 1975, Vol. 2, p. 563-564. - CA, 1975, 83, 170518. PA, 1975, 78, 86079.

Exp.; T; T_d; d^9.

3.1.208. Friebel, C., Reinen, D., Ligand field and ESR spectroscopic study of the Jahn-Teller effect of silver(2+) ions in fluoridic coordination. [Germ.]. - Z. Anorg. Allg. Chem., 1975, 413, N1, 51-60. - CA, 1975, 82, 177747.

Exp.; E - e; d^9; 2.1.

3.1.209. Groot, M. S., de, Hesselmann, I. A. M., Reinders, F. J., Paramagnetic resonance of phosphorescent tetramethylpyrazine. - Mol. Phys., 1975, 29, N1, 37-48. - PA, 1975, 78, 9317.

Exp.; 2.2.

3.1.210. Jesion, A., Shing, Y. H., Walsh, D., Suppression of the trigonal crystal field splitting of titanium(3+) ions in alum. - In: Magn. Resonance & Related Phenomena: Proc. 18th Congr. AMPERE, 1974. Amsterdam, 1975, Vol. 2, p. 561-562. - CA, 1975, 83, 170517. PA, 1975, 78, 86078.

Exp.; Γ_8; C_{3v}; d^1.

3.1.211. Kaneshima Nobuo, Yahara Itsuo, Kubo Hidenori, Hirakawa Kazuyoschi, Superhyperfine interactions of ligand nuclei in Cu^{2+} compounds. I. Theory. [Jap.]. - Kyushu Daigaku Kogaku Shuho, Technol. Rep. Kyushu Univ., 1975, 48, N5, 641-646. - CA, 1977, 86, 98188. PA, 1976, 79, 40400.

Theor.; E - e; O_h; d^9.

3.1.212. Lowther, J. E., Symmetric Jahn-Teller distortions around superoxide(-) ion in alkali halides. - Chem. Phys. Lett., 1975, 35, N1, 136-137. - CA, 1975, 83, 170209. PA, 1975, 78, 81361.

Theor.

3.1.213. Raizman, A., Suss, J. T., EPR study of ruthenium(3+) and rhodium(2+) in single crystals of calcium oxide. - In: Magn. Resonance & Related Phenomena: Proc. 18th Congr. AMPERE, 1974, Amsterdam, 1975, Vol. 1, p. 121-122. - CA, 1975, 83, 185919. PA, 1975, 78, 77384.

Exp.; T; E; O_h; d^5; d^7.

3.1.214. Setser, G. G., Investigation by electron paramagnetic resonance of the Jahn-Teller effect in nickel(3+)-doped lithium niobate(V) and copper(2+)-doped lithium niobate(V). - 1975. - 245 p. - [Order N 75-22059.]. - Diss. Abstr. Int. B, 1975, 36, N4, 1800-1801. - CA, 1975, 83, 211072.

Exp.; d^7; d^9.

1976

3.1.215. Abou-Ghantous, M., Clark, I. A., Moore, W. S., The properties of the ion chromium(5+) in alumina. II. The thermally detected EPR spectrum and the Jahn-Teller theory. - J. Phys. C, 1976, 9, N10, 1965-1973. - CA, 1976, 85, 133531. PA, 1976, 79, 61444.

Exp.; T; O_h; C_{3v}; d^1.

3.1.216. Bhattacharyya, B. D., Tunnel splitting in a trigonal O^- center. - Phys. Status Solidi b, 1976, 74, N1, K53-K56.

Theor.; T - t; O_h; p^5.

3.1.217. Bhattacharyya, B. D., A new aspect of Jahn-Teller effect in the low-spin ferrihemoglobin azide. - Phys. Status Solidi b, 1976, 74, N2, 695-700. - CA, 1976, 85, 12073. PA, 1976, 79, 53561.

Theor.; T; C_{4v}; d^5.

3.1.218. Bir, G. L., Static Jahn-Teller effect on an E state ion located in a tetragonal crystalline field. [Russ.]. - Fiz. Tverd. Tela, 1976, 18, N6, 1627-1630. - CA, 1976, 85, 101387. PA, 1977, 80, 24539.

Theor.; E - e; O_h; D_{4h}; d^9.

3.1.219. Canters, G. W., Jansen, G., Noort, M., Van der Waals, J. H., High resolution Zeeman experiments on singlet, triplet, and quartet states of metalloporphins. - J. Phys. Chem., 1976, 80, N20, 2253-2259. - CA, 1976, 85, 151412. RZK, 1977, 5B146.

Exp.; E - $(b_1 + b_2)$; D_{4h}.

3.1.220. Clerjaud, B., Gelineau, A., Jahn-Teller effect in the ground state of nickel(+) in zinc sulfide and zinc selenide. - In: Magn. Resonance & Related Phenomena: Proc. 19th Congr. AMPERE. Heidelberg; Geneva, 1976, p. 503-506. - CA, 1977, 87, 31549.

Theor.; T - e; T_d; d^9.

3.1.221. De Jong, H. J., Glasbeek, M., Chromium(5+) ion in strontium titanate(IV): an example of a static Jahn-Teller effect in a d^1 system. - Solid State Commun., 1976, 19, N12, 1197-1200. - CA, 1976, 85, 151496. PA, 1976, 79, 83967.

Exp.; T - e; O_h; d^1.

3.1.222. Jesion, A., Shing, Y. H., Walsh, D., Anisotropic EPR spectrum of the Jahn-Teller system Cu^{2+} in zinc bromate hexahydrate. - J. Phys. C, 1976, 9, N8, L219-L222. - PA, 1976, 79, 53937.

Exp.; E - e; O_h; d^9.

3.1.223. Mier-Maza, R., The superhyperfine structure of a system exhibiting the dynamic Jahn-Teller effect: scandium(2+) ion-doped calcium fluoride. - 1976. - 231 p. - [Order N 76-21696.]. Diss. Abstr. Int. B, 1976, 37, N4, 1764. CA, 1977, 86, 10454.

Exp.; E - e; T_d; d^1.

3.1.224. Pontnau, J., Adde, R., Analysis of the ground-state spin-Hamiltonian parameters and electric field effect for the $3d^2$ ions (V^{3+}, Cr^{4+}) in corundum. - Phys. Rev. B, 1976, 14, N9, 3778-3792. - Bibliogr.: 54 ref. CA, 1977, 86, 24169. PA, 1977, 80, 15801.

Theor.; E - e; C_{3v}; d^2; 3.5.

3.1.225. Rao, P. Sambasiva, Subramanian, S., Jahn-Teller tunneling in copper(II)-doped hexaimidazole zinc(II) dichloride tetrahydrate. - J. Magn. Resonance, 1976, 22, N2, 191-206. - CA, 1976, 85, 101733.

Exp.; E - e; O_h; d^9.

3.1.226. Teodorescu, M., Manifestations of the Jahn-Teller effect in ESR spectra of paramagnetic species with an electronic doublet ground state. [Rom.]. - Stud. & Cercet. Fiz., 1976, 28, N3, 233-249. - CA, 1976, 85, 26828. PA, 1976, 79, 68260.

Theor.; E - e.

3.1.227. Vaysleyb, A. V., Rozenfel'd, Yu. B., Tsukerblat, B. S., Electron paramagnetic resonance in an orbital doublet with a weak Jahn-Teller interaction. [Russ.]. - Fiz. Tverd. Tela, 1976, 18, N7, 1864-1873. - CA, 1976, 85, 101821. PA, 1977, 80, 28427.

Theor.; E - e.

3.1.228. Ziatdinov, A. M., Zaripov, M. M., Yablokov, Yu. V., The depression of the dynamic Jahn-Teller effect. - Phys. Status Solidi b, 1976, 78, N2, K69-K71. - CA, 1977, 86, 63091. PA, 1977, 80, 16164.

Exp.; E - e; O_h; d^9.

1977

3.1.229. Aldous, R., Baker, J. M., EPR in a new photoexcited center of terbium in calcium fluoride. - J. Phys. C, 1977, 10, N23, 4837-4841. - CA, 1978, 88, 161065. PA, 1978, 81, 19319.

Exp.; O_h.

3.1.230. Ball, D., Lowther, J. E., Jahn-Teller effects and hyperfine parameters of V^{4+} in tetragonally distorted tetrahedral sites. - Phys. Lett. A, 1977, 61, N5, 333-335. - CA, 1977, 87, 60384. PA, 1977, 80, 69276.

Theor.; E - e; T_d; d^1.

3.1.231. Bertini, I , Gatteschi, D., Scozzafava, A., Jahn-Teller distortions of tris(ethylenediamine)copper(II) complexes. - Inorg. Chem., 1977, 16, N8, 1973-1976. - CA, 1977, 87, 60348.

Exp.; E - e; D_3; d^9.

3.1.232. Bhattacharyya, B. D., Jahn-Teller effect of an orthorhombic center in MgO. - Acta Phys. Pol. A, 1977, 51, N6, 859-863. - CA, 1977, 87, 76058. PA, 1977, 80, 66140.

Theor.; T; O_h; d^1.

3.1.233. Boate, A. R., Morton, J. R., Preston, K. F., ESR spectra of arsenic hexafluoride(-2) and antimony hexafluoride(-2). - Chem. Phys. Lett., 1977, 50, N1, 65-69. - CA, 1977, 87, 159655. PA, 1977, 80, 81416.

Exp.; T; O_h.

3.1.234. Dey, D. K., Ghoshal, A. K., Sthanapati, J., Pal, A. K., Jahn-Teller effects in EPR of a square planar copper complex. - Phys. Lett. A, 1977, 62, N4, 265-266. - CA, 1977, 87, 125058. PA, 1977, 80, 80994.

Exp.; E - (b_1 + b_2); D_4h; d^9.

3.1.235. Dixon, J. M., The dynamic Jahn-Teller effect of an electronic Γ_8 quartet resulting from strong coupling to Γ_5 lattice modes in the fine structure of erbium(3+) in palladium-erbium. - J. Phys. C, 1977, 10, N6, 833-849. - CA, 1977, 87, 93083. PA, 1977, 80, 35895.

Theor.; Γ_8 - t.

3.1.236. Dixon, J. M., Crystal fields and dynamic Jahn-Teller coupling of an electronic quartet to Γ_5 lattice modes in the fine structure of erbium(3+) in palladium-erbium. - In: Proc. 2nd Int. Conf. Cryst. Field. Eff. Met. Alloys, New York, 1977, p. 89-93. - CA, 1977, 87, 159671.

Theor.; Γ_8 - t.

3.1.237. Ferrante, R. F., Wilkerson, J. L., Graham, W. R. M., Weltner, W., Jr., ESR spectra of the MnO, MnO_2, MnO_3, and MnO_4 molecules at 4°K. - J. Chem. Phys., 1977, 67, N12, 5904-5913. - CA, 1978, 88, 97056. RZK, 1978, 12B255.

Exp.; T; T_d.

3.1.238. Henke, W., Reinen, D., Spectroscopic studies on the Jahn-Teller effect of the copper(2+) ion in the terpyridine complexes $Cu(terpy)_2X_2 \cdot nH_2O$ [X = NO_3^-, ClO_4^-, Br^-]. [Germ.]. - Z. Anorg. Allg. Chem., 1977, 436, 187-200. - CA, 1978, 88, 96855.

Exp.; E - e; O_h; d^9; 2.1.

3.1.239. Kim, H., Lange, J., Superhyperfine splitting and the dynamic Jahn-Teller effect for the ferrous ion in potassium magnesium fluoride. - Phys. Rev. Lett., 1977, 39, N8, 501-504. - CA, 1977, 87, 125055. PA, 1977, 80, 85555.

Exp.; T - t; O_h; d^6.

3.1.240. Kooter, J. A., Canters, G. W., Van der Waals, J. H., Electron spin resonance of the lowest triplet state of palladium-porphine in a n-octane crystal at 1.3 K. - Mol. Phys., 1977, 33, N6, 1545-1564. - PA, 1977, 80, 58050.

Exp.; E - $(b_1 + b_2)$; D_{4h}.

3.1.241. Krebs, J. J., Stauss, G. H., EPR of $Cr(3d^3)$ in GaAs - evidence for strong Jahn-Teller effects. - Phys. Rev. B, 1977, 15, N1, 17-22. - CA, 1977, 86, 113374. PA, 1977, 80, 43875.

Exp.; E - e; T_d; d^3.

3.1.242. Krebs, J. J., Stauss, G. H., EPR of chromium(2+)$(3d^4)$ in gallium arsenide: Jahn-Teller distortion and photoinduced charge conversion. - Phys. Rev. B, 1977, 16, N3, 971-973. - CA, 1977, 87, 175371.

Exp.; T - e; T_d; d^4.

3.1.243. Luz, Z., Raizman, A., Suss, J. T., Oxygen-17 superhyperfine structure of Rh^{2+} Jahn-Teller ions in MgO. - Solid State Commun., 1977, 21, N8, 849-852. - PA, 1977, 80, 32253.

Exp.; E - e; O_h; d^7.

3.1.244. Moreno, M., The measurement of covalency in tetragonal complexes. - J. Phys. C, 1977, 10, N8, L183-L186. - PA, 1977, 80, 47781.

Exp.; E - $(b_1 + b_2)$; D_{4h}; d^9.

3.1.245. Mukai Kazuo, Yorimitsu Koichi, Mishina Tadashi, ESR study of the tert-pentyl derivative of Yang's biradical. - Bull. Chem. Soc. Jpn., 1977, 50, N9, 2471-2472. - CA, 1977, 87, 183627.

Exp.; E - e; D_{3h}.

3.1.246. Raizman, A., Suss, J. T., Low, W., Superhyperfine structure of magnesium-25 in the EPR spectrum of iridium(2+) ion in magnesium oxide. - Physica B + C, 1977, 86-88, N3, 1229-1230. - CA, 1977, 86, 180218. PA, 1977, 80, 69683.

Exp.; E - e; O_h; d^7.

3.1.247. Reedijk, J., Octahedral copper(II) complexes having tetragonal compression. - Chem. Weekbl. Mag., 1977, March, p. 97. - CA, 1978, 88, 15221.

Exp.; E - e; O_h; d^9; 2.1.

3.1.248. Stauss, G. H., Krebs, J. J., Henry, K. L., EPR study of iron(3+) and chromium(2+) indium phosphide. - Phys. Rev. B, 1977, 16, N3, 974-977. - CA, 1977, 87, 175372.

Exp.; T - e; O_h; d^4; d^5.

3.1.249. Surendran, K. K., Lingam, K. V., Rao, M. J., Bhattacharyya, B. N., Jahn-Teller effect of copper(2+) ion: An EPR study. - In: Proc. Nucl. Phys. & Solid State Phys. Symp., 1977, Vol. 20C, p. 570-572. - CA, 1979, 90, 130090.

Exp.; E - e; D_{3h}; d^9.

3.1.250. Van Ooijen, J. A. C., Van der Put, P. J., Reedijk, J., Compressed tetragonal geometry in Cu(II) doped dichlorobis(pyrazole)-cadmium. - Chem. Phys. Lett., 1977, 51, N2, 380-382. - CA, 1977, 87, 209175. PA, 1978, 81, 6568.

Exp.; d^9.

3.1.251. Vaysleyb, A. V., Rozenfel'd, Yu. B., Tsukerblat, B. S., Paraelectric and paramagnetic absorption in trigonal exchange clusters with weak Jahn-Teller coupling. [Russ.]. - In: Fiz. Mat. Metody Koord. Khim.: Tez. Dokl. 6 Vsesoyuz. Soveshch. Kishinev, 1977, p. 75-76.

Theor.; E - e; 3.5.

3.1.252. Wood, J. S., Keijzers, C. P., De Boer, E., The ESR spectra of the pure and zinc doped hexakis(pyridine N-oxide) cuprate $Cu(C_5H_5NO)_6{}^{2+}$ ion. - Chem. Phys. Lett., 1977, 51, N3, 489-492. - CA, 1978, 88, 30020. PA, 1978, 81, 10754.

Exp.; O_h; d^9.

1978

3.1.253. Bates, C. A., Jahn-Teller effects in paramagnetic crystals. - Phys. Rep., 1978, 35, N3, 187-304. - Bibliogr.: 372 ref. CA, 1978, 88, 201098.

Rev.

3.1.254. Bersuker, I. B., Vekhter, B. G., Polinger, V. Z., Jahn-Teller effect in EPR. [Russ.]. - In: Probl. Magn. Rezonansa. Moscow, 1978, p. 31-48. - Bibliogr.: 54 ref. CA, 1979, 90, 212403. RZF, 1978, 12D781.

Rev.

3.1.255. Bill, H., Schwan, H., Sigmund, E., Jahn-Teller model of the S^- ion in KCl and KBr crystals. - C.R. Seances Soc. Phys. et Hist. Natur. Geneve, 1978, 13, N1, 23-30. - PA, 1979, 82, 82567.

Theor.; T - t; O_h; p^5.

3.1.256. De Jong, H. J., Glasbeek, M., Electric field effects in EPR of the $SrTiO_3:Cr^{5+}$ Jahn-Teller system. - Solid State Commun., 1978, 28, N8, 683-687. - CA, 1979, 90, 130096. PA, 1979, 82, 28599. RZF, 1979, 7D720.

Exp.; T - e; O_h; d^1.

3.1.257. Galindo, S., Owen, J., Murrieta, S. H., Transferred hyperfine interactions for ferrous ion in potassium trifluoromagnesate(I-). - J. Phys. C, 1978, 11, N2, L73-L75. - CA, 1978, 89, 14378. PA, 1978, 81, 31105.

Exp.; E - e; O_h; d^6.

3.1.258. Holuj, F., ESR of Cu^{2+} in $Ca(OD)_2$: secondary spectra. - J. Magn. Resonance, 1978, 30, N2, 343-349. - CA, 1978, 89, 120336. PA, 1978, 81, 88174.

Exp.; d^9.

3.1.259. Kooter, J. A., Canters, G. W., Van der Waals, J. H., ODMR of the triplet state of palladium porphine. - Semicond. & Insul., 1978, 4, N3/4, 249-254. - CA, 1979, 90, 46318. PA, 1979, 82, 19809.

Exp.; E - $(b_1 + b_2)$; D_{4h}.

3.1.260. Krap, C. J., Glasbeek, M., Van Voorst, J. D. W., Phonon-assisted conversion between the Jahn-Teller states of the phosphorescent F center in calcium oxide: optically detected magnetic resonance under uniaxial stress. - Phys. Rev. B, 1978, 17, N1, 61-68. - CA, 1978, 88, 161089.

Exp.; T - e; O_h; $2p^1$; 2.1; 2.2; 3.3.

3.1.261. Krishnan, V. G., Electron spin resonance studies on single crystals of copper(2+)-doped monopyrazine zinc sulfate trihydrate. - J. Phys. C, 1978, 11, N16, 3493-3498. - CA, 1979, 90, 130071. PA, 1978, 81, 84478.

Exp.; T; T_d; d^9.

3.1.262. Le Si Dang, Merle d'Aubigne, Y., Romestain, R., Magnetic resonance in relaxed excited states A_x and A_t of Ga^+ in alkali halides. - Solid State Commun., 1978, 26, N8, 413-416. - CA, 1978, 89, 97528. PA, 1978, 81, 72437.

Exp.; T - (e + t_2); O_h; 2s2p.

3.1.263. Le Si Dang, Merle d'Aubigne, Y., Rasoloarison, Y., Romestain, R., Uniaxial stress effects on EPR lines of $^3T_{1u}$ in CaO:F. - Semicond. & Insul., 1978, 3, N2/3, 149-150. - PA, 1978, 81, 48040.

Exp.; T - e; O_h; $2p^1$.

3.1.264. Loubser, J. H. N., Van Wyk, J. A., Electron spin resonance in the study of diamond. - Rep. Prog. Phys., 1978, 41, N8, 1201-1248. - PA, 1978, 81, 95521.

Rev.; 3.2.

3.1.265. Lowther, J. E., Stoneham, A. M., Theoretical implications of the stress and magnetic field splitting of the GR1 line in diamond. - J. Phys. C, 1978, 11, N11, 2165-2169. - CA, 1978, 89, 206618. PA, 1978, 81, 60095.

Theor.; T; T_d; 2.3; 2.4.

3.1.266. Setser, G. G., Estle, T. L., Jahn-Teller effect in EPR spectra: Multistate theory for ^2E orbital states in cubic symmetry. - Phys. Rev. B, 1978, 17, N3, 999-1014. - CA, 1978, 88, 179829. PA, 1978, 81, 56369.

Theor.; E - e; O_h; d^9.

3.1.267. Van der Linde, R. H., Ammerlaan, C. A. J., Optically induced divacancy reorientations in silicon. - Semicond. & Insul., 1978, 4, N1-2; 139-150. - CA, 1978, 89, 138060. PA, 1978, 81, 80216.

Exp.; T_d; 2.1; 2.4; 3.3.

1979

3.1.268. Astruc, D., Hamon, J. R., Althoff, G., Roman, E., Batail, P., Michaud, P., Mariot, J. P., Varret, F., Cozak, D., Design, stabilization, and efficiency of organometallic "electron reservoirs": 19-Electron sandwiches η^5-$C_5R_5Fe^I$-η^6-C_6R_6', a key class active in redox catalysis. - J. Am. Chem. Soc., 1979, 101, N18, 5445-5447. - CA, 1980, 92, 163230.

Exp.; d^7; 3.3; 3.4.

3.1.269. Bertini, I., Gatteschi, D., Scozzafava, A., Six-coordinate copper complexes with $g_\parallel < g_\perp$ in the solid state. - Coord. Chem. Rev., 1979, 29, N1, 67-84. - CA, 1979, 91, 167558.

Rev.; E - e; O_h; d^9; 4.1; 4.4

3.1.270. Bugay, A. A., Vikhnin, V. S., Kustov, V. E., Influence of axial stress on the EPR spectrum of the Mn^0 Jahn-Teller center in Si. [Russ.]. - Fiz. Tverd. Tela, 1979, 21, N7, 2022-2027. - CA, 1979, 91, 131709. PA, 1980, 83, 44911. RZF, 1979, 10D690.

Exp.; T - (e + t); T_d.

3.1.271. Canters, G. W., Kooter, J. A., Van Dijk, N., Van der Waals, J. H., Zeeman experiments on the orbitally degenerate phosphorescing triplet and quartet states of palladium and copper porphin. - J. Lumin., 1979, 18-19, N1, 196-200. - CA, 1979, 90, 129906. PA, 1979, 82, 33786.

Exp.; E - $(b_1 + b_2)$; D_4h; 2.2.

3.1.272. Hammons, J. H., Bernstein, M., Myers, R. J., Electron spin resonance study of the radical anions of substituted cycloloctatetraenes: The effects of Jahn-Teller distortions and vibronic mixing. - J. Phys. Chem., 1979, 83, N15, 2034-2040. - CA, 1979, 91, 90717.

Exp.

3.1.273. Hoffmann, S. K., Goslar, J. R., Dynamical effects in EPR spectra of Cu(II)-doped tetraamminecadmium chloride and acetate. - Acta Phys. Pol. A, 1979, 55, N4, 471-479. - CA, 1979, 91, 46870. PA, 1979, 82, 61204.

Exp.; T; T_d; d^9.

3.1.274. Kool, T. W., Glasbeeck, M., V^{4+} in $SrTiO_3$: a Jahn-Teller impurity. - Solid State Commun., 1979, 32, N11, 1099-1101. - CA, 1980, 92, 85438. PA, 1980, 83, 29608.

Exp.; T - e; O_h; d^1.

3.1.275. Kooter, J. A., Van der Waals, J. H., The metastable triplet state of zinc porphin and magnesium porphin: a study by ESR in an n-octane crystal at 1.4 K. - Mol. Phys., 1979, 37, N4, 997-1013. - CA, 1979, 91, 184483.

Exp.; E - $(b_1 + b_2)$; D_{4h}.

3.1.276. Kooter, J. A., Van der Waals, J. H., Knop, J. V., Calculation of the spin density and spin-spin dipolar interaction in the triplet state of metal porphins. - Mol. Phys., 1979, 37, N4, 1015-1036.

Theor.; E - $(b_1 + b_2)$; D_{4h}.

3.1.277. Lipatov, V. D., Quasistatic Jahn-Teller effect in a $AgCl:Pd^+$ EPR spectrum. [Russ.]. - Fiz. Tverd. Tela, 1979, 21, N10, 2922-2926. - CA, 1980, 92, 13314. RZF, 1980, 2D689.

Exp.; O_h.

3.1.278. Mehran, F., Stevens, K. W. H., Plaskett, T. S., Dynamic interactions in $EuAsO_4$(Gd) and $EuVO_4$(Gd). - Phys. Rev. B, 1979, 20, N5, 1817-1822. - CA, 1979, 91, 219864. PA, 1980, 83, 7162.

Exp.; D_{4h}.

3.1.279. Narayana, M., Sivarama, Sastry G., ESR and ground state wave function of Cu^{2+} in $Mg(H_2O)_6H_2EDTA$. - J. Phys. C, 1979, 12, N4, 695-701. - PA, 1979, 82, 56605. RZF, 1979, 10D700.

Exp.; E - e; O_h; d^9.

3.1.280. Pradilla, S. J., Chen, H. W., Koknat, F. W., Fackler, J. P., Jr., Structure and electron paramagnetic resonance spectrum of the product of the reaction of aqueous pyridine with copper(II) hexafluoroacetylacetonate: Tetrakis(pyridine)bis(trifluoroacetato)-copper(II). - Inorg. Chem., 1979, 18, N12, 3519-3522. - CA, 1979, 91, 221628.

Exp.; E - e; O_h; d^9; 5.2.

3.1.281. Raizman, A., Schoenberg, A., Suss, J. T., Jahn-Teller effect in the EPR spectrum of Pt^{3+} in MgO. - Phys. Rev. B, 1979, 20, N5, 1863-1866. - CA, 1979, 91, 201736. PA, 1980, 83, 7160.

Exp.; E - e; O_h.

3.1.282. Schaafsma, T. J., Van der Bent, S. J., Kooyman, R. P. H., Optically detected magnetic resonance of the triplet state of pheophytins. - In: Magn. Resonance & Related Phenomena: Proc. 20th Congr. AMPERE, Tallin, 1978, Berlin, etc., 1979, p. 160.

Exp.; E - $(b_1 + b_2)$; D_{4h}; 3.3.

3.1.283. Shen, L. N., Estle, T. L., The effects of applied stress on the electron paramagnetic resonance spectra of Ni^{3+} in Al_2O_3. - J. Phys. C, 1979, 12, N11, 2119-2132. - CA, 1979, 91, 219834. PA, 1979, 82, 82862.

Exp.; E - e; O_h; C_{3v}; d^7.

3.1.284. Sorokin, M. V., Chirkin, G. K., Manifestation of the pseudo-Jahn-Teller effect in the EPR spectra of Cu^{2+} in Nd_4Br. [Russ.]. - Fiz. Tverd. Tela, 1979, 21, N10, 2987-2993. - CA, 1979, 91, 219913. RZF, 1980, 2D687.

Theor.; D_{4h}; d^9; 1.5.

3.1.285. Van der Waals, J. H., van Dorp, W. G., Schaafsma, T. J., Electron spin resonance of porphyrin excited states. - In: The Porphyrins. New York, 1979, Vol. 4, p. 257-312.

Rev.; E - $(b_1 + b_2)$; D_{4h}.

3.1.286. Yamaga Mitsuo, Yoshioka Hide, Jahn-Teller type self-trapping of holes in mixed crystals $AgBr_{1-x}Cl_x$. - J. Phys. Soc. Jpn., 1979, 46, N5, 1538-1545. - CA, 1979, 91, 65762. PA, 1979, 82, 65709. RZF, 1979, 11D950.

Exp.; D_{4h}; d^9.

3.1.287. Yamaga Mitsuo, Hayashi Yoshikazu, Yoshioka Hide, Jahn-Teller effect in the ESR spectrum of the complex $(AgBr_6)^{4-}$ in mixed crystals $AgBr_{1-x}Cl_x$. - J. Phys. Soc. Jpn., 1979, 47, N2, 677-678. - CA, 1979, 91, 165979. PA, 1979, 82, 91294.

Exp.; E - e; O_h; d^9.

See also: 1.1.-19, 20, 30, 47, 68; 1.2.-1, 10; 1.4.-2, 10, 12, 13; 1.7.-2, 17, 18, 22; 1.8.-8; 2.1.-20, 51, 84, 116, 173, 174, 202, 228, 273; 2.2.-52, 155; 2.3.-41, 58, 73, 115, 123; 2.4.-28, 35, 51, 70, 126; 2.6.-1, 18; 2.7.-32, 55, 58, 68, 74, 82; 3.3.-3, 5, 6, 7, 11, 12, 20, 32; 3.5.-39, 60; 3.6.-9, 20, 24; 4.1.-50, 75; 4.3.-10, 14, 37; 4.4.-11, 12, 26, 41, 44, 48, 56, 58, 67, 80, 81, 83, 90, 92, 93, 94, 105, 114, 130, 131, 150, 154, 156, 157, 166, 167; 4.6.-38, 53, 100, 101, 103, 111, 112, 158, 160, 162, 165, 166, 168, 174, 186; 5.1.-10, 58; 5.2.-15, 17, 25, 31, 53; 5.4.-19, 26, 50, 51, 55, 69.

3.2. EPR Line Shape. Temperature Dependence. Magnetic Relaxation

More complete investigations of the vibronic effects by means of EPR are possible when the temperature dependence of the spectrum is taken into account. Using the spin Hamiltonian presentation, the temperature dependence of the EPR line position is interpreted as due to temperature-dependent g-factors [3.2.-26, 52, 115, 119, 125, 129, etc.]. A number of works are devoted to the investigation of the so-called transition from the static to the Jahn-Teller effect dynamic [3.2.-2, 54, 67, 72, 96, 97, 103, 111, 121, 150].

The anisotropic low-temperature Jahn-Teller EPR spectrum becomes gradually isotropic when the temperature increases. This may be due to either the population of the near-lying excited isotropic singlet and/or relaxation. The latter is considered in the papers [3.2.-1, 7, 11, 16, 17, 19, 33, 39, 74, 93, 98, 99, 147, 148]. Motional narrowing of the lines is studied in Refs. [3.2.-4, 51, 82, 104, etc.]. For line shape investigations, see Ref. [3.2.-24, 69, 142, 145].

An important problem here (as in other related problems) is whether interaction with only the first coordination sphere (the cluster model) is sufficient to explain the origin of the temperature dependence of the spectrum. Discussions of this problem and models going beyond the cluster one are given in the works [3.2-89, 101] (see also [3.1.-193, 253, 254]).

1940

3.2.1. Van Vleck, J. H., Paramagnetic relaxation times for titanium and chrome alum. - Phys. Rev., 1940, 57, 426-427. - CA, 1940, 34, 6143[3].

Theor.; O_h.

1952

3.2.2. Bleaney, B., Bowers, K. D., The cupric ion in a trigonal crystalline electric field. - Proc. Phys. Soc. London, Ser. A, 1952, 65, 667-668.

Exp.; E - e; O_h; C_{3v}; d^9.

1953

3.2.3. Bijl, D., Rose-Innes, A. C., Temperature change in the paramagnetic resonance spectrum of copper lanthanum nitrate. - Proc. Phys. Soc. London, Ser. A, 1953, 66, N11, 954-956.

Exp.; E - e; O_h; d^9.

1958

3.2.4. Sack, R. A., A contribution to the theory of the exchange narrowing of spectral lines. - Mol. Phys., 1958, 1, N1, 163-167.

Theor.; E; T.

1961

3.2.5. McConnell, H. M., Spin-orbit coupling in orbitally degenerate states of aromatic ions. - J. Chem. Phys., 1961, 34, N1, 13-16. - CA, 1961, 55, 119458c.

Theor.; E; D_{6h}.

3.2.6. Orton, J. W., Anzins, P., Griffiths, J. H. E., Vertz, J. E., Electron spin resonance studies of impurity ions in magnesium oxide. - Proc. Phys. Soc. London, 1961, 78, N502, 554-568. - CA, 1962, 56, 3042a. RZF, 1962, 7V235.

Exp.; E - e; O_h; d^9.

1962

3.2.7. Vekhter, B. G., Bersuker, I. B., The paramagnetic relaxation in free complexes taking into account external perturbations. [Russ.]. - In: Tez. Dokl. Soveshch. Primeneniyu Fiz. Metodov Issled. Kompleks. Soedin. Kishinev, 1962, p. 19.

Theor.

1963

3.2.8. Groot, M. S., de, Van der Waals, J. H, Paramgnetic resonance in phosphorescent aromatic hydrocarbons. III. Conformational isomerism in benzene and triptycene. - Mol. Phys., 1963, 6, N6, 545-562.

Exp.; D_{6h}.

3.2.9. Höchli, U., Leifson, O. S., Müller, K. A., Spin-lattice relaxation of Ni^{3+} in Al_2O_3. - Helv. Phys. Acta, 1963, 36, N5, 484.

Exp.; O_h; C_{3v}; d^7.

3.2.10. Low, W., Rubins, R. S., Paramagnetic resonance of iron-group and rare earth impurities in calcium oxide. - In: Paramagnetic Resonance: Proc. 1st Int. Conf., Jerusalem, 1962. New York; London, 1963, Vol. 1, p. 79-89. - CA, 1964, 60, 3647h.

Exp.; O_h; 3.1.

3.2.11. Vekhter, B. G., Bersuker, I. B., The EPR line broadening and relaxation times in transition metal octahedral complexes taking into account the inversion splitting. [Russ.]. - In: Tez. Dokl. 3 Soveshch. Kvant. Khim. Kishinev, 1963, p. 10.

Theor.; O_h.

1964

3.2.12. Allen, H. C., Jr., Kokoszka, G. F., Inskeep, R. G., Electron paramagnetic resonance (EPR) spectrum of some tris complexes of Cu^{++}. - J. Am. Chem. Soc., 1964, 86, N6, 1023-1025. - CA, 1964, 60, 11513h.

Exp.; E - e; O_h; d^9.

3.2.13. Höchli, U., Müller, K. A., Resonance relaxation of Pt^{3+} in Al_2O_3. - Helv. Phys. Acta, 1964, 37, N3, 209.

Exp.; E - e; O_h; C_{3v}; d^7.

3.2.14. Höchli, U., Müller, K. A., Observation of the Jahn-Teller splitting of three-valent d^7 ions via Orbach relaxation. - Phys. Rev. Lett., 1964, 12, N26, 730-733. - CA, 1964, 61, 7853h. RZF, 1965, 3D476.

Exp.; E - e; O_h; d^7.

3.2.15. Wysling, P., Resonance relaxation of Ni^{3+} in MgO and CaO. - Helv. Phys. Acta, 1964, 37, N7/8, 629.

Exp.; E - e; O_h; d^7.

1965

3.2.16. Bersuker, I. B., Vekhter, B. G., On the mechanism of spin-lattice relaxation in systems with inversion splitting. [Russ.]. - Fiz. Tverd. Tela, 1965, 7, N4, 1231-1233. - RZF, 1965, 8D452.

Theor.; E - e; T - t.

3.2.17. Bersuker, I. B., Vekhter, B. G., The inversion broadening and paramagnetic relaxation times in systems with electronic E- and T_2-terms. [Russ.]. - Izv. Akad. Nauk Mold.SSR, Ser. Khim. i Biol. Nauk, 1965, N11, 3-13. - RZF, 1967, 2D415.

Theor.; E - e; T - t.

3.2.18. Höchli, U. T., Müller, K. A., Wysling, P., Paramagnetic resonance and relaxation of Cu^{2+} and Ni^{3+} in MgO and CaO: the determination of Jahn-Teller energy splittings. - Phys. Lett., 1965, 15, N1, 5-6. - CA, 1965, 62, 12593b. RZF, 1965, 9D431.

Exp.; E - e; O_h; d^9.

3.2.19. Jones, M. T., Spin-lattice relaxation on hexakis(trifluoromethyl)benzene anion radical. - J. Chem. Phys., 1965, 42, N11, 4054-4055. - CA, 1965, 63, 6506e.

Exp.; E - e; D_{6h}.

3.2.20. Stoneham, A. M., The theory of the spin-lattice relaxation of copper in a Tutton salt crystal. - Proc. Phys. Soc. London, 1965, 85, N1, 107-117. - RZF, 1965, 6D454.

Theor.; E - e; T - e; d^9.

3.2.21. Wysling, P., Müller, K. A. Höchli, U., Paramagnetic resonance and relaxation of Ag^{2+} and Pd^{3+} in MgO and CaO. - Helv. Phys. Acta, 1965, 38, N4, 358.

Exp.; E - e; O_h; d^7; d^9.

1966

3.2.22. Burnham, D. C., Temperature dependence of the Jahn-Teller effect in AgCl:Cu^{2+}. - Bull. Am. Phys. Soc., 1966, 11, N3, 186.

Exp.; E - e; O_h; d^9.

3.2.23. Höchli, U. T., EPR of V^{2+} in CaF_2. - Bull. Am. Phys. Soc., 1966, 11, N3, 203.

Exp.; T; O_h; d^3.

3.2.24. Kivelson, D., Theory of ESR linewidths for benzene negative ion radicals. - J. Chem. Phys., 1966, 45, N2, 751-752.

Theor.; E - e.

3.2.25. Merritt, E. R., Sturge, M. D., Transition from the static to dynamic Jahn-Teller effect in the ESR of Al_2O_3:Ni^{3+}. - Bull. Am. Phys. Soc., 1966, 11, N5, 202.

Exp.; E - e; O_h; C_{3v}; d^7.

3.2.26. Reddy, T. Ramasubba, Srinivasan, R., Temperature dependence of g values in tetragonal ionic Cu^{++} salts. - Phys. Lett., 1966, 22, N2, 143-144. - CA, 1966, 65, 11465f. RZF, 1966, 12E563.

Exp.; E - e; O_h; d^9.

3.2.27. Tucker, E. B., Paramagnetic spin-phonon interaction in crystals. - In: Physical Acoustics. London, 1966, Vol. 4A, p. 47-112.

Rev.

1967

3.2.28. Boettger, H., Jahn-Teller effects studied on spin 1 ions. - Phys. Status Solidi, 1967, 23, N1, 325-333. - CA, 1968, 68, 7960.

Theor.; O_h.

3.2.29. Kokoszka, G. F., Reimann, C. W., Allen, H. C., Jr., Gordon, G., Optical and magnetic measurements on single crystals of copper(II)-doped tris(phenanthroline)zinc(II) nitrate dihydrate. - Inorg. Chem., 1967, 6, N9, 1657-1661. - CA, 1967, 67, 103839.

Exp.; E - e; O_h; d^9; 2.1.

3.2.30. Loubser, J. H. N., van Ryneveld, W. P., The dynamic Jahn-Teller and other effects in the high-temperature electron spin resonance spectrum of nitrogen in diamond. - Brit. J. Appl. Phys., 1967, 18, N7, 1029-1031. - CA, 1967, 67, 59383.

Exp.; T_d.

3.2.31. Novak, P., On the inversion splitting in octahedrally coordinated E electronic states. - Phys. Status Solidi, 1967, 23, N1, K45-K47. - Solid State Abstr., 1967, 8, 64054.

Theor.; E - e; O_h.

3.2.32. Reinberg, A. R., Estle, T. L., Electron-paramagnetic resonance studies of the fluoride donor in beryllium oxide. - Phys. Rev., 1967, 160, N2, 263-273. - CA, 1967, 67, 103831.

Exp.; O_h.

3.2.33. Stevens, K. W. H., The theory of paramagnetic relaxation. - Rep. Prog. Phys., 1967, 30, N1, 189-226. - PA, 1968, 71, 9415. RZF, 1968, 3D440.

Rev.

1968

3.2.34. Freed, J. H., Kooser, R. G., T_1/T_2 and spin relaxation in the benzene anion. - J. Chem. Phys., 1968, 49, N10, 4715-4716. - CA, 1969, 70, 42674.

Exp.; E - (b_1 + b_2); E - e; D_{6h}.

3.2.35. Lee, K. P., Walsh, D., Spin-lattice relaxation of cuprous ion in double nitrate. - Phys. Lett. A, 1968, 27, N1, 17-18. - CA, 1968, 69, 63436.

Exp.; E - e; O_h; d^9.

3.2.36. Müller, K. A., Berlinger, W., Slonczewski, J. C., Rubins, R. S., Dynamical Jahn-Teller effect in the paramagnetic resonance of a nearly degenerate impurity. - Bull. Am. Phys. Soc., 1968, 13, N3, 433-434.

Exp.; E - e; O_h.

3.2.37. Stevens, K. W. H., Walsh, D., The ground state of the ferrous ion in alumina and calculation of spin-lattice relaxation. - J. Phys. C, 1968, 1, N6, 1554-1562. - PA, 1969, 72, 12315.

Theor.

3.2.38. Williams, F. I. B., Breen, D. P., Spin-lattice relaxation of a Jahn-Teller system. - Bull. Am. Phys. Soc., 1968, 13, N8, 458.

Exp.; E - e; O_h; d^9.

1969

3.2.39. Bates, C. A., Dixon, J. M., Relaxation processes in irradiated rubies. - J. Phys. C, 1969, 2, N12, 2225-2237. - PA, 1970, 73, 31576.

Theor.; O_h; C_{3v}; d^3; d^4.

3.2.40. Boatner, L. A., Dischler, B., Herrington, J. R., Estle, T. L., Study of the dynamic Jahn-Teller effect by observation of the EPR of $SrCl_2$:La^{2+}. - Bull. Am. Phys. Soc., 1969, 14, N3, 355.

Exp.; E - e; O_h; d^1.

3.2.41. Breen, D. P., Krupka, D. C., Williams, F. I. B., Relaxation in a Jahn-Teller system. I. Copper in an octahedral water coordination. - Phys. Rev., 1969, 179, N2, 241-254. - CA, 1969, 70, 119955. PA, 1969, 72, 31159.

Exp.; E - e; O_h; d^9.

3.2.42. Holton, W. C., de Wit, M., Watts, R. K., Estle, T. L., Paramagnetic copper centers in ZnS. - J. Phys. & Chem. Solids, 1969, 30, N4, 963-977.

Exp.; d^9.

3.2.43. Lee, Ker-ping, Spin-lattice relaxation in the presence of Jahn-Teller instability. - 1969. - Diss. Abstr. Int. B, 1970, 30, N7, 3343-3344. CA, 1970, 72, 138046.

Exp.

3.2.44. Sadlej, A. J., Witkowski, A., Temperature effects in EPR spectra of the tropylium radical in crystalline matrices. I. Intramolecular vibronic coupling. - Acta Phys. Pol. A, 1969, 36, N6, 971-986. - CA, 1970, 73, 61102.

Theor.; E - e; D_{7h}.

3.2.45. Sadlej, A. J., Witkowski, A., Temperature effects in EPR spectra of the tropylium radical in crystals matrices. II. Influence of the vibronic coupling due to intermolecular vibrations. - Acta Phys. Pol. A, 1969, 36, N6, 987-994. - CA, 1970, 73, 61103.

Theor.; E - e; D_{7h}.

3.2.46. Williams, F. I. B., Krupka, D. C., Breen, D. P., Relaxation in a Jahn-Teller system. II. - Phys. Rev., 1969, 179, N2, 255-272. - CA, 1969, 70, 119556. PA, 1969, 72, 31160.

Theor.; E - e.

1970

3.2.47. Bates, C. A., Bentley, J. P., Jones, B. F., Moore, W. S., Lattice-ion interactions of Ti^{3+} in corundum. III. Analysis of the electron paramagnetic resonance line-shape. - J. Phys. C, 1970, 3, N3, 570-578. - PA, 1970, 73, 51387.

Exp.; T; O_h; d^1.

3.2.48. Borcherts, R. H., Kanzaki, H., Abe, H., EPR spectrum of a Jahn-Teller system, NaCl:Cu^{+2}. - Phys. Rev. B, 1970, 2, N1, 23-27. - CA, 1971, 74, 70084. PA, 1970, 73, 76224. RZF, 1971, 4D660.

Exp.; E - e; O_h; d^9.

3.2.49. Cox, A. F. J., Hagston, W. E., Correlation between infrared absorption and paramagnetic copper centers in zinc and cadmium sulfide. - J. Phys. C, 1970, 3, N9, 1954-1962. - CA, 1970, 73, 125266. RZF, 1971, 2D533.

Theor.; E - e; O_h; d^9; 2.7.

3.2.50. Englman, R., Temperature change in the electron-spin resonance of Cu^{2+} in calcium oxide. - Phys. Lett. A, 1970, 31, N8, 473-474. - CA, 1970, 73, 9258.

Theor.; E - e; O_h; d^9.

3.2.51. Herrington, J. R., Estle, T. L., Boatner, L. A., Dischler, B., Averaging by relaxation and dynamic Jahn-Teller effect. - Phys. Rev. Lett., 1970, 24, N18, 984-986. - CA, 1970, 73, 9500. PA, 1970, 73, 44981.

Exp.; E - e; T_d; d^1.

3.2.52. Slonczewski, J. C., Müller, K. A., Berlinger, W., Dynamic Jahn-Teller effect of an impurity in a spontaneously distorted crystal. - Phys. Rev. B, 1970, 1, N9, 3445-3551. - CA, 1970, 73, 50585. PA, 1970, 73, 56673. RZF, 1970, 12D626.

Exp.; O_h; d^7.

3.2.53. Wagner, G. R., Murphy, J., Castle, J. G., Jr., Spin-relaxation studies of paramagnetic defects in II-VI crystals. - 1970. - 157 p. - U.S. Clearinghouse Fed. Sci. Tech. Inform., AD 713864. - CA, 1971, 75, 56470.

Exp.; d^4; d^7.

3.2.54. Wilson, R. G., Holuj, F., Hedgecock, N. E., Jahn-Teller effect in the electron paramagnetic resonance of Cu^{++} in $Ca(OH)_2$. - Phys. Rev. B, 1970, 1, N9, 3609-3613. - CA, 1970, 73, 71954. PA, 1970, 73, 54152. RZF, 1970, 12D635.

Exp.; d^9.

1971

3.2.55. Herrington, J. R., Electron paramagnetic resonance investigation of the dynamic Jahn-Teller effect exhibited by lanthanum activated strontium chloride: A d^1 configuration in eightfold coordination. - 1971. - 180 p. - [Order N 71-26296.]. Diss. Abstr. Int. B, 1971, 32, N4, 2351-2352. - CA, 1972, 76, 66100.

Exp.; E - e; O_h; d^1.

3.2.56. Herrington, J. R., Estle, T. L., Boatner, L. A., Electron-paramagnetic resonance investigation of the dynamic Jahn-Teller effect in $SrCl_2:La^{2+}$. - Phys. Rev. B, 1971, 3, N9, 2933-2945. - CA, 1971, 74, 148969. PA, 1971, 74, 34093.

Exp.; E - e; O_h; d^1.

3.2.57. Koch, R. C., EPR study of the dynamic Jahn-Teller effect in tris(octamethylpyrophosphoramide)copper(II) perchlorate. - 1971. - 224 p. - [Order N 72-15478.]. Diss. Abstr. Int. B, 1972, 32, N11, 6277-6278. CA, 1972, 77, 107458.

Exp.; E - e; D_3; d^9.

3.2.58. Lee, K. P., Walsh, D., Theory of tunneling assisted electron spin-lattice relaxation. - Can. J. Phys., 1971, 15, N12, 1620-1629. - CA, 1971, 75, 56377. PA, 1971, 74, 49409. RZF, 1972, 1E1163.

Theor.; E - e; T - t.

3.2.59. Moss, R. E., Perry, A. J., g-Factor deviations in degenerate aromatic hydrocarbon radicals: The benzene anion. - Mol. Phys., 1971, 22, N5, 789-798. - CA, 1972, 76, 76669. RZF, 1972, 7D207.

Theor.; E - e; D_{6h}.

3.2.60. Rumin, N., Walsh, D., Lee, K. P., Low-temperature paramagnetic properties of Cu^{2+} and Ti^{3+} in octahedral environments. - J. Phys. [France], 1971, 32, Colloq. N1, Pt. 2, 946-947. - CA, 1971, 75, 28072. CA, 1974, 80, 8830.

Exp.; T - t; E - e; O_h; d^1; d^9.

3.2.61. Suss, J. T., Raizman, A., Szapiro, S., Lows, W., Jahn-Teller effect of 4d ions in single crystals of MgO. - In: Abstr. 4th Int. Symp. Magn. Resonance. Rehovot, 1971, 1 p. - PA, 1972, 75, 7121.

Exp.; O_h; d^7.

1972

3.2.62. Cianchi, L., Mancini, M., Ion-phonon interaction in paramagnetic crystals. - Riv. Nuovo Cimento, 1972, 2, N1, 25-87. - Bibliogr.: 90 ref.

Rev.

3.2.63. Shul'man, L. A., Podzyarey, G. A., Dynamic Jahn-Teller effect in impurity atoms of nitrogen in diamond. [Russ.]. - In: Kristallokhim. Tugoplavkikh Soedineniy. Kiev, 1972, p. 182-190. - CA, 1973, 78, 64909. RZK, 1972, 19B560.

Exp.; T - t; T_d.

3.2.64. Suss, J. T., Raizman, A., Szapiro, S., Low, W., Jahn-Teller effect of $4d^7$ ions in single crystals of MgO. - J. Magn. Resonance, 1972, 6, N3, 438-443. - CA, 1972, 76, 133774. RZF, 1972, 11D623.

Exp.; E - e; O_h; d^7.

3.2.65. Wilkinson, E. L., Energy of the first excited electronic level (s) of Fe^{2+} in CaO. - Phys. Rev. B, 1972, 6, N7, 2517-2521. - PA, 1972, 75, 72478. RZF, 1973, 3D815.

Exp.; T; O_h; d^6.

1973

3.2.66. Aiki Kunio, Narita Koziro, Tanabe Motoko, ESR study of Tb^{3+} and Pr^{3+} in La_2O_2S single crystals. - J. Phys. Soc. Jpn., 1973, 35, N3, 745-749. - CA, 1973, 79, 110041. PA, 1973, 76, 68835.

Exp.; C_{3v}.

3.2.67. Boatner, L. A., Reynolds, R. W., Abraham, M. M., Chen, Y., Transition from static to dynamic Jahn-Teller effects as exhibited by the EPR spectra of silver(2+) ion in strontium oxide, calcium, and magnesium oxide. - Phys. Rev. Lett., 1973, 31, N1, 7-10. - CA, 1973, 79, 59780. PA, 1973, 76, 53016.

Exp.; E - e; O_h; d^9.

3.2.68. Herrington, J. R., Estle, T. L., Boatner, L. A., Electron-paramagnetic-resonance investigation of the dynamic Jahn-Teller effect in $SrCl_2:Y^{2+}$ and $SrCl_2:Sc^{2+}$. - Phys. Rev. B, 1973, 7, N7, 3003-3013. - CA, 1973, 78, 153493. PA, 1973, 76, 33807. RZF, 1973, 11D1008.

Exp.; E - e; O_h; d^1.

3.2.69. Le Si Dang, Buisson, R., Williams, F. I. B., Relaxation and line braodening by phon-assited tunneling in $Cu^{2+}:ZnSiF_6 \cdot 6H_2O$. - In: Magn. Resonance & Related Phenomena: Proc. 17th Congr. AMPERE, Turku, 1972. Amsterdam; London, 1973, p. 389-392.

Exp.; E - e; O_h; d^9.

3.2.70. Lijphart, E. E., De Vroomen, A. C., Poulis, N. J., Electron spin-lattice relaxation in the copper Tutton salts. - In: Magn. Resonance & Related Phenomena: Proc. 17th Congr. AMPERE, Turku, 1972. Amsterdam; London, 1973, p. 397-400. - CA, 1974, 80, 150826.

Exp.; E - e; O_h; d^9.

3.2.71. Moore, W. S., Bates, C. A., Al-Sharbati, T. M., Electric-field-induced thermally detected EPR of non-Kramers ions in Al_2O_3. - J. Phys. C, 1973, 6, N10, L209-L214. - PA, 1973, 76, 44319.

Exp.; O_h; C_{3v}; d^4; d^6.

3.2.72. Pradilla-Sorzano, J., Fackler, J. P., Jr., Base adducts of β-ketoenolates. II. Single-crystal electron paramagnetic resonance and optical studies of copper(II)-doped cis-bis(hexafluoroacetylacetonato)bis(pyridine)zinc(II), Cu-Zn$(F_6acac)_2(py)_2$. - Inorg. Chem., 1973, 12, N5, 1182-1189. - CA, 1973, 78, 153526.

Exp.; E - e; O_h; d^9.

3.2.73. Rumin, N., Vincent, C., Walsh, D., Paramagnetic resonance and relaxation of the Jahn-Teller complex (hexaaquatitanium)-(3+). - Phys. Rev. B, 1973, 7, N5, 1811-1816. - CA, 1973, 78, 142171. PA, 1973, 76, 34197.

Exp.; T - e; O_h; d^1.

3.2.74. Shing, Y. H., Vincent, C., Walsh, D., Spin-lattice relaxation anisotropy for a dynamic Jahn-Teller system. - Phys. Rev. Lett., 1973, 31, N17, 1036-1038. - CA, 1973, 79, 151320. PA, 1973, 76, 75877. RZF, 1974, 4D718.

Theor.; T - e; O_h; d^1.

3.2.75. Teller, J., Vencent, C., Walsh, D., Spin-lattice relaxation in a Jahn-Teller system. [Fr.]. - Can. J. Phys., 1973, 29, N24, 36. - PA, 1974, 77, 41251.

Exp.; d^9.

3.2.76. Vincent, C., Spin-lattice relaxation of a ^2E Jahn-Teller system. - 1973. - Diss. Abstr. Int. B, 1974, 34, N10, 5141. CA, 1974, 81, 129508.

Exp.; E - e.

1974

3.2.77. Bates, C. A., Moore, W. S., Al-Sharbati, T. M., Steggles, P., Gavaix, A., Vasson, A., Vasson, A.-M., Cross-relaxation between Kramers and non-Kramers ions by a quadrupole-electric field interaction. - J. Phys. C, 1974, 7, N5, L83-L87.

Exp.; O_h; C_{3v}; d^3; d^6.

3.2.78. Berti, M., Cianchi, L., Mancini, M., Spina, G., Inconsistency of Orbach's procedure in the evaluation of g shifts of rare earth ions: the case of Ce^{3+} in ethyl sulfates. - Lett. Nuovo Cimento, 1974, 9, N3, 75-78. - CA, 1974, 80, 101763.

Theor.; E - e; C_{3v}.

3.2.79. Bill, H., Silsbee, R. H., Dynamical Jahn-Teller and reorientation effect in the EPR spectrum of oxygen(-) doped calcium fluoride. - Phys. Rev. B, 1974, 10, N7, 2697-2709. - CA, 1975, 82, 9708. PA, 1975, 78, 18820.

Theor.; T - e; O_h; p^5.

3.2.80. Herrington, J. R., Boatner, L. A., Aton, T. J., Estle, T. L., Electron paramagnetic resonance investigation of the dynamic Jahn-Teller effect for scandium(2+) in barium fluoride, strontium fluoride, and calcium fluoride. - Phys. Rev. B, 1974, 10, N3, 833-843. - CA, 1974, 81, 113270. PA, 1974, 77, 79589.

Exp.; E - e; O_h; d^1.

3.2.81. Lee, K., Walsh, D., Electron spin-lattice relaxation measurements of the Jahn-Teller complex hexaaquocopper(2+) ion. - Phys. Status Solidi b, 1974, 62, N2, 689-696. - CA, 1974, 80, 139180. PA, 1974, 77, 37482.

Exp.; E - e; O_h; d^9.

3.2.82. Le Si Dang, Buisson, R., Williams, F. I. B., Dynamics of an octahedral Cu^{2+} Jahn-Teller system: Consequences on its electron spin resonance. - J. Phys. [France], 1974, 35, N1, 49-65. - CA, 1974, 80, 76348. PA, 1974, 77, 20844. RZF, 1974, 5D648.

Exp.; E - e; O_h; d^9.

3.2.83. Pradilla-Sorzano, J., Fackler, J. P., Jr., Base adducts of β-ketoenolates. VII. EPR studies of some fluxional 1,1,1,6,6,6-hexafluoro-2,4-pentanedionato Cu(II) complexes. - Inorg. Chem., 1974, 13, N1, 38-44. - CA, 1974, 80, 54294.

Exp.; E - e; O_h; d^9.

3.2.84. Reynolds, R. W., Boatner, L. A., Abraham, M. M., Chen, Y., EPR investigation of intermediate Jahn-Teller coupling effects for copper(2+) ion in magnesium oxide and calcium oxide. - Phys. Rev. B, 1974, 10, N9, 3802-3817. - CA, 1975, 82, 49590. PA, 1975, 78, 22145.

Exp.; E - e; O_h; d^9.

3.2.85. Schoenberg, A., Suss, J. T., Luz, Z., Low, W., Dynamic Jahn-Teller effect in the EPR spectrum of nickel(+) and nickel(3+) ions in magnesium oxide. - Phys. Rev. B, 1974, 9, N5, 2047-2050. - CA, 1974, 81, 8108. PA, 1974, 77, 53310. RZF, 1975, 7D725.

Exp.; E - e; O_h; d^7; d^9.

3.2.86. Shing, Y. H., Vincent, C., Walsh, D., Energy splittings of the $^2T_2$ vibronic state of titanium(3+) in methylammonium alum: Comments. - Phys. Rev. B, 1974, 9, N1, 340-341. - CA, 1974, 80, 114613.

Exp.; T; d^1.

3.2.87. Silver, B. L., Getz, D., ESR of hexaaquocopper(II). II. Quantitative study of the dynamic Jahn-Teller effect in copper-doped zinc Tutton's salt. - J. Chem. Phys., 1974, 61, N2, 638-650. - CA, 1974, 81, 97429. PA, 1974, 77, 76148.

Exp.; E - e; O_h; d^9.

3.2.88. Vasson, A.-M., Vasson, A., Gavaix, A., Yalcin, P., Steggles, P., Moore, W. S., Bates, C. A., The determination of zero-field splittings from measurements of spin-lattice relaxation times. - J. Phys. Lett. [France], 1974, 35, N5, L73-L76.

Exp.; O_h; C_{3v}; d^3; d^6.

1975

3.2.89. Bates, C. A., Steggles, P., Jahn-Teller theory and calculation of the relaxation time and resonance line shape for iron(2+) ions in aluminum oxide. - J. Phys. C, 1975, 8, N14, 2283-2299. - CA, 1975, 83, 123770. PA, 1975, 78, 73041. RZF, 1975, 12E85.

Theor.; T; O_h; C_{3v}; d^6; 1.4.

3.2.90. Bates, C. A., Steggles, P., Gavaix, A., Vasson, A., Vasson, A.-M., Cross-relaxation to fast relaxing ions in alumina. - J. Phys. C, 1975, 8, N14, 2300-2316. - PA, 1975, 78, 77376.

Exp.; O_h; C_{3v}; d^3; d^5; d^6.

3.2.91. Bates, C. A., Gavaix, A., Steggles, P., Vasson, A., Vasson, A.-M., Cross-relaxation measurements and their interpretation. - In: Magn. Resonance & Related Phenomena: Proc. 18th Congr. AMPERE, 1974. Amsterdam, 1975, Vol. 2, p. 417-418. - PA, 1976, 79, 2381.

Exp.; E - e; O_h; C_{3v}; d^1; d^2; d^3.

3.2.92. Bertini, I., Gatteschi, D., Scozzafava, A., Single crystal ESR study of the $Cu(en)_3SO_4$ complex. - Inorg. Chim. Acta, 1975, 11, N2, L17-L19. - CA, 1975, 82, 49593.

Exp.; E - e; D_3; d^9.

3.2.93. Gill, J. C., The establishment of thermal equilibrium in paramagnetic crystals. - Rep. Prog. Phys., 1975, 38, N1, 91-150. - Bibliogr.: p. 146-150.

Theor.

3.2.94. Jesion, A., Shing, Y. H., Walsh, D., Ground vibronic levels of titanium alum. - Phys. Rev. Lett., 1975, 35, N7, 51-52. - PA, 1975, 78, 62280.

Exp.; T - e; O_h; d^1.

3.2.95. Moore, W. S., Al-Sharbati, T. M., Clark, I. A., Knowles, A. P., Experimental techniques for thermal detection of strongly-coupled ions. - In: Magn. Resonance & Related Phenomena: Proc. 18th Congr. AMPERE, 1974. Amsterdam, 1975, Vol. 2, p. 559-560. - PA, 1975, 78, 86060.

Exp.

3.2.96. Reynolds, R. W., Boatner, L. A., Dynamic intermediate, and static Jahn-Teller effect in the EPR spectra of ^2E orbital states. - Phys. Rev. B, 1975, 12, N11, 4735-4754. - CA, 1976, 84, 82170. PA, 1976, 79, 31368.

Theor.; E - e; O_h.

3.2.97. Setser, G. G., Barksdale, A. O., Estle, T. L., Jahn-Teller effect in EPR spectra: Three-state theory for ^2E orbital states in cubic symmetry. - Phys. Rev. B, 1975, 12, N11, 4720-4734. - CA, 1976, 84, 67477. PA, 1976, 79, 31367.

Theor.; E - e; O_h; d^9.

3.2.98. Vaughan, R. A., Electron spin-lattice relaxation. - Magn. Resonance Rev., 1975, 4, N1, 25-62. - Bibliogr.: 158 ref. CA, 1976, 84, 186700.

Rev.

1976

3.2.99. Bates, C. A., Gavaix, A., Steggles, P., Vasson, A., Vasson, A.-M., Non-resonant cross-relaxation and its occurrence in vanadium-doped rubies. - J. Phys. C, 1976, 9, N12, 2413-2427. - PA, 1976, 79, 72880.

Theor.; E - e; O_h; C_{3v}.

3.2.100. Bates, C. A., Gavaix, A., Steggles, P., Vasson, A., Vasson, A.-M., Non-resonant cross-relaxation to ^5E Jahn-Teller ions in ruby. - In: Magn. Resonance & Related Phenomena: Proc. 19th Congr. AMPERE. Heidelberg; Geneva, 1976, p. 507-510. - CA, 1977, 86, 197416.

Exp.; E - e; O_h; C_{3v}; d^3; d^4.

3.2.101. Bates, C. A., Gavaix, A., Steggles, P., Szymczak, H., Vasson, A., Vasson, A.-M., A study of spin-phonon interactions by spin-lattice relaxation processes. - In: Phonon Scattering in Solids: Proc. 2nd Int. Conf., 1975. New York; London, 1976, p. 214-216.

Exp.; E - e; O_h; C_{3v}; d^7; 1.7.

3.2.102. Bir, G. L., The Jahn-Teller effect on impurity centers in semiconductors. [Russ.]. - Zh. Eksp. & Teor. Fiz., 1976, 18, N6, 1622-1626.

Exp.; E - e; O_h; d^9.

3.2.103. Driessen, W. L., The Jahn-Teller effect in hexakis-(hydrocyanic acid)copper(II)-bis(tetrachloroindate(III)). - Inorg. & Nucl. Chem. Lett., 1976, 12, N11, 873-875. - CA, 1977, 86, 48935. RZK, 1977, 10B229.

Exp.; E - e; O_h; d^9.

3.2.104. Englman, A., A motional narrowing diagnostic for ESR lines in a vibronic system. - J. Phys. Lett. [France], 1976, 37, N10, 261-263. - RZF, 1977, 3D668.

Theor.; E - e; O_h.

3.2.105. Gavaix, A., Vasson, A., Vasson, A.-M., Bates, C. A., Steggles, P., Resonant cross-relaxation between Cr^{3+} and V^{3+} in Al_2O_3. - J. Phys. & Chem. Solids, 1976, 37, N11, 1051-1057. - PA, 1976, 79, 91325.

Exp.; O_h; C_{3v}; d^2; d^3.

3.2.106. Jesion, A., Microwave spectroscopy of Jahn-Teller crystals. - 1976. - Diss. Abstr. Int. B, 1977, 37, N11, 5729-5730. CA, 1979, 91, 148863.

Exp.; T - (e + t); O_h; d^1; d^9.

3.2.107. Konning, A. T., Svare, I., Jahn-Teller motion in copper bromate. - Phys. Scr., 1976, 14, N1-2, 79-80. - CA, 1976, 85, 184492. PA, 1976, 79, 93922. RZK, 1977, 12B247.

Exp.; E - e; O_h; d^9.

3.2.108. Maksimov, N. G., Anufrienko, V. F., The state of copper(II) ions and their distribution in cubic magnesium oxide, cadmium oxide, and calcium oxide according to EPR data. [Russ.]. - Dokl. Akad. Nauk SSSR, 1976, 228, N6, 1391-1394. - CA, 1976, 85, 151967.

Exp.; E - e; O_h; d^9.

3.2.109. Sambasiva, R. P., Subramanian, S., Jahn-Teller tunneling in copper(II)-doped hexaimidazole zinc(II) dichloride tetrahydrate. - J. Magn. Resonance, 1976, 22, N2, 191-206. - RZK, 1976, 22B226.

Exp.; E - e; O_h; d^9.

3.2.110. Zaritsky, I. M., Bratus, V. Ya., Vikhnin, V. S., Vishnevskiy, A. S., Konshits, A. A., Ustintsev, V. M., Spin-lattice relaxation of a nitrogen Jahn-Teller center in diamond. [Russ.]. - Fiz. Tverd. Tela, 1976, 18, N11, 3226-3230. - CA, 1977, 86, 63081. PA, 1977, 80, 66412.

Exp.; T - t; T_d.

1977

3.2.111. Boatner, L. A., Reynolds, R. W., Chen, Y., Abraham, M. M., Static, quasistatic, and quasidynamic Jahn-Teller effect in the EPR spectra of silver(2+) ion in strontium oxide, calcium oxide, and magnesium oxide. - Phys. Rev. B, 1977, 16, N1, 86-106. - CA, 1977, 87, 125026. PA, 1977, 80, 85561.

Exp.; E - e; O_h; d^9.

3.2.112. Bontemps-Moreau, N., Boccara, A. C., Thibault, P., Laser selective excitation and double resonance in Jahn-Teller system: F center in calcium oxide. - Phys. Rev. B, 1977, 16, N5, 1822-1827. - CA, 1977, 87, 209150.

Exp.; T - e; O_h; 2p.

3.2.113. Clerjaud, B., Gelineau, A., Strong spin-lattice coupling of Kramers doublets. - Phys. Rev. B, 1977, 16, N1, 82-85. - CA, 1977, 87, 109045. PA, 1977, 80, 85181.

Theor.; T - e; T_d; d^9.

3.2.114. Jesion, A., Shing, Y. H., Walsh, D., EPR and spin-lattice relaxation of the Jahn-Teller system: Copper(2+) in zinc bromate hydrate ($Zn(BrO_3)_2 \cdot 6H_2O$). - Phys. Rev. B, 1977, 16, N7, 3012-3015. - CA, 1977, 87, 209137.

Exp.; E - e; O_h; d^9.

3.2.115. Kurzynski, M., Explicit temperature dependence of spin Hamiltonian parameters for Jahn-Teller ions. - Acta Phys. Pol. A, 1977, 52, N5, 647-656. - CA, 1978, 88, 29690. PA, 1978, 81, 30891.

Theor.; E - e; T - (e + t).

3.2.116. Le Si Dang, Romestain, R., Merle d'Aubigne, Y., Fukuda Atsuo, Jahn-Teller effect in an orbital triplet coupled to both E_g and T_{2g} modes of vibration: Experimental evidence for the coexistence of tetragonal and trigonal minima. - Phys. Rev. Lett., 1977, 38, N26, 1539-1543. - CA, 1977, 87, 60372. PA, 1977, 80, 66441.

Exp.; T - (e + t); O_h; sp; 1.2.

3.2.117. Petrashen', V. E., Yablokov, Yu. V., Davidovich, R. L., Study of the $K_2Zn(ZrF_6)_2 \cdot 6H_2O$ crystals with use of impurity ions of the iron group. [Russ.]. - In: Fiz. Mat. Metody Koord. Khim.: Tez. Dokl. 6 Vsesoyuz. Soveshch. Kishinev, 1977, p. 59.

Exp.; E - e; O_h; d^9.

3.2.118. Raizman, A., Suss, J. T., Low, W., Quadrupole interaction and static Jahn-Teller effect in the EPR spectra of iridium-(2+) in magnesium oxide and calcium oxide. - Phys. Rev. B, 1977, 15, N11, 5184-5196. - CA, 1977, 87, 60394. PA, 1977, 80, 72842.

Exp.; E - e; O_h; d^7.

3.2.119. Sastry, S. G., Rao, M. J., Lingam, K. V., Bhattacharya, B.N., Jahn-Teller effect in copper-doped zinc tris(ethylenediamine)-chloride. - In: Proc. Nucl. Phys. & Solid State Phys. Symp., 1977, Vol. 20C, p. 573-575. - CA, 1979, 90, 94963.

Exp.; E - e; O_h; d^9; 3.1.

3.2.120. Ziatdinov, A. M., Zaripov, M. M., Yablokov, Yu. V., Davidovich, R. L., The nature of the Jahn-Teller EPR spectra of the copper(II) ion in $ABF_6 \cdot 6H_2O$-type crystals. [Russ.]. - Kazan', 1977. - 61 p. - Deposited Doc., VINITI, 1977, N3550-77 Dep. CA, 1979, 90, 130104. RZF, 1978, 1D734.

Exp.; E - e; O_h; d^9.

1978

3.2.121. Ammeter, J. H., EPR of orbitally degenerate sandwich compounds. - J. Magn. Resonance, 1978, 30, N2, 299-325. - CA, 1978, 89, 138045. PA, 1978, 81, 86893.

Exp.; E - e; D_{5h}; d^5; d^7.

3.2.122. Bugay, A. A., Vikhnin, V. S., Kustov, V. E., Maksimenko, V. M., Krylikovskiy, B. K., Spin-lattice relaxation of a Jahn-Teller chromium(0) center in silicon. [Russ.]. - Zh. Eksp. & Teor. Fiz., 1978, 74, N6, 2250-2258. - CA, 1978, 89, 97523. PA, 1978, 81, 72402.

Exp.; T - e; T - t; T - (e + t); T_d.

3.2.123. Cibert, J., Edel, P., Merle d'Aubigne, Y., Romestain, R., Relaxation in the $^3T_{1u}$ state of F centers in CaO. - Semicond. & Insul., 1978, 3, N2/3, 163-164.

Exp.; T; O_h; 2p.

3.2.124. Clerjaud, B., Influence des interactions vibroniques sur les properiétés des ions d^9 dans les composes II-VI de structure blende: Diss. [Fr.]. - Paris, 1978. - 140 p.

The Influence of Vibronic Interactions on the Properties of d^9 Ions Doped in II-VI Compounds with the Zinc Blende Structure.

Exp.; T; T_d; d^9.

3.2.125. Grunin, V. S., Patrina, I. B., Davtyan, G. D., EPR of Cr^{5+} and Mo^{5+} ions in anatase single crystals. [Russ.]. - Fiz. Tverd. Tela, 1978, 20, N5, 1556-1558. - CA, 1978, 89, 68111. RZF, 1978, 10D663.

Exp.; T; d^1.

3.2.126. Hayashi Makoto, Nakagawa Hideyuki, Matsumoto Hiroaki, The Jahn-Teller center in cadmium halide crystals. I. Copper(II) chloride-doped cadmium chloride. - Fukui Daigaku Kogakubu Kenkyu Hokoku, 1978, 26, N1, 15-27. - CA, 1979, 90, 112640.

Exp.; E - e; O_h; d^9.

3.2.127. Le Si Dang, Merle d'Aubigne, Y., Rasoloarison, Y., Uniaxial stress effect and spin-lattice coupling in the $^3T_{1u}$ relaxed excited states of F centers in CaO. - J. Phys. [France], 1978, 39, N7, 760-770. - CA, 1978, 89, 82601. PA, 1978, 81, 72405.

Exp.; T - e; O_h; 2p.

3.2.128. Petrashen, V. E., Yablokov, Yu. V., Davidovich, R. L., Electron paramagnetic resonance study of copper(2+) in potassium zinc hexafluorozirconate hexahydrate ($K_2Zn(ZrF_6)_2 \cdot 6H_2O$). - Phys. Status Solidi b, 1978, 88, N2, 439-443. - CA, 1978, 89, 171451. PA, 1978, 81, 80201.

Exp.; E - e; O_h; d^9.

3.2.129. Radhakrishna, S., Rao, T. B., ESR studies of hexaaquacopper(2+) ion in magnesium sulfate heptahydrate. - J. Magn. Resonance, 1978, 32, N1, 71-81. - CA, 1979, 90, 46337. PA, 1979, 82, 38377.

Exp.; E - e; O_h; d^9; 3.1.

3.2.130. Vasson, A.-M., Vasson, A., Bates, C. A., Resonant cross-relaxation to Cr^{4+} in ruby. - J. Phys. [France], 1978, 39, N6, 1005-1006. - CA, 1979, 90, 78844. PA, 1979, 82, 65723.

Exp.; O_h; C_{3v}; d^2; d^3.

3.2.131. Vikhnin, V. S., Reorientation and spin-lattice relaxation caused by a tunnel-controlled process. [Russ.]. - Fiz. Tverd. Tela, 1978, 20, N5, 1340-1346. - CA, 1978, 89, 52478.

Theor.

3.2.132. Vikhnin, V. S., Sochava, L. S., Tolparov, Yu. N., The manifestation of the off-center ion tunneling in the EPR spectra. [Russ.]. - Fiz. Tverd. Tela, 1978, 20, N8, 2412-2419.

Exp.; E - e; O_h; C_{3v}; d^9; 3.5.

3.2.133. Yamaga Mitsuo, Hayashi Yoshikazu, Yoshioka Hide, Jahn-Teller and reorientation effects in the ESR spectrum of the self-trapped holes in mixed crystals silver bromide chloride ($AgBr_{1-x}Cl_x$). - J. Phys. Soc. Jpn., 1978, 44, N1, 154-161. - CA, 1978, 88, 112939. PA, 1978, 81, 27092.

Exp.

3.2.134. Yamaga Mitsuo, Yoshioka Hide, Jahn-Teller and reorientation effects in the ESR spectrum of the self-trapped holes in mixed crystals silver bromide chloride ($AgBr_{1-x}Cl_x$). II. - J. Phys. Soc. Jpn., 1978, 44, N6, 1901-1908. - CA, 1978, 89, 120300. PA, 1978, 81, 64273.

Exp.

1979

3.2.135. Bates, C. A., Vasson, A., Vasson, A.-M., Studies of $3d^4$ ions in Al_2O_3 by cross relaxation. - J. Phys. [France], 1979, 40, N10, 1075-1085. - CA, 1979, 91, 201709.

Exp.; O_h; C_{3v}; d^4.

3.2.136. Bates, C. A., Wardlaw, R., An analysis of the effects of stress on the EPR and Raman spectra of Ni^{3+}:Al_2O_3. - J. Phys. C, 1979, 12, N11, 2133-2142. - CA, 1979, 91, 201670. PA, 1979, 82, 82863.

Exp.; E - e; O_h; C_{3v}; d^7; 2.8.

3.2.137. Bencini, A., Benelli, C., Gatteschi, D., Sacconi, L., ESR spectra of halo[1,1,1-tris(diphenylphosphinomethyl)ethane]nickel-(I) complexes. - Inorg. Chim. Acta, 1979, 37, N2, 195-199. - RZF, 1980, 5D640.

Exp.; E - e; O_h; D_3; d^9; 2.1.

3.2.138. Bermudez, V. M., McClure, D. S., Spectroscopic studies of the two-dimensional magnetic insulators chromium trichloride and chromium tribromide. II. - J. Phys. & Chem. Solids, 1979, 40, N2, 149-173. - CA, 1979, 91, 99308.

Exp.; T.

3.2.139. Bugai, A. A., Vikhnin, V. S., Kustov, V. E., Maksimenko, V. M., Spin lattice relaxation in the case of dynamical Jahn-Teller effect (Cr^0 and Mn^0 in Si). - In: Magn. Resonance and Related Phenomena: Proc. 20th Congr. AMPERE, Tallin, 1978, Berlin, etc., 1979, p. 272.

Exp.; T - (e + t); T_d.

3.2.140. Bugay, A. A., Vikhnin, V. S., Kustov, V. E., The spin lattice relaxation of the Jahn-Teller centers Cr^0 and Mn^0 in silicium: the coexistence of minima of different symmetry. [Russ.]. - Fiz. Tverd. Tela, 1979, 21, N11, 3332-3339. - CA, 1980, 92, 68498. PA, 1980, 83, 80602.

Exp.; T - (e + t); T_d; 1.2.

3.2.141. Glasbeek, M., De Jong, H. J., Koopmans, W. E., Nonlinear Jahn-Teller coupling and local dynamics of $SrTiO_3$:Cr^{5+} near the structural phase transition point. - Chem. Phys. Lett., 1979, 66, N1, 203-206. - CA, 1979, 91, 201722. PA, 1980, 83, 2625.

Exp.; T - (e + t); O_h; d^1; 4.4.

3.2.142. Khlopin, V. P., Polinger, V. Z., Bersuker, I. B., The Jahn-Teller effect for an icosahedral T_2 term in EPR spectra. - In: Magn. Resonance & Related Phenomena: Proc. 20th Congr. AMPERE, Tallin, 1978. Berlin, etc., 1979, p. 266.

Theor.; T - v; I; I_h.

3.2.143. Krebs, J. J., Stauss, G. H., Effects of uniaxial stress and temperature variation on the Cr^{2+} center in GaAs. - Phys. Rev. B, 1979, 20, N2, 795-800. - CA, 1980, 92, 13276. PA, 1979, 82, 95636.

Exp.; E - e; T - e; O_h; d^4.

3.2.144. Miyanaga Takeschi, EPR Studies of strong Jahn-Teller coupling systems $CdCl_2:Ag^{2+}$ and $CdBr_2:Ag^{2+}$. - J. Phys. Soc. Jpn., 1979, 46, N1, 167-175. - PA, 1979, 82, 28605.

Exp.; E - e; O_h; d^9.

3.2.145. Shen, L. N., Estle, T. L., An investigation of the nature of the Jahn-Teller effect for Ni^{3+} in Al_2O_3 by electron paramagnetic resonance. - J. Phys. C, 1979, 12, N11, 2103-2118. - CA, 1979, 91, 219833. PA, 1979, 82, 82564.

Exp.; E - e; O_h; C_{3v}; d^7.

3.2.146. Sochava, L. S., Tolparov, Yu. N., Vikhnin, V. S., Off-center impurity ions in crystal as studied by EPR. - In: Magn. Resonance & Related Phenomena: Proc. 20th Congr. AMPERE, Tallin, 1978. Berlin, etc., 1979, p. 320.

Exp.; O_h; d^5; d^7; d^9.

3.2.147. Vasson, A., Vasson, A.-M., Bates, C. A., Studies of E-type Jahn-Teller ions by resonant cross relaxations. - In: Magn. Resonance & Related Phenomena: Proc. 20th Congr. AMPERE, Tallin, 1978. Berlin, etc., 1979, p. 267.

Exp.; E - e; O_h; C_{3v}; d^3; d^4; d^5; d^7.

3.2.148. Vikhnin, V. S., Spin-lattice relaxation of tunneling impurities. - In: Magn. Resonance & Related Phenomena: Proc. 20th Congr. AMPERE, Tallin, 1978. Berlin, etc., 1979, p. 145.

Theor.

3.2.149. Vikhnin, V. S., Sochava, L. S., Tolparov, Yu. N., Noncentral Jahn-Teller ion: Localization in energy minimum with C_1 symmetry. [Russ.]. - Fiz. Tverd. Tela, 1979, 21, N6, 1789-1797. - CA, 1979, 91, 100079. PA, 1980, 83, 25198.

Exp.; E - e; O_h; C_{3v}; d^9.

3.2.150. Wood, J. S., De Boer, E., Keijzers, C. P., Nature of the Jahn-Teller effect in the $Cu(C_5H_5NO)_6^{2+}$ ion. - Inorg. Chem., 1979, 18, N3, 904-906. - CA, 1979, 90, 130081.

Exp.; E - e; O_h; d^9.

See also: 1.3.-102; 1.7.-32, 47; 3.1.-1, 10, 37, 57, 62, 64, 71, 92, 119, 159, 184, 254; 3.5.-28; 4.4.-55, 128.

3.3. NMR, ENDOR, and Other Double Resonances

The first publications on this subject appeared in the middle sixties. These are experimental works in which the Jahn-Teller effect is used in order to explain the origin of line broadening [3.3.-13] and local static and dynamic distortions [3.3.-4, 11, 15, 18, 24, 25]. The high lability and fast exchange of ligands in complexes in solutions, observed by NMR, is discussed in Ref. [3.3.-19]. The line splitting of ligand NMR spectra in some transition metal hexafluorides is attributed to the pseudo-Jahn-Teller effect [3.3.-16, 21, 26, 27, 31]. The influence of the Jahn-Teller effect on spin-lattice relaxation, as well as the role of random strain, were studied by means of ENDOR [3.3.-1, 3, 5, 6, 8, 20] and optically detected EPR [3.3.-28, 29] (see also [3.1.-86, 131, 148, 262, 275]).

1964

3.3.1. Watkins, G. D., Corbett, J. W., Defects in irradiated silicon: electron paramagnetic resonance and electron nuclear double resonance of the Si - E center. - Phys. Rev., 1964, 134, N5A, 1359-1377. - CA, 1964, 60, 152222a. RZF, 1964, 11D376.

Exp.; T_d.

1965

3.3.2. Simza, Z., The distribution of valence in manganese ferrites. - Czech. J. Phys., 1965, 15, N6, 435-437. - CA, 1966, 64, 4286h.

Theor.; O_h; d^4; d^5; d^6.

3.3.3. Watkins, G. D., A review of EPR studies in irradiated silicon. - In: Radiation Damage Semicond.: 7th Int. Conf. Phys.

Semicond., Paris-Rayaumont, 1964. Paris, 1965, Vol. 3, p. 97-113. - CA, 1966, 65, 14687a. RZF, 1965, 11D426.

Rev.

1967

3.3.4. Kubo Takeji, Abe Hisashi, Hirai Akira, Manganese-55 nuclear magnetic resonance of manganese(III) ions located at β-sites in manganese ferrite single crystal. - J. Phys. Soc. Jpn., 1967, 23, N1, 124-125. - CA, 1968, 68, 82944.

Exp.; O_h; d^4.

3.3.5. Watkins, G. D., Defects in irradiated silicon: electron-paramagnetic-resonance and electron-nuclear double resonance. - Phys. Rev., 1967, 155, N3, 802-815.

Exp.; E - e; C_{3v}.

1968

3.3.6. Elkin, E. L., Watkins, G. D., Defects in irradiated silicon: electron paramagnetic resonance and electron-nuclear double resonance [ENDOR] of the arsenic- and antimony-vacancy pairs. - Phys. Rev., 1968, 174, N3, 881-897. - CA, 1968, 69, 112023.

Exp.; E - e; C_{3v}.

3.3.7. Noack, M., Gordon, G., Oxygen-17 NMR and copper EPR linewidths in aqueous solutions of copper(II) ion and 2,2'-dipyridine. - J. Chem. Phys., 1968, 48, N6, 2689-2699. - CA, 1968, 69, 6953.

Exp.; E - e; O_h; d^9; 3.2.

1969

3.3.8. Borcherts, R. H., Lohr, L. L., Jr., Optical, EPR, and ENDOR [electron nuclear double resonance] studies of $CdF_2:V^{3+},V^{2+}$. - J. Chem. Phys., 1969, 50, N12, 5262-5265. - CA, 1969, 71, 44312. PA, 1969, 72, 47385.

Exp.; E - e; O_h; d^2; d^3; 2.1; 3.1.

3.3.9. Kubo Takeji, Hirai Akira, Abe Hisashi, Manganese-55 nuclear magnetic resonance of the manganese(III) ion located at the β-site in manganese ferrite single crystal anisotropic hyperfine field due to the local Jahn-Teller distortion. - J. Phys. Soc. Jpn., 1969, 26, N5, 1094-1109. - CA, 1969, 71, 26412. PA, 1969, 72, 43299.

Exp.; E - e; O_h; d^4.

1970

3.3.10. Chase, L. L., Microwave-optical double resonance of the metastable $4f^6 5d$ level of Eu^{2+} in the fluorite lattices. - Phys. Rev. B, 1970, 2, N7, 2308-2318. - CA, 1971, 74, 70055. PA, 1971, 74, 5898. RZF, 1971, 7D552.

Exp.; E - e; O_h; $4f^6 5d$.

3.3.11. Mizoguchi Mogiri, Tasaki Akira, Mn^{55} NMR results for $(NiMn_2O_4)_{1-x}(MnFe_2O_4)_x$. - J. Phys. Soc. Jpn., 1970, 29, N5, 1382-1383. - CA, 1971, 74, 81534. PA, 1971, 74, 19454.

Exp.; d^3; d^4; d^5.

1971

3.3.12. Chan, I. Y., Van Dorp, W. G., Schaafsma, T. J., Van der Waals, J. H., Lowest triplet state of zinc porphyrin. I. Modulation of its phosphorescence by microwaves. - Mol. Phys., 1971, 22, N5, 741-751. - CA, 1972, 76, 78892.

Exp.; E - $(b_1 + b_2)$; D_{4h}.

1972

3.3.13. Le Dang Khoi, Rotter, M., Krishnan, R., Hyperfine interaction of Cr^{3+} and Cu^{2+} ions in Cr- and Cu-doped YIG. - Phys. Status Solidi a, 1972, 12, N2, 569-574. - CA, 1972, 77, 107359. RZF, 1973, 2D691.

Exp.; E - e; O_h; C_{3v}; d^3; d^9.

1973

3.3.14. Harris, C. B., Glasbeek, M., Hensley, E. B., Phosphorescence microwave double resonance spectroscopy in ionic solids and its application to coherent Jahn-Teller states in F-centers of calcium oxide. - Phys. Rev. Lett., 1973, 33, N9, 537-540. - CA, 1974, 81, 113106. PA, 1974, 77, 73049.

Exp.; T; O_h; 2p; 2.2.

3.3.15. Skjaeveland, S. M., Svare, I., Jahn-Teller effect in $FeSiF_6 \cdot 6H_2O$ observed with proton resonance. - Phys. Norv., 1973, 7, N2, 105. - PA, 1974, 77, 16991.

Exp.; O_h; d^6.

1974

3.3.16. Afanas'ev, M. L., Lundin, A. G., NMR and Jahn-Teller effect. [Russ.]. - In: Fiz. Mat. Metody Koord. Khim.: Tez. Dokl. 5 Vsesoyuz. Soveshch. Kishinev, 1974, p. 114-115. - CA, 1975, 83, 18444.

Theor.

3.3.17. Le Dang Khoi, Veillet, P., Krishnan, R., Nuclear resonance study of the distribution of manganese ions on various sites of yttrium garnet. [Fr.]. - C.R. Hebd. Seances Acad. Sci. Ser. B, 1974, 278, N24, 1047-1050. - PA, 1974, 77, 64268.

Exp.; d^3; d^5.

3.3.18. Skjaeveland, S. M., Svare, I., Motion in fluorosilicates studied with nuclear magnetic resonance. - Phys. Scr., 1974, 10, N5, 273-276. - CA, 1975, 82, 105077. PA, 1975, 78, 15443.

Exp.

3.3.19. West, R. J., Lincoln, S. F., Nitrogen-14 magnetic resonance study of the exchange of acetonitrile on copper(II) complexes. - J. Chem. Soc. Dalton Trans., 1974, N3, 281-284. - RZF, 1974, 7D651.

Exp.; E - e; O_h; d^9.

1975

3.3.20. Watkins, G. D., Defects in irradiated silicon: EPR and electron-nuclear double resonance of interstitial boron. - Phys. Rev. B, 1975, 12, N12, 5824-5839. - CA, 1976, 84, 67959.

Exp.; T_d; 3.1.

1977

3.3.21. Afanas'ev, M. L., Lundin, A. G., Seryshev, S. A., The NMR of ^{19}F at high hydrostatic pressures and the pseudo-Jahn-Teller effect in WF_6 and KPF_6. [Russ.].- In: Fiz. Mat. Metody Koord. Khim.: Tez. Dokl. 6 Vsesoyuz. Soveshch. Khishinev, 1977, p. 90.

Exp.; 1.5.

3.3.22. Doddrell, D. M., Bendall, M. R., Pegg, D. T., Healy, P. C., Gregson, A. K., Temperature dependence of proton spin-lattice relaxation times in some paramagnetic transition metal acetylacetonate complexes: The possible influence of the Jahn-Teller effect on electron spin relaxation. - J. Am. Chem. Soc., 1977, 99, N4, 1281-1282. - CA, 1977, 86, 154774.

Exp.; E ; T; O_h; d^4; 3.2.

3.3.23. Kubo Hidenori, Kaneshima Nobuo, Matsuki Nobuyuku, Hirakawa Kazuyoshi, Dynamic Jahn-Teller effect in $K_2Mn_{1-x}Cu_xF_4$. [Jap.]. - Kyushu Daigaki Kogaku Shuho, Technol. Rep. Kyushu Univ., 1977, 50, N4, 437-444. - CA, 1979, 90, 177697. PA, 1978, 81, 19347.

Exp.; E - b_1; d^4; d^9; 3.3.

3.3.24. Le Dang Khoi, Veillet, P., Walker, P. J., Nuclear magnetic resonance studies of chromium-53 in a single crystal of chromium rubidium chloride ($CrRb_2Cl_4$). - J. Phys. C, 1977, 10, N22, 4593-4597. - CA, 1978, 88, 143857.

Exp.; O_h; d^4.

3.3.25. Le Dang Khoi, Veillet, P., Renard, J.-P., NMR study in a single crystal of the one-dimensional antiferromagnet $CsCuCl_3$. - Solid State Commun., 1977, 24, N4, 313-316. - RZF, 1978, 6D843.

Exp.; E - e; C_{3v}; d^9.

3.3.26. Lundin, A. G., Afanas'ev, M. L., Investigation of octahedral complexes by NMR. [Russ.]. - In: Fiz. Mat. Metody Koord. Khim.: Tez. Dokl. 6 Vsesoyuz. Soveshch. Kishinev, 1977, p. 9.

Theor.; O_h.

3.3.27. Zeer, E. P., Afanas'ev, M. L., Bondarenko, V. S., Lybzikov, G. F., Lundin, A. G., The pseudo-Jahn-Teller effect and the mobility in hexafluotitanates of pyridinium and zincum. [Russ.]. - In: Fiz. Mat. Metody Koord. Khim.: Tez. Dokl. 6 Vsesoyuz. Soveshch. Kishinev, 1977, p. 89-90.

Exp.; O_h.

1978

3.3.28. Cibert, J., Edel, P., Merle d'Aubigne, Y., Romestain, R., Optical detection of magnetic resonance in the $^3T_{1u}$ state of F centers in calcium oxide. - Semicond. & Insul., 1978, 3, N2/3, 137-148. - CA, 1978, 88, 200527.

Exp.; T; O_h; 3.2.

3.3.29. Glasbeek, M., Van Voorst, J. D. W., Influence of random intrinsic strain upon zero-field double resonance spectra of the CaO F-center in its excited Jahn-Teller states. - Phys. Rev. B, 1978, 17, N12, 4895-4907. - CA, 1978, 89, 138086. PA, 1978, 81, 84499.

Exp.; T - e; O_h.

3.3.30. Latas, K. J., Nishimura, A. M., Solvent effects upon the phosphorescent triplet states of cyanopyridine and pyridinecarboxaldehyde. - J. Photochem., 1978, 9, N6, 577-580. - CA, 1979, 90, 71444.

Exp.; D_{6h}.

1979

3.3.31. Afanas'ev, M. L., Lundin, A. G., Seryshev, S. A., Fluorine-19 NMR at high hydrostatic pressures and the pseudo-Jahn-Teller effect in KPF_6 and $(ND_4)_2TiF_6$. [Russ.]. - Koord. Khim., 1979, 5, N10, 1453-1456. - CA, 1979, 91, 219892. RZK, 1980, 2B981.

Exp.; O_h; 1.5.

3.3.32. Ugolev, I. I., Makatun, V. N., Potapovich, A. K., Shneerson, V. L., The manifestation of dynamical effectsin the EPR spectra of Cu^{2+} in some salt hydrates. [Russ.]. - Zh. Prikl. Spectrosk., 1979, 31, N5, 840-843. - CA, 1980, 92, 119088.

Exp.; d^9.

See also: 1.1.-30; 2.1.-280; 3.1.-156, 211, 255; 3.4.-27; 4.2.-79; 4.4.-146, 156; 5.1.-10, 54; 5.2.-18.

3.4. Mössbauer Spectrosopy

The first work on Mössbauer spectroscopy (or nuclear gamma resonance (NGR)), in which the observed quadrupole splitting is attributed to the Jahn-Teller effect, was published in 1963 [3.4.-1]. The same assumption with regard to the system $MgO:Fe^{2+}$ was made in Ref. [3.4.-5]. In these works the temperature dependence of the quadrupole splitting was related to tunneling transitions between the Jahn-Teller configurations giving different contributions to the electric field gradient on the nucleus. Quantitative relations for the NGR band shape, based on the Kubo-Andersen model, are given in Refs. [3.4-12, 13]. An alternative interpretation is given in Ref. [3.4-8]. The works [3.4.-10, 11, 15, 16, 18] (and also [3.4.-22, 23]) are devoted to vibronic effects in the NGR of systems with E and T terms of different multiplicity. In some papers the possible observation of vibronic reduction of the electric field gradient [3.4-7, 22], as well as spin-orbit interaction [3.4.-28], is noted.

In the seventies the study of electric and magnetic hyperfine interactions in NGR spectra taking into account vibronic effects advanced [3.4.-30, 31, 32, 33]. In Ref. [3.4.-34] the influence of the Jahn-Teller effect on the NGR line of hemoglobin is discussed. In a series of papers [3.4-28, 36, 37] the vibronic effects in the $CaO:Fe^{2+}$ system studied by IR and NGR spectroscopy are elucidated.

No works on NGR studied by methods going beyond the cluster model have been reported till now.

The article [3.4.-42] on NGR in vibronic systems is included in this subsection.

1963

3.4.1. Bemski, G., Fernandes, J. C., Isomer shift of Fe^{57} in InAs. - Phys. Lett., 1963, 6, N1, 10-11. - CA, 1963, 59, 12275h.

Exp.; d^6.

3.4.2. Tanaka Midori, Mizoguchi Tadashi, Aiyama Yoshimichi, Mössbauer study of ^{57}Fe in distorting spinels. - J. Phys. Soc. Jpn., 1963, 18, N7, 1089. - CA, 1965, 62, 1236a.

Exp.; d^6.

1964

3.4.3. Bacchella, G. L., Imbert, P., Meriel, P., Martel, E., Pinot, M., Study by x-rays and Mössbauer effect of phase changes in iron chromites. [Fr.]. - Bull. Soc. Sci. Bretagne, 1964, 39, 121-127. - CA, 1966, 65, 6367h.

Exp.; T; T_d; d^6; 4.6.

1965

3.4.4. Gerard, A., Use of the Mössbauer effect to study the Jahn-Teller distortion in iron chromites. - Bull. Soc. Belge Phys., 1965, N6, 379-388. - CA, 1966, 65, 6466b.

Exp.; T; O_h; T_d; d^5; d^6.

1966

3.4.5. Pipkorn, D. N., Leider, H. R., Mössbauer effect in $MgO:Fe^{2+}$: Quadrupole splitting at low temperature. - Bull. Am. Phys. Soc., 1966, 11, N1, 49.

Exp.; T; O_h; d^6.

1967

3.4.6. Bornaz, M., Filoti, G., Gelberg, A., Rosenberg, M., Mössbauer study of the Jahn-Teller effect in $Fe_{0.3}Mn_{2.7}O_4$. - Phys. Lett. A, 1967, 24, N9, 449-450. - CA, 1967, 67, 69297.

Exp.; 4.6.

3.4.7. Chappert, J., Frankel, R. B., Blum, N. A., Iron(I) and iron(II) hyperfine fields in magnesium oxide and calcium oxide. - Phys. Lett. A, 1967, 25, N2, 149-150. - CA, 1968, 68, 73934.

Exp.; T; O_h; d^6; d^7; 1.4.

3.4.8. Ham, F. S., Mössbauer spectrum of Fe^{2+} in MgO. - Phys. Rev., 1967, 160, N2, 328-333. - RZF, 1968, 3E808.

Theor.; T; O_h; d^6; 3.2.

3.4.9. Ono, K., Chandler, L., Ito, A., A Mössbauer study of natural spinel crystals. - Phys. Lett. A, 1967, 24, N5, 273-274.

Exp.; T; O_h; d^6; 4.6.

1968

3.4.10. Bersuker, I.B., Ogurtsov, I. Ya., The hyperfine interaction in impurity centers and complexes with degenerate electronic terms. [Russ.]. - Fiz. Tverd. Tela, 1968, 10, 12, 3651-3660.

Theor.; T; O_h; T_d.

3.4.11. Bersuker, I. B., Ogurtsov, I. Ya., The hyperfine interactions in molecular systems taking into account the Jahn-Teller effect. [Russ.]. - In: Conf. republ. de chimie fizica generala si aplicata: Resum. lucr. Bucuresti, 1968, p. 157.

Theor.; T - t_2; O_h; T_d; d^5.

3.4.12. Hartmann-Boutron, F., Effects of quadrupole coupling fluctuations on the Mössbauer spectra of ions in crystals. [Fr]. - J. Phys. [France], 1968, 29, N1, 47-56. - CA, 1968, 69, 14708.

Theor.

3.4.13. Tjon, J. A., Blume, M., Mössbauer spectra in a fluctuating environment. II. Randomly varying electric field gradients. - Phys. Rev., 1968, 165, N2, 456-461.

Theor.

1969

3.4.14. Bancroft, G. M., Mays, M. J., Prater, B. E., Single line Mössbauer spectrum of $[Fe(NH_3)_6]^{2+}$. - Chem. Phys. Lett., 1969, 4, N5, 248-250. - CA, 1970, 72, 26887.

Exp.; T; O_h; d^6.

3.4.15. Bersuker, I. B., Ogurtsov, I. Ya., Jahn-Teller effect and inversion (tunneling) splitting in Mössbauer spectra of coordination systems. - In: Abstr. Papers 22nd Int. Congr. Pure and Appl. Chem. Sydney, 1969, p. 92.

Theor.; T - t.

3.4.16. Bersuker, I. B., Ogurtsov, I. Ya., Polinkovskii, E. I., On the possible splitting of the Mössbauer line in the absence of static distortions. [Russ.]. - Izv. Akad. Nauk Mold.SSR. Ser. Biol. i Khim. Nauk, 1969, N4, 70-73. - RZF, 1970, 5E344.

Theor.

3.4.17. Bornaz, M., Filoti, G., Gelberg, A., Rosenberg, M., Mössbauer study of iron manganites. - J. Phys. C, 1969, 2, N6, 1008-1015. - CA, 1969, 71, 44206. PA, 1969, 72, 37701.

Exp.

3.4.18. Ogurtsov, I. Ya., Nekotoryye Voprosy Teorii Sverkh-tonkikh Vzaimodeystviy v Mossbaurovskikh Spektrakh: Avtoref. Dis. ... Kand Fiz.-Mat. Nauk. [Russ.]. - Kishinev, 1969. - 14 p. - Kishinev Univ.

Some Aspects of the Theory of Hyperfine Interactions in Mössbauer Spectra.

Theor.; T - t; O_h; T_d.

1971

3.4.19. Aleksandrov, A. Yu., Ionov, S. P., Pritchard, A. M., Gol'danskiy, V. I., Mössbauer spectra and the pseudo-Jahn-Teller effect in M_2SbCl_6[M = NH_4, Rb, Cs]. [Russ.]. - Pis'ma, Zh. Eksp. & Teor. Fiz., 1971, 13, N1, 13-15. - CA, 1971, 74, 132826. PA, 1971, 74, 41521. RZF, 1971, 5E649.

Exp.

3.4.20. Filoti, G., Gelberg, A., Gomolea, V., Rosenberg, M., Mössbauer investigation of Jahn-Teller distortion in spinelic manganites. - In: Hyperfine Interactions Excited Nucl.: Proc. Conf., 1970. New York, etc., 1971, Vol. 3, p. 834-835. - CA, 1972, 77, 68276. RZF, 1972, 6E765.

Exp.; d^0; d^5; d^7; d^{10}.

3.4.21. Gerard, A., Imbert, P., Prange, H., Varret, F., Wintenberger, M., Fe^{2+} impurities, isolated and in pairs, in ZnS and CdS studied by the Mössbauer effect. - J. Phys. & Chem. Solids, 1971, 32, N9, 2091-2100.

Exp.; E - e; T_d; d^6.

1973

3.4.22. Bersuker, I. B., Borshch, S. A., Ogurtsov, I. Ya., Hyperfine quadrupole splitting and Mössbauer spectra for an orbital doublet. - Phys. Status Solidi b, 1973, 59, N2, 707-714.

Theor.; E - e.

3.4.23. Bersuker, I. B., Borshch, S. A., Ogurtsov, I. Ya., Quadrupole splitting in nuclear gamma resonance spectra of orbital degenerated E-term systems. - In: Abstr. 11th Eur. Congr. Mol. Spectrosc. Tallin, 1973, 273(C7). - PA, 1973, 76, 75931.

Theor.; E - e.

3.4.24. Borshch. S. A., Sverkhtonkaya Struktura Messbaurovskikh Spektrov v Peremennykh Pol'akh: Avtoref. Dis. ... Kand. Fiz.-Mat. Nauk. [Russ.]. - Kishinev, 1973. - 14 p. - Kishinev Univ.

The Hyperfine Structure of the Mössbauer Spectra in Variable Fields.

Theor.; E - e.

3.4.25. Regnard, J. R., Mössbauer study of iron-57 in doped calcium oxide crystals. - Solid State Commun., 1973, 12, N3, 207-209. - CA, 1973, 78, 104124. PA, 1973, 76, 24639.

Exp.; T; O_h; d^5; d^6; d^7.

1974

3.4.26. Ham, F. S., Jahn-Teller effects in Mössbauer spectroscopy. - J. Phys. [France], 1974, 35, Colloq. N6, 121-130. - CA, 1975, 82, 131644.

Theor.; T; E; O_h; T_d; d^6.

3.4.27. Love, J. C., Obenshain, F. E., Nickel-61: Mössbauer studies of substituted nickel spinels. - In: Magn. & Magn. Mater.: 19th AIP Annu. Conf., Boston, 1973. New York, 1974, Pt. 2, p. 513-517. - CA, 1974, 81, 43758.

Exp.; O_h; d^8; 3.3.

3.4.28. Regnard, J. R., Chappert, J., Ribeyron, A., Mössbauer and far infrared evidence for a dominant E_g vibronic coupling in iron(2+) ion-doped calcium oxide. - Solid State Commun., 1974, 15, N9, 1539-1542. - CA, 1975, 82, 24053. PA, 1975, 78, 10252. RZF, 1975, 3D627.

Exp.; T - e; O_h; d^6.

3.4.29. Triplett, B. B., Dixon, N. S., Boolchand, P., Hanna, S. S., Bucher, E., Low-temperature Mössbauer studies of thulium compounds. - J. Phys. [France], 1974, 35, Colloq. N6, 653-657. - CA, 1975, 82, 131669.

Exp.; D_{4h}; 4.6.

1975

3.4.30. Belyaev, L. M., Dmitrieva, T. V., Lyubutin, I. S., Mazhara, A. P., Fedorov, V. E., Features of electric and magnetic hyperfine interactions of the iron-57 nuclei in chalcogenide spinels. [Russ.]. - Zh. Eksp. & Teor. Fiz., 1975, 68, N3, 1176-1182. - CA, 1975, 83, 18501. PA, 1975, 78, 47085. RZF, 1975, 7E198.

Exp.; E - e; T_d; d^6; 4.6.

3.4.31. Filoti, G., Mössbauer studies on Jahn-Teller distorted compounds. [Rom.]. - Stud. & Cercet. Fiz., 1975, 27, N7, 665-702. - CA, 1975, 83, 170482. PA, 1976, 79, 26788.

Exp.

3.4.32. Spender, M. R., Morrish, A. H., Mössbauer study of some sulfur spinels. - In: Proc. 5th Int. Conf. Mössbauer Spectrosc., 1973. Prague, 1975, Vol. 1/3, p. 125-129. - CA, 1977, 87, 192910.

Exp.; E - e; T; O_h; d^3; d^6; 4.6.

1976

3.4.33. Brossard, L., Oudet, H., Gibart, P., Distribution of strains at Fe^{2+}_A site in the thiospinel iron(II) scandium tetrasulfide. - J. Phys. [France], 1976, 37, Colloq. N6, 23-28. - CA, 1977, 87, 93078.

Exp.; d^6.

3.4.34. Cianchi, L., Mancini, M., Spina, G., A comment on the role of the Jahn-Teller effect in oxy-hemoglobin. - Lett. Nuovo Cimento. 1976, 15, N8, 263-264. - PA, 1976, 79, 42502.

Theor.; T - e; O_h; d^6.

3.4.35. Pollak, H., Quartier, R., Bruyneel, W., Walter, P., Jahn-Teller coherent transition and other properties of iron(3+) borate. - J. Phys. [France], 1976, 37, Colloq. N6, 589-590. - CA, 1977, 86, 130638.

Exp.; d^5.

3.4.36. Ray, T., Regnard, J. R., Study of the vibronic effects from the Zeeman-field-induced Mössbauer spectra of Fe^{2+} ions in octahedral symmetry. I. The effective Hamiltonian formalism. - Phys. Rev. B, 1976, 14, N5, 1796-1804. - CA, 1976, 85, 184476. PA, 1977, 80, 8476.

Theor.; T; O_h; d^6; 1.4.

3.4.37. Regnard, J. R., Ray, T., Study of the vibronic effects from the Zeeman-field-induced Mössbauer spectra of Fe^{2+} ions in octahedral symmetry. II. Cases of $KMgF_3:Fe^{2+}$ and $CaO:Fe^{2+}$. - Phys. Rev. B, 1976, 14, N5, 1805-1810.- PA, 1977, 80, 8477.

Exp.; T; O_h; d^6.

1977

3.4.38. Garcin, C., Imbert, P., Jehanno, G., Low-temperature Mössbauer study of iron(2+) ion impurities in zinc sulfide: source and absorber experiments. - Solid State Commun., 1977, 21, N6, 545-549. - CA, 1977, 86, 148422. PA, 1977, 80, 28500.

Exp.; d^6; 1.4.

3.4.39. Griebler, W. D., Pebler, J., Schmidt, K., Babel, D., Mössbauer spectroscopic studies of the high/low spin iron(II) compound $CsFe^HFe^L(CN)_6$. [Germ.]. - Z. Naturforsch. b, 1977, 32, N9, 992-997. - CA, 1977, 87, 191679.

Exp.; d^6.

1978

3.4.40. Buvnik, V. M., Ikonnikov, V. P., Kokov, I. T., Petrov, M. I., Study of alkali metal tetrachloriferrate crystals by Mössbauer spectra of iron-57. [Russ.]. - In: Yader. Magn. Rezonans v Kristallakh. Krasnoyarsk, 1978, p. 131-134. - CA, 1980, 92, 31539. RZK, 1979, 18B607.

Exp.; d^6.

3.4.41. Van Diepen, A. M., Lotgering, F. K., Preparation and Mössbauer spectra of CoV_2O_4 spinel doped with Fe^{2+} on tetrahedral sites. - Solid State Commun., 1978, 28, N11, 951-955. - PA, 1979, 82, 28660.

Exp.; E - e; T_d; d^6.

1979

3.4.42. Asaji Tetsuo, Ikeda Ryuichi, Nakamura Daiyu, Nitrogen-14 nuclear quadrupole resonance in some hexanitro complexes of the type $R_2PbCu(NO_2)_6$. - Z. Naturforshch. b, 1979, 34, N12, 1722-1728. - CA, 1980, 92, 138169.

Exp.; E - e; O_h; d^9; 4.6.

3.4.43. Bacci, M., Cerdonio, M., Vitale, S., Ground and low-lying electronic states of oxyhemoglobin from temperature dependent

Mössbauer and susceptibility data. - Biophys. Chem., 1979, 10, N1, 113-117. - CA, 1979, 91, 104027.

Theor.; $E - b_1$; C_{4v}; d^6; 3.6.

3.4.44. Deshpande, P. D., Khandekar, P. V., Umadikar, P. H., Diwan, A. J., Mössbauer study of spinel system $ZnMn_{1-x}Cr_xFeO_4$. - Pramāna, A J. Phys., 1979, 13, N6, 607-6h0. - PA, 1980, 83, 29664.

Exp.; d^4; 4.6.

3.4.45. Garcin, C., Gerard, A., Imbert, P., Jehanno, G., Abnormal populations and vibronic properties of the electronic levels of Fe^{2+} ion in ZnS:^{57}Co Mössbauer sources. - J. Phys. [France], 1979, 40, Colloq. N2, 413-414. - CA, 1979, 90, 177729. PA, 1980, 83, 15912.

Exp.; $E - e$; T_d; d^6.

3.4.46. Grandjean, F., Charge transfer chromium copper oxide ($CuCr_2O_4$): a Mössbauer study of iron-57 below the tetragonal-cubic transfer. - J. Phys. C, 1979, 12, N21, 4601-4610. - CA, 1980, 92, 112087.

Exp.; d^5; d^6.

See also: 1.1.-29, 30; 2.7.-32; 3.1.-105, 121, 268; 3.6.-34; 4.4.-3, 6, 7, 18, 38, 39, 49, 50, 70, 98, 118, 121, 131, 132, 150, 162, 165; 4.6.-51, 70, 184; 5.1.-58.

3.5. Acoustic Properties. Acoustic, Paraelastic and Paraelectric Resonances

In the first works on these topics [3.5.-1, 2, 3] the resonance absorption of ultrasound due to transitions between tunneling energy levels was predicted. Experimentally the acoustic losses were studied in Ref. [3.5.9]. Later, absorption and scattering of sound in crystals with Jahn-Teller impurity centers was investigated in a series of papers [3.5.-38, 50, 56, 58, 61, 62, 65]. In these works the parameters of tunneling splitting, as well as relaxation times, were also obtained.

A large number of studies were devoted to acoustic paramagnetic resonance in Jahn-Teller systems [3.5.-6, 11-13, 15-17, 19-35, 37, 39, 40, 45, 49, 51, 52, 54, 55, 57, 60, 63, 64]. The applicability of acoustic paramagnetic resonance to the investigation of spin-phonon interactions and the necessity to take into account the Jahn-Teller effect in the interpretation of experimental results were noted in the review article [3.5.-34]. Some particular problems of acoustic paramagnetic resonance were also considered: influence of random strain [3.5.-12, 22, 26] and external stress [3.5.-54], manifestation of vibronic reduction [3.5.-17, 25, 35], comparison of the cluster and multimode approaches [3.5.-49, 60, 64], problem of spin-phonon interaction, relaxation, and line shape [3.5.-41, 45] (see also the review article [3.5.-42]), and evaluation of vibronic constants and tunneling splitting magnitudes from experimental data [3.5.-45, 52, 55, 57, 63, etc.].

Besides the acoustic paramagnetic resonance method, the paraelectric [3.5.-10, 59, 67], nuclear acoustic [3.5.-44], and paraelastic [3.5.-36] resonances were also used to study vibronic effects in Jahn-Teller systems.

1963

3.5.1. Bersuker, I. B., Strong resonant absorption of ultrasound in octahedral complexes of transition metals with inversion splitting. [Russ.]. - Zh. Eksp. & Teor. Fiz., 1963, 44, N5, 1577-1582.

Theor.

3.5.2. Bersuker, I. B., The principle of a polarizer and analyzer of ultrasonic vibrations. [Russ.]. - Akust. Zh., 1963, 9, N3, 378-379.

Theor.

1964

3.5.3. Bersuker, I. B., "Inversion" sound absorption by hydrated Cu salt single crystals in a magnetic field. [Russ.]. - Fiz. Tverd. Tela, 1964, 6, N2, 436-439. - CA, 1964, 60, 15325g.

Theor.; E - e; O_h; d^9; 1.3.

1965

3.5.4. Fraser, D. B., Gyorgy, E. M., Le Craw, R. C., Remeika, J. P., Schnettler, F. J., Van Uitert, L. G., New acoustic and magnetic properties of YIG [yttrium iron garnet] and YAG [yttrium aluminum garnet] with small Mn and Ni additions. - J. Appl. Phys., 1965, 36, N3, 1016-1017. - CA, 1965, 62, 14013b.

Exp.; T; O_h; T_d; d^4; d^8.

3.5.5. Gyorgy, E. M., Sturge, M. D., Fraser, D. B., Le Craw, R. C., Observation of Jahn-Teller tunneling by acoustic loss. - Phys. Rev. Lett., 1965, 15, N1, 19-22. - CA, 1965, 63, 7667f. RZF, 1966, 1D404.

Exp.; E - e; O_h; d^4; d^7.

1966

3.5.6. Fletcher, J. R., Marshall, F. G., Rampton, V. W., Rowell, P. M., Stevens, K. W. H., Microwave ultrasonic paramagnetic resonance of the Cr^{2+} ion in magnesium oxide. - Proc. Phys. Soc. London, 1966, 88, N1, 127-130. - CA, 1966, 64, 18724. RZF, 1967, 6D409.

Exp.; O_h; d^4; 3.3.

3.5.7. Gyorgy, E. M., Le Craw, R. C., Sturge, M. D., Influence of Jahn-Teller ions on the acoustic and magnetic properties of YIG. - J. Appl. Phys., 1966, 37, N3, 1303-1309. - CA, 1966, 64, 15149f. RZF, 1966, 10E650.

Exp.; d^4; 3.6.

1967

3.5.8. Fraser, D. B., Gyorgy, E. M., Le Craw, R. C., Schnettler, F. J., Temperature-stable sonic transmission elements comprising crystalline materials containing Jahn-Teller ions: Pat. USA N3296555. - Appl. Oct. 8, 1964; (Cl.331-157). - Jan. 3, 1967. - 5p. - CA, 1967, 66, 50273.

Exp.; E; T; O_h; D_{4h}; d^3; d^4; d^8; d^9.

3.5.9. Sturge, M. D., Krause, J. T., Gyorgy, E. M., Le Craw, R. C., Merritt, F. R., Acoustic behavior of the Jahn-Teller ion Ni^{3+} in Al_2O_3. - Phys. Rev., 1967, 155, N2, 218-224. - CA, 1967, 67, 15145. RZF, 1968, 1E421.

Exp.; E - e; O_h; C_{3v}; d^7.

3.5.10. Williams, F. I. B., Paraelectric resonance of praseodymium in yttrium ethylsulfate. - Proc. Phys. Soc. London, 1967, 91, Pt. 1, N571, 111-123. - CA, 1967, 66, 119653.

Exp.

1968

3.5.11. Marshall, F. G., Rampton, V. W., Acoustic paramagnetic resonance spectrum of chromous ions in magnesium oxide. - J. Phys. C, 1968, 1, N3, 594-598. - CA, 1968, 69, 63457.

Exp.; O_h; d^4.

1969

3.5.12. Fletcher, J. R., Stevens, K. W. H., Jahn-Teller effect of octahedrally coordinated $3d^4$ ions. - J. Phys. C, 1969, 2, N3, 444-456. - CA, 1969, 70, 90842. PA, 1969, 72, 24346.

Theor.; O_h; d^4.

3.5.13. Pointon, A. J., Taylor, R. G. F., The acoustic paramagnetic resonance of chromous ions in magnesium oxide. - Phys. Lett. A, 1969, 28, N7, 535-536.

Exp.; O_h; d^4.

3.5.14. Vekhter, B. G., Piezoelectric effect in systems with a multivalley potential. [Russ.]. - In: Tez. Dokl. 2 Vsesoyuz. Konf. Teor. Tverd. Tela. Moscow, 1969, p. 49.

Theor.

1970

3.5.15. Anderson, R. S., Bates, C. A., The effects of electric fields on the APR spectra of $Cr^{2+}:Al_2O_3$. - J. Phys. C, 1970, 3, N7, 1139-1143. - PA, 1970, 73, 62749.

Exp.; O_h; C_{3v}; d^4.

3.5.16. Anderson, R. S., Brabin-Smith, R. G., Rampton, V. W., The acoustic paramagnetic resonance of X-irradiated ruby. - J. Phys. C, 1970, 3, N12, 2379-2386. - PA, 1971, 74, 31123.

Exp.; O_h; C_{3v}; d^4.

3.5.17. Bates, C. A., Goodfellow, L. C., Stevens, K. W. H., The theory of acoustic paramagnetic resonance of V^{3+} in MgO. - J. Phys. C, 1970, 3, N8, 1831-1838. - RZF, 1971, 2D591.

Theor.; T; O_h; d^2; 1.3; 1.4.

3.5.18. Hendriks, J. H., Dijksta, F. J., Acoustic relaxation in manganese ferrites ($Mn_xFe_{3-x}O_4$) at temperatures between 90 and 400°K for $1.0 < x < 1.6$. - J. Phys. & Chem. Solids, 1970, 31, N12, 2793-2795. - CA, 1971, 74, 116169. RZF, 1971, 7E1153.

Exp.

3.5.19. Locatelli, M., Jausaud, P., De Goer, A. M., Acoustic paramagnetic resonance spectrum of Al_2O_3:Mn. [Fr.]. - C.R. Hebd. Seances Acad. Sci. Ser. B, 1970, 270, N9, 620-623. - CA, 1970, 72, 138226.

Exp.; E - e; O_h; C_{3v}; d^4.

3.5.20. Rabin-Smith, R. G., The study of some iron group transition ions by acoustic paramagnetic resonance: Thesis. - Univ. Nottingham. - PA, 1970, 73, 76221.

Exp.; O_h; d^2; d^4.

1971

3.5.21. Bates, C. A., Goodfellow, L. C., Jaussaud, P., Smith, W., Jahn-Teller theory of acoustic paramagnetic resonance in aluminum oxide. - In: Magn. Resonance & Related Phenomena: Proc. 16th Congr. AMPERE, 1970. Bucharest, 1971, p. 258-264. - CA, 1973, 78, 9923.

Theor.; O_h; C_{3v}; d^2; d^4.

3.5.22. Ham, F. S., Acoustic paramagnetic resonance spectrum of Cr^{2+} in magnesium oxide. - Phys. Rev. B, 1971, 4, N11, 3854-3869. - CA, 1972, 76, 8678. PA, 1971, 74, 83797.

Theor.; E - e; O_h; d^4.

3.5.23. Ray, T., Dynamical Jahn-Teller effects in V^{3+} ion in a magnesium oxide crystal. - Solid State Commun., 1971, 9, N12, 911-915. - CA, 1971, 75, 82251. PA, 1971, 74, 59443.

Theor.; T; O_h; d^2.

1972

3.5.24. Abou-Ghantous, M., Dusdier, F., Locatelli, M., Acoustic paramagnetic resonance of V^{4+} ions in Al_2O_3. - Phys. Lett. A, 1972, 39, N1, 53-54.

Exp.; T; O_h; C_{3v}; d^1.

3.5.25. Abou-Ghantous, M., Bates, C. A., Acoustic paramagnetic resonance of V^{4+} ion in Al_2O_3. - In: Introduct. Abstr. Int. Conf. Phonon Scattering in Solids, Paris. Saclay, 1972, p. 266-271. - PA, 1973, 76, 2347.

Exp.; T; O_h; d^1.

3.5.26. Anderson, R. S., Bates, C. A., Jaussaud, P. C., The properties of $(3d)^4$ ions in corundum. I. Acoustic paramagnetic resonance spectra and their interpretation. - J. Phys. C, 1972, 5, N23, 3397-3413. - PA, 1973, 76, 8427.

Exp.; E - e; O_h; C_{3v}; d^4.

3.5.27. Anderson, R. S., Bates, C. A., Rampton, V. W., Jaussaud, P. C., The properties of $(3d)^4$ ions in corundum. II. Effects of electric fields on the acoustic paramagnetic resonance spectra and lineshape analysis. - J.Phys. C, 1972, 5, N23, 3414-3428. - PA, 1973, 76, 8428.

Exp.; E - e; O_h; C_{3v}; d^4.

3.5.28. Bates, C. A., Jones, S. C., Rampton, V. W., Steggles, P., Ferrous ions in ruby crystals. - J. Phys. C, 1972, 5, N11, L136-L138.

Exp.; T; O_h; C_{3v}; d^6.

3.5.29. Buisson, R., Nahmani, A., Influence of strains on paramagnetic resonance spectrum for strongly coupled ions: Application to V^{3+} in magnesium oxide. - Phys. Rev. B, 1972, 6, N7, 2648-2659. - CA, 1972, 77, 107353. PA, 1972, 75, 72495.

Theor.; T - e; O_h; d^2.

3.5.30. Rampton, V. W., Bates, C. A., Fletcher, J. R., Jones, S. C., Acoustic paramagnetic resonance at high magnetic fields from $(3d)^4$ ions. - In: Introduct. Abstr. Int. Conf. Phonon Scattering in Solids, Paris. Saclay, 1972, p. 243-246. - PA, 1973, 76, 2346.

Exp.; O_h; C_{3v}; d^4.

3.5.31. Ray, T., Dynamical Jahn-Teller effects for a V^{3+} ion in a MgO crystal. - Phys. Rev. B, 1972, 5, N5, 1758-1772. - CA, 1972, 76, 92232. PA, 1972, 75, 20147. RZF, 1972, 9E756.

Theor.; E; O_h; d^2.

3.5.32. Stevens, K. W. H., Spin-phonon interactions. - In: Abstr. Int. Conf. Phonon Scattering in Solids, Paris. Saclay, 1972, p. 217-223. - PA, 1973, 76, 1753.

Theor.; 3.2.

1973

3.5.33. Bates, C. A., Jones, S. C., Rampton, V. W., Steggles, P., An investigation of relaxation processes in ruby by APR. - In: Magn. Resonance & Related Phenomena: Proc. 17th Congr. AMPERE, Turku, 1972. Amsterdam; London, 1973, p. 75. - PA, 1973, 76, 2373.

Exp.; O_h; C_{3v}; d^5.

3.5.34. Bates, C. A., Jahn-Teller effects and APR [acoustic paramanetic resonance] spectra of $3d^n$ ion impurities in crystals. [Pol.]. - Postepy Fiz., 1973, 24, N6, 613-629. - CA, 1974, 80, 114216. PA, 1974, 77, 28695. RZF, 1974, 6D619.

Rev.

3.5.35. Ganapol'skiy, E. M., The acoustic paramagnetic resonance of Fe^{2+} and Fe^{3+} in the GaAs. [Russ.]. - Fiz. Tverd. Tela, 1973, 15, N2, 368-375.

Exp.; E - e; T_d; d^6; 1.4.

3.5.36. Glinchuk, M. D., Karmazin, A. A., Paraelastic resonance of Jahn-Teller and off-center ions. - Phys. Status Solidi a, 1973, 15, N2, K145-K147. - CA, 1973, 78, 130352. RZF, 1973, 9E762.

Theor.; E - e.

3.5.37. Goodfellow, L. C., The theory of acoustic paramagnetic resonance for the V^{3+} ion in MgO and Al_2O_3: Thesis. - Univ. Nottingham, 1973. - PA, 1974, 77, 3898.

Theor.; T; O_h; C_{3v}; d^2.

3.5.38. Lange, J. N., Dynamic Jahn-Teller effect for Cr^{2+} in MgO: hypersonic attenuation. - Phys. Rev. B, 1973, 8, N12, 5999-6009. - CA, 1974, 80, 112847. PA, 1974, 77, 44858. RZF, 1974, 9E496.

Exp.; E - e; O_h; d^4

1974

3.5.39. Abou-Ghantous, M., Bates, C. A., Clark, I. A., Fletcher, J. R., Jaussaud, P. C., Moore, W. S., Study of the Jahn-Teller orbital doublet: Nickel(3+) ion in aluminum oxide. - J. Phys. C, 1974, 7, N15, 2707-2720. - CA, 1974, 81, 113290. RZF, 1975, 1E164.

Exp.; E - e; O_h; C_{3v}; d^7.

3.5.40. Ganapol'skiy, E. M., On the state of chromium atom impurities in gallium arsenide. [Russ.]. - Fiz. Tverd. Tela, 1974, 16, N10, 2886-2893. - PA, 1975, 78, 62282. RZF, 1975, 2E24.

Exp.; T - (e + t); T_d; d^4.

3.5.41. Obata Yukio, On the spin-lattice interaction. II. [Jap.]. - Kotai Butsuri, Solid State Phys., 1974, 9, N1, 25-31. - RZF, 1974, 7D537.

Rev.; 3.2.

3.5.42. Obata Yukio, On the spin-lattice interaction. III. [Jap.]. - Kotai Butsuri, Solid State Phys., 1974, 9, N4, 179-181. - RZF, 1975, 1E52.

Rev.; 3.2.

3.5.43. Rampton, V. W., The acoustic paramagnetic resonance of rare-earth-doped calcium fluoride. - J. Phys. C, 1974, 7, N23, 4346-4354. - PA, 1975, 78, 13139.

Exp.; E - e; O_h.

3.5.44. Shutilov, V. A., Issledovaniya Akusticheskogo Yadernogo Rezonanca v Kristallakh: Avtoref. Dis. ... Dokt. Fiz.-Mat. Nauk. [Russ.]. - Leningrad, 1974. - 34 p. - Leningrg. Univ.

Investigation of Acoustic Nuclear Resonance in Crystals.

Exp.

1975

3.5.45. Abou-Ghantous, M., Bates, C. A., Fletcher, J. R., Jaussaud, P. C., Nickel(3+) ion-doped aluminum oxide Jahn-Teller system. - J. Phys. C, 1975, 8, N21, 3641-3652. - CA, 1976, 84, 24171. PA, 1976, 79, 14366.

Exp.; E - e; O_h; C_{3v}; d^7.

3.5.46. Ganapol'skiy, E. M., Structure of the energy spectrum and electron-phonon interaction of iron impurity atoms in corundum. [Russ.]. - Fiz. Tverd. Tela, 1975, 17, N1, 67-75. - PA, 1976, 79, 2375. RZF, 1975. 5D708.

Exp.; T; O_h; d^6; 3.2.

3.5.47. Ganapol'skiy, E. M., Giperzvukovyye Issledovaniya Akusticheskogo Paramagnitnogo Rezonansa Primesnykh Tsentrov v Kristallakh: Avtoref. Dis. ... Dokt. Fiz.-Mat. Nauk. [Russ.]. - Kazan', 1975, - 46 p. - Kazansk. Univ.

Hypersonic Investigations of Acoustic Paramagnetic Resonance of Impurity Centers.

Exp.

3.5.48. Lange, J. N., Strongly coupled impurities in MgO: Acoustic paramagnetic resonance of Fe^{2+} and Cr^{2+}. - Phys. Rev. B, 1975, 12, N1, 226-238.

Exp.; E; T; O_h; d^5; d^6.

1976

3.5.49. Abou-Ghantous, M., Bates, C. A., Goodfellow, L. C., Jahn-Teller effects in $(3d)^2$ ions in Al_2O_3. - J. Phys. & Chem. Solids, 1976, 37, N11, 1059-1068. - CA, 1977, 86, 98509. PA, 1976, 79, 93920.

Exp.; T; O_h; C_{3v}; d^2.

3.5.50. King, P. J., Oates, S. G., A hypersonic attenuation study of V^{3+} in magnesium oxide. - J. Phys. C, 1976, 9, N2, 389-400. - PA, 1976, 79, 30721.

Exp.; O_h; d^2.

3.5.51. Lange, J., Dynamic Jahn-Teller effect for Cr^{2+} in MgO: Acoustic paramagnetic resonance. - Phys. Rev. B, 1976, 14, N11, 4791-4802. - CA, 1977, 86, 63117. PA, 1977, 80, 20693.

Exp.; O_h; d^4.

3.5.52. Lange, J., Guha, S., Temperature dependence of the acoustic paramagnetic resonance spectra for chromium(2+) in magnesium oxide and potassium magnesium fluoride. - In: Phonon Scattering in Solids: Proc. 2nd Int. Conf., 1975. New York; London, 1976, p. 175-177. - CA, 1977, 86, 197347.

Exp.; E - e; O_h; d^4.

3.5.53. Nahmani, A., Buisson, R., Acoustic paramagnetic resonance of Ti^{2+} in CaO. - Solid State Commun., 1976, 18, N3, 297-300.

Exp.; T - e; O_h; d^2.

3.5.54. Rampton, V. W., Shellard, I. J., Stress-induced changes in the acoustic paramagnetic resonance of chromous ions in magnesium oxide. - In: Phonon Scattering in Solids: Proc. 2nd Int. Conf., 1975. New York; London, 1976, p. 178-180. - CA, 1977, 86, 180047.

Exp.; E - e; O_h; d^4.

1977

3.5.55. Guha, S., Lange, J., Dynamic Jahn-Teller effect for the chromous ion in potassium trifluoromagnesate. - Phys. Rev. B, 1977, 15, N9, 4157-4167. - CA, 1977, 87, 31598. PA, 1977, 80, 66416.

Exp.; E - e; O_h; d^4.

3.5.56. Ivanov, M. A., Fishman, A. Ya., Ultrasonic absorption in dielectrics with orbitally degenerate impurities of the E-type. [Russ.]. - In: Fiz. Mat. Metody Koord. Khim.: Tez. Dokl. 6 Vsesoyuz. Soveshch. Kishinev, 1977, p. 124.

Theor.; E - e.

3.5.57. Sasaki Ken, Obata Yukio, Theory of paramagnetic acoustic resonance absorption by U^{4+} ions in cubic crystalline field. - J. Phys. Soc. Jpn., 1977, 42, N1, 43-49. - CA, 1977, 86, 132413.

Theor.; T - (e + t); O_h; f^2.

3.5.58. Uchastkin, V. I., Low-temperature acoustical losses in high-molecular-weight compounds and Jahn-Teller effect. [Russ.]. - In: Mater. 9 Vsesoyuz. Akust. Konf. Sekts. V. Moscow, 1977, p. 135-138. - CA, 1978, 88, 153193. RZF, 1978, 2E1812.

Exp.

3.5.59. Vaysleyb, A. V., Rozenfel'd, Yu. B., Tsukerblat, B. S., Paraelectric resonance in an orbital doublet with weak Jahn-Teller interaction. [Russ.]. - Fiz. Tverd. Tela, 1977, 19, N5, 1484-1486. - PA, 1978, 81, 15137.

Theor.; E - e.

1978

3.5.60. Bates, C. A., Jahn-Teller study of Cr^{2+}:Al_2O_3. - J. Phys. C, 1978, 11, N16, 3447-3460. - CA, 1979, 90, 112566. PA, 1978, 81, 84172.

Theor.; E - e; O_h; C_{3v}; d^4; 3.1.

3.5.61. Ivanov, M. A., Fishman, A. Ya., Acoustic absorption in crystals with orbitally degenerate impurity centers. [Russ.]. - Fiz. Tverd. Tela, 1978, 20, N7, 2148-2158. - PA, 1979, 82, 37765. RZF, 1978, 10E396.

Theor.; E - e; 3.2.

3.5.62. Kim, H., Lange, J., Acoustical relaxation due to the ferrous ion in $KMgF_3$. - Phys. Rev. B, 1978, 15, 1961-1965. - PA, 1979, 82, 4375.

Exp.; T - t_2; O_h; d^6.

3.5.63. Shellard, I. J., Acoustic paramagnetic spectra of Cr^{2+} in MgO and CaO: Thesis. - Univ. Nottingham, 1978. - PA, 1979, 82, 87313.

Exp.; E - e; O_h; d^4.

1979

3.5.64. Bates, C. A., Maynard, C. M., Rampton, V. W., Shellard, I. J., The Cr^{2+}:CaO Jahn-Teller system. - J. Phys. C, 1979, 12, N17, 3561-3569. - CA, 1980, 92, 85413. PA, 1979, 82, 99756.

Exp.; E - e; O_h; C_{3v}; d^4.

3.5.65. Tokumoto Hiroshi, Ishiguro Takehiko, Observation of the dynamic Jahn-Teller effect for Cr^{2+} centers in GaAs. - In: Phys. Semicond.: Invited and Contributed Papers 14th Int. Conf., Edinburgh, 1978. Bristol; London, 1979, p. 299-302. - CA, 1979, 90, 159294. PA, 1980, 83, 6798.

Exp.; E - e; T_d; d^4; 3.2.

3.5.66. Tokumoto Hiroshi, Ishiguro Takehiko, Ultrasonic study of dynamic behavior of Jahn-Teller distorted Cr^{2+} centers in GaAs. - J. Phys. Soc. Jpn., 1979, 46, N1, 84-91. - CA, 1979, 90, 96460. PA, 1979, 82, 28048.

Exp.; E - e; T_d; d^4; 3.2.

3.5.67. Vikhnin, V. S., Paraelastic resonance of the nonstationary tunneling states. [Russ.]. - Ukr. Fiz. Zh., 1979, 24, N12, 1802-1805.

Theor.; 3.2.

See also: 1.7.-18, 47; 3.1.-123, 251; 3.2.-71; 4.1.-72, 92, 102, 103, 121.

3.6. Magnetic Susceptibility

The magnetic susceptibility was the first physical method used to verify the predictions of the Jahn-Teller theorem [3.6.-1, 2], but intensive work on this topic began only in the seventies. Using vibronic effects, the anisotropy of magnetic properties in crystals with Jahn-Teller impurities [3.6.-5, 7, 8] and the magnetic susceptibility of oxide solid solutions [3.6.-15] and of some manganese [3.6.-11], cobalt [3.6.-16], and uranium [3.6.-19] compounds were considered. For the magnetic properties of cooperative Jahn-Teller systems, see Section 4. For cobaltocenes ($^2E_{1g}$ term) a combined investigation of absorption spectra, EPR, and magnetic susceptibility was carried out [3.1.-143]. The theoretical aspects of the problem are given in Refs. [3.6.-1, 2, 3, 14, 16, 19, 30].

1937

3.6.1. Van Vleck, J. H., On the magnetic susceptibilities of Ti, V, and Cr alum. - Phys. Rev., 1937, 52, 246.

Theor.

1939

3.6.2. Van Vleck, J. H., The magnetic behavior of vanadium, titanium, and chrome alum. - J. Chem. Phys., 1939, 7, N1, 61-71. - CA, 1939, 33, 1561^2.

Theor.

1958

3.6.3. Van Vleck, J. H., II. Energetics of complexes: The magnetic behavior of regular and inverted crystalline levels. - Discuss. Faraday Soc., 1958, 26, N1, 96-102. - CA, 1959, 53, 15779.

Theor.

1960

3.6.4. Cooke, A. H., Lazenby, R., Leask, J. M., Magnetic susceptibilities of erbium and holmium ethyl sulfates. - 1960. - 3 p. - U.S. Dept. Com. Office Techn. Serv., PB, Rep. 155474. - CA, 1963, 58, 2001e.

Exp.; T.

1961

3.6.5 Baltzer, P. K., Contribution to magnetic anisotropy from cations which locally distort the crystal lattice. - J. Phys. Soc. Jpn., 1961, 17, Suppl., B-1, 192-195.

Exp.; T; T_d; d^8.

1964

3.6.6. Figgis, B. N., Lewis, J., Mabbs, F., Webb, G. A., Trigonal or tetragonal distortions and the magnetic properties of transition metal ions with triply orbitally degenerate ground terms. - Nature, 1964, 203, N4950, 1138-1141. - PA, 1965, 68, 5875.

Exp.; T; T_d; d^8.

3.6.7. Gyorgy, E. M., Schnettler, F. J., Induced anisotropy in YIG with Si and Mn additions. - J. Appl. Phys., 1964, 35, N3, 1067-1068. - CA, 1964, 60, 8758e.

Exp.; E; T; O_h; d^4.

3.6.8. Gyorgy, E. M., Schnettler, F. J., Induced magnetic anisotropy in ferrite single crystals. - J. Appl. Phys., 1964, 35, N5, 1648-1649. - CA, 1964, 61, 162b.

Exp.; T; T_d; O_h; d^4; d^8.

1966

3.6.9. Tulupov, V. A., Shigorin, D. N., Verein, N. V., On the structure of π-complexes formed by ions of transition group elements with olefins. I. np^6-$nd^{10} \rightarrow \pi^*$ promotion. [Russ.]. - Zh. Fiz. Khimii, 1966, 40, N5, 1020-1025. - CA, 1966, 65, 11581h.

Theor.

1967

3.6.10. Burns, R. G., Strens, R. G. J., Structural interpretation of polarized absorption spectra of the aluminum-iron-manganese-chromium epidotes. - Mineral. Mag., 1967, 36, N278, 204-226. - CA, 1968, 68, 73721.

Exp.; d^4; d^5; 2.1.

1969

3.6.11. Pal, A. K., Mitra, S. N., Sen Gupta, P., Paramagnetic studies on single crystal manganese(III) acetylacetonate. - In: Proc. 13th Nucl. Phys. & Solid State Phys. Symp., 1969, Vol. 3, p. 196-201. - CA, 1970, 73, 114520.

Exp.; d^4.

1970

3.6.12. Bhattacharyya, B. D., Role of the Jahn-Teller effect in the ligand field theory of the magnetic behavior of tetramethylammonium tetrachloroferrate. - In: Proc. 14th Nucl. Phys. & Solid State Phys. Symp., Roorkee, 1969. Bombay, 1970, Vol. 3, p. 395-398. - CA, 1971, 75, 124441.

Theor.; E - e; T_d; d^6.

3.6.13. Bhattacharyya, B. D., Role of the Jahn-Teller effect in the ligand field theory of the magnetic behavior of $(NMe_4)_2FeCl_4$. - Phys. Status Solidi, 1970, 38, N2, K157-K161. - CA, 1970, 73, 19747. PA, 1971, 74, 2642.

Theor.; E - e; T_d; d^6.

3.6.14. Sasaki Ken, Obata Yukio, Studies of the dynamical Jahn-Teller effect on static magnetic susceptibility. - J. Phys. Soc. Jpn., 1970, 28, N5, 1157-1167. - CA, 1970, 73, 8742. PA, 1970, 73, 51165.

Theor.; T - e.

3.6.15. Syono Yasuhiko, Magnetic oxide solid solutions. [Jap.]. - Nippon Kessho Gakkaishi, J. Crystallogr. Soc. Jpn., 1970, 12, N4, 220-231. - CA, 1972, 76, 19479.

Rev.

1971

3.6.16. Bhattacharyya, B. D., Lattice-ion interactions of Co^{2+} ion in an octahedral environment. - Phys. Status Solidi b, 1971, 48, N1, K87-K91. - PA, 1972, 75, 7018.

Theor.; O_h; d^7; 5.3.

3.6.17. Bhattacharyya, B. D., Lattice-ion interactions of Fe^{2+} ion in a tetrahedral environment. - J. Phys. & Chem. Solids, 1971, 32, N10, 2357-2362. - PA, 1971, 74, 80355.

Theor.; E - e; T_d; d^6; 2.1.

3.6.18. Chandler, P. E., Jahn-Teller effects in orbital triplet: Thesis. - Univ. Nottingham, 1971. - PA, 1972, 75, 40586.

Theor.; T.

3.6.19. Sasaki Ken, Obata Yukio, Theory of dynamical Jahn-Teller effect on magnetic susceptibility and its application to the uranium dioxide-thorium dioxide system. - J. Phys. [France], 1971, 32, Colloq. N1, Pt. 2, 739-740. - CA, 1971, 75, 27209. CA, 1974, 80, 42272.

Theor.; T - e.

1972

3.6.20. Bhattacharyya, B. D., On lattice-ion interactions of octahedrally coordinated trisacetylacetonato vanadium(III). - In: Abstr. Nuclear Phys. & Solid State Phys. Symp. Bombay, 1972, 1 p. - PA, 1972, 75, 36793.

Theor.; T - e; O_h; d^2; 2.1; 3.1.

3.6.21. Kahn, O., Kettle, S. F. A., Influence of vibronic coupling on the paramagnetism of a cubic complex in $^2T_2$ electronic state. [Fr.]. - Theor. Chim. Acta, 1972, 27, N3, 187-196. - CA, 1973, 78, 9431.

Theor.; T - e; O_h; C_{3v}.

3.6.22. Prabhackaran, C. P., Patel, C. C., Jahn-Teller distortion in monodentate ligand complexes of manganese(III). - J. Inorg. & Nucl. Chem., 1972, 34, N7, 2371-2374. - CA, 1972, 77, 81737.

Exp.; E - e; O_h; 2.7.

1973

3.6.23. Bernier, J. C., Kahn, O., Magnetic behavior of vanadium hexacarbonyl. - Chem. Phys. Lett., 1973, 19, N3, 414-417. - PA, 1973, 76, 36703.

Exp.; T; O_h.

3.6.24. Bhattacharyya, B. D., Dynamic Jahn-Teller effect in $3d^1$ ions. - In: Proc. 17th Nucl. Phys. & Solid State Phys. Symp., 1972. Bombay, 1973, Vol. C, p. 689-692. - CA, 1974, 80, 126091.

Exp.; T - (e + t_2); O_h; d^1; 3.1.

3.6.25. Jansen, M., Hoppe, R., Sodium chloride structure family: Crystal structure of sodium manganate(III). [Germ.]. - Z. Anorg. Allg. Chem., 1973, 399, N2, 163-169. - CA, 1973, 79, 119344.

Exp.; T; E; d^4.

3.6.26. Kahn, O., Kettle, S. F. A., Vibronic coupling and paramagnetic anisotropy in a cubic complex $^2T_2$ subjected to a trigonal field. [Fr.]. - Theor. Chim. Acta, 1973, 29, N4, 359-374. - CA, 1973, 79, 71596.

Theor.; T - e; O_h; C_{3v}.

1974

3.6.27. Kahn, O., Paramagnetism and orbital reduction. [Fr.]. - Chim. Phys. & Phys. - Chim. Biol., 1974, 71, N4, 581-586. - CA, 1974, 81, 18437.

Theor.; O_h; 1.4.

3.6.28. Novak, P., Effective Hamiltonian for calculation of the magnetocrystalline anisotropy of magnetic insulators. - In: Tr. Mezhdunar. Konf. Magn. 1973. Moscow, 1974, Vol. 5, p. 449-453. - CA, 1976, 84, 53081.

Theor.; d^9.

1975

3.6.29. Alpert, Y., Banerjee, R., Magnetic susceptibility measurements of deoxygenated hemoglobins and isolated chains. - Biochem. & Biophys. Acta, 1975, 405, N1, 144-154.

Exp.; C_{4v}; C_{2v}; d^6.

3.6.30. Chandler, P. E., Magnetic susceptibility and the Jahn-Teller effect for a $T_2 \times E$ ion. - J. Phys. C, 1975, 8, N3, 316-320. - PA, 1975, 78, 33429.

Theor.; T - e.

1976

3.6.31. Bhattacharyya, B. D., Jahn-Teller effect in cobalt-substituted ferrites. - In: Abstr. 2nd Int. Conf. Ferrites, Bellevue, 1976, p. 15-16. - (Europhys. Conf. Abstr., 1976; Vol. 2B). - CA, 1977, 86, 82723.

Exp.; E - e; O_h; T_d; d^7.

3.6.32. Walz, F., Rivas, J., Magnetic aftereffects in manganese ferrites. - Phys. Status Solidi a, 1976, 37, N1, 151-160. - PA, 1976, 79, 91300.

Exp.; 3.2.

1977

3.6.33. Bacci, M., Vibronic coupling effect on the paramagnetism of deoxygenated hemoglobin. - Chem. Phys. Lett., 1977, 48, N1, 184-186.

Theor.; E - $(b_1 + b_2)$; C_{4v}; d^6.

1978

3.6.34. Bacci, M., Vibronic coupling in deoxyhemoglobin and deoxymyoglobin. - J. Chem. Phys., 1978, 68, N11, 4907-4911. - RZF, 1978, 11I380.

Theor.; E - $(b_1 + b_2)$; C_{4v}; d^6; 3.4.

See also: 1.1.-30; 1.4.-17; 1.7.-16; 2.1.-20, 228, 237; 2.7.-71; 3.1.-105, 143; 3.4.-43; 4.2.-79; 4.3.-34; 4.4.-4, 7, 8, 9, 10, 14, 16, 17, 22, 26, 27, 35, 36, 37, 46, 52, 57, 64, 65, 74, 87, 96, 102, 106, 107, 124, 129, 132, 133, 134, 149, 151, 160, 164a; 4.5.-13, 28; 4.6.-141; 5.1.-58; 5.2.-25, 45; 5.4.-45, 55.

3.7. Impurity Heat Capacity and Thermoconductivity. Phonon Scattering. Electroconductivity. Electric Properties

The influence of strong vibronic coupling on the thermal properties of paramagnetic salts was first noted in Ref. [3.7.-1]. In Refs. [3.7.-2, 4] anomalies in the temperature dependence of heat capacity and heat conductivity were considered as due to the interactions of Jahn-Teller clusters with lattice vibrations at temperatures comparable with the vibronic energy level separation (tunneling splitting). Heat conductivity features resulting from phonon scattering [3.7.-8, 13, 17, 18, 41], influence of magnetic fields and uniaxial stress [3.7.-29, 34, 35], random strain [3.7.-45], etc., were also considered. Disagreement of the theory with experimental data, even taking into account multimode effects, was noted in Ref. [3.7.-41]. Several works have been reported on the electric resistance of vibronic systems [3.7.-3, 11, 12, 37].

Concerning electric properties, it was reported that dipolar instability was due to the pseudo-Jahn-Teller effect and its influence on the temperature dependence of molecular polarizabilities [3.7.-7, 24, 43] (see also [1.5.-14, 19, 25] and Section 4). The possibilities of a phonon maser effect in Jahn-Teller paramagnetic crystals has been suggested [3.7.-14].

1963

3.7.1. Persico, F., Stevens, K. W. H., Tucker, J. W., Some consequences of strong spin-lattice interactions. - Phys. Lett., 1963, 5, N1, 16-17. - CA, 1963, 59, 8279f. PA, 1963, 66, 18139.

Theor.; 3.2.

1964

3.7.2. Vekhter, B. G., Bersuker, I. B., The interaction of the inversion levels of transition metal complexes with lattice vibrations. [Russ.]. - In: Tez. Dokl. 4 Soveshch. Teorii Poluprovodnikov. Kishinev, 1964, p. 13.

Theor.

1965

3.7.3. Aoki Ikuo, Kameyama Ikuji, Hinata Noriyasu, Ono Tadashi, Yanachi Tadao, Yamaneshi Terumi, Tetragonal distortion of $CuFe_xMn_{2-x}O_4$. [Jap.]. - Chiba Daigaku Bunri Gakubu Kiyo Shizen Kagaku, J. College Arts and Sciences, Chiba Univ. Natural Sci. Ser., 1965, 4, N3, 253-261. - CA, 1966, 65, 16232f.

Exp.

3.7.4. Bersuker, I. B., Vekhter, B. G., Anomalies of heat capacity and heat conductivity in crystals of complex compounds. [Russ.]. - Fiz. Tverd. Tela, 1965, 7, N5, 1569-1570.

Theor.

1967

3.7.5. Stevens, K. W. H., Van Eekelen, H. A. M., Thermodynamic effects of spin-phonon coupling. - Proc. Phys. Soc. London, 1967, 92, N3, 680-696. - PA, 1968, 71, 8636.

Theor.; 3.2.

1968

3.7.6. Bates, C. A., Dixon, J. M., Fletcher, J. R., Stevens, K. W. H., The E-type orbitals of octahedrally coordinated ions. - J. Phys. C, 1968, 1, N4, 859-871. - PA, 1968, 71, 48339.

Theor.; E - e; O_h; 1.2; 1.7.

1969

3.7.7. Bersuker, I. B., Dipole instability and dipole moments of symmetrical molecular systems. - In: Abstr. Papers 22nd Int. Congr. Pure and Appl. Chem. Sydney, 1969, p. 16.

Theor.

3.7.8. McConachie, M. A., Resonant phonon scattering at low temperatures: Thesis. - Univ. Nottingham, 1969. - PA, 1970, 73, 71877.

Exp.; O_h; d^4.

3.7.9. Sheard, F. W., Negative thermal expansion due to paramagnetic ions in cubic crystals. - Phys. Lett. A, 1969, 30, N3, 156-157.

Theor.

1970

3.7.10. Challis, L. J., De Goer, A. M., Evidence for a Jahn-Teller effect in d^4 ions in aluminum oxide. - Phys. Lett. A, 1970, 31, N8, 463-468. - CA, 1970, 73, 20224. PA, 1970, 73, 41075.

Exp.; E - e; O_h; C_{3v}.

3.7.11. Crevecoeur, C., De Wit, H. J., Electrical conductivity of Li-doped MnO. - J. Phys. & Chem. Solids, 1970, 31, N4, 783-791. - CA, 1970, 72, 137572. PA, 1970, 73, 44204.

Exp.; O_h.

3.7.12. Kumar Naresh, Attractive electron-electron interaction induced by the dynamical Jahn-Teller effect involving degenerate impurity orbitals: theory of the impurity enhancement of the superconductive transition temperature in dilute alloys. - Phys. Rev. B, 1970, 2, N7, 2500-2511. - CA, 1971, 74, 15832. PA, 1971, 74, 9282.

Theor.; E; 4.5.

1971

3.7.13. Challis, L. J., De Goer, A. M., Evidence from thermal conductivity measurements for a Jahn-Teller effect for Mn^{3+} and Cr^{2+} in Al_2O_3. - In: Proc. 12th Int. Conf. Low Temp. Phys., Kyoto, 1970. Tokyo, 1971, p. 788. - RZF, 1974, 6E945.

Exp.; O_h; C_{3v}; d^4.

3.7.14. Lee, K. P., Phonon generation in Jahn-Teller type paramagnetic crystals. - Phys. Canada, 1971, 27, N4, 49. - PA, 1971, 74, 65983.

Theor.; E - e.

3.7.15. Sheard, F. W., Anomalous thermal expansion due to paramagnetic impurity ions. - J. Phys. [France], 1971, 32, Colloq. N1, 939-940. - CA, 1974, 80, 7908.

Theor.

3.7.16. Ward, R. G., Thermal conduction across Kapitza boundaries and in solids at low temperatures: Thesis. - Univ. Nottingham. - PA, 1971, 74, 55853.

Exp.; O_h; C_{3v}; d^4; 4.5.

1972

3.7.17. Challis, L. J., De Goer, A. M., Guckelsberger, K., Slack, G. A., An investigation of the ground state of Cr^{2+} in MgO based on thermal conductivity measurements. - Proc. R. Soc. London. Ser. A, 1972, 330, N1580, 29-58. - PA, 1972, 75, 72238.

Exp.; T - t_2; O_h; d^4.

3.7.18. Challis, L. J., Halbo, L., Evidence for a Jahn-Teller effect in p-Ge from magnetothermal conductivity measurements. - Phys. Rev. Lett., 1972, 28, N13, 816-819. - CA, 1972, 76, 133124. PA, 1972, 75, 33185. RZF, 1972, 8E930.

Exp.; T_d.

3.7.19. Fletcher, J. R., Blumson, J., Sheard, F. W., Phonon scattering by Jahn-Teller defects. - In: Introduct. Abstr. Int. Conf. Phonon Scattering in Solids, Paris. Saclay, 1972, p. 228-232. - PA, 1973, 76, 1755.

Theor.; E.

3.7.20. De Goer, A. M., Thermal conductivity at low temperatures of single crystal Al_2O_3:Ni [the dynamical Jahn-Teller effect]. [Fr.]. - In: Introduct. Abstr. Int. Conf. Phonon Scattering in Solids, Paris. Saclay, 1972, p. 260-265. - PA, 1973, 76, 1845.

Exp.; E - e; O_h; d^7.

3.7.21. De Goer, A. M., Devismes, N., Phonon scattering by vanadium ions in Al_2O_3. - J. Phys. & Chem. Solids, 1972, 33, N9, 1785-1793. - PA, 1972, 75, 68214.

Exp.; O_h; C_{3v}; d^1.

3.7.22. Halbo, L., Challis, L. J., The anisotropy of the magnetothermal conductivity of p-Si and p-Ge [and the Jahn-Teller effect in bound acceptor states]. - In: Introduct. Abstr. Int. Conf. Phonon Scattering in Solids, Paris. Saclay, 1972, p. 139-149. - PA, 1973, 76, 1842.

Exp.; T_d.

1973

3.7.23. Anderson, B. R., Challis, L. J., The measurement of the trigonal splitting of Fe^{2+} in Al_2O_3 by a frequency crossing technique using thermal phonons. - J. Phys. C, 1973, 6, N13, L266-L270.

Exp.; T; O_h; C_{3v}; d^6.

3.7.24. Bersuker, I. B., Ogurtsov, I. Ya., Shaparev, Yu. V., The temperature dependence of the mean dipole moment of symmetrical molecular systems. [Russ.]. - Teor. & Eksp. Khim., 1973, 9, N4, 451-459. - CA, 1974, 80, 19674.

Theor.

1974

3.7.25. Hu, P., Naryanamurti, V., Dynes, R. C., Heat pulses and Jahn-Teller tunneling in Al_2O_3:Ni^{3+}. - Bull. Am. Phys. Soc., 1974, 19, N3, 337.

Exp.; O_h; C_{3v}; d^7.

3.7.26. Locatelli, M., De Goer, A. M., Phonon scattering by Ni^{3+} ions in Al_2O_3. - Solid State Commun., 1974, 14, N2, 111-117.

Exp.; E - e; O_h; C_{3v}; d^7.

3.7.27. Nandini, R., Kumar Naresh, Resistance minimum in dilute magnetic alloys in the absence of impurity moments. - Phys. Status Solidi b, 1974, 63, N1, 377-384. - CA, 1974, 81, 7280. PA, 1974, 77, 48960.

Theor.

3.7.28. White, G. K., Sheard, F. W., The thermal expansion at low temperatures of UO_2 and UO_2/ThO_2. - J. Low Temp. Phys., 1974, 14, N5/6, 445-457. - CA, 1974, 80, 126090. RZF, 1974, 9E1081.

Exp.

1975

3.7.29. Anderson, B. R., Challis, L. J., An experimental investigation of frequency-crossing effects in phonon scattering by Fe^{2+} ions in Al_2O_3. - J. Phys. C, 1975, 8, N10, 1484-1494.

Exp.; T; O_h; C_{3v}; d^6.

1976

3.7.30. Devismes, N., De Goer, A. M., A study of phonon-spin interaction in the system $3d^1$ in Al_2O_3. - In: Phonon Scattering in Solids: Proc. 2nd Int. Conf., 1975. New York; London, 1976, p. 208-210. - CA, 1977, 87, 12467.

Exp.; T - e; O_h; C_{3v}; d^1.

3.7.31. Kinder, H., Dietsche, W., Phonon spectroscopy in Al_2O_3 doped with transition metal impurities. - In: Phonon Scattering in Solids: Proc. 2nd Int. Conf., 1975. New York; London, 1976, p. 199-201. - CA, 1977, 86, 180048.

Exp.; O_h; C_{3v}.

3.7.32. Locatelli, M., De Goer, A. M., Phonon scattering by Ni^{3+} ions in Al_2O_3. - In: Phonon Scattering in Solids: Proc. 2nd Int. Conf., 1975. New York; London, 1976, p. 193-195. - CA, 1977, 87, 12464.

Exp.; O_h; C_{3v}; d^7.

3.7.33. Lopez, A. R., Dixon, G. S., Thermal conductivity of $KMgF_3:Cr^{2+}$. - Phys. Lett. A, 1976, 58, N4, 267-268. - PA, 1976, 79, 90989.

Exp.; E - e; O_h; d^4.

3.7.34. Patel, J. L., Wigmore, J. K., Heat pulse investigations of chromium(2+)-doped magnesium oxide. - In: Phonon Scattering in Solids: Proc. 2nd Int. Conf., 1975. New York; London, 1976, p. 187-189. - CA, 1977, 87, 13687.

Exp.; O_h; d^4.

3.7.35. Rivallin, J., Salce, B., De Goer, A. M., The effect of uniaxial stress on thermal conductivity at low temperatures of Al_2O_3 and MgO doped with $3d^n$ ions. - Solid State Commun., 1976, 19, N1, 9-13. - PA, 1976, 79, 68705.

Exp.; E - e; O_h; C_{3v}; d^1; d^4; d^7.

3.7.36. Rivallin, J., Salce, B., Contribution to the study of $3d^4$ ions in aluminum oxide and magnesium oxide by low temperature thermal conductivity measurements under uniaxial stress. - In: Phonon Scattering in Solids: Proc. 2nd Int. Conf., 1975. New York; London, 1976, p. 184-186. - CA, 1977, 87, 75779.

Exp.; E - e; O_h; C_{3v}; d^1; d^4; d^7.

3.7.37. Salce, B., Determination of the linear Jahn-Teller coupling constant of nickel(3+) in aluminum oxide by low temperature thermal conductivity measurements under uniaxial stress. - In: Phonon Scattering in Solids: Proc. 2nd Int. Conf., 1975. New York; London, 1976, p. 196-198. - CA, 1977, 86, 179693.

Exp.; E - e; O_h; C_{3v}; d^7.

3.7.38. Van Vechten, J. A., Thurmond, C. D., Comparison of theory with quenching experiments for the entropy and enthalpy of vacancy formation in Si and Ge. - Phys. Rev. B, 1976, 14, N8, 3551-3557. - CA, 1977, 86, 34951. PA, 1977, 80, 11872.

Exp.; T_d.

1977

3.7.39. Altukhov, V. I., Zavt, G. S., The resonance scattering of phonons on paramagnetic ions. [Russ.]. - Fiz. Tverd. Tela, 1977, 19, N4, 1057-1064. - CA, 1977, 87, 29970.

Theor.; T; O_h; d^4; 3.2.

3.7.40. Patel, J. L., Wigmore, J. K., Jahn-Teller parameters of chromium(2+)-doped magnesium oxide determined using heat pulses. - J. Phys. C, 1977, 10, N11, 1829-1842. - CA, 1977, 87, 175758. PA, 1977, 80, 62876.

Exp.; O_h; d^4.

1978

3.7.41. Devismes, N., De Goer, A. M., Phonon scattering by V^{3+} ions in Al_2O_3: Comparison between theory and experiment. - J. Phys. C, 1978, 11, N18, 3805-3815. - CA, 1979, 90, 160391. PA, 1978, 81, 95116.

Exp.; O_h; C_{3v}; d^2.

3.7.42. Ghazi, A. A., Phonon spectroscopy of transition metal ions using frequency crossing techniques: Thesis. - Univ. Nottingham, 1978. - PA, 1979, 82, 42193.

Exp.; O_h; C_{3v}; d^2; d^4.

1979

3.7.43. Bogomolov, V. N., Kudinov, E. K., Pavlova, T. M., Petranovskiy, V. P., Appearance of large temperature-activated dipole moments of sulfur clusters in zeolite cavities. [Russ.]. - Pis'ma Zh. Eksp. & Teor. Fiz., 1979, 30, N7, 409-411. - CA, 1979, 91, 220956. PA, 1980, 83, 20575.

Exp.

3.7.44. Challis, L. J., Ghazi, A. A., Maxwell, K. J., An investigation of the ground state of Mn^{3+} in MgO based on thermal conductivity measurements. - J. Phys. C, 1979, 12, N2, 303-310. - CA, 1979, 91, 129810. PA, 1979, 82, 42497.

Exp.; O_h; d^4.

3.7.45. Salce, B., De Goer, A. M., Studies of the $Ni^{3+}:Al_2O_3$ system by thermal conductivity measurements. - J. Phys. C, 1979, 12, N11, 2081-2101. - CA, 1970, 91, 217865. PA, 1979, 82, 82395.

Exp.; O_h; C_{3v}; d^7.

See also: 1.7.-34, 39, 44; 2.2.-134; 2.3.-88; 2.8.-15; 3.1.-127; 3.6.-22; 4.1.-4, 13, 153, 156; 4.2.-22, 53, 64, 79, 81, 82, 86, 95, 100, 101, 102, 105; 4.3.-4, 7, 43; 4.4.-13, 25, 29, 35, 36, 37, 41, 61, 64, 68, 79, 88, 98, 99, 107, 115, 124, 160, 166a; 4.5.-11; 4.6.-62; 4.7.-1, 2, 3, 4, 5, 7, 9, 11, 12, 13, 14, 16, 17, 19, 21, 22, 26, 27, 29, 39, 40; 5.1.-32.

4. COOPERATIVE JAHN-TELLER EFFECT. STRUCTURAL PHASE TRANSITIONS. FERROELECTRICITY. CRYSTAL CHEMISTRY*

It took about 20 years to advance from the first papers of Jahn and Teller [1.1.-3, 5] in 1937 to the systematic investigation of the properties of isolated Jahn-Teller molecules. Then another 15 years elapsed till the beginning of investigations of the vibronic coupling effects in crystals that contain electronic degenerate ions as sublattices. In such crystals at low temperature one may expect the ordering of local Jahn-Teller distortions, leading in the case of ferro-ordering to the bulk deformation of the whole crystal. If the temperature is increased to such an extent that the thermal fluctuations destroy the correlation between local Jahn-Teller distortions, the latter become independent (disordered), and the crystal deformation disappears - the structural phase transition takes place, induced by the electron-lattice (vibronic) interaction. This situation is called the cooperative Jahn-Teller effect.

At present the cooperative Jahn-Teller effect theory is one of the few consistent microscopic theories of phase transitions. The great interest in this theory is also due to the fact that crystals with the cooperative Jahn-Teller effect possess quite a number of unusual special properties that are due to the essential coupling of the electron and lattice subsystems.

The first attempts to reveal the correlation between the vibronic coupling and the crystal structure were undertaken by Dunitz and Orgel [4.1.1; 4.6.2] as early as in 1952, followed by the papers of Woitowich [4.1.2] and Kanamori [4.1.3]. While the papers [4.1.-1; 2; 4.6.-2, 3] were mainly of a qualitative nature, the approach of Kanamori [4.1.3] formed the basis of the quantitative theoretical description of the cooperative Jahn-Teller effect. Nevertheless, after the paper [4.1.3] about 10 years passed before interest in the cooperative Jahn-Teller effect surged: in 1971-1975 more than 200 theoretical and experimental papers were published concerning different aspects of the effect. In the beginning this activity

*The introductory reviews for this section were written by B. G. Vekhter.

was restricted mainly to investigation of rare earth compounds: vanadates, arsenates, and phosphates. Investigation of these compounds has several advantages compared to other crystals, including optical transparency in the visible region and relatively weak vibronic coupling of degenerate electronic states of rare earth ions to the lattice vibrations, leading to low temperatures of phase transitions. Investigation of this class of crystals had a marked influence on studies of the cooperative Jahn-Teller effect. It proved the existence of the effect in a large number of crystals and indicated the different possibilities of theoretical and experimental study of this phenomenon.

Presently there are some review papers devoted to separate parts of the cooperative Jahn-Teller effect. First, there is the general review of Gehring and Gehring [4.1.76], the reviews on acoustic [4.1.97] and optic [4.3.69] properties, and some review papers in the books "Electron-Phonon Interactions and Phase Transitions" (edited by T. Riste, Plenum Press, New York-London, 1977) and "Structural Phase Transitions and Soft Modes" (edited by E. J. Samuelsen, E. Anderson, and J. Feder, Oslo, Tromösö, 1972).

Below, the classification of papers on the cooperative Jahn-Teller effect and analysis of its development are rather arbitrary.

4.1. General Theory. Vibronic Modes. Acoustic Anomalies

First there is the previously mentioned fundamental paper of Kanamori [4.1.3] (see also Kataoka and Kanamori [4.1.341]), the two large and detailed papers of Englman and Halperin [4.1.-17, 26] on the microscopic picture of the cooperative Jahn-Teller effect and perovskites, and the main paper of Elliott et al. [4.3.23] on the theoretical and experimental study of the effect in rare earth vanadates, and arsenates, and phosphates.

The phase transitions in Jahn-Teller crystals, induced by the electron-phonon coupling, may be considered by two different (in principle) approaches. In the first approach the phase transition within the electron subsystem, caused by the interaction of localized electronic states of the Jahn-Teller sublattice ions through the phonon field, are considered. This approach appears sufficient to describe the thermodynamic properties and optical absorption spectra. However, it is the phonon acoustic mode that becomes soft at structural phase transition. Its investigation by ultrasonic, light, and neutron scattering requires taking into account of the dynamic electron-phonon coupling and the occurrence of mixed vibronic modes. The dynamics in systems with the cooperative Jahn-Teller effect was considered in Refs. [4.3.23; 4.1.-42, 56, 68, 70, 118, 119, 120, 152, 153; 4.4.103]. Investigations of the vibronic mode dispersion in the wide range of wave number values by neutron scattering (Ref. [4.6.124] must also be mentioned. Sublattices with a complex scheme of electronic levels were studied in Ref. [4.1.83]; the peculiarities of the spectra of elementary excitations were considered in [4.1.-62, 86]. Most of the investigations were devoted to the study of the cooperative Jahn-Teller effect leading to ferro-ordering of local Jahn-Teller distortions in the low-symmetric phase. More complex ordered structures were discussed in Refs. [4.1.-100, 111, 128, 151]. A remarkable softening of the elastic modulus due to the correlation of local Jahn-Teller distortions takes place even when the structural phase transition is not realized [4.1.127]. Investigations of the influence of the electric field on the soft modulus and of the magnetoelastic properties of cooperative Jahn-Teller effect systems were made in the papers [4.1.-158, 159] and [4.1.-53, 60, 78; 4.4.91] respectively; these effects are caused by the field rearrangement of the active electronic levels.

1958

4.1.1. Orgel, L. E., Ferroelectricity and the structure of transition-metal oxides. - Discuss. Faraday Soc., 1958, N26, 138-144.

Theor.; 4.2.

1959

4.1.2. Wojtowicz, P. J., Theoretical model for tetragonal-to-cubic phase transformations in transition metal spinels. - Phys. Rev., 1959, 116, N1, 32-45. - CA, 1960, 54, 13798i. RZF, 1960, N9, 20155.

Theor.; O_h; D_{4h}.

1960

4.1.3. Kanamori, Junjiro, Crystal distortion in magnetic compounds. - J. Appl. Phys., 1960, 31, N5, Suppl., 14-23. - CA, 1961, 55, 1173d. RZF, 1961, 2E584.

Theor.; E - e; O_h; d^4; d^9.

1963

4.1.4. Rosenberg, M., Nicolau, P., Manaila, R., Pausescu, P., Preparation, electrical conductivity and tetragonal distortion of some magnite systems. - J. Phys. & Chem. Solids, 1963, 24, N12, 1419-1434.

Exp.; E - e; O_h; d^4; 4.7.

4.1.5. Wold, A., Arnott, R. J., Whipple, F., Goodenough, J. B., Crystallographic transitions in several chromium spinel systems. - J. Appl. Phys., 1963, 34, N4, 1085-1086. - CA, 1963, 58, 9691e.

Exp.; T; T_d; d^7; d^8; d^9; 4.6.

1964

4.1.6. Slonczewski, J. C., Phase transformations induced by the Jahn-Teller effect. - Bull. Soc. Sci. Bretagne, 1964, 39, 53-58. - CA, 1966, 65, 3087e. PA, 1968, 71, 35076. RZF, 1966, 12E362.

Theor.; O_h.

1965

4.1.7. Sträsler, S., Kittel, C., Degeneracy and the order of the phase transformation in the molecular-field approximation. - Phys. Rev., 1965, 139, N3, A758-A760. - RZF, 1966, 1E216.

Theor.

1966

4.1.8. Kino Yoshihiro, Miyahara Syohei, Crystal deformation of copper and nickel chromite system. - J. Phys. Soc. Jpn., 1966, 21, N12, 2732-2733. - CA, 1967, 66, 119562.

Exp.; T; T_d; d^8; d^9; 4.6.

1967

4.1.9. Sarfatt, J., Stoneham, A. M., The Goldstone theorem and the Jahn-Teller effect. - Proc. Phys. Soc. London. Ser. A, 1967, 91, N1, 214-221. - CA, 1967, 67, 5815.

Theor.; E - e; 1.1.

1968

4.1.10. Allen, S. J., Spin-lattice interaction in UO_2. I. Ground-state and spin-wave excitations. - Phys. Rev., 1968, 166, N2, 530-539. - CA, 1968, 68, 91700.

Theor.; T - (e + 2t); O_h; $5f^2$.

4.1.11. Allen, S. J., Jr., Spin-lattice interaction in UO_2. II. Theory of the first-order phase transition. - Phys. Rev., 1968, 167, N2, 492-498. - Solid State Abstr., 1969, 8, 73257.

Theor.; T - (e + 2t); O_h; $5f^2$.

4.1.12. Brandt, O. G., Walker, C. T., Ultrasonic attenutation and elastic constants for uranium dioxide. - Phys. Rev., 1968, 170, N2, 528-541. - CA, 1968, 69, 22203.

Exp.; T - (e + 2t); O_h; $5f^2$; 3.5.

4.1.13. Goodenough, J. B., Spin-orbit-coupling effects in transition-metal compounds. - Phys. Rev., 1968, 171, N2, 466-479. - CA, 1968, 69, 38870.

Theor.; E; T; O_h; T_d; C_{3v}; D_{3h}.

4.1.14. Kanamori, Junjiro, Kataoka, Mitsuo, Itoh, Yoshimasa, Cooperative Jahn-Teller distortion in mixed chromites. - J. Appl. Phys., 1968, 39, N2, 688-689. - CA, 1968, 68, 73068. PA, 1968, 71, 35052.

Theor.

4.1.15. Stevens, K. W. H., Chromous ions in magnesium oxide. - J. Appl. Phys., 1968, 39, N2, 686-687. - PA, 1968, 71, 53053.

Theor.

1970

4.1.16. Cracknell, A. P., Cracknell, M. F., Davies, B. L., Landau theory of second-order phase transitions and the magnetic phase transitions in some antiferromagnetic oxides. - Phys. Status Solidi, 1970, 39, N2, 463-469. - CA, 1970, 73, 40106.

Theor.; 4.4.

4.1.17. Englman, R., Halperin, B., Cooperative dynamic Jahn-Teller effect. I. Molecular field treatment of spinels. - Phys. Rev. B, 1970, 2, N1, 75-94. - CA, 1971, 74, 47413. PA, 1970, 73, 75980.

Theor.; E; O_h; d^4; d^9.

4.1.18. Halperin, B., Englman, R., Elementary excitation in cooperative Jahn-Teller systems. - Solid State Commun., 1970, 8, N19, 1555-1558. - PA, 1970, 73, 79221.

Theor.; E; D_{4h}; d^4.

4.1.19. Novak, P., Aspects of the cooperative Jahn-Teller effect in spinel and perovskite systems. - J. Phys. & Chem. Solids, 1970, 31, N1, 125-130. - CA, 1970, 72, 59154. PA, 1970, 73, 28551.

Theor.; E; O_h.

4.1.20. Novak, P., Collective excitations in spinel systems exhibiting the cooperative Jahn-Teller effect. - Czech. J. Phys. B, 1970, 20, N2, 196-201. - PA, 1970, 73, 37615.

Theor.

4.1.21. Stoneham, A. M., Bullough, R., Mean field models for martensitic and cooperative Jahn-Teller transformations. - J. Phys. C, 1970, 3, N12, L195-L197. - CA, 1972, 77, 156453. PA, 1971, 74, 30654. RZF, 1971, 8E374.

Theor.

1971

4.1.22. Elliott, R. J., Gehring, G. A., Malozemoff, A. P., Smith, S. R. P., Staude, W. S., Tyte, R. N., Theory of cooperative Jahn-Teller distortions in dysprosium vanadate and terbium vanadate. - J. Phys. C, 1971, 4, N9, L179-L184. - CA, 1971, 75, 69114. PA, 1971, 74, 55783.

Theor.; E; D_4h.

4.1.23. Elliott, R. J., Young, A. P., Smith, S. R. P., Acoustic anomalies in Jahn-Teller coupled systems. - J. Phys. C, 1971, 4, N15, L317-L322. - CA, 1972, 76, 6778. PA, 1971, 74, 83134.

Theor.; E; D_4h.

4.1.24. Englman, R., Elementary excitations of ordered Jahn-Teller centers. - In: Light Scattering in Solids: Proc. 2nd Int. Conf. Paris, 1971, p. 366-369. - CA, 1973, 78, 21788. PA, 1972, 75, 19991.

Rev.; 4.3.

4.1.25. Gorodetsky, G., Lüthi, B., Wanklyn, B. M., Elastic properties of a magnetically controllable Jahn-Teller distortion in $DyVO_4$. - Solid State Commun., 1971, 9, N23, 2157-2160. - CA, 1972, 76, 51133. PA, 1972, 75, 9802. RZF, 1972, 4E830.

Exp.; E; D_4h.

4.1.26. Halperin, B. I., Englman, R., Cooperative dynamic Jahn-Teller effect. II. Crystal distortion in perovskites. - Phys. Rev. B, 1971, 3, N5, 1698-1708. - CA, 1971, 74, 104039. PA, 1971, 74, 26344.

Theor.; E; O_h; d^4; d^9.

4.1.27. Nogues, M., Poix, P., Jahn-Teller type deformation in the mixed oxide $ZnMn_2O_4$. [Fr.]. - C.R. Hebd. Seances Acad. Sci. Ser. C, 1971, 272, N14, 1318-1320. - CA, 1971, 75, 41652. PA, 1971, 74, 56160.

Exp.; T; O_h; d^4.

4.1.28. Pytte, E., Structural phase transition in spinels induced by the Jahn-Teller effect. - Phys. Rev. B, 1971, 3, N10, 3503-3509. - CA, 1971, 75, 11458. PA, 1971, 74, 36725. RZF, 1971, 12E636.

Theor.; E; O_h; d^4; d^9.

4.1.29. Pytte, E., Stevens, K. W. H., Tunneling model of phase changes in tetragonal rare-earth crystals. - Phys. Rev. Lett., 1971, 27, N13, 862-865. - PA, 1971, 74, 76887.

Theor.; E; D_{4h}.

4.1.30. Pytte, E., Mechanisms of structural phase transitions. - In: Struct. Phase Transitions Soft Modes: Proc. NATO Adv. Study Inst. Norway, 1971, p. 133. - CA, 1975, 82, 163115.

Rev.

4.1.31. Stevens, K. W. H., Ion-lattice interactions in rare-earth compounds. - J. Phys. C, 1971, 4, N15, 2297-2303. - PA, 1971, 74, 83623.

Theor.; E; D_{4h}.

4.1.32. Syono Yasuhiko, Fukai Yuh, Ishikawa Yoshikazu, Anomalous elastic properties of Fe_2TiO_4. - J. Phys. Soc. Jpn., 1971, 31, N2, 471-476.

Exp.; d^6; 4.4.

4.1.33. Gorodetsky, G., Kino, Y., Lüthi, B., Elastic anomalies near cooperative Jahn-Teller transitions. - In: Introduct. Abstr. Int. Conf. Phonon Scattering in Solids, Paris. Saclay, 1972, p. 330-332. - PA, 1973, 76, 2014.

Exp.

4.1.34. Kataoka, Mitsuo, Kanamori, Junjiro, A theory of the cooperative Jahn-Teller effect: Crystal distortion in $Cu_{1-x}Ni_xCr_2O_4$ and $Fe_{1-x}Ni_xCr_2O_4$. - J. Phys. Soc. Jpn., 1972, 32, N1, 113-134. - CA, 1972, 76, 78645. PA, 1972, 75, 16411. RZF, 1972, 6E755.

Theor.

4.1.35. Levy, P. M., Chen, H. H., Quadrupole coupling in the rare-earth pnictides. - In: Magn. and Magn. Mater.: 17th AIP Annu. Conf., Chicago, 1971. New York, 1972, Pt. 1, p. 373-377. - CA, 1972, 77, 40775. RZF, 1973, 1E819.

Theor.; 4.4.

4.1.36. Lüthi, B., Sound propagation effects near phase transitions in crystals. - IEEE Trans. Magn., 1972, 8, N3, 581. - PA, 1973, 76, 46709.

Theor.; E; D_4h.

4.1.37. Massard, P., Bernier, J. C., Michel, A., Jahn-Teller effect in the Ta_2CrO_6-$TaCrO_4$ system. [Fr.]. - J. Solid Chem., 1972, 4, N2, 269-274. - CA, 1972, 76, 78741. PA, 1972, 75, 26095.

Exp.; d^3; d^4.

4.1.38. Melcher, R. L., Scott, B. A., Soft acoustic mode at the cooperative Jahn-Teller phase transiton in $DyVO_4$. - Phys. Rev. Lett., 1972, 28, N10, 607-610. - CA, 1972, 76, 104060. PA, 1972, 75, 25662. RZF, 1972, 7E487.

Theor.; E; D_{4h}.

4.1.39. Melcher, R. L., Scott, B. A., Soft acoustic mode and the strain dependence of the g factor in $DyVO_4$. - In: Introduct. Abstr. Int. Conf. Phonon Scattering in Solids, Paris. Saclay, 1972, p. 333-337. - PA, 1973, 76, 1761.

Exp.; E; D_4h.

4.1.40. Pytte, E., Structural transitions in rare-earth vanadates. - Acta Crystallogr. A, 1972, 28, N4, Suppl., S189. - PA, 1973, 76, 24266.

Theor.; E; D_4h.

4.1.41. Samuelsen, E. J., A theory to describe Jahn-Teller induced phase transition in spinels. - Phys. Norv., 1972, 6, N3/4, 207. - PA, 1973, 76, 41014.

Theor.; E; d^4; d^6; d^9.

4.1.42. Sandercock, J. R., Palmer, S. B., Elliott, R. J., Hayes, W., Smith, S. R. P., Young, A. P., Brillouin scattering, ultrasonic and theoretical studies of acoustic anomalies in crystals showing Jahn-Teller phase transitions. - J. Phys. C, 1972, 5, N21, 3126-3146. - CA, 1972, 77, 169861. PA, 1972, 75, 84038. RZF, 1973, 4E644.

Exp.; E; D_{4h}; 4.3.

4.1.43. Sandercock, J. R., Acoustic mode anomalies in the Jahn-Teller phase transitions in dysprosium vanadate and terbium vanadate. - Solid State Commun., 1972, 11, N5, 729-731. - CA, 1972, 77, 144914. PA, 1972, 75, 71681.

Exp.; E; D_4h; 4.3.

4.1.44. Sivardiere, J., Theory of phase transitions in rare-earth vanadates. - Phys. Rev. B, 1972, 6, N11, 4284-4293.

Theor.; E; D_4h; 4.4.

4.1.45. Thomas H., Müller, K. A., Theory of a structural phase transition induced by the Jahn-Teller effect. - Phys. Rev. Lett., 1972, 28, N3, 820-823. - PA, 1972, 75, 32757. RZF, 1972, 8E893.

Theor.; E; O_h; T_d; C_3v.

4.1.46. Viliani, G., Thermodynamical properties of Jahn-Teller induced phase transitions in rare-earth compounds. - Lett. Nuovo Cimento, 1972, 4, N13, 641-642. - CA, 1972, 77, 106305. PA, 1972, 75, 64806. RZF, 1972, 11E400.

Theor.; E; D_4h; 4.7.

1973

4.1.47. Chakraverty, B. K., Macroscopic strain-coupling induced symmetry change in Fe_3O_4. - Solid State Commun., 1973, 12, N5, 317-319. - PA, 1973, 76, 27222. RZF, 1973, 7E447.

Theor.; T - (e + t_2); O_h.

4.1.48. Feder, J., Pytte, E., Statistical mechanics of the collective Jahn-Teller phase transition. - Phys. Rev. B, 1973, 8, N8, 3978-3981. - PA, 1974, 77, 32652. RZF, 1974, 6E515.

Theor.

4.1.49. Halperin, B., Structural phase transitions induced by a pseudo-Jahn-Teller effect. - Phys. Rev. B, 1973, 7, N2, 894-897. - CA, 1973, 78, 76052. PA, 1973, 76, 18089. RZF, 1973, 7E446.

Theor.; 1.5.

4.1.50. Harley, R. T., Hayes, W., Perry, A. M., Smith, S. R. P., The phase transitions of $PrAlO_3$. - J. Phys. C, 1973, 6, N14, 2382-2400.

Exp.; (E + T) - t; O_h; 1.5; 1.6; 4.3.

4.1.51. Kino, Y., Lüthi, B., Mullen, M. E., Elastic properties and cooperative Jahn-Teller effect in nickel chromite. - Solid State Commun., 1973, 12, N4, 275-277. - CA, 1973, 78, 115512. PA, 1973, 76, 24181. RZF, 1973, 7E494.

Exp.; T - e; T_d; d^8.

4.1.52. Levy, P. M., A theoretical study of the elastic properties of dysprosium antimonide. - J. Phys. C, 1973, 6, N24, 3545-3556. - PA, 1974, 77, 16541.

Theor.

4.1.53. Melcher, R. L., Pytte, E., Scott, B. A., Phonon instabilities in $TmVO_4$. - Phys. Rev. Lett., 1973, 31, N5, 307-310. PA, 1973, 76, 55069. RZF, 1973, 12E877.

Theor.; E; D_{4h}.

4.1.54. Melcher, R. L., The anomalous elasticity of materials undergoing cooperative Jahn-Teller phase transitions. - In: Ultrason. Symp. Proc., Monterey, Calif. 1973. New York, 1973, p. 293-298. - CA, 1976, 85, 70882. RZF, 1974, 11E597.

Exp.; E; D_{4h}.

4.1.55. Moran, T. J., Thomas, R. L., Levy, P. M., Chen, H.H., Elastic properties of DySb near the magnetic and structural phase transition. - Phys. Rev. B, 1973, 7, N7, 3238-3241. - PA, 1973, 76, 33591.

Exp.; E - e; O_h.

4.1.56. Pytte, E., Dynamics of Jahn-Teller phase transitions. - Phys. Rev. B, 1973, 8, N8, 3954-3959. - CA, 1974, 80, 30832. PA, 1974, 77, 3261. RZF, 1974, 6E514.

Theor.

4.1.57. Ray, D. K., Young, A. P., A theory for the elastic properties of dysprosium antimonide. - J. Phys. C, 1973, 6, N23, 3353-3358. - RZF, 1974, 4E1080.

Theor.; E - e; O_h.

4.1.58. Stevens, K. W. H., Pytte, E., Phase changes in cubic rare-earth crystals. - Solid State Commun., 1973, 13, N1, 101-104. - CA, 1973, 79, 71611. PA, 1973, 76, 52976.

Theor.

4.1.59. Stinchcombe, R. B., Ising model in a transverse field. I. Basic theory. - J. Phys. C, 1973, 6, N15, 2459-2483. - PA, 1973, 76, 58113.

Theor.; E.

4.1.60. Vekhter, B. G., Kaplan, M. D., The elastic modulus in crystals with Jahn-Teller cations. [Russ.]. - In: Ref. Dokl. 8 Vsesoyuz. Akust. Konf. Moscow, 1973, p. 212.

Theor.; E; D_{4h}; 4.4.

1974

4.1.61. Chakraverty, B. K., Verwey ordering on magnetite as a cooperative Jahn-Teller transition. - Solid State Commun., 1974, 15, N8, 1271-1275. - CA, 1974, 81, 178093. PA, 1974, 77, 82374.

Theor.; d^5; d^6.

4.1.62. Gehring, G. A., The cooperative Jahn-Teller effect for several active ions in the unit cell. - J. Phys. C, 1974, 7, N21, L379-L383. - PA, 1975, 78, 6732. RZF, 1975, 4E223.

Theor.

4.1.63. Kanamori, Junjiro, Cooperative Jahn-Teller effect. [Jap.]. - Nippon Butsuri Gakkaishi, Proc. Phys. Soc. Jpn., 1974, 29, N5, 420-423. - CA, 1974, 81, 43305. RZF, 1975, 1E53.

Rev.

4.1.64. Khomskii, D. I., Kugel, K. I., Superexchange-induced orbital ordering and magnetic structure of insulators with Jahn-Teller ions. - In: Tr. Mezhdunar. Konf. Magn. MKM-1973. Moscow, 1974, Vol. 3, p. 59-63. - CA, 1976, 84, 37989.

Theor.

4.1.65. Kjems, J. K., Shirane, G., Cooperative Jahn-Teller phase transitions in $PrAlO_3$. - Phys. Rev. B, 1974, 10, N6, 2512-2534. - PA, 1975, 78, 18564.

Exp.; 4.6.

4.1.66. Kroese, C. J., Maaskant, W. J. A., Relation between the high-temperature and room-temperature structures of cesium trichlorocuprate(II). - Chem. Phys., 1974, 5, N2, 224-233. - CA, 1975, 82, 50002.

Exp.; E; O_h; d^9; 4.6.

4.1.67. Page, J. H., Rosenberg, H. M., Investigation of elastic constants in cooperative Jahn-Teller system. - In: Satellite Symp. 8th Int. Congr. Acoustics on Microwave Acoustics. Lancaster, 1974, p. 141-144. - PA, 1975, 78, 54311.

Exp.; E; D_{4h}; 4.4.

4.1.68. Pytte, E., Dynamics of Jahn-Teller phase transitions. - Ferroelectrics, 1974, 7, N1/4, 193. - PA, 1975, 78, 15520.

Theor.

4.1.69. Scott, J. F., Soft-mode spectroscopy: experimental studies of structural phase transitons. - Rev. Mod. Phys., 1974, 46, N1, 83-128. - PA, 1974, 77, 56851.

Rev.; 4.3.

4.1.70. Thomas, H., Jahn-Teller effect as a mechanism for structural phase transitions. - In: Anharmonic Lattices, Structural Transitions and Melting. Leiden; Noordhoff, 1974, p. 213-229. (NATO Adv. Study Inst. Ser., Ser. E, 1974, N1.) - CA, 1976, 84, 23874.

Theor.

4.1.71. Uffer, L. F., Levy, P. M., Sablik, M. J., Tricriticality of HoSb. - Solid State Commun., 1974, 15, N2, 191-194. - CA, 1974, 81, 128900. RZF, 1975, 2E1378.

Theor.; O_h.

4.1.72. Vekhter, B. G., Kaplan, M. D., The electron-phonon modes and the sound absorption in crystals with cooperative Jahn-Teller effect. [Russ.]. - In: Tez. Dokl. 8 Vsesoyuz. Soveshch. Kvant. Akust. Akustoelektron. Kazan', 1974, p. 189.

Theor.; 4.4.

1975

4.1.73. Aleksandrov, K. S., Zinenko, V. I., Phase transitions of displacive type. [Russ.]. - In: Fazovye Perekhody Kristallakh. Krasnoyarsk, 1975, p. 3-67.

Rev.; D_{4h}.

4.1.74. Bersuker, I. B., Polinger, V. Z., Rhombic ordering in crystals with a cooperative Jahn-Teller effect. [Russ.]. - Izv. Akad. Nauk SSSR. Ser. Fiz., 1975, 39, N4, 655-658. - PA, 1975, 78, 58656. RZF, 1975, 8E723.

Theor.; T - $(e + t_2)$; O_h; T_d; 4.2.

4.1.75. Bhattacharyya, B. D., Phase transition in the spinel $ZnMn_2Se_4$. - In: Abstr. Conf. Phase Transitions and Equilibria, Bangalore. New Delhi, 1975, p. 22-23. - PA, 1976, 79, 26696.

Exp.; O_h; 3.1; 4.4.

4.1.76. Gehring, G. A., Gehring, K. A., Cooperative Jahn-Teller effects. - Rep. Prog. Phys., 1975, 38, N1, 1-89. - CA, 1975, 82, 163130. PA, 1975, 78, 29601. RZF, 1975, 9E94.

Rev.

4.1.77. Kanamori, Junjiro, Spontaneous distortion of crystals. [Jap.]. - Kobutsugaku Zasshi, J. Mineral. Soc. Jpn., 1975, 12, N2, 61-72. - CA, 1976, 85, 184893.

Rev.

4.1.78. Kaplan, M. D., Vekhter, B. G., Electron-phonon modes and sound propagation in crystals with a cooperative Jahn-Teller effect. [Russ.]. - Fiz. Tverd. Tela, 1975, 17, N1, 76-80. - CA, 1975, 82, 103340. PA, 1976, 79, 1731. RZF, 1975, 5E71.

Theor.; 4.4.

4.1.79. Mullen, M. E., Ultrasonic investigations of the Jahn-Teller effect. - 1975. - 188 p. - [Order N75-17464.] - Diss. Abstr. Int. B, 1975, 36, N2, 770. CA, 1975, 83, 183603.

Exp.

4.1.80. Sturge, M. D., Cohen, E., Van Uitert, L. G., Van Stapele, R. P., Microscopic order parameters in $PrAlO_3$. - Phys. Rev. B, 1975, 11, N12, 4768-4779. - PA, 1975, 78, 76798.

Exp.

4.1.81. Tindemans-van Eijndhoven, J. C. M., Kroese, C. J., Local modes in the theory of lattice vibrations: derivation and application. - J. Phys. C, 1975, 8, N23, 3963-3974. - PA, 1976, 79, 13641.

Theor.

4.1.82. Vekhter, B. G., Effekt Jahna-Tellera v Molekulakh i Kristallakh: Avtoref. Dis. ... Dokt. Fiz.-Mat. Nauk. [Russ.]. - Kishinev, 1975. - 25 p. - Akad. Nauk SSSR. Inst. Khim. Fiz.

The Jahn-Teller Effect in Molecules and Crystals.

Rev.; 1.3; 1.4; 1.8.

4.1.83. Young, A. P., Excitations in Jahn-Teller coupled systems with complicated electronic level schemes. - J. Phys. C, 1975, 8, N19, 3158-3170. - CA, 1975, 83, 210831. PA, 1976, 79, 1977.

Theor.

1976

4.1.84. Battison, J. E., Leask, M. J. M., Lowry, J. B., Kasten, A., First order phase transitions in thulium arsenate. - J. Phys. C, 1976, 9, N12, 2295-2303. - CA, 1976, 85, 152056. PA, 1976, 79, 72351.

Exp.; D_{4h}.

4.1.85. Cowan, W. B., Zuckerman, M. J., Anharmonic effects in a Jahn-Teller phase transition. - Solid State Commun., 1976, 16, N2, 207-210. - CA, 1975, 82, 105284. PA, 1975, 78, 18396. RZF, 1975, 7E697.

Theor.; E; O_h; d^9.

4.1.86. Ergun, H. B., Gehring, K. A., Gehring, G. A., Jahn-Teller induced Davydov splitting in terbium vanadate(V). - J. Phys. C, 1976, 9, N6, 1101-1115. - CA, 1976, 85, 26952. PA, 1976, 79, 45779.

Exp.; D_{4h}.

4.1.87. Fernandez, J., Tello, M. J., Peraza, J., Bocanegra, E. H., New high temperature phase transitions in cesium cupric chloride. - Mater. Res. Bull., 1976, 11, N9, 1161-1167. - CA, 1976, 85, 152082. PA, 1976, 79, 93586.

Exp.; E; O_h; d^9.

4.1.88. Gehring, G. A., Phonons in cooperative systems. - In: Phonon Scattering in Solids: Proc. 2nd Int. Conf., 1975. New York; London, 1976, p. 231-236. - CA, 1977, 86, 179627.

Rev.

4.1.89. Harrowfield, B. V., Comments on the paper "Planar dynamic Jahn-Teller effect in nitrocomplexes." - Solid State Commun., 1976, 19, N10, 983-984. - CA, 1976, 85, 133620. PA, 1976, 79, 79518.

Theor.; E - e; O_h; d^9.

4.1.90. Höck, K.-H., Thomas, H., Local Jahn-Teller effect at a structural phase transition. - In: Crystal Field Eff. in Met. & Alloys: Proc. 2nd Int. Conf., Zurich, 1976. New York; London, p. 184-188.

Theor.; E; O_h.

4.1.91. Hutchings, M. T., Scherm, R., Smith, S. R. P., Neutron scattering investigation of the central mode and acoustic phonon anomaly arising from the Jahn-Teller phase transition in terbium vanadate(V). - In: Magn. & Magn. Mater.: 21st AIP Annu. Conf., Philadelphia, 1975. New York, 1976, N29, p. 372-378. - CA, 1976, 85, 114975. PA, 1976, 79, 87041.

Exp.; D_{4h}.

4.1.92. Kaplan, M. D., Magnitoakusticheskiye Svoystva Kristallov s Kooperativnym Effektom Jahn-Tellera: Avtoref. Dis. ...Kand. Fiz.-Mat. Nauk. [Russ.]. - Leningrad, 1976. - 21 p. - Leningr. Univ.

Magnetoacoustical Properties of Crystals with Cooperative Jahn-Teller Effect.

Rev.; 4.4.

4.1.93. Kino Yoshihiro, The elastic properties in the cooperative Jahn-Teller phase transition. [Jap.]. - Nippon Butsuri Gakkaishi, Proc. Phys. Soc. Jpn., 1976, 31, N3, 198-200. - CA, 1976, 85, 70726.

Rev.

4.1.94. Kurzynski, M., Effects of electron-"breathing" mode coupling: A theory of the three phase transitions of $LaCoO_3$. - J. Phys. C, 1976, 9, N20, 3731-3745.

Theor.

4.1.95. Lage, E. J. S., Stinchcombe, R. B., Transverse Ising model with substitutional disorder - an effective medium theory. - J. Phys. C, 1976, 9, N17, 3295-3306.

Theor.

4.1.96. Mehran, F., Stevens, K. W. H., Plaskett, T. S., Random strain effects in the cooperative Jahn-Teller system $TmVO_4$. - Phys. Rev. Lett., 1976, 37, N21, 1403-1406. - CA, 1977, 86, 10465. PA, 1977, 80, 12200.

Exp.; E; D_{4h}; 3.1; 4.4.

4.1.97. Melcher, R. L., The anomalous elastic properties of materials undergoing cooperative Jahn-Teller phase transitions. - In: Phys. Acoustics. New York, 1976, Vol. 12, p. 1-77. - Bibliogr.: 106 ref. CA, 1977, 86, 147879.

Rev.

4.1.98. Noda, Y., Mori, M., Yamada, Y., Successive Jahn-Teller transitions and the canted pseudospin state in potassium lead copper nitrite ($K_2PbCu(NO_2)_6$). - Solid State Commun., 1976, 19, N11, 1071-1074. - CA, 1976, 85, 133425. PA, 1976, 79, 79708.

Exp.; E; O_h; d^9; 4.6.

4.1.99. Ray, D. K., Occurrence of a distortive transition at a temperature higher than the magnetic one: case of CeAg. - Solid State Commun., 1976, 19, N11, 1053-1057. - PA, 1976, 79, 79551.

Theor.; 4.4.

4.1.100. Schröder, G., Thomas, H., Structural phase transition induced by the Jahn-Teller effect: Antiferrodistortive ordering. - Z. Phys. B, 1976, 25, N4, 369-380. - CA, 1977, 86, 36502. PA, 1977, 80, 11657.

Theor.; E.

4.1.101. Thomas, H., Mean field theory of phase transitions. - In: Proc. Int. School of Physics "Enrico Fermi." Amsterdam, 1976, p. 36-44. - PA, 1976, 79, 89618.

Theor.

4.1.102. Vekhter, B. G., Kaplan, M. D., Shutilov, V. A., Nuclear dipolar spin-phonon interaction in crystals with cooperative Jahn-Teller effect. [Russ.]. - In: Tez. Dokl. 9 Vsesoyuz. Konf. Akustoelektronike i Kvant. Akustike. Moscow, 1976, p. 89-90.

Theor.; $E - b_1$; D_{4h}; 3.5; 4.4.

4.1.103. Vekhter, B. G., Kaplan, M. D., Shutilov, V. A., Dipolar mechanism of nuclear spin-phonon coupling and absorption of sound in crystals with cooperative Jahn-Teller effect. [Russ.]. - Fiz. Tverd. Tela, 1976, 18, N6, 1734-1736. - CA, 1976, 85, 114236. PA, 1977, 80, 24753.

Theor.; $E - b$; D_{4h}; 3.5; 4.4.

1977

4.1.104. Brühl, S., Sigmund, E., Renormalization group approach to a one-dimensional cooperative T - e Jahn-Teller system. - Z. Naturofrsch. a, 1977, 32, N1, 111-112. - CA, 1977, 86, 146212.

Theor.; T - e.

4.1.105. Buynov, N. S., Nagibarov, V. R., Solovarov, N. K., A phonon superradiation phase transition. [Russ.]. - Ukr. Fiz. Zh., 1977, 22, N1, 151-152.

Theor.

4.1.106. Elliott, R. J., The cooperative Jahn-Teller effect and related problems. - Physica B + C, 1977, 86-88, N3, 1118-1121. - CA, 1977, 87, 90808. PA, 1977, 80, 62878.

Theor.

4.1.107. Hirotsu Shunsuke, Jahn-Teller induced phase transition in cesium copper chloride ($CsCuCl_3$): structural phase transition with helical atomic displacements. - J. Phys. C, 1977, 10, N7, 967-985. - CA, 1977, 87, 109655. PA, 1977, 80, 43301.

Theor.; E - e; O_h; d^9.

4.1.108. Höck, K.-H., Thomas, H., Local Jahn-Teller effect at a structural phase transition. - In: Electron-Phonon Interactions and Phase Transitions. New York; London, 1977, p. 271-276. (NATO Adv. Study Inst. Ser., Ser. B, 1977, N29.) - CA, 1979, 90, 127898.

Theor.; E - e; 1.4; 1.7.

4.1.109. Huberman, B. A., Martin, R. M., Dynamic long-range forces and Z branches at phase transitions. - Phys. Rev. Lett., 1977, 39, N8, 478-481. - PA, 1977, 80, 84887.

Theor.; E; D_{4h}.

4.1.110. Ivanov, M. A., Mitrofanov, V. Ya., Fishman, A. Ya., On the possibility of spin-glass-type states in systems with orbital degenerate impurities. [Russ.]. - In: Fiz. Mat. Metody Koord. Khim.: Tez. Dokl. 6 Vsesoyuz. Soveshch. Kishinev, 1977, p. 128.

Theor.

4.1.111. Khomskiy, D. I., Cooperative Jahn-Teller transition with tripling of the period: $CsCuCl_2$. [Russ.]. - Pis'ma Zh. Eksp. & Teor. Fiz., 1977, 25, N12, 579-582. - CA, 1977, 87, 92895. PA, 1977, 80, 76557.

Theor.; E - e; O_h; d^9.

4.1.112. Mehran, F., Plaskett, T. S., Stevens, K. W. H., Jahn-Teller induced random strain effects in thulium phosphate. - Phys. Rev. B, 1977, 16, N1, 1-3. - CA, 1977, 87, 109042. PA, 1977, 80, 85565.

Exp.; D_4h; 3.1.

4.1.113. Ostrovskiy, V. S., Kharkyanen, V. N., Orbital ordering in Jahn-Teller crystals with direct Coulomb interaction. [Russ.]. - Fiz. Tverd. Tela, 1977, 19, N6, 1545-1551. - CA, 1977, 87, 90954. PA, 1978, 81, 26833.

Theor.; E - e; T - e; T - t.

4.1.114. Page, J. H., Rosenberg, H. M., An investigation of the elastic constants of thulium vanadate at 9 GHz. - J. Phys. C, 1977, 10, N11, 1817-1827. - CA, 1977, 87, 175823. PA, 1977, 80, 62576.

Exp.; E - (b_1 + b_2); D_4h; 4.7.

4.1.115. Paquet, D., Cooperative pseudo-Jahn-Teller model for the sequence of ferroelastic transitions in barium sodium niobate. - J. Chem. Phys., 1977, 66, N3, 886-895. - CA, 1977, 86, 99230. PA, 1977, 80, 35689.

Theor.; O_h; 1.5.

4.1.116. Paquet, D., Cooperative pseudo-Jahn-Teller model of the sequence of ferroelastic transitions in barium sodium niobate. - In: Electron-Phonon Interactions and Phase Transitions. New York; London, 1977, p. 337-344. (NATO Adv. Study Inst. Ser., Ser. B, 1977, N29.) - CA, 1979, 90, 129699.

Theor.; D_4h.

4.1.117. Schneck, J., Primot, J., Von der Mühll, R., Pavez, J., New phase transition with increasing symmetry on cooling in barium sodium niobate. - Solid State Commun., 1977, 21, N1, 57-60. - CA, 1977, 86, 81937.

Exp.; C_4v.

4.1.118. Smith, S. R. P., Hutchings, M. T., The central peak in $TbVO_4$. - In: Electron-Phonon Interactions and Phase Transitions. New York; London, 1977, p. 327-330. (NATO Adv. Study Inst. Ser., Ser. B, 1977, N29.)

Exp.; E - b_1; D_{4h}; 2.6.

4.1.119. Stinchcombe, R. B., Pseudo-spin approach to structural phase transitions. - In: Electron-Phonon Interactions and Phase Transitions. New York; London, 1977, p. 209-244. (NATO Adv. Study Inst. Ser., Ser. B, 1977, N29.)

Rev.

4.1.120. Thomas, H., Theory of Jahn-Teller transitions. - In: Electron-Phonon Interactions and Phase Transitions. New York; London, 1977, p. 245-270. (NATO Adv. Study Inst. Ser., Ser. B, 1977, N29.) - CA, 1979, 90, 130745.

Rev.

4.1.121. Vekhter, B. G., Kaplan, M. D., Shutilov, V. A., Interference nuclear acoustic resonance in crystals with a cooperative Jahn-Teller effect. [Russ.]. - Zh. Eksp. & Teor. Fiz., 1977, 72, N1, 354-362. - CA, 1977, 86, 113370. PA, 1977, 80, 28460. RZF, 1977, 6D629.

Theor.; 3.5.

4.1.122. Yamada, Y., Electron-phonon interactions and charge ordering in insulators. - In: Electron-Phonon Interactions and Phase Transitions. New York; London, 1977, p. 370-392. (NATO Adv. Study Inst. Ser., Ser. B, 1977, N29.)

Rev.; E - e; O_h; d^9.

1978

4.1.123. Belov, K. P., Vekhter, B. G., Goryaga, A. N., Kaplan, M. D., Mirzoakhmedov, Kh. M., Mukhtarov, N. M., Nedbay, A. I., Khromova, N. N., Shutilov, V. A., The ultrasonic velocity and absorption anomalies in Jahn-Teller spinels. [Russ.]. - In: Akust. Spektrosk. Kvantovaya Akustika. Akustoelektronika: Mater. 2 Vsesoyuz. Simp. Akust. Spektrosk. Tashkent, 1978, p. 155-158.

Exp.

4.1.124. Brühl, S., Wagner, M., Cooperative Jahn-Teller systems of T nature: electron-phonon excitation branches. - In: Proc. Int. Conf. Lattice Dynamics. Paris, 1978, p. 215-217. - CA, 1979, 90, 92668. PA, 1978, 81, 63823.

Theor.; T - e; T - t.

4.1.125. Feiner, L. F., Van Stapele, R. P., Degenerate electron states and structural phase transitions. - Ned. Tijdschr. Natuurkd. A, 1978, 44, N3, 111-114. - CA, 1979, 90, 144337.

Theor.

4.1.126. Ghatak, S. K., Ray, D. K., Tannous, C., Studies of the structural stability of an electronic system with a two-fold degenerate band. - Phys. Rev. B, 1978, 18, N10, 5379-5384. - CA, 1979, 90, 113210.

Theor.; E - e; O_h.

4.1.127. Harley, R. T., Manning, D. I., Jahn-Teller induced elastic constant changes in $TmPO_4$. - J. Phys. C, 1978, 11, N12, L633-L636. - CA, 1979, 90, 76663. PA, 1978, 81, 79657.

Exp.; D_{4h}.

4.1.128. Höck, K.-H., Schröder, G. R., Thomas, H., Order-disorder structural phase transitions described by the three-state Potts model. - Z. Phys. B, 1978, 30, N4, 403-413. - CA, 1978, 89, 138571. RZF, 1979, 1E707.

Theor.; E - e.

4.1.129. Ionova, G. V., Makarov, E. F., Pachev, O. M., Ionov, S. P., Analysis of charge-orbital ordering in compounds with mixed valence. - Phys. Status Solidi b, 1978, 85, N2, 683-691.

Theor.

4.1.130. Ivanov, M. A., Mitrofanov, V. Ya., Fishman, A. Ya., Thermodynamics of interacting orbital degenerate impurities. [Russ.]. - Fiz. Tverd. Tela, 1978, 20, N10, 3023-3032. - CA, 1979, 90, 13093. PA, 1979, 82, 69477.

Theor.

4.1.131. Junker, W., Wagner, M., High temperature series expansions in cooperative Jahn-Teller T-systems. - Physica A, 1978, 94, N3/4, 385-402. - RZF, 1979, 4E125.

Theor.; T - e; T - t.

4.1.132. Kashida Shoji, Kaga Hiroyuki, Effect of hydrostatic pressure on the successive phase transitions in dipotassium lead hexanitrocuprate. - J. Phys. Soc. Jpn., 1978, 44, N3, 930-932. - CA, 1978, 89, 15035. PA, 1978, 81, 43120.

Theor.; E; O_h; d^9.

4.1.133. Kashida Shoji, Successive Jahn-Teller phase transitions in $K_2PbCu(NO_2)_6$. - J. Phys. Soc. Jpn., 1978, 1978, 45, N2, 414-421. - CA, 1978, 89, 138566. PA, 1978, 81, 79762.

Theor.; E - e; O_h; d^9.

4.1.134. Kashida Shoji, Ultrasonic velocities of $K_2PbCu(NO_2)_6$ near the cooperative Jahn-Teller phase transitions. - J. Phys. Soc. Jpn., 1978, 45, N6, 1874-1879. - PA, 1979, 82, 23680.

Exp.; E; O_h; d^9.

4.1.135. Mori Masahiro, Noda Yukio, Yamada Yasusada, Successive Jahn-Teller phase transitions in $Cs_2PbCu(NO_2)_6$. - Solid State Commun., 1978, 27, N7, 735-737. - CA, 1979, 90, 14809. PA, 1978, 81, 95154.

Exp.; E - e; O_h; d^9; 4.6.

4.1.136. Rao, C. N. R., Rao, K. J., Phase Transitions in Solids: An Approach to the Study of the Chemistry and Physics of Solids. - New York, etc.: McGraw-Hill, 1978. - XI, 330 p. - CA, 1978, 89, 121383.

Theor.

4.1.137. Reinen, D., Steffens, F., Structure and bonding in transition metal fluorides $M^{II}Me^{IY}F_6$: A. Phase transitions. [Germ.]. - Z. Anorg. Allg. Chem., 1978, 441, 63-82. - CA, 1978, 89, 155871.

Exp.

4.1.138. Saban, A. Ya., Dipole and quadrupole vibrations in terbium vanadate ($TbVO_4$)-type crystals. [Russ.]. - In: Materialy 5 Konf. Molodykh Uchenykh L'vov. Fil. Mat. Fiziki Inst. Mat. Akad. Nauk. L'vov, 1978, p. 98-101. - Deposited Doc., VINITI, 1978, N3778-78. - Dep. Rukopisi, 1979, N3, p. 24. CA, 1980, 92, 32198.

4.1.139. Vekhter, B. G., Kaplan, M. D., Electric field effects in elastic properties of $DyVO_4$. - In: Akust. Spektrosk. Kvantovaya Akustika. Akustoelektronika: Mater. 2 Vsesoyuz. Simp. Akust. Spektrosk. Tashkent, 1978.

Theor.; 4.4.

4.1.140. Yamada Yasusada, Electron-phonon interactions and structural phase transitions. [Jap.]. - Kotai Butsuri, Solid State Phys., 1978, 13, N5, 251-264. - Bibliogr.: 64 ref. CA, 1979, 90, 46648. PA, 1979, 82, 4384.

Rev.

1979

4.1.141. Antoshina, L. G., Vekhter, B. G., Kaplan, M. D., Mirzoakhmedov, Kh., Ultrasonic spectroscopy of spinels with Jahn-Teller ions. - In: Tez. Dokl. 6 Vsesoyuz. Simp. Spektrosk. Kristallov, Aktivir. Ionami Redkozem. Perekhodnykh Metallov. Krasnodar, 1979, p. 35.

Exp.

4.1.142. Belov, K. P., Vekhter, B. G., Goryaga, A. N., Kaplan, M. D., Mirzoakhmedov, Kh. M., Khromova, N. N., Shutilov, V. A., Display of a cooperative Jahn-Teller effect in acoustic properties of polycrystalline spinels. [Russ.]. - Fiz. Tverd. Tela, 1979, 21, N6, 1679-1682. - CA, 1979, 91, 112589. PA, 1980, 83, 24538.

Exp.; d^8.

4.1.143. Brühl, S., Dynamics of cooperative Jahn-Teller T-systems. I. Vibronic excitation branches. - Z. Phys. B, 1979, 35, N3, 287-298. - CA, 1979, 91, 219490. PA, 1980, 83, 10934. RZF, 1980, 3E62.

Theor. T - (e + t).

4.1.144. Brühl, S., Dynamics of cooperative Jahn-Teller T-systems. II. Elastic properties. - Z. Phys. B, 1979, 37, N3, 231-241. - RZF, 1980, 9E311.

Theor.; T - e.

4.1.145. Höck, K.-H., Thomas, H., Discontinuous freezing-out of local order. - Z. Phys. B, 1979, 32, N3, 323-326. - CA, 1979, 90, 113183. PA, 1979, 82, 42204.

Theor.; E - e; O_h.

4.1.146. Iqbal, Z., Basic concepts of structural phase transitions. - In: Magn. Resonance Phase Transitions. New York. 1979, p. 1-22. - CA, 1980, 92, 50127.

Rev.

4.1.147. Ivanov, M. A., Mitrofanov, V. Ya., Fal'kovskaya, L. D., Fishman, A. Ya., Effect of internal strain on thermodynamic properties of systems with a structural Jahn-Teller transition. [Russ.]. - Fiz. Tverd. Tela, 1979, 21, N8, 2426-2429. - CA, 1979, 91, 166591.

Theor.

4.1.148. Kristofel', N. N., Local phase transition in a center with quasidegenerate levels and temperature effects in lattice dynamics. [Russ.]. - Fiz. Tverd. Tela, 1979, 21, N3, 895-900. - CA, 1979, 91, 30725.

Theor.

4.1.149. Kristofel', N. N., The illustration of the local relaxational phase transitions in the anisotropic impurity centers. [Russ.]. - Izv. Akad. Nauk Est.SSR. Fiz. Mat., 1979, 28, N3, 243-248.

Theor.

4.1.150. Kristofel', N. N., On the influence of the soft mode on the local Jahn-Teller effect. [Russ.]. - Opt. & Spektrosk., 1979, 47, N2, 609-611. - PA, 1980, 83, 59937.

Theor.

4.1.151. Lee, B. S., Helical ordering in the Jahn-Teller cooperative phase transition. - J. Phys. C, 1979, 12, N5, 855-863. - CA, 1979, 91, 81689. PA, 1979, 82, 46778.

Theor.

4.1.152. Leung Kok Ming, Huber, D. L., Low-frequency dynamics in cooperative Jahn-Teller systems. - Phys. Rev. B, 1979, 19, N11, 5483-5494. - CA, 1979, 91, 114584. PA, 1979, 82, 82565.

Theor.

4.1.153. Leung Kok Ming, Low-frequency dynamics in cooperative Jahn-Teller systems. - 1979. - 108 p. - [Order N79-27186.]. Diss. Abstr. Int. B, 1980, 40, N7, 3241. CA, 1980, 92, 155194.

Theor.

4.1.154. Leung Kok Ming, Huber, D. L., Ultrasonic attenuation near the 208 K phase transition of $RbAg_4I_5$. - Phys. Rev. Lett., 1979, 42, N7, 452-456. - PA, 1979, 82, 33085.

Theor.

4.1.155. Leung Kok Ming, Huber, D. L., Ultrasonic attenuation in a cooperative Jahn-Teller system: TmZn. - J. Appl. Phys., 1979, 50, N3, 1831-1833. - CA, 1979, 91, 30861. PA, 1979, 82, 78053.

Theor.

4.1.156. Nusser, H., Wagner, M., Cluster expansions in cooperative Jahn-Teller T-systems. - Physica A, 1979, 98, N1/2, 118-138. - CA, 1980, 92, 28857. PA, 1980, 83, 4661.

Theor.; T - e; T - t.

4.1.157. Srinivasan, V., Srivastava, K. N., A soft mode in the Jahn-Teller effect. - J. Phys. C, 1979, 12, N10, L367-L369. - CA, 1979, 91, 202888. PA, 1979, 82, 69221.

Theor.

4.1.158. Vekhter, B. G., Kaplan, M. D., Novosel'skiy, I. A., Effect of an electric field on the elastic properties of dysprosium vanadate ($DyVO_4$). [Russ.]. - Fiz. Tverd. Tela, 1979, 21, N5, 1386-1391. - CA, 1979, 91, 100123. PA, 1980, 83, 34123.

Theor.; 4.2.

4.1.159. Vekhter, B. G., Kaplan, M. D., New type of a structural Jahn-Teller phase transition in dysprosium vanadate ($DyVO_4$) induced by an electric field. [Russ.]. - Pis'ma Zh. Eksp. & Teor. Fiz., 1979, 29, N3, 173-176. - CA, 1979, 90, 160914. PA, 1979, 82, 46802.

Theor.; D_{4h}; 4.2.

4.2. Dielectric Properties. Vibronic Ferro- and Antiferroelectrics

The softening of the elastic modulus is a characteristic property of the Jahn-Teller crystals. Special interest attaches to the case where Jahn-Teller distortions lead to dipole moment formation (dipolar instability) [1.5.-14, 19, 25]. For centrosymmetric lattice sites this is possible only as a result of the vibronic mixing of electronic states of opposite parity. In this case there are large anomalies of dielectric properties, the acoustic anomalies being considerably smaller. This situation formed the basis of the so-called vibronic theory of ferroelectricity [4.2.-1, 2, 4, 10, 16, 18, 49, 84]; the last two papers [4.2.-49, 84] are review articles. In these works the conditions under which the vibronic dipolar instability of crystal lattices exists were revealed, and different mechanisms of high-temperature phase stabilization were considered. For a recent refinement of the theory, see the papers [4.2.-62, 78, 79].

Essential anomalies of both elastic and dielectric susceptibilities can be expected in noncentrosymmetric crystals. For these the cooperative Jahn-Teller effect may lead to ferroelectric phase transitions, provided the vibronic coupling with odd optical modes predominates. The characteristic features of such transitions were discussed in Refs. [4.2-52, 75, 105]. Even if the electron-phonon coupling with the acoustic mode predominates, resulting in the structural ordered (nondipolar) phase, large anomalies of the dielectric constant besides the softening of the elastic modulus may occur [4.2.-81, 83, 96, 105]. Together with studies on the influence of electric fields, this allows one to differentiate between the contributions of intra- and intersublattice couplings to the mean field, and to estimate the constant of electron-polarization interaction [4.1.-158, 159].

1964

4.2.1. Englman, R., Microscopic theory of ionic dielectrics. - 1964. - 71 p. - AEC Accession N 14129, Rep. No. IA-994. - CA, 1965, 63, 9170c.

Theor.

4.2.2. Sinha, K. P., Sinha, A. P. B., Role of Jahn-Teller effects in the origin of ferroelectricity: occurrence of ordered phase in perovskite-type ferroelectrics. - Indian J. Pure & Appl. Phys., 1964, 2, N3, 91-94. - RZF, 1964, 11E417.

Theor.; O_h; d^o.

1966

4.2.3. Bersuker, I. B., Origin of ferroelectricity in perovskite-type crystals. - Phys. Lett., 1966, 20, N6, 589-590. - CA, 1966, 65, 136f.

Theor.; O_h; d^o; 1.5.

4.2.4. Bersuker, I. B., Vekhter, B. G., Three-dimensional alternation of bonds in perovskite type crystal lattice. [Russ.]. - In: Tez. Dokl. 4 Vsesoyuz. Soveshch. Kvant. Khim. Kiev, 1966, p. 58.

Theor.; O_h; d^o; 1.5.

4.2.5. Brout, R., Müller, K. A., Thomas, H., Tunnelling and collective excitations in a microscopic model of ferroelectricity. - Solid State Commun., 1966, 4, N10, 507-511. - CA, 1967, 66, 23357.

Theor.

4.2.6. Ching Te Li, Electronic spontaneous polarization of O^{2-} in the ABO_3-type crystals. [Chinese]. - Wu Li Hsueh Pao, 1966, 22, N2, 188-196. - CA, 1966, 65, 11542g.

Theor.; O_h; d^o.

4.2.7. Englman, R., Vibronic effect in the temperature dependence of the soft frequency of $BaTiO_3$. - In: Proc. Int. Meet. Ferroelectricity. Prague, 1966, Vol. 1, p. 51-54. - RZF, 1968, 5E615.

Theor.; O_h; d^o; 1.5.

4.2.8. Havinga, E. E., Ferroelectric perovskites containing manganese ions. - Philips Res. Rep., 1966, 21, N1, 49-62. - CA, 1966, 64, 18642d.

Exp.

4.2.9. Sabane, C. D., Sinha, A. P. B., Biswas, A. B., Electrical properties of copper manganite. - Indian J. Pure & Appl. Phys., 1966, 4, N5, 187-190. - CA, 1966, 65, 12974d.

Exp.; O_h; 4.6.

1967

4.2.10. Bersuker, I. B., Vekhter, B. G., The interband interaction and spontaneous polarization of crystal lattices. [Russ.]. - Fiz. Tverd. Tela, 1967, 9, N9, 2652-2660. - CA, 1968, 68, 73694.

Theor.; O_h; d^o.

4.2.11. Englman, R., Microscopic theory of ferroelectricity. - 1967. - 91 p. - U.S. Air Force Syst. Command, Air Force Mater. Lab., Tech. Rep., AFML, TR-67-256. - CA, 1970, 73, 29863.

Theor.; O_h; d^o; 1.5.

4.2.12. Kristofel', N. N., Konsin, P. I., On the possibility of the ferroelectric phase transitions in connection with electron-phonon interaction. [Russ.]. - Izv. Akad. Nauk Est. SSR, Fiz., Mat., 1967, 16, N4, 431-439. - CA, 1968, 68, 63620.

Theor.; 1.5.

4.2.13. Kristofel', N. N., Konsin, P., Pseudo-Jahn-Teller effect and second order phase transitions in crystals. - Phys. Status Solidi, 1967, 21, N7, K39-K43. - CA, 1967, 67, 69049.

Theor.; O_h; d^o; 1.5.

1968

4.2.14. Bersuker, I. B., Vekhter, B. G., The "interband" theory of spontaneous polarization of crystals and ferroelectric phase transitions. [Russ.]. - In: Tez. Dokl. 6 Vsesoyuz. Mezhvuz. Konf. Segnetoelektrichestvu. Riga, 1968, p. 165.

Theor.; 1.5.

4.2.15. Konsin, P. I., Kristofel', N. N., On the dependence of the ferroelectric soft mode frequencies on the electric field. [Russ.]. - Fiz. Tverd. Tela, 1968, 10, N7, 2250-2252. - RZF, 1968, 11E631.

Theor.; 4.1.

4.2.16. Kristoffel, N., Konsin, P., Displacive vibronic phase transitions in narrow-gap semiconductors. - Phys. Status Solidi, 1968, 28, N2, 731-739. - RZF, 1968, 12E709.

Theor.; 1.5; 4.1.

1969

4.2.17. Bersuker, I. B., Vekhter, B. G., Danil'chuk, G. S., Kremenchugskiy, L. S., Muzalevskiy, A. A., Rafalovich, M. L., On the origin of spontaneous polarization and ferroelectric phase transitions in TGS-type crystals. [Russ.]. - Fiz. Tverd. Tela, 1969, 11, N9, 2452-2458. - CA, 1969, 71, 117618.

Theor.

4.2.18. Bersuker, I. B., Vekhter, B. G., The "interband" theory of spontaneous polarization of crystals and ferroelectric phase transitions. [Russ.]. - Izv. Akad. Nauk SSSR, Ser. Fiz., 1969, 33, N2, 199-203. - CA, 1969, 70, 91649. RZF, 1969, 8E777.

Theor.; O_h; d^o; 1.5.

4.2.19. Bersuker, I. B., Vekhter, B. G., Muzalevskiy, A. A., On the role of the quadratic terms of the electron-vibrational interactions in the formation of the dipolar instability of polyatomic systems. [Russ.]. - Izv. Akad. Nauk Mold.SSR, Ser. Biol. Khim. Nauk, 1969, N4, 60-63.

Theor.; 1.5.

4.2.20. Bersuker, I. B., Vekhter, B. G., "Interband" interaction and spontaneous polarization of crystals. - In: Abstr. 2nd Int. Meet. Ferroelectricity. Kyoto, 1969, p. 274-275.

Theor.; O_h; d^o; 1.5.

4.2.21. Bokov, V. A., Grigoryan, N. A., Bryzhina, M. F., Kazaryan, V. S., Ferroelectric and magnetic properties of (1-x)Bi-MnO_3-x$PbTiO_3$ solid solutions. [Russ.]. - Izv. Akad. Nauk SSSR, Ser. Fiz., 1969, 33, N7, 1164-1167. - CA, 1969, 71, 75478.

Exp.; O_h; d^0; d^3.

4.2.22. Konsin, P. I., Kristofel', N. N., The temperature dependence of specific heat at vibronic phase transitions. [Russ.]. - Izv. Akad. Nauk Est.SSR, Fiz., Mat., 1969, 18, N3, 344-346. - CA, 1970, 72, 16255.

Theor.; 1.5. 4.7.

4.2.34. Konsin, P. I., Kristofel', N. N., The vibronic phase transitions of crystals taking into account the degeneracy of the participating bands and vibration modes. [Russ.]. - Izv. Akad. Nauk Est.SSR, Fiz., Mat., 1969, 18, N4, 438-444. - CA, 1970, 72, 94048.

Theor.; O_h; d^0; 1.5.

4.2.24. Kristofel', N. N., Konsin, P. I., On the electron-phonon nature of ferroelectric phase transitions. [Russ.]. - Izv. Akad. Nauk SSSR, Ser. Fiz., 1969, 33, N1, 187-191.

Theor.

4.2.25. Venevtsev, Yu. N., Analysis of conditions for the formation of a spontaneous polarized state in perovskite. [Russ.]. - Izv. Akad. Nauk SSSR, Ser. Fiz., 1969, 33, N7, 1125-1132. - CA, 1969, 71, 75467.

Rev.

1970

4.2.26. Nardelli, G. F., Off-center ions in crystal lattices.- 1970. - 14 p. - U.S.Clearinghouse Fed. Sci. Tech. Inform. AD N722859. - CA, 1971, 75, 92344.

Theor.; 1.5.

1971

4.2.27. Bersuker, I. B., Vekhter, B. G., Musalevski, A. A., On the origin of the spontaneous polarization and the isotope effects in ferroelectric KH_2PO_4 crystals. - Phys. Status Solidi b, 1971, 45, N1, K25-K27.

Theor.

4.2.28. Bersuker, I. B., Vekhter, B. G., Muzalevski, A. A., On the microtheory of spontaneous polarization and phase transitions in crystals of perovskite-type and of ones with tetrahedral structural units. - In: Abstr. 2nd Eur. Meet. Ferroelectricity. Dijon, 1971, p. 8-9.

Theor.; O_h; d^o; 1.5.

4.2.29. Bersuker, I. B., Vekhter, B. G., Rafalovich, M. L., Instability and distortions of coordination compound chain structures in electronic degenerate states. - In: Proc. 3rd Conf. Coord. Chem. Bratislava, 1971, p. 31-38. - CA, 1972, 76, 105660.

Theor.

4.2.30. Konsin, P. I., Kristofel', N. N., Additional features of ferroelectric phase transitions in the vibronic model. [Russ.]. - Izv. Akad. Nauk Est.SSR, Fiz., Mat., 1971, 20, N1, 37-47. - CA, 1971, 75, 11977.

Theor.; 1.5. 4.4.

4.2.31. Kristofel, N. N., Electron-phonon interaction and the soft vibration modes active in phase transitions. - In: Light Scattering in Solids: Proc. 2nd Int. Conf. Paris, 1971, p. 384-387. - CA, 1973, 78, 35270.

Theor.; 1.5.

4.2.32. Kristofel, N. N., Konsin, P., Theory of ferroelectric phase transitions with vibronic mechanism. - Ann. Univ. Turku. Ser. A1, 1971, N146, 14 p. - CA, 1971, 75, 113732.

Theor.; 1.5; 4.3.

4.2.33. Kristofel', N. N., Konsin, P. I., On the theory of interband mechanisms of ferroelectric phase transitions. [Russ.]. - Izv. Akad. Nauk SSSR, Ser. Fiz., 1971, 35, N9, 1770-1775. - CA, 1972, 76, 38758.

Theor.; 1.5.

4.2.34. Kristofel', N. N., Konsin, P. I., Theory of vibronic phase transition in wide-gap ferroelectrics. [Russ.]. - Fiz. Tverd. Tela, 1971, 13, N9, 2513-2520. - CA, 1971, 75, 134525.

Theor.; O_h; d^o.

4.2.35. Kristofel', N. N., Konsin, P. I., Effect of impurities on the Curie point of a ferroelectric with a wide gap. [Russ.]. - Fiz. Tverd. Tela, 1971, 13, N12, 3513-3516. - CA, 1972, 76, 78148.

Theor.

1972

4.2.36. Bersuker, I. B., Vekhter, B. G., Musalevski, A. A., On the microtheory of spontaneous polarization and phase transitions in crystals of perovskite-type and of ones with tetrahedral structural units. - J. Phys. [France], 1972, 33, N2, 139-140.

Theor.; O_h; d^o.

4.2.37. Bersuker, I. B., Vekhter, B. G., Muzalevskiy, A. A., On the microtheory of ferroelectricity. [Russ.]. - In: Fiz. i Khim. Tverd. Tela: Sb. Nauch. Tr. Moscow, 1972, N2, p. 4-15. - RZF, 1972, 7E822.

Theor.; O_h; d^o.

4.2.38. Bersuker, I. B., Vekhter, B. G., The temperature dependence of the band structure in semiconductor-ferroelectrics. [Russ.]. - In: Tez. Dokl. 2 Seminara Poluprovodnikam-Segnetoelektrikam. Rostov n/Donu, 1972, p. 14.

Theor.

4.2.39. Do Chan Quat, Kopaev, Yu. V., On the instability of a metal two-band model relative to annihiliation scattering. [Russ.]. - Fiz. Tverd. Tela, 1972, 14, N9, 2603-2611. - CA, 1973, 78, 7899. RZF, 1973, 4E1827.

Theor.

4.2.40. Konsin, P. I., Kristofel', N. N., Curie point of a wide-gap ferroelectric in relation to the nonequilibrium concentration of carriers. [Russ.]. - Kristallografiya, 1972, 17, N4, 712-715. - CA, 1972, 77, 119592.

Theor.; O_h; d^o.

4.2.41. Konsin, P. I., On the possibility of the exciton-phonon structural phase transitions. [Russ.]. - Fiz. Tverd. Tela, 1972, 14, N4, 1144-1147. - RZF, 1972, 8E996.

Theor.

4.2.42. Konsin, P. I., Kristofel', N. N., Phase transitions in ferroelectrics with hydrogen bonds. [Russ.]. - Fiz. Tverd. Tela, 1972, 14, N10, 2873-2879. - CA, 1973, 78, 35273. RZF, 1973, 2E1055.

Theor.

1973

4.2.43. Bercha, D. M., Baletskiy, D. Yu., On the nature of nonferroelectric phase transitions in chain crystals. [Russ.]. - Fiz. Tverd. Tela, 1973, 15, N1, 275-277. - CA, 1973, 78, 116201. RZF, 1973, 6E355.

Theor.

4.2.44. Bersuker, I. B., Vekhter, B. G., Rafalovich, M. L., The spontaneously polarized ferro and antiferro states of chain structures taking into account vibronic interactions. [Russ.]. - In: Fiz. i Khim. Tverd. Tela: Sb. Nauch. Tr. Moscow, 1973, N4, p. 4-11. - RZF, 1974, 4E1183.

Theor.

4.2.45. Bersuker, I. B., Vekhter, B. G., Rafalovich, M. L., Ferro- and antiferroelectric properties of chain structures taking into account vibrational interactions. [Russ.]. - Fiz. Tverd. Tela, 1973, 15, N3, 946-948. - CA, 1973, 78, 165823. PA, 1973, 76, 72586. RZF, 1973, 7E934.

Theor.

4.2.46. Bersuker, I. B., Vekhter, B. G., Muzalevski, A. A., Ferro- and antiferroelectric ordered states in the vibronic theory: Influence of electronic band structure. - In: Abstr. 3rd Int. Meet. Ferroelectricity. Edinburgh, 1973, p. 86.

Theor.

4.2.47. Konsin, P. I., Kristofel', N. N., The temperature dependence of the forbidden band in wide-gap ferroelectrics. [Russ.]. - Izv. Akad. Nauk Est.SSR, Fiz. Mat., 1973, 22, N2, 173-178. - CA, 1973, 79, 58983.

Theor.; O_h; d^o.

4.2.48. Kristofel', N. N., Konsin, P. I., Interband theory of ferroelectricity. [Russ.]. - In: Titanat Bariya: Mater. Semin. Posvyashch. 25-letiyu Otkrytiya Segnetoelektr. Svoystv Titanata Bariya, 1970. Moscow, 1973, p. 11-19. - CA, 1974, 80, 8181. RZF, 1973, 9E821.

Rev.

4.2.49. Kristofel', N. N., Konsin, P. I., Electron-phonon interaction, microscopic mechanism, and properties of ferroelectric phase transitions. [Russ.]. - In: Poluprovodniki-Segnetoelektriki. Rostov n/Donu, 1973, p. 42-60.

Theor.

4.2.50. Kristofel, N. N., Konsin, P. I., Electron-phonon interaction, microscopic mechanism, and properties of ferroelectric phase transitions. - Ferroelectrics, 1973, 6, N1/2, 3-12. - Bibliogr.: 48 ref. CA, 1974, 81, 112373.

Rev.

4.2.51. Vekhter, B. G., Dipole moments and ferroelectric phase transitions in crystals with Jahn-Teller centers. [Russ.]. - Fiz. Tverd. Tela, 1973, 15, N2, 509-513. - CA, 1973, 78, 141500. PA, 1973, 76, 75950. RZF, 1973, 6E868.

Theor.; T_d.

4.2.52. Vekhter, B. G., Bersuker, I. B., On the temperature dependence of the band structure in semiconductor-ferroelectrics. - Ferroelectrics, 1973, 6, N1, 13-14.

Theor.

4.2.53. Vishnevskiy, I. I., Skripak, V. N., The behavior of the lattice heat conductivity in the region of the phase transition. [Russ.]. - In: Teplofiz. Svoystva Tverd. Veshchestv. Moscow, 1973, p. 44-48. - RZF, 1974, 1E987.

Theor.

1974

4.2.54. Bersuker, I. B., Vekhter, B. G., Muzalevskii, A. A., Interband vibronic interaction theory for ferro- and antiferroelectric ordered states in rock-salt-type crystals. - Ferroelectrics, 1974, 6, N3/4, 197-202. - CA, 1974, 81, 112600. RZF, 1975, 2E1119.

Theor.

4.2.55. Bersuker, I. B., Vekhter, B. G., Muzalevskii, A. A., Ferro- and antiferroelectric ordered states in the vibronic theory: Influence of electronic band structure. - Ferroelectrics, 1974, 8, N1/2, 465. - RZF, 1975, 5E1359.

Theor.

4.2.56. Chanussot, G., Static pseudo-Jahn-Teller effect at point defects in irradiated ferroelectric crystals (perovskite structures). II. Theoretical. - Ferroelectrics, 1974, 8, N3/4, 671-683. - CA, 1975, 83, 20301. PA, 1975, 78, 29995. RZF, 1975, 9E1541.

Theor.

4.2.57. Konsin, P. I., Concentration dependences of the Curie-Weiss constant and Curie-Weiss point of the ferroelectric solutions of the (Ba, Sr)TiO_3 type. [Russ.]. - Fiz. Tverd. Tela, 1974, 16, N9, 2709-2712. - CA, 1974, 81, 178620. RZF, 1975, 1E1065.

Theor.; O_h; d^o.

4.2.58. Kristofel', N. N., The atomic model of the vibronic ferroelectric-semiconductor with inter-electron interaction. [Russ.]. - Fiz. Tverd. Tela, 1974, 16, N9, 2535-2540.

Theor.

1975

4.2.59. Bersuker, I. B., Vekhter, B. G., Zenchenko, V. P., Ferroelectric phase transitions in magnetic fields. - In: Proc. 3rd Eur. Meet. Ferroelectricity. Zurich, 1975, K3.

Theor.

1976

4.2.60. Bersuker, I. B., Vekhter, B. G., Zenchenko, V. P., Ferroelectric phase transitions in magnetic fields. - Ferroelectrics, 1976, 13, N1/4, 373-376. - RZF, 1977, 5E1533.

Theor.

4.2.61. Chanussot, G., Inhomogeneous instabilities in an illuminated crystal of doped $BaTiO_3$. - Ferroelectrics, 1976, 13, N1/4, 313-315. - CA, 1977, 86, 82445. PA, 1976, 79, 93751.

Theor.; O_h; d^o; d^5; d^7; d^9.

4.2.62. Girshberg, Ya. G., Tamarchenko, V. I., The instability and phase transition in systems with interband interaction. [Russ.]. - Fiz. Tverd. Tela, 1976, 18, N4, 1066-1072.

Theor.

4.2.63. Girshberg, Ya. G., Fazovyy Perekhod i Temperaturnaya Zavisimost' Kriticheskogo Kolebaniya i Parametra Poryadka v Sistemakh s Mezhzonnoy Svyazyu: Avtoref. Dis. ... Kand. Fiz. Mat. Nauk. [Russ.]. - Leningrad, 1976. - 12 p. - Leningr. Ped. Inst.

The Phase Transition and Temperature Dependence of the Critical Vibraton and the Order Parameter in Systems with Interband Coupling.

Theor.

4.2.64. Harrison, J. P., Hessler, J. P., Taylor, D. R., One- and three-dimensional antiferroelectric ordering in $PrCl_3$. - Phys. Rev. B, 1976, 14, N7, 2979-2982. - CA, 1976, 85, 185427.

Exp.

4.2.65. Itskovich, M., Konsin, P. I., Kristofel', N. N., Effect of impurity carriers on the ferroelectric properties of small-gap semiconductors. [Russ.]. - Izv. Akad. Nauk Est.SSR, Fiz. Mat., 1976, 25, N3, 269-273. - CA, 1976, 85, 185414.

Theor.

4.2.66. Konsin, P. I., Kristoffel, N. N., Ferroelectric semimetal-dielectric phase transition induced by the interband electron-phonon interaction. - Ferroelectrics, 1976, 13, N1/4, 393-395. - CA, 1977, 86, 99671.

Theor.

4.2.67. Konsin, P. I., Ferroelectric phase transitions in systems with broad electronic bands caused by interband electron-phonon interaction. - Phys. Status Solidi b,.1976, 76, N2, 487-495. - CA, 1976, 85, 115393.

Theor.

4.2.68. Konsin, P. I., Ferroelectric phase transition of the semimetal-dielectric type. [Russ.]. - Fiz. Tverd. Tela, 1976, 18, N3, 701-705. - CA, 1976, 84, 143539.

Theor.

4.2.69. Konsin, P. I., Vibronnaya i Dinamicheskaya Teoriya Strukturnykh Fazovykh Perekhodov: Avtoref. Dis. ... Dokt. Fiz.-Mat. Nauk. [Russ.]. - Tartu, 1976. - 31 p. - Inst. Phys. Akad. Nauk Est.SSR.

The Vibronic and Dynamic Theory of the Structural Phase Transitions.

Rev.

4.2.70. Kristofel', N. N., Konsin, P. I., The vibronic theory of ferroelectricity. [Russ.]. - Usp. Fiz. Nauk, 1976, 120, N4, 507-510.

Rev.

4.2.71. Kristofel', N. N., Konsin, P. I., On the matrix element of the interband electron-phonon interaction of the vibronic theory of ferroelectrics. [Russ.]. - Izv. Akad. Nauk Est.SSR, Fiz. Mat., 1976, 25, N1, 23-28.

Theor.

4.2.72. Natori Akiko, Displacive phase transition in narrow-gap semiconductors. - J. Phys. Soc. Jpn., 1976, 40, N1, 163-171.

Theor.

4.2.73. Perron-Simon, A., Ravez, J., Hagenmuller, P., Effect on the ferroelectric Curie temperature of phase with the "tetragonal tungsten bronze" structure of introducing a cation showing a Jahn-Teller effect in the octahedral site. [Fr.]. - C.R. Hebd. Seances Acad. Sci. Ser. C, 1976, 283, N2, 33-35. - CA, 1976, 85, 200795.

Exp.

4.2.74. Vekhter, B. G., Kaplan, M. D., Ferroelectric phase transitions in crystals with cooperative Jahn-Teller effect. [Russ.]. - Fiz. Tverd. Tela, 1976, 18, N3, 784-788. - PA, 1977, 80, 8532.

Theor.

1977

4.2.75. Konsin, P. I., On the possibility of a ferroelectric phase transition induced by impurity electrons. [Russ.]. - Izv. Akad. Nauk Est.SSR, Fiz. Mat., 1977, 26, N4, 411-416.

Theor.

4.2.76. Kristofel', N. N., On tunneling effects in vibronic ferroelectrics. [Russ.]. - Fiz. Tverd. Tela, 1977, 19, N3, 775-780. - RZF, 1977, 8E1431.

Theor.

4.2.77. Mailyan, G. L., Plakida, N. M., Fluctuations of order parameter and anharmonic interaction in the vibronic model of ferroelectrics. - Phys. Status Solidi b, 1977, 80, N2, 543-547.

Theor.

4.2.78. Mailyan, G. L., On the phase transition temperature in ferroelectric semiconductors. - Solid Commun., 1977, 24, N9, 611-614.

Theor.

4.2.79. Taylor, D. R., Harrison, J. P., Cooperative Jahn-Teller ordering of electric dipoles in non-Kramers doublets. - Ferroelectrics, 1977, 16, N1/4, 253-255. - CA, 1978, 88, 161084.

Exp.; C_{3h}.

4.2.80. Taylor, D. R., Harrison, J. P., McColl, D. B., Cooperative Jahn-Teller ordering in praseodymium compounds. - Physica B + C, 1977, 86-88, N3, 1164-1165. - CA, 1977, 86, 198646. PA, 1977, 80, 66144.

Exp.; C_{3h}.

4.2.81. Taylor, D. R., Electric susceptibility studies of cooperative Jahn-Teller ordering in rare-earth crystals. - In: Electron-Phonon Interactions and Phase Transitions. New York; London, 1977, p. 297-301. (NATO Adv. Study Inst. Ser., Ser. B, 1977, N29.) - CA, 1979, 90, 131217.

Rev.; 4.7.

4.2.82. Unoki Hiromi, Sakudo Tsunetaro, Dielectric anomaly and improper antiferroelectricity at the Jahn-Teller transitions in rare-earth vanadates. - Phys. Rev. Lett., 1977, 38, N3, 137-140. - PA, 1977, 80, 24802.

Exp.; D_{2d}; D_{4h}.

1978

4.2.83. Bersuker, I. B., Vekhter, B. G., The vibronic theory of ferroelectricity. - Ferroelectrics, 1978, 19, N3/4, 137-150. - Bibliogr.: 65 ref. CA, 1979, 90, 178613. PA, 1979, 82, 19854.

Rev.

4.2.84. Chanussot, G., Physical models for the photoferroelectric phenomena. - Ferroelectrics, 1978, 20, N1/2, 37-50. - CA, 1979, 90, 160538. PA, 1979, 82, 15407.

Theor.

4.2.85. Chaudhuri, B. K., Static and dynamic properties of the ferroelectric phase transition in hydrogen-bonded KDP and its isomorphs. - Indian J. Pure & Appl. Phys., 1978, 16, N9, 831-835. - CA, 1979, 90, 160913.

Theor.

4.2.86. Harley, R. T., Taylor, D. R., Dielectric anomalies at the cooperative Jahn-Teller phase transitions in $DyAsO_4$ and $DyVO_4$. - In: Proc. Int. Conf. Lattice Dynamics, 1977. Paris, 1978, p. 218-219. - CA, 1979, 90, 79668. PA, 1978, 81, 64312.

Exp.; D_{2d}.

4.2.87. Konsin, P. I., Kristoffel, N. N., Soft modes in the vibronic theory of antiferroelectric and structural modulated phase transitions. - In: Proc. Int. Conf. Lattice Dynamics, 1977. Paris, 1978, p. 666-668. - CA, 1979, 90, 79726.

Theor.

4.2.88. Konsin, P. I., Structural phase transitions of antiferroelectric and displacive modulated types caused by electron-phonon interaction. - Phys. Status Solidi b, 1978, 86, N1, 57-66.

Theor.

4.2.89. Konsin, P. I., Kristoffel, N. N., Impurities and ferroelectric properties of semiconductors. - Ferroelectrics, 1978, 18, N1/3, 121-126. - CA, 1978, 89, 139082.

Theor.

4.2.90. Kristoffel, N. N., Konsin, P. I., Some new results of the vibronic theory of ferroelectricity. - Ferroelectrics, 1978, 21, N1/4, 477-479. - CA, 1979, 90, 213928. RZF, 1979, 7E1745.

Theor.; O_h; d^o.

4.2.91. Kristofel', N. N., The dependence of the conduction electron plasma frequency of the vibronic ferroelectrics on the spontaneous distortion of the lattice. [Russ.]. - Izv. Akad. Nauk Est.SSR, Fiz. Mat., 1978, 27, N3, 373-375.

Theor.

4.2.92. Kristofel', N. N., Rebane, L. A., Impurity centers and phase transitons. [Russ.]. - In: Tr. Inst. Fiz. i Astron. Akad. Nauk Est.SSR, 1978, Vol. 48, p. 64-84.

Theor.

4.2.93. Ouedraogo, A., Dehaut, B., Chanussot, G., Photoferroelectric properties of Cu-doped $BaTiO_3$ crystals. - J. Phys. [France], 1978, 39, N12, L179-L182. - CA, 1978, 89, 52209. PA, 1978, 81, 60400.

Exp.; E - e; O_h; d^0; d^9; 4.3.

4.2.94. Unoki Hiromi, Sakudo Tsunetaro, Jahn-Teller phase transition and improper antiferroelectricity in dysprosium vanadate. [Jap.]. - Kotai Butsuri, Solid State Phys., 1978, 13, N3, 173-179. - CA, 1979, 90, 46967. PA, 1978, 81, 80294.

Exp.; D_{2d}.

4.2.95. Vekhter, B. G., Kaplan, M. D., Dielectric properties of centrosymmetric orthovanadates with a cooperative Jahn-Teller effect. [Russ.]. - Fiz. Tverd. Tela, 1978, 20, N5, 1433-1437. - CA, 1978, 89, 69264. PA, 1979, 82, 28694. RZF, 1978, 9E688.

Theor.

4.2.96. Vekhter, B. G., Zenchenko, V. P., Bersuker, I. B., Phase transitions in large gap ferroelectrics in magnetic fields. - Ferroelectrics, 1978, 20, N3/4, 163-164. - RZF, 1979, 7E1747.

Theor.

1979

4.2.97. Bersuker, I. B., The Jahn-Teller effect and vibronic interactions in ferroelectrics. [Russ.]. - In: Tez. Dokl. 9 Vsesoyuz. Soveshch. Segnetoelektrichestvu. Rostov n/Donu, 1979, p. 10.

Theor.

4.2.98. Konsin, P. I., Kristoffel, N. N., The vibronic theory of structural phase transitions and properties of narrow-gap semiconductors of the (Pb, Sn)Te type. - In: Phys. Semicond.: Invited and Contributed Papers 14th Int. Conf., Edinburgh, 1978. Bristol; London, 1979, p. 453-456. - CA, 1979, 90, 196298. RZF, 1980, 6E1632.

Theor.

4.2.99. Kristofel', N. N., Gulbis, A. V., The possibility of a proper anomalous photovoltaic effect in vibronic ferroelectrics. [Russ.]. - Izv. Akad. Nauk Est.SSR, Fiz. Mat., 1979, 28, N3, 268-271.

Theor.; 4.3.

4.2.100. Page, J. H., Smith, S. R. P., Taylor, D. R., Harley, R. T., Dielectric studies and interpretation of the first-order cooperative Jahn-Teller phase transition in $DyAsO_4$. - J. Phys. C, 1979, 12, N23, L875-L881. - PA, 1980, 83, 19940.

Exp.

4.2.101. Pelikh, L. N., Gurskas, A. A., Jahn-Teller phase transition in potassium dysprosium tungsten $KDy(WO_4)_2$. [Russ.]. - Fiz. Tverd. Tela, 1979, 21, N7, 2136-2138. - CA, 1979, 91, 132306. PA, 1980, 83, 55012.

Exp.

4.2.102. Vekhter, B. G., Kaplan, M. D., Microscopic theory of Jahn-Teller ferro- and antiferroelastics. [Russ.]. - Izv. Akad. Nauk SSSR, Ser. Fiz., 1979, 43, N8, 1586-1592. - CA, 1979, 91, 202464. RZF, 1979, 11E686.

Theor.

4.2.103. Vekhter, B. G., Zenchenko, V. P., Bersuker, I. B., The vibronic ferroelectrics in electric and magnetic fields. [Russ.]. - In: Tez. Dokl. 9 Vsesoyuz. Sovesch. Segnetoelektrichestvu. Rostov n/Donu, 1979, Pt. 1, p. 35.

Theor.; 4.4.

4.2.104. Vekhter, B. G., Kaplan, M. D., The dynamical theory of dielectric anomalies in Jahn-Teller crystals. [Russ.]. - In: Tez. Dokl. 9 Vsesoyuz. Soveshch. Segnetoelektrichestvu. Rostov na/Donu, 1979, Pt. 1, p. 46.

Theor.

4.2.105. Vekhter, B. G., Kaplan, M. D., The microscopic theory of Jahn-Teller ferro- and antiferroelectrics. - In: Abstr. 4th Eur. Meet. Ferroelectricity. Ljubljana, 1979, p. 27.

Theor.; 4.1.

4.2.106. Vekhter, B. G., Zenchenko, V. P., Bersuker, I. B., Temperature dependence of the band structure for vibronic ferroelectrics. - In: Abstr. 4th Eur. Meet. Ferroelectricity. Ljubljana, 1979, p. 28.

Theor.

See also: 4.1.-1, 74, 82, 115, 122; 4.3.-1, 8, 11, 44, 50, 54, 59, 63, 72, 73; 4.4.-119a, 122, 124, 137, 144a, 164a, 166a.

4.3. Optical Manifestations

Optical spectroscopic investigations have been widely and successfully used for studying the cooperative Jahn-Teller effect, especially in compounds which are transparent to the visible light (see reviews [4.3.-57, 69]). In optical absorption and Raman spectra the cooperative Jahn-Teller effect manifests itself in the temperature-dependent splitting of the corresponding spectral lines at $T < T_C$. Since the phase transition in Jahn-Teller crystals is due to the electron state degeneracy, its spectral manifestations are essentially dependent on external perturbations that remove the electronic degeneracy. This circumstance was already outlined in the first investigations [4.3.-10, 19, 23, 24] and used to prove the Jahn-Teller origin of the phase transitions. In these (and later) studies the influence of magnetic fields and stress upon the optical spectra was examined; the piezospectroscopic experiments allow one to differentiate between the electron-deformation and electron-phonon contributions to the constant of intersite interaction. The sattelites of the special lines caused by the intersite vibronic coupling were investigated in Refs. [4.3.-35, 52, 60]. In Raman scattering experiments the splitting of degenerate phonon modes [4.3.23] was also observed. The possibility of using birefringence investigations for the study of cooperative Jahn-Teller effect problems, in particular for the determination of the critical exponent, was illustrated in Refs. [4.3.-42, 46].

1949

4.3.1. Kay, H. F., Vousden, P., Symmetry changes in barium titanate at low temperatures and their relation to its ferroelectric properties. - Philos. Mag., 1949, 40, 1019-1040. - CA, 1950, 44, 1774h.

Exp.; O_h; C_{4v}; d^0; 4.2.

1965

4.3.2. Banerjee, R. S., Basu, S., Crystal spectra of chromium trisbiguanide. - J. Inorg. & Nucl. Chem., 1965, 27, N11, 2452-2453. - CA, 1966, 64, 5929d.

Exp.; T; O_h.

4.3.3. Dingle, R., The polarized single crystal spectrum of ammonium pentafluoromanganate(III). - Inorg. Chem., 1965, 4, N9, 1287-1290.

Exp.; E - e; O_h; d^4.

1966

4.3.4. Greskovich, C., Stubican, V. S., Divalent chromium in magnesium-chromium spinels. - J. Phys. & Chem. Solids, 1966, 27, N9, 1379-1384. - CA, 1966, 65, 11515e.

Exp.; T; E; O_h; d^4.

4.3.5. Stubican, V. S., Greskovich, C., Jahn-Teller distortion induced by tetrahedral-site Cr^{2+} ions in spinels. - J. Am. Ceram. Soc., 1966, 49, N9, 518-519. - CA, 1967, 66, 14613.

Exp.; T; E; O_h; d^4.

4.3.6. Reinen, D., The Jahn-Teller effect of Cu^{2+} ions in oxidic solids. - In: Proc. 10th Int. Conf. Coord. Chem. Tokyo, 1967, p. 243-244.

Exp.; E - e; O_h; d^9.

1967

4.3.7. Siratori Kiiti, Effect of crystal deformaion on the lattice vibration of oxide spinels. - J. Phys. Soc. Jpn., 1967, 23, N5, 948-954. - RZF, 1968, 6D582.

Exp.; d^0-d^{10}.

1969

4.3.8. Chisler, E. V., Raman spectrum study of phase transition in sodium nitrate crystal. [Russ.]. - Fiz. Tverd. Tela, 1969, 11, N5, 1272-1281. - CA, 1969, 71, 44146. PA, 1970, 73, 4514.

Exp.; E - e; D_{3h}.

4.3.9. Kanturek, J., Simsa, Z., Remark on the infrared spectra of manganese ferrites. - Phys. Status Solidi, 1969, 36, N1, K47-K49. - CA, 1970, 72, 26658. PA, 1970, 73, 21477.

Exp.; E - e; O_h; d^4.

1970

4.3.10. Cooke, A. H., Ellis, C. J., Gehring, K. A., Leask, M. J. M., Martin, D. M., Wanklyn, B. M., Wells, M. R., White, R. L., Observation of a magnetically controllable Jahn-Teller distortion in dysprosium vanadate at low temperatures. - Solid State Commun., 1970, 8, N9, 689-692. - CA, 1970, 73, 49424. PA, 1970, 73, 44728.

Exp.; D_{2d}; 4.4.

4.3.11. Pisarev, R. V., Druzhinin, V. V., Nesterova, N. N., Prokhorova, S. D., Andreeva, G. T., Optical absorption of ferroelectrical iron boracites. - Phys. Status Solidi, 1970, 40, N2, 503-512. - CA, 1970, 73, 71572. RZF, 1971, 3D470.

Exp.; T; E; O_h; d^6; 4.2.

1971

4.3.12. Becker, P. J., Laugsch, J., Spectroscopic confirmation of a magnetically controllable Jahn-Teller distortion in dysprosium vanadate. - Phys. Status Solidi b, 1971, 44, N2, K109-K112. - CA, 1971, 74, 132687. PA, 1971, 74, 45646.

Exp.; D_{2d}; 4.4.

4.3.13. Elliott, R. J., Electron-phonon coupled modes in Jahn-Teller systems. - In: Light Scattering in Solids: Proc. 2nd Int. Conf. Paris, 1971, p. 354-357. - CA, 1973, 78, 35643. PA, 1972, 75, 20066.

Theor.; 2.8.

4.3.14. Ellis, C. J., Gehring, K. A., Leask, M. J. M., White, R. L., Spectroscopic properties of dysprosium vanadate. - J. Phys. [France], 1971, 32, Colloq. N1, 1024-1025. - CA, 1974, 80, 21046.

Exp.; D_{2d}; 3.1.

4.3.15. Gehring, K. A., Malozemoff, A. P., Staude, W., Tyte, R. N., Observation of magnetically controllable distortion in $TbVO_4$ by optical spectroscopy. - Solid State Commun., 1971, 9, N9, 511-514.

Exp.; 4.4.

4.3.16. Harley, R. T., Hayes, W., Smith, S. R. P., Raman study of phase transitions in rare-earth vanadates. - Solid State Commun., 1971, 9, N9, 515-517. - CA, 1971, 75, 26320.

Exp.; 2.8.

4.3.17. Harley, R. T., Hayes, W., Smith, S. R. P., Raman studies of Jahn-Teller induced phase transitions in rare-earth compounds. - In: Light Scattering in Solids: Proc. 2nd Int. Conf. Paris, 1971, p. 357-363. - CA, 1973, 78, 9792. PA, 1972, 75, 16534. RZF, 1973, 5E642.

Exp.; 2.8.

4.3.18. Klein, L., Wüchner, W., Kahle, H. G., Schopper, H. C., Study of a phase transition in $TbAsO_4$. - Phys. Status Solidi b, 1971, 48, N2, K139-K141.

Exp.; 4.4.

1972

4.3.19. Becker, P. J., Leask, M. J. M., Tyte, R. N., Optical study of the cooperative Jahn-Teller transition in thulium vanadate, $TmVO_4$. - J. Phys. C, 1972, 5, N15, 2027-2036. - CA, 1972, 77, 95043. PA, 1972, 75, 65269. RZF, 1973, 1E821.

Exp.; $E - b_1$; D_4h; 4.4.

4.3.20. Becker, P. J., Leask, M. J. M., Maxwell, K. J., Tyte, R. N., Jahn-Teller induced magnon side bands in thulium vanadate. - Phys. Lett. A, 1972, 41, N3, 205-207. - CA, 1972, 77, 170730. PA, 1972, 75, 81377. RZF, 1973, 3E1489.

Exp.; $E - b_1$; D_4h; 4.4.

4.3.21. Böhm, W., Herb, R., Kahle, H. G., Kasten, A., Laugsch, J., Wüchner, W., Spectroscopic confirmation of a crystallographic phase transition and of antiferromagnetic ordering in $TbPO_4$. - Phys. Status Solidi b, 1972, 54, N2, 527-536. - CA, 1973, 78, 64215.

Exp.; $E - (b_1 + b_2)$; D_4h; 4.4.

4.3.22. D'Ambrogio, F., Brüesch, P., Kalbfleisch, H., Raman and infrared measurements and the study of phase transitions in $DyVO_4$. - Phys. Status Solidi b, 1972, 49, N1, 117-124. - CA, 1972, 76, 65890.

Exp.; 2.8.

4.3.23. Elliott, R. J., Harley, R. T., Hayes, W., Smith, S. R. P., Raman scattering and theoretical studies of Jahn-Teller induced phase transition in some rare-earth compounds. - Proc. R. Soc. London. Ser. A, 1972, 328, N1573, 217-266. - CA, 1972, 76, 160296. PA, 1972, 75, 48198.

Exp.; 2.8.

4.3.24. Gehring, G. A., Malozemoff, A. P., Staude, W., Tyte, R. N., Effect of uniaxial stress on the optical spectrum of $DyVO_4$. - J. Phys. & Chem. Solids, 1972, 33, N7, 1499-1510. - PA, 1972, 75, 51338.

Exp.

4.3.25. Harley, R. T., Hayes, W., Smith, S. R. P., Raman scattering investigations of Jahn-Teller induced phase transitions in $TuAsO_4$ and $TuVO_4$. - J. Phys. C, 1972, 5, N12, 1501-1510. - CA, 1972, 77, 81548. PA, 1972, 75, 54436. RZF, 1972, 12E1297.

Exp.; E - b_1; D_{4h}; 2.8.

4.3.26. Lee, J. N., Moos, H.W., Spectroscopic study of magnetic behavior and Jahn-Teller distortions in the even-electron system terbium phosphate. - Phys. Rev. B, 1972, 5, N9, 3645-3654. - CA, 1972, 76, 160506. PA,1972, 75, 36779.

Exp.; 4.4.

4.3.27. Wüchner, W., Böhm, W., Kahle, H. G., Kasten, A., Laugsch, J., Spectroscopic properties of $TbAsO_4$. - Phys. Status Solidi b, 1972, 54, N1, 273-283. - PA, 1973, 76, 2560. RZF, 1973, 4D619.

Exp.

1973

4.3.28. Commander, R. J., Finn, C. B. P., Investigation of the middle infrared absorption spectrum of $DyVO_4$ at low temperatures. - J. Phys. C, 1973, 6, N1, L29-L31. - CA, 1973, 78, 77647.

Exp.

4.3.29. Grefer, J., Reinen, D., Low temperature reflectance spectra of octahedrally coordinated copper(2+) ions in the ligand field range. [Germ.]. - Z. Naturforsch. a, 1973, 28, N3/4, 464-471. - CA, 1973, 79, 36769. PA, 1973, 76, 36877. RZF, 1973, 9E744.

Exp.; E - e; O_h; d^9.

4.3.30. McCaffery, A. J., Rowan, P. D., Shatwell, R. A., The detection of crystallographic phase transitions by linear dichroism. - J. Phys. C, 1973, 6, N21, L387-L389. - RZF, 1974, 4D953.

Exp.; 2.4.

4.3.31. Nouet, J., Tomas, D. J., Scott, J. F., Light scattering from magnons and excitons in rubidium trifluorocobaltate(II). Phys. Rev. B, 1973, 7, N11, 4874-4883. - CA, 1973, 79, 11317.

Exp.; E - e; O_h; d^7; 4.4.

4.3.32. Reinen, D., Grefer, J., Jahn-Teller effect of copper-(II) ions in tetrahedral oxygen coordination. I. Ligand field spectra of copper(II)-containing oxidic lattices. [Germ.]. - Z.

Naturforsch. a, 1973, 28, N7, 1185-1192. - CA, 1974, 80, 8581. PA, 1973, 76, 61052. RZF, 1973, 12D641.

Exp.; T; T_d; d^9.

4.3.33. Zvyagin, A. I., Stetsenko, T. S., Yurko, V. G., Vayshnoras, R. A., Low-temperature phase transition in potassium dysprosium molybdate [$KDy(MoO_4)_2$] induced by a cooperative Jahn-Teller effect. [Russ.]. - Pis'ma Zh. Eksp. & Teor. Fiz., 1973, 17, N4, 190-193. - CA, 1973, 78, 153413. PA, 1973, 76, 43875. RZF, 1973, 7E500.

Exp.; 4.4.

4.3.34. Zvyagin, A. I., Stetsenko, T. S., El'chaniova, S. D., Pshisukha, A. M., Yurko, V. G., Vayshnoras, R. A., Pelikh, L. N., Kobets, M. I., Optical, resonance, and magnetic properties of single crystals of the $KY(MoO_4)_2$-$KDy(MoO_4)_2$ system. [Russ.]. - In: Fiz. Kondens. Sostoyaniya. Kiev, 1973, p. 3-37. (Tr. Fiz. - Tekhn. Inst. Nizk. Temp. Akad. Nauk Ukr.SSR; Vol. 26.) - CA, 1975, 83, 88217. RZF, 1974, 10E826.

Exp.; 4.4.

1974

4.3.35. Becker, P.-J., Laugsch, J., Sidebands in the spectrum of thulium arsenate. - In: Tr. Mezhdunar. Konf. Magn. MKM - 1973. Moscow, 1974, Vol. 3, p. 78-82. - CA, 1976, 84, 10662. RZF, 1974, 10E1396.

Exp.; 4.4.

4.3.36. Fleury, P. A, Lazay, P. D., Van Uitert, L. G., Brillouin-scattering evidence for a new phase transition in perovskite crystals: $PrAlO_3$. - Phys. Rev. Lett., 1974, 33, N8, 492-495.

Exp.; 4.1.

4.3.37. Friebel, C., Refinement of the structure of Ba_2CuF_6. [Germ.]. - Z. Naturforsch. b, 1974, 29, N9/10, 634-641.

Exp.; E - e; O_h; d^9; 3.1.

4.3.38. Harley, R. T., Hayes, W., Perry, A. M., Smith, S. R. P., Elliott, R. J., Savill, I. D., Cooperative Jahn-Teller effects in the mixed crystals terbium gadolinium vanadate ($Tb_pGd_{1-p}VO_4$) and dysprosium yttrium vanadate ($Dy_pY_{1-p}VO_4$). - J. Phys. C, 1974, 7, N17, 3145-3160. - CA, 1974, 81, 143629. PA, 1974, 77, 79298. RZF, 1975, 3E725.

Exp.; 2.8.

4.3.39. Harley, R. T., Hayes, W., Perry, A. M., Smith, S. R. P., Jahn-Teller phase transitions in $Tb_cGd_{1-c}VO_4$. - Solid State Commun., 1974, 14, N6, 521-524. - CA, 1974, 80, 150719. PA, 1974, 77, 37010. RZF, 1974, 10E649.

Exp.; 2.8.

4.3.40. Lewis, J. F. L., Prinz, G. A., Far infrared spectroscopy of terbium(III) phosphate. - Phys. Rev. B, 1974, 10, N7, 2892-2899. - CA, 1975, 82, 9595.

Exp.

4.3.41. Wappler, D., Spectroscopic investigations on dysprosium arsenate. [Germ.]. - Phys. Kondens. Mater., 1974, 17, N2, 113-124. - CA, 1974, 80, 114343. PA, 1974, 77, 20779.

Exp.

1975

4.3.42. Becker, P. J., Gehring, G. A., Linear birefringence in magnetic and cooperative Jahn-Teller rare earth salts. - Solid State Commun., 1975, 16, N6, 795-798. - CA, 1975, 82, 162063. PA, 1975, 78, 33809.

Theor.; 4.4.

4.3.43. Bowen, H. K., Adler, D., Auker, B. H., Electrical and optical properties of FeO. - J. Solid State Chem., 1975, 12, N3/4, 355-359. - PA, 1975, 78, 26113.

Exp.; d^6; 4.2.

4.3.44. Chanussot, G., Voisin, J., Dehaut, B., Godefroy, G., Optical study of the Jahn-Teller structure in doped barium titanate. [Fr.]. - Rev. Phys. Appl. [France], 1975, 10, N6, 405-407. - CA, 1976, 84, 83130.

Exp.; O_h; d^0; d^6; d^7; d^9.

4.3.45. Harley, R. T., Hayes, W., Perry, A. M., Smith, S. R. P., Raman study of excitations in praseodymium orthoaluminate. - J. Phys. C, 1975, 8, N8, L123-L125. - CA, 1975, 83, 35220. PA, 1975, 78, 47182.

Exp.; 2.8.

4.3.46. Harley, R. T., Macfarlane, R. M., Determination of the critical exponent β in terbium vanadate ($TbVO_4$) and dysprosium vanadate ($DyVO_4$) using linear birefringence. - J. Phys. C, 1975, 8, N21, L451-L455. - CA, 1976, 84, 24612. PA, 1976, 79, 18697.

Exp.

4.3.47. Hirotsu Shunsuke, Optical and thermal properties of cesium copper trichloride and its phase transition near 423°K. - J. Phys. C, 1975, 8, N1, L12-L16. - CA, 1975, 82, 147322.

Exp.; E - e; d^9; 4.7.

4.3.48. Lyons, K. B., Birgeneau, R. J., Blount, E. I., Van Uitert, L. G., Electronic excitations in $PrAlO_3$. - Phys. Rev. B, 1975, 11, N2, 891-900.

Exp.; 2.8.

1976

4.3.49. Battison, J. E., Kasten, A., Leask, M. J. M., Lowry, J. B., Maxwell, K. J., Effects of uniaxial stress on thulium vanadate and thulium arsenate. - J. Phys. C, 1976, 9, N7, 1345-1350. - PA, 1976, 79, 49374.

Exp.; E - b_1; D_{4h}.

4.3.50. Chanussot, G., Optical response of ferroelectrics driven by Jahn-Teller effects. - Ferroelectrics, 1976, 10, N1/4, 149-151. - CA, 1977, 86, 49672. PA, 1976, 79, 62157.

Exp.; d^0; d^5; d^7; d^9; 4.7.

4.3.51. Colwell, P. J., Rahn, L. A., Walker, C. T., Raman scattering from antiferromagnetic UO_2. - In: Light Scattering in Solids: Proc. 3rd Int. Conf., 1975. Paris, 1976, p. 239-243. - CA, 1976, 85, 133466.

Exp.; O_h; d^1; 2.8.

4.3.52. Gehring, G. A., Theory of side-bands in cooperative Jahn-Teller material. - Phys. Status Solidi b, 1976, 78, N1, 211-218. - CA, 1977, 86, 10161. PA, 1977, 80, 1398.

Theor.

4.3.53. Sigmund, E., The T - t Jahn-Teller system: solution of the fundamental combinatorial problem and its application to moments and zero phonon line. - Z. Naturforsch. a, 1976, 31, N8, 904-910. - CA, 1976, 85, 151065. PA, 1976, 79, 83726.

Theor.; T - t; 4.1.

1977

4.3.54. Chanussot, G., Fridkin, V. M., Godefroy, G., Jannot, B., The photoinduced Rayleigh scattering in $BaTiO_3$ crystals showing the bulk photovoltaic effect. - Appl. Phys. Lett., 1977, 31, N1, 3-4. - CA, 1977, 87, 60092. PA, 1977, 80, 77226.

Exp.

4.3.55. Ferre, J., Regis, M., Farge, Y., Kleemann, W., Magneto-optical properties of the 2d-ferromagnet: dipotassium copper(II) fluoride. - Physica B + C, 1977, 89, 181-184. - CA, 1977, 87, 15146. PA, 1977, 80, 52062.

Exp.; D_{2h}; d^9; 2.4; 4.4.

4.3.56. Glynn, T. J., Harley, R. T., Macfarlane, R. M., A study of the cooperative Jahn-Teller effect in rare-earth vanadates by linear birefringence. I. $TbVO_4$ and $Tb_cGd_{1-c}VO_4$. - J. Phys. C, 1977, 10, N15, 2937-2946. - PA, 1977, 80, 76550.

Exp.

4.3.57. Harley, R. T., Optical studies of Jahn-Teller transitions. - In: Electron-Phonon Interactions and Phase Transitions. New York; London, 1977, p. 277-296. (NATO Adv. Study Inst. Ser., Ser. B, 1977, N29.) - CA, 1979, 90, 129604.

Rev.

4.3.58. Iqbal, Z., Raman scattering and the electronically-induced phase transition in $K_3Fe(CN)_6$ at 130 K. - J. Phys. C, 1977, 10, N18, 3533-3543. - CA, 1978, 88, 97614. PA, 1977, 80, 90142.

Exp.; 2.8.

4.3.59. Kristofel', N. N., Gulbis, A. V., The spontaneous birefringence in a wide-gap vibronic ferroelectric. [Russ.]. - Fiz. Tverd. Tela, 1977, 19, N10, 3071-3074. - CA, 1977, 87, 208781.

Theor.; O_h; d^0; 4.2.

4.3.60. Laugsch, J., Gehring, G. A., Optical side bands in a cooperative Jahn-Teller material: $TmAsO_4$. - Phys. Status Solidi b, 1977, 81, N1, 167-178. - PA, 1977, 80, 52123.

Exp.

4.3.61. Salamon, M. B., Jahn-Teller-like model of the 208 K phase transition in the solid electrolyte $RbAg_4I_5$. - Phys. Rev. B, 1977, 15, N4, 2236-2241. CA, 1977, 86, 148025. PA, 1977, 80, 35694.

Exp.

4.3.62. Zvyagin, A. I., El'chaninova, S. D., Pelikh, L. N., Anomalous effect of Eu^{3+} ion impurity on the structural phase transition temperature in $CsDy(MoO_4)_2$. (Short communication.) [Russ.]. - Fiz. Nizk. Temp., 1977, 3, N3, 533-535. - CA, 1977, 87, 60959.

Exp.

1978

4.3.63. Andrievskiy, B. V., Romanyuk, N. A., Optical characteristics of Rochelle salt crystals in the 4-22 eV region. [Russ.]. - Ukr. Fiz. Zh., 1978, 23, N8, 1351-1354. - CA, 1978, 89, 188203. PA, 1978, 81, 84550. RZF, 1979, 1D1118.

Exp.

4.3.64. El'chaninova, S. D., Zvyagin, A. I., Phase transitions in $CsDy(MoO_4)_2$ doped with trivalent ion impurities. [Russ.]. - Fiz. Nizk. Temp., 1978, 4, N11, 1465-1470. - CA, 1979, 90, 130929.

Exp.

4.3.65. Ferre, J., Regis, M., Optical determination of the Cu-Mn exchange interaction in $K_2Cu_{1-x}Mn_xF_4$. - Solid State Commun., 1978, 26, N4, 225-228. - PA, 1978, 81, 72551.

Exp.; E; T; O_h; d^5; d^9.

4.3.66. Köhler, P., Massa, W., Reinen, D., Hofmann, B., Hoppe, R., Jahn-Teller effect of manganese(3+) ions in octahedral fluorine coordination: Ligand field spectroscopic and magnetic studies. [Germ.]. - Z. Anorg. Allg. Chem., 1978, 446, N1, 131-158. - CA, 1979, 90, 94667.

Exp.; E; O_h; d^4.

4.3.67. Skorobogatova, I. V., Zvyagin, A. I., Low temperature phase transitions in $RbDy(MoO_4)_2$ and $KDy(WO_4)_2$. [Russ.]. - Fiz. Nizk. Temp., 1978, 4, N6, 800-804. - PA, 1978, 81, 87787.

Exp.

4.3.68. Swanson, B. I., Lucas, B. C., Ryan, R. R., Structural phase transformations and temperature dependent Raman spectra of $Cs_2LiFe(CN)_6$. - J. Chem. Phys., 1978, 69, N10, 4328-4334. - PA, 1979, 82, 11416.

Exp.; 2.8; 4.6.

4.3.69. Vekhter, B. G., Kaplan, M. D., Spectroscopy of Jahn-Teller crystals. [Russ.]. - In: Spektrosk. Kristallov. Leningrad, 1978, p. 149-159. - CA, 1979, 91, 11699. RZK, 1979, 7B747.

Rev.; 2.8.

1979

4.3.70. Domann, G., Kasten, A., Optical study of the cooperative Jahn-Teller distortion in dysprosium vanadate ($DyVO_4$) and dysprosium arsenate ($DyAsO_4$). - J. Magn. & Magn. Mater., 1979, 13, N1/2, 167-170. - CA, 1980, 92, 31236. PA, 1980, 83, 15240.

Exp.

4.3.71. Ferre, J., Regis, M., Farge, Y., Kleemann, W., Magneto-optical properties of the two-dimensional ferromagnet K_2CuF_4. I. Magnetic circular and linear dichroism. - J. Phys. C, 1979, 12, N13, 2671-2688. - CA, 1980, 92, 32924. PA, 1979, 82, 87419.

Exp.; E; T; O_h; d^9; 2.4; 4.4.

4.3.72. Fujita Ikuo, Temperature effects on the doublet structure of the 5.76 eV absorption band in Tl-doped KDP single crystals. - J. Phys. Soc. Jpn., 1979, 46, N6, 1889-1893. - CA, 1979, 91, 65713. PA, 1979, 82, 78713.

Exp.

4.3.73. Gulbis, A. V., Kristofel', N. N., Estimation of the light-induced refractive index change in a vibronic ferroelectric. [Russ.]. - Izv. Akad. Nauk Latv.SSR, Ser. Fiz. Tekh. Nauk, 1979, N2, 119-121. - CA, 1979, 91, 11513.

Theor.; 4.2.

4.3.74. Reinen, D., Friebel, C., Local cooperative Jahn-Teller interactions in model structures: Spectroscopic and structural evidence. - In: Structure and Bonding. Berlin, etc., 1979, Vol. 37, p. 1-60. - CA, 1979, 91, 219451.

Rev.; 4.6.

4.3.75. Reut, E. G., Manifestation of a static Jahn-Teller effect in intrinsic luminescence of crystals with scheelite structure. [Russ.]. - Izv. Akad. Nauk SSSR, Ser. Fiz., 1979, 43, N6, 1186-1193. - CA, 1979, 91, 114673. RZF, 1979, 10D807.

Exp.

4.3.76. Strauss, E., Gerhardt, V., Gebhardt, W., Exciton and exciton-magnon emission in $KMnF_3$ and $RbMnF_3$. - J. Lumin., 1979, 18-19, N1, 151-153. - CA, 1979, 90, 129901. PA, 1979, 82, 33779.

Exp.

See also: 4.2.-3, 32, 40, 56, 61, 84, 93, 99.

4.4. Magnetic Properties. Mössbauer Effect

Electronic degeneracy in one of the crystal sublattices has a great influence on magnetic properties. Even in the absence of the Jahn-Teller effect (e.g., in the case of very weak vibronic coupling), the exchange interaction between electronic degenerate states results in correlated magnetic and orbital orderings. On the other hand, in the absence of exchange interaction, the vibronic coupling and the consequent structural phase transition complicate essentially the magnetic behavior of Jahn-Teller crystals. Modification of exchange interaction in Jahn-Teller crystals was examined in detail only for the case of cubic orbital doublets, where the spin-orbital coupling vanishes [4.4.-43, 60, 62, 95, 117]. In crystals with cooperative Jahn-Teller effect (and without exchange) the vibronic coupling causes a great reduction of the orbital contribution to the magnetic properties and the occurrence of large magnetic anisotropy [4.4.-4, 15, 25, 30; 4.6.21]. In crystals where both the exchange and vibronic interaction are essential, a specific mutual quenching of distortive and magnetic orderings occurs [4.4.-59, 72; 4.1.-10, 11, 15]. This plays a role, in particular, in the above-mentioned influence of magnetic fields on the structural Jahn-Teller phase transitions and the magnetostriction effects [4.4.-22, 26; 4.4.-91, 111, 113]. The cooperative Jahn-Teller effect in low-lying excited states with spin multiplicity different from that of the ground state may essentially affect the magnetic properties, leading to the low-spin → high-spin transitions [4.4.-72, 82, 163].

1960

4.4.1. Goodenough, J. B., Direct cation-cation interactions in several oxides. - Phys. Rev., 1960, 117, N6, 1442-1451.

Exp.

1962

4.4.2. Miyahara Shohei, Jahn-Teller distortion in magnetic spinels. - J. Phys. Soc. Jpn., 1962, 17, Suppl. B-1, 181-184. - CA, 1962, 57, 14567g.

Exp.; d^9.

4.4.3. Yamadaya Tokio, Mitui Tadayasu, Okada Takuija, Shikazono Naomoto, Hamaguchi Yoshikazu, A Mössbauer study of Jahn-Teller crystal distortion in copper ferrite. - J. Phys. Soc. Jpn., 1962, 17, N12, 1897-1898. - CA, 1963, 58, 13220c.

Exp.

1963

4.4.4. Krupicka, S., Origin of the magnetic anisotropy in the $Mn_xFe_{3-x}O_4$ system. - Phys. Status Solidi, 1963, 3, N3, K118-K121. - CA, 1963, 59, 4651h.

Theor.; O_h; d^4.

1966

4.4.5. Blasse, G., Antiferromagnetism of the spinal $LiCuVO_4$. - J. Phys. & Chem. Solids, 1966, 27, N3, 612-613. - CA, 1966, 64, 16806g.

Exp.; E; O_h; d^9; 4.6.

4.4.6. Imbert, P., Study of the crystallographic and magnetic transitions in the spinel FeV_2O_4 by the Mössbauer effect. [Fr.]. - C.R. Hebd. Seances Acad. Sci. Ser. B, 1966, 263, N13, 767-770. - CA, 1967, 66, 6622.

Exp.; d^6.

4.4.7. Tanaka Midori, Tokoro Tetsuzi, Jahn-Teller effects on Mössbauer spectra of ^{57}Fe in $FeCr_2O_4$ and FeV_2O_4. - J. Phys. Soc. Jpn., 1966, 21, N2, 262-267. - CA, 1966, 64, 12039b.

Exp.; T_d; d^6.

1968

4.4.8. Bhattacharyya, B. D., Datta, S. K., Role of Jahn-Teller effect in the ligand field theory of copper hexafluorosilicate hexahydrate. - Indian J. Phys., 1968, 42, N3, 181-197. - CA, 1969, 70, 62484. PA, 1970, 73, 2151.

Exp.; E - e; O_h; d^9.

4.4.9. Iida, S., Mizushima, K., Yamada, N., Iizuka, T., Details of the electronic relaxation in $Mn_xFe_{3-x}O_{4+\gamma}$. - J. Appl. Phys., 1968, 39, N2, 818-820. - CA, 1968, 68, 100124.

Exp.; d^4.

4.4.10. Krupicka, S., Zaveta, K., Magnetic aftereffects in ferrimagnetic oxidic spinels. - J. Appl. Phys., 1968, 39, N2, 930-938. - CA, 1968, 68, 91214.

Exp.

4.4.11. Friebel, C., Reinen, D., ESR measurements on Cu^{2+} containing oxide systems. - In: Proc. 11th Int. Conf. Coord. Chem., Jerusalem, 1968. London, 1969, p. 597-598.

Exp.; E - e; O_h; d^9; 3.1.

4.4.12. Friebel, C., Reinen, D., ESR studies on the Jahn-Teller effect of the Cu^{2+} ions in oxidic perovskites. [Germ.]. - Z. Naturforsch. a, 1969, 24, N10, 1518-1525. - CA, 1970, 72, 26898. PA, 1970, 73, 17056. RZF, 1970, 5D668.

Exp.; E - e; O_h; d^9; 3.1.

1970

4.4.13. Becker, P. J., Dummer, G., Kahle, H. G., Klein, L., Muller-Vogt, G., Schopper, H. C., Investigation of the magnetic properties of $DyVO_4$ at low temperatures. - Phys. Lett. A, 1970, 31, N9, 499-500.

Exp.; D_{2d}.

4.4.14. Gerber, R., Elbinger, G., Contribution of Fe^{2+}, Mn^{3+}, and Fe^{3+} ions to the magnetic anisotropy of $Mg_xMn_{0.6}Fe_{2.4-x}O_4$. - J. Phys. C, 1970, 3, N6, 1363-1375. - CA, 1970, 73, 50102.

Exp.; d^4.

4.4.15. Goodenough, J. B., Menyuk, N., Dwight, K., Kafalas, J. A., Effects of hydrostatic pressure and of Jahn-Teller distortions on the magnetic properties of $RbFeF_3$. - Phys. Rev. B, 1970, 2, N11, 4640-4645. - CA, 1971, 74, 69470. PA, 1971, 74, 37411. RZF, 1971, 7E1166.

Exp.

4.4.16. Matsumoto Gen, $(La_{1-x}Ca_x)MnO_3$. I. Magnetic structure of $LaMnO_3$-. J. Phys. Soc. Jpn., 1970, 29, N3, 606-615. - CA, 1970, 73, 103576.

Exp.; E; O_h; d^4.

4.4.17. Novak, P., Dynamic Jahn-Teller effect and magnetic anisotropy in $Mn_xFe_{3-x}O_4$ systems. - Czech. J. Phys., 1970, 20, N3, 269-276. - CA, 1970, 72, 126707. PA, 1970, 73, 51229.

Theor.; d^4.

1971

4.4.18. Filoti, G., Gelberg, A., Gomolea, V., Rosenberg, M., Mössbauer study of the Jahn-Teller effect in spinels. - In: Proc. Conf. Application Mössbauer Effect, Tikany, 1969. Budapest, 1971, p. 593. - PA, 1971, 74, 49476.

Exp.

4.4.19. Goodenough, J. B., Menyuk, N., Dwight, K., Kafalas, J. A., Effects of hydrostatic pressure and of Jahn-Teller distortions on the magnetic properties of rubidium iron trifluoride. - J. Phys. [France], 1971, 32, Colloq. N1, Pt. 2, 621-624. - CA, 1971, 1975, 27228. CA, 1974, 80, 20807.

Exp.

4.4.20. Ham, F. S., Jahn-Teller effects and magnetism. - J. Phys. [France], 1971, 32, Colloq. N1, Pt. 2, 952-957. - CA, 1971, 75, 42229. CA, 1974, 80, 42266.

Theor.

4.4.21. Hirakawa Kinshiro, Yamada Isao, Kurogi Yukihisa, Linea chain antiferromagnet potassium copper fluoride ($KCuF_3$) and ferromagnet potassium copper fluoride (K_2CuF_4). - J. Phys. [France], 1971, 32, Colloq. N1, Pt. 2, 890-891. - CA, 1974, 80, 8379.

Exp.; C_{3v}; d^9.

4.4.22. Hudson, R. P., Mangum, B. W., Effects of applied magnetic fields on cooperative Jahn-Teller transition temperatures of dysprosium arsenate and dysprosium vanadate. - Phys. Lett. A, 1971, 36, N3, 157-158. - CA, 1972, 76, 8193.

Exp.

4.4.23. Ishikawa Yoshikazu, Syono Yashuhiko, Giant magnetostriction due to Jahn-Teller distortion in ulvo-spinel (Fe_2TiO_4). - Phys. Rev. Lett., 1971, 26, N21, 1335-1338. - CA, 1971, 75, 27282. PA, 1971, 74, 45494.

Exp.; d^6.

4.4.24. Ishikawa Yoshikazu, Syono Yashuhiko, Giant magnetostriction and magnetic anisotropy of Fe_2TiO_4. - J. Phys. Soc. Jpn., 1971, 31, N2, 461-570.

Exp.; d^6.

4.4.25. Kahle, H. G., Klein, L., Muller-Vogt, G., Schopper, H. C., Magnetic properties of $DyAsO_4$ at low temperatures. - Phys. Status Solidi b, 1971, 44, N2, 619-626.

Exp.

4.4.26. Mangum, B. W., Lee, J. N., Moos, H. W., Magnetically controllable cooperative Jahn-Teller distortion in thulium arsenate. - Phys. Rev. Lett., 1971, 27, N22, 1517-1520. - CA, 1972, 76, 39703. PA, 1972, 75, 4459.

Exp.

4.4.27. Novak, P., Roskovec, V., Simsa, Z., Brabers, V. A. M., Effect of Mn^{3+} ion orbit lattice coupling on magnetic anisotropy of $Mn_xFe_{3-x}O_4$ system. - J. Phys. [France], 1971, 32, Colloq. N1, Pt. 1, 59-61. - CA, 1971, 75, 27091.

Exp.; O_h; d^4.

4.4.28. Novak, P., Contribution of octahedrally coordinated Mn^{3+} ion to magnetic torque. - Czech. J. Phys., Sect. B, 1971, 21, N11, 1198-1212. - CA, 1972, 76, 51903.

Exp.; O_h; d^4.

4.4.29. Pincus, P., Instability of the uniform antiferromagnetic chain. - Solid State Commun., 1971, 9, N22, 1971-1973. - PA, 1972, 75, 4401.

Theor.

4.4.30. Schopper, H. C., Becker, P. J., Böhm, W., Dummer, G., Kahle, H. G., Investigation of the magnetic properties of $TbPO_4$ at low temperatures. - Phys. Status Solidi b, 1971, 46, N2, K115-K116.

Exp.; 4.7.

4.4.31. Will, G., Schäfer, W., The magnetic structure of the antiferromagnetic $DyVO_4$. - J. Phys. C, 1971, 4, N7, 811-819.

Exp.; 4.6.

4.4.32. Will, G., Schäfer, W., Scharenberg, W., Göbel, H., Magnetic structures and phase transitions in $DyVO_4$, $DyAsO_4$, and $DyPO_4$. [Germ.]. - Z. Angew. Phys., 1971, 32, N2, 122-127.

Exp.; 4.6.

4.4.33. Will, G., Scharenberg, W., Schäfer, W., Bargonth, M. O., Siedel, I., Structure analysis by thermal neutron diffraction. [Germ.]. - 1971. - 33 p. - Rep. BMBW-FBK-71-15. - CA, 1972, 77, 119344.

Rev.

4.4.34. Wright, J. C., Moos, H. W., Spectroscopic study of magnetic phenomena in $DyAsO_4$ and $DyVO_4$. - Phys. Rev. B, 1971, 4, N1, 163-166.

Exp.; 4.3.

1972

4.4.35. Bucher, E., Birgeneau, R. J., Maita, J. P., Felcher, G. P., Brun, T. O., Magnetic and structural phase transition in DySb. - Phys. Rev. Lett., 1972, 28, N12, 746-749. - RZF, 1972, 8E1319.

Exp.; 4.7.

4.4.36. Cooke, A. H., Swithenby, S. J., Wells, M. R., The properties of thulium vanadate: Example of molecular field behavior. - Solid State Commun., 1972, 10, N3, 265-268. - CA, 1972, 76, 119095. PA, 1972, 75, 22665.

Exp.; D_{4h}.

4.4.37. Cooke, A. H., Cooperative Jahn-Teller transitions in the rare earth vanadates. - In: Abstr. Conf. Low Temp. Phys., Freudenstadt. Geneva, 1972, p. 30. - PA, 1972, 75, 43926.

Exp.; D_{4h}; 4.3; 4.7.

4.4.38. Filoti, G., Gelberg, A., Gomolea, V., Rosenberg, M., Mössbauer investigation of Jahn-Teller distortion in manganites. - Int. J. Magn., 1972, 2, N2, 65-69. - CA, 1972, 77, 27064. PA, 1972, 75, 36829.

Exp.; O_h; d^o; d^7.

4.4.39. Filoti, G., Gelberg, A., Rosenberg, M., Spanu, V., Telnic, P., Mössbauer study of Jahn-Teller distortion in $Co_xMn_{2.9-x}Fe_{0.1}O_4$. - Phys. Status Solidi a, 1972, 14, N1, K91-K94. - CA, 1973, 78, 35946. PA, 1973, 76, 8492.

Exp.; O_h; d^7.

4.4.40. Gehring, G. A., Malozemoff, A. P., Staude, W., Tyte, R. N., Theory of uniaxial stress in the cooperative Jahn-Teller distortion of dysprosium vanadate. - J. Phys. & Chem. Solids, 1972, 33, N7, 1487-1498. - CA, 1972, 77, 54256. PA, 1972, 75, 51292.

Theor.

4.4.41. Harrowfield, B. V., Weber, R., Evidence of phase transitions in $K_2PbCu(NO_2)_6$: [Heat capacity and e.s.r.]- Phys. Lett. A, 1972, 38, N1, 27-28. - PA, 1972, 75, 15967.

Exp.; E - e; O_h; d^9; 4.7; 3.1.

4.4.42. Klein, L., Kahle, H. G., Schopper, H. C., Walter, H., Magnetically controllable phase transition in terbium arsenate. - Int. J. Magn., 1972, 3, N1-3, 17-21. - CA, 1972, 77, 170520.

Exp.

4.4.43. Kugel', K. I., Khomskiy, D. I., Superexchange ordering of degenerate orbitals and magnetic structure of dielectrics with Jahn-Teller ions. [Russ.]. - Pis'ma Zh. Eksp. & Teor. Fiz., 1972, 15, N10, 629-632. - CA, 1972, 77, 94741. PA, 1972, 75, 68565. RZF, 1972, 9E276.

Theor.; D_{4h}.

4.4.44. Nogues, M., Poix, P., Cooperative Jahn-Teller effect in the zinc manganite-zinc stannate system. [Fr.].- Ann. Chim. [France], 1972, 7, N5, 301-314. - CA, 1973, 78, 116252. RZF, 1973, 7E497. RZK, 1973, 14B840.

Exp.; E - e; O_h; d^4; 4.6.

4.4.45. Novak, P., Contribution of cuprous ion to magnetocrystalline anisotropy. - Int. J. Magn., 1972, 2, N4, 177-181. - RZK, 1973, 8B678.

Exp.; d^4; d^9.

4.4.46. Novak, P., Contribution of tetrahedral Ni^{2+} ion to magnetocrystalline anisotropy. - Czech. J. Phys. Sect. B, 1972, 22, N11, 1133-1154.

Theor.; T; T_d; d^8.

4.4.47. Plaskin, P. M., Stonfer, R. C., Mathew, M., Palenik, G. J., A novel antiferromagnetic oxo-bridged manganese complex. - J. Am. Chem. Soc., 1972, 94, N6, 2121-2122. - PA, 1972, 75, 44303.

Exp.; d^3; d^4.

4.4.48. Reinen, D., Friebel, C., Reetz, K. P., Cooperative Jahn-Teller effects of Cu^{2+} ions in hexanitro complexes of the type $M'_2[M''Cu(NO_2)_6]$ (M' = K, Tl, Rb, Cs; M'' = Ca, Sr, Pb, Ba) and investigations of Ni^{2+} analogs. [Germ.]. - J. Solid State Chem., 1972, 4, N1, 103-114. - CA, 1972, 76, 65761. PA, 1972, 75, 19943. RZF, 1972, 4E818.

Exp.; E - e; O_h; d^9; 4.3.

4.4.49. Spender, M. R., Morrish, A. H., A low temperature transition in $FeCr_2S_4$. - Solid State Commun., 1972, 11, N10, 1417-1421. - CA, 1973, 78, 35943. PA, 1973, 76, 5151.

Exp.; d^6.

4.4.50. Spender, M. R., Morrish, A. H., Mössbauer study of the ferrimagnetic spinel $FeCr_2S_4$. Dynamic Jahn-Teller effect. - Can. J. Phys., 1972, 50, N11, 1125-1138. - PA, 1972, 75, 57331.

Exp.; d^6.

1973

4.4.51. Belov, K. P., Goryaga, A. N., Antoshina, L. G., The magnetic ordering of tetragonally distorted copper ferrite $CuFe_2O_4$. [Russ.]. - Fiz. Tverd. Tela, 1973, 15, N10, 2895-2898. - CA, 1974, 80, 20853. PA, 1974, 77, 57342. RZF, 1974, 2E1224.

Exp.; E - e; O_h; d^9.

4.4.52. Belov, K. P., Goryaga, A. N., Antoshina, L. G., Korayem, T., The magnetic state of the Cr^{3+} ion in the $CuCr_2O_4$ compound. [Russ.]. - Fiz. Tverd. Tela, 1973, 15, N12, 3687-3688. - CA, 1974, 80, 75829. RZF, 1974, 4E1504.

Exp.; E - e; O_h; d^9.

4.4.53. Dmitrieva, N. V., Shur, Ya. S., The temperature dependence of magnetic properties of Co-substituted Ni-Zn ferrites. [Russ.]. - Fiz. Tverd. Tela, 1973, 15, N12, 3513-3519. - CA, 1974, 80, 75886. PA, 1974, 77, 73001. RZF, 1974, 4E1541.

Exp.; T_d; d^6.

4.4.54. Gehring, K. A., Cooperative Jahn-Teller effects in magnetic materials. - In: Magn. & Magn. Mater.: 18th AIP Annu. Conf., Denver, Colo., 1972. New York, 1973, Pt. 2, p. 1648-1663. - CA, 1973, 79, 24172. PA, 1973, 76, 68556. RZF, 1973, 9E1101.

Theor.; E - b_1; D_{4h}.

4.4.55. Harrowfield, B. V., Pilbrow, J. R., The Jahn-Teller effect in $K_2PbCu(NO_2)_6$. - J. Phys. C, 1973, 6, N4, 755-766, - CA, 1973, 78, 129300. PA, 1973, 76, 30687. RZF, 1973, 8D667.

Exp.; E - e; O_h; d^9.

4.4.56. Harrowfield, B. V., Dempster, A. J., Freeman, T. E., Pilbrow, J. R., Phase transitions and the Jahn-Teller effect in $Tl_2PbCu(NO_2)_6$. - J. Phys. C, 1973, 6, N12, 2058-2065. - CA, 1973, 79, 58634. PA, 1973, 76, 50136. RZF, 1973, 12E557.

Exp.; E - e; O_h; d^9.

4.4.57. Ihrig, H., Vigren, D. T., Kübler, J., Methfessel, S., Cubic-to-tetragonal transformation and susceptibility in lanthanum-silver-indium ($LaAg_xIn_{1-x}$) alloys. - Phys. Rev. B, 1973, 8, N10, 4525-4533. - CA, 1974, 80, 65032.

Exp.; E - e; O_h; d^9.

4.4.58. Kambara, T., Haas, W. J., Spedding, F. H., Good, R. H., Jr., Theory of crystal distortion and Zeeman effects in rare-earth compounds with zircon structure. - Phys. Rev. B, 1973, 7, N8, 3945-3953. - CA, 1973, 78, 153333. PA, 1973, 76, 38802.

Theor.; D_{4h}.

4.4.59. Kaplan, M. D., Vekhter, B. G., Magnetic properties of systems with a cooperative Jahn-Teller effect. [Russ.]. - Fiz. Tverd. Tela, 1973, 15, N12, 3675-3676. - CA, 1974, 80, 75830. PA, 1974, 77, 79500. RZF, 1974, 5E1286.

Theor.; D_{4h}.

4.4.60. Khomskii, D. I., Kugel, K. I., Orbital and magnetic structure of two-dimensional ferromagnets with Jahn-Teller ions. - Solid State Commun., 1973, 13, N7, 763-766. - CA, 1973, 79, 150917. PA, 1974, 77, 10273.

Theor.; E - e; O_h; d^9.

4.4.61. Klein, L., Magnetic and crystallographic properties of terbium arsenate at low temperatures. - Int. J. Magn., 1973, 5, N1-3, 231-235. - CA, 1974, 80, 53819.

Exp.

4.4.62. Kugel', K. I., Khomskiy, D. I., Crystal structure and magnetic properties of substances with orbital degeneracy. [Russ.]. -

Zh. Teor. & Eksp. Fiz., 1973, 64, N4, 1429-1439. - PA, 1974, 77, 79506.

Theor.

4.4.63. Leask, M. J. M., Maxwell, K. J., Tyte, R. N., Becker, P. J., Kasten, A., Wüchner, W., Magnetic field induced crystallographic domains in rare earth vanadates and arsenates. - Solid State Commun., 1973, 13, N6, 693-695. - PA, 1973, 76, 72495. RZF, 1974, 3E642.

Exp.; 2.4.

4.4.64. Lüthi, B., Muller, M. E., Andres, K., Bucher, E., Maita, J. P., Experimental investigation of the cooperative Jahn-Teller effect in TmCd. - Phys. Rev. B, 1973, 8, N6, 2639-2648. - PA, 1974, 77, 32649. RZF, 1974, 5E936.

Exp.; E - e; O_h; 4.2.

4.4.65. Mitra, S. N., Sengupta, P., Jahn-Teller anisotropy of paramagnetism in single crystal of $Cu(en)_3SO_4$. - Indian J. Phys., 1973, 47, N2, 79-88. - CA, 1973, 78, 152970. PA, 1973, 76, 50341. RZF, 1973, 8E1051.

Exp.; E - e; O_h; D_3; d^9.

4.4.66. Obi, Y., Saito, H., Crystal distortion and magnetic properties of some spinels containing copper. - Phys. Status Solidi a, 1973, 16, N1, K9-K12. - PA, 1973, 76, 30632.

Exp.; E - e; O_h; d^9; 4.6.

4.4.67. Reinen, D., Backes, G., Cooperative Jahn-Teller distortions in complexes $A_2^IB^{II}Co(NO_2)_6$. - In: Proc. 15th Int. Conf. Coord. Chem. Moscow, 1973, p. 9.

Exp.; O_h; d^7.

4.4.68. Schmidbauer, E., Magnetoresistance anomalies at the Curie point of some iron-chromium spinels. - Solid State Commun., 1973, 12, N6, 507-509. - CA, 1973, 78, 153019. PA, 1973, 76, 30462.

Exp.

4.4.68a. Ting, C. S., Ganguly, A. K., Zeyer, R., Birman, J. L., Influence of a strong magnetic field on the tetragonal phase of V_3Si-type compounds. - Phys. Rev. B, 1973, 8, N4, 3665-3674.

Theor.

4.4.69. Uffer, L. F., Levy, P. M., Chen, H.H., Cooperative Jahn-Teller effect in dysprosium antimonide. - In: Magn. & Magn. Mater.: 18th AIP Annu. Conf., Denver, Colo., 1972. New York, 1973, Pt. 1, p. 553-558. - CA, 1973, 79, 23734. PA, 1973, 76, 44022. RZF, 1973, 9E1138.

Exp.

4.4.70. Van Diepen, A. M., Van Stapele, R. P., Ordered local distortions in cubic $FeCr_2S_4$. - Solid State Commun., 1973, 13, N10, 1651-1653.

Exp.; E - e; T_d; d^6.

4.4.71. Vekhter, B. G., Kaplan, M. D., Structural phase transitions and orbital magnetism in crystals containing Jahn-Teller ions. [Russ.]. - Fiz. Tverd. Tela, 1973, 15, N7, 2013-2017. - CA, 1973, 79, 97932. PA, 1974, 77, 37176. RZF, 1973, 11E1288. RZK, 1973, 24B1074.

Theor.; E - b_1; D_{4h}; 4.1.

4.4.72. Vekhter, B. G., Kaplan, M. D., Orbital magnetism and structural phase transitions in crystals with Jahn-Teller ions. [Russ.]. - In: Tez. Dokl. Mezhdunar. Konf. Magn. MKM-1973. Moscow, 1973, p. 147.

Theor.; D_{4h}; 4.1.

4.4.73. Vekhter, B. G., Kaplan, M. D., Orbital magnetism and structural phase transitions in crystals with Jahn-Teller ions. - Phys. Lett. A, 1973, 43, N4, 389-390. - CA, 1973, 79, 24918. PA, 1973, 76, 34053. RZF, 1973, 8E1098.

Theor.; D_{4h}; 4.1.

4.4.74. Wood, V. E., Austin, A. E., Collings, E. W., Brog, K. C., Magnetic properties of heavy rare earth orthomanganites. - J. Phys. & Chem. Solids, 1973, 34, N5, 859-868. - CA, 1973, 78, 153021.

Exp.; d^4.

4.4.75. Wüchner, W., Laugsch, J., Observation of induced magnetism and magnetic ordering in terbium arsenate by optical spectroscopy. - Int. J. Magn., 1973, 5, N1-3, 181-185. - CA, 1974, 80, 53925. PA, 1974, 77, 6725.

Exp.

4.4.76. Zeiger, H. J., Pratt, G. W., Magnetic Interactions in Solids. - Oxford: Clarendon Press, 1973. - XV, 660 p. - RZF, 1974, 6E1313.

Pt. 4.11. The Jahn-Teller Effect in Crystals, p. 208-266.

Rev.

1974

4.4.77. Ablov, A. V., Bersuker, I. B., Vekhter, B. G., Tsukerblat, B. S., Magneto-structural phase transitions in Fe^{3+} coordination compounds. [Russ.]. - In: Fiz. Mat. Metody Koord. Khim.: Tez. Dokl. 5 Vsesoyuz. Soveshch. Kishinev, 1974, p. 151-152.

Theor.; d^5.

4.4.78. Arai, K. I., Tsuya, N., Observation of magnetostriction in Cu ferrite single crystals. - Phys. Status Solidi b, 1974, 66, N2, 547-552. - PA, 1975, 78, 18806.

Exp.; d^5; d^9.

4.4.79. Belov, K. P., Goryaga, A. N., Antoshina, L. G., Anomalous behavior of the electrical properties of tetragonal-distorted copper ferrite. [Russ.]. - Fiz. Tverd. Tela, 1974, 16, N8, 2446-2447. - PA, 1975, 78, 42252. RZF, 1974, 11E1234.

Exp.; E - e; O_h; d^9.

4.4.80. Cohen, E., Sturge, M. D., Birgeneau, R. J., Blount, E. I., Van Uiter, L. G., Kjems, J. K., Internal displacement order parameter below the 151-K phase transition in $PrAlO_3$. - Phys. Rev. Lett., 1974, 32, N5, 232-235. - PA, 1974, 77, 24992. RZF, 1974, 7E610.

Exp.

4.4.81. Cowan, W. B., Vincent, C., Walsh, D., Some Jahn-Teller effects for Cu^{2+} in LMN(^2F). - In: Tr. Mezhdunar. Konf. Magn. MKM-1973. Moscow, 1974, Vol. 3, p. 72-76. - CA, 1976, 84, 24149. RZF, 1974, 9E1410.

Exp.; E - e; O_h; d^9; 3.2.

4.4.82. Iida, S., Yamamoto, M., Umemura, S., Detailed study on the electronic transition of Fe_3O_4. - In: Magn. & Magn. Mater.: 19th AIP Annu. Conf., Boston, 1973. New York, 1974, Pt. 2, p. 913-917. - RZF, 1974, 11E639.

Exp.; E - e; T_d; d^6; 4.6.

4.4.83. Kasten, A., Becker, P. J., Spectroscopic investigation of a metamagnetic phase transition in $DyVO_4$. - J. Phys. C, 1974, 7, N17, 3120-3132. - RZF, 1975, 3E1725.

Exp.

4.4.84. Kataoka Mitsuo, Theory of the giant magnetostriction in Fe_2TiO_4. - J. Phys. Soc. Jpn., 1974, 36, N2, 456-463. - CA, 1974, 80, 88859. PA, 1974, 77, 45159.

Theor.

4.4.85. Krupicka, S., Roskovec, V., Zounova, F., Nevriva, M., Anisotropy and relaxation in $Mn_{1+x}Cr_{2-x}O_4$. - In: Tr. Mezhdunar. Konf. Magn. MKM-1973. Moscow, 1974, Vol. 1, p. 223-226. - CA, 1976, 85, 115704. RZF, 1975, 7E1393.

Exp.; O_h; d^4.

4.4.86. Mullen, M. E., Lüthi, B., Wang, P. S., Bucher, E., Longinotti, L. D., Maita, J. P., Ott, H. R., Magnetic-ion-lattice interaction: Rare-earth antimonides. - Phys. Rev. B, 1974, 10, N1, 186-199.

Exp.

4.4.87. Nogues, M., Poix, P., Contribution to the study of the ferrimagnetic properties of the system $(Mn_3O_4)_t(Mn_2SnO_4)_{1-t}$. [Fr.].- Solid State Commun., 1974, 15, N3, 463-470. - PA, 1974, 77, 70148.

Exp.

4.4.88. Ott, H. R., Andres, K., Low temperature thermal expansion and magnetostriction of TmCd. - Solid State Commun., 1974, 15, N8, 1341-1345. - CA, 1974, 81, 178994. PA, 1974, 77, 82598.

Exp.

4.4.89. Pytte, E., Magnetic field dependence of the Jahn-Teller transition in $DyVO_4$. - Phys. Rev. B, 1974, 9, N3, 932-941. - CA, 1974, 80, 138616. PA, 1974, 77, 63882. RZF, 1974, 10E646.

Theor.; 4.1.

4.4.90. Reinen, D., Friebel, C., Propach, V., High and low spin behavior of Ni(III) ions in octahedral coordination. [Germ.]. - Z. Anorg. Allg. Chem., 1974, 408, N2, 187-204. - CA, 1974, 81, 161744.

Exp.; E - e; O_h; d^7.

4.4.91. Vekhter, B. G., Kaplan, M. D., Magnetoelastic interactions in crystals with cooperative Jahn-Teller effect. [Russ.]. - Fiz. Tverd. Tela, 1974, 16, N6, 1630-1634. - CA, 1974, 81, 70198. PA, 1975, 78, 37975. RZF, 1974, 11E1391.

Theor.; E - b_1; D_4h; 4.1.

1975

4.4.92. Friebel, C., Jahn-Teller induced phase transitions in the hexanitrocuprates $A_2PbCu(NO_2)_6$ [A = potassium, rubidium, cesium]. [Germ.]. - Z. Anorg. Allg. Chem., 1975, 417, N3, 197-212. - CA, 1976, 84, 37428.

Exp.; E - e; O_h; d^9.

4.4.93. Friebel, C., Gonzalez, G. M. J., Cooperative Jahn-Teller coupling in mercury tetraisothiocyanato cuprate(II) crystals. [Germ.]. - Z. Naturforsch. b, 1975, 30, N11/12, 883-888. - CA, 1976, 84, 67467.

Exp.; E - e; O_h; d^9.

4.4.94. Friebel, C., EPR and ligand field spectroscopic studies of Cu(II) compounds. [Germ.]. - Z. Naturforsch. b, 1975, 30, N11/12, 970-972. - CA, 1976, 84, 67468.

Exp.; E - e; O_h; d^9; 3.1.

4.4.95. Inagaki Satoru, Effect of orbital degeneracy and intraatomic exchange on the occurrence of ferromagnetism. - J. Phys. Soc. Jpn., 1975, 39, N3, 596-604.

Theor.

4.4.96. Kaji Gentaro, Tasaki Akira, Siratori Kiiti, Magnetic annealing in lithium ferrite-manganite. - J. Phys. Soc. Jpn., 1975, 39, N2, 292-294. - CA, 1975, 83, 125290.

Exp.; d^4.

4.4.97. Kugel', K. I., Khomskiy, D. I., Exchange interaction in systems with triple orbital degeneracy. [Russ.]. - Fiz. Tverd. Tela, 1975, 17, N2, 454-461. - CA, 1975, 82, 175436. PA, 1976, 79, 5994. RZF, 1975, 6E75.

Theor.; T; O_h.

4.4.98. Lotgering, F. K., Van Diepen, A. M., Olijhoek, J. F., Mössbauer spectra of iron-chromium sulfospinels with varying metal ratio. - Solid State Commun., 1975, 17, N9, 1149-1153. - CA, 1976, 84, 37082.

Exp.; E - e; T_d; d^6.

4.4.99. MacPherson, J. W., Wang Yung-li, Magnetic field effects in the cooperative Jahn-Teller system $TbVO_4$. - J. Phys. & Chem. Solids, 1975, 36, N6, 493-499. - CA, 1975, 83, 52241. PA, 1975, 78, 50354. RZF, 1975, 11E116.

Theor.

4.4.100. McPherson, J. W., The structural and magnetic phase transitions in the cooperative Jahn-Teller systems: terbium vanadate(V) and terbium arsenate. - 1975. - 170 p. - [Order N76-2669.]. Diss. Abstr. Int. B, 1976, 36, N8, 4051. CA, 1976, 84, 129689.

Rev.; Theor.

4.4.101. Novak, P., Structural phase transitions controlled by a magnetic field. [Czech.]. - Cesk. Gas. Fis., 1975, 25, N6, 631. - PA, 1976, 79, 40159.

Theor.

4.4.102. Padel, L., Poix, P., Crystallographic and magnetic properties of $Ba_2(UCu)O_6$. [Fr.]. - C.R. Hebd. Seances Acad. Sci. Ser. C, 1975, 281, N1, 19-21. - CA, 1975, 83, 171158. PA, 1976, 79, 1606.

Exp.

4.4.103. Paul, R., Stinchcombe, R. B., Thermodynamic and magnetic properties of a cooperative spin-phonon coupled system. - Phys. Rev. B, 1975, 11, N1, 325-336. - CA, 1975, 82, 103961. PA, 1975, 78, 33275. RZF, 1975, 6E147.

Theor.

4.4.104. Schaak, G., Magnetic-field dependent phonon states in paramagnetic CeF_3. - Solid State Commun., 1975, 17, N4, 505-509.

Exp.; D_{3d}; f^1.

4.4.105. Schwab, G., EPR of Gd ions in Jahn-Teller-distorted rare-earth crystals $TbAsO_4$, $TmAsO_4$, and $TmVO_4$. - Phys. Status Solidi b, 1975, 68, N1, 359-367. - PA, 1975, 78, 50863.

Exp.; E - b_1; D_{4h}; 3.1.

4.4.106. Zvyagin, A. I., El'chaninova, S. D., Stetsenko, T. S., Pelikh, L. N., Khats'ko, E. N., Low-temperature structural phase transition in cesium dysprosium molybdate. [Russ.]. - Fiz. Nizk. Temp., 1975, 1, N1, 79-82. - RZF, 1975, 5E748.

Exp.

1976

4.4.107. Andres, K., Wang, P. S., Wong, Y. H., Lüthi, B., Ott, H. R., Jahn-Teller transition in $PrCu_2$. - In: Magn. & Magn. Mater.: 1st Joint MMM-Intermag. Conf., Pittsburgh, 1976. AIP Conf. Proc., 1976, N34, p. 222-223. - CA, 1977, 86, 82677. PA, 1977, 80, 20276.

Exp.; 4.1; 4.2.

4.4.108. Becker, P. J., Magnetostriction and linear magnetic birefringence in the Jahn-Teller coupled system thulium arsenate with effective spin S = I. [Germ.]. - Phys. Status Solidi b, 1976, 74, N1, 285-295. - CA, 1976, 84, 171612. PA, 1976, 79, 45566. RZK, 1976, 18B608.

Exp.; 4.3.

4.4.109. Bhattacharyya, B. D., Jahn-Teller effect in iridium-(4+)-doped nickel iron oxide ($NiFe_2O_4$) crystals. - In: Abstr. 2nd Int. Conf. Ferrites. Bellevue, 1976, p. 17-18. - (Europhys. Conf. Abstr., 1976; Vol. 2B). - CA, 1977, 86, 114526.

Exp.; O_h.

4.4.110. Buisson, G., Tchéou, F., Sayetat, F., Scheunemann, K., Crystallographic and magnetic properties of $TbCrO_4$ at low temperature (X-ray and neutron experiments). - Solid State Commun., 1976, 18, N7, 871-875. - CA, 1976, 84, 187755.

Exp.

4.4.111. Cooke, A. H., Davidson, M. M., England, N. J., Leask, M. J. M., Lowry, J. B., Tropper, A. C., Wells, M. R., The magnetic, spectroscopic, and thermal properties of $KDyMo_2O_8$. - J. Phys. C, 1976, 9, N20, L573-L578. - CA, 1977, 86, 99931.

Exp.; 4.3; 4.7.

4.4.112. Cyrot, M., Lacroix, L.-C. C., Orbital degeneracy and magnetic and crystallographic structure. [Fr.]. - J. Phys. [France], 1976, 37, Colloq. N10, 183-188. - RZF, 1977, 4E26.

Theor.; d^9.

4.4.113. Faber, J., Jr., Lander, G. H., Cooper, B. R., Magnetoelastic interactions in UO_2. - In: Magn. & Magn. Mater.: 21st AIP Annu. Conf., Philadelphia, 1975. New York, 1976, p. 379-381. - PA, 1976, 79, 87617. RZF, 1977, 6E1606.

Theor.

4.4.114. Friebel, C., Propach, V., Reinen, D., Copper(2+) ions in compressed coordination octahedrons - spectroscopic studies of the Jahn-Teller effect of copper(2+) in barium copper zinc fluoride ($Ba_2Cu_xZn_{1-x}F_6$) solid solutions. [Germ.]. - Z. Naturforsch. b, 1976, 31, N12, 1574-1584. - CA, 1977, 86, 113331.

Exp.; d^9.

4.4.115. Gehring, G. A., Kahle, H. G., Nägele, W., Simon, A., Wüchner, W., Two-level singlet magnetism in $TbAsO_4$ and $TbVO_4$. - Phys. Status Solidi b, 1976, 74, N1, 297-309.

Exp.; 4.3; 4.7.

4.4.116. Ihrig, H., Methfessel, S., Influence of the martensitic structure transformation in $LaAg_xIn_{1-x}$ on the magnetic properties of substituted Ce. - Z. Phys. B, 1976, 24, N4, 385-389. - CA, 1976, 85, 115818. PA, 1976, 79, 79995.

Exp.

4.4.117. Kugel', K. I., Khomskiy, D. I., A double Heisenberg model in a magnetic field and the metamagnetism of Jahn-Teller systems. [Russ.]. - Pis'ma, Zh. Eksp. & Teor. Fiz., 1976, 23, N5, 264-267. - PA, 1976, 79, 53836.

Theor.

4.4.118. Lee, J. D., Schroeer, D., Mössbauer study of $Fe_{1-x}V_{2-x}O_4$ spinels for the determination of cation distributions and magnetic structure. - J. Phys. & Chem. Solids, 1976, 37, N8, 739-746. - CA, 1976, 85, 201199. PA, 1976, 79, 72923.

Exp.; d^5; d^6.

4.4.119. McPherson, J. W., Wang Yung-li, Induced moment magnetism in the cooperative Jahn-Teller systems terbium arsenate ($TbAsO_4$) and terbium vanadate ($TbVO_4$). - J. Phys. & Chem. Solids, 1976, 37, N2, 143-150. - CA, 1976, 84, 83350. PA, 1976, 79, 18445.

Exp.

4.4.119a. Murase, K., Sugai, S., Takaoka, S., Katayama, S., Study of the phase transition of IV-VI compound alloy semiconductors. - In: Phys. Semiconduct.: Proc. 13th Int. Conf. Rome. Amsterdam, 1976, p. 305-308. - CA, 1977, 87, 60945.

4.4.120. Stevens, K. W. H., Exchange interactions in magnetic insulators. - Phys. Rep., 1976, 24, N1, 1-75. - RZF, 1976, 8E1479.

4.4.121. Van Diepen, A. M., Lotgering, F. K., Olijhoek, J. F., Effect of small variations of the Fe/Cr ratio on the ^{57}Fe Mössbauer spectra of $FeCr_2S_4$. - J. Magn. & Magn. Mater., 1976, 3, N1/2, 117-119. - CA, 1976, 85, 133637. PA, 1976, 79, 80198.

Exp.; E - e; T_d; d^6.

4.4.122. Vekhter, B. G., Zenchenko, V. P., Bersuker, I. B., The influence of magnetic fields on the phase transition in vibronic ferroelectrics. [Russ.]. - Fiz. Tverd. Tela, 1976, 18, 2325-2330. - RZF, 1977, 1E1304.

Theor.; 4.2.

1977

4.4.123. Bowden, G. J., Day, R. K., On the magnetic anomly in erbium-iron(1:3) compound at 47 K. - J. Phys. F, 1977, 7, N1, 191-197. - CA, 1977, 86, 198925.

Exp.

4.4.124. Folinsbee, J. T., Tapster, P. R., Taylor, D. R., Mroczkowski, S., Wolf, W. P., Jahn-Teller antiferroelectric ordering in praseodymium hydroxide. - Solid State Commun., 1977, 24, N8, 499-500. - CA, 1978, 88, 30953. PA, 1978, 81, 10859.

Exp.; 4.2.

4.4.125. Harley, R. T., Perry, C. H., Richter, W., Magnetic field dependence of the cooperative Jahn-Teller distortion in dysprosium vanadate. - J. Phys. C, 1977, 10, N8, L187-L189. - CA, 1977, 87, 108767. PA, 1977, 80, 4777.

Exp.

4.4.126. Khomskii, D. I., Kugel, K. I., Degenerate Hubbard model in a magnetic field: Application to Jahn-Teller system. - Phys. Status Solidi b, 1977, 79, N2, 441-450. - PA, 1977, 80, 28358.

Theor.

4.4.127. Komarov, A. G., Korovin, L. I., Kudinov, E. K., Spectrum of excitations in Jahn-Teller crystals in the presence of biquadratic exchange. [Russ.]. - In: XXX Gertsenov. Chteniya: Teor. Fiz. i Astron. Leningrad, 1977, p. 17-21. - CA, 1978, 89, 12358. RZF, 1977, 12E20.

Theor.; 4.1.

4.4.128. Mehran, F., Stevens, K. W. H., Plaskett, T. S., Interplay of dipolar and random strain effects in the cooperative Jahn-Teller system $TmAsO_4$. - Solid State Commun., 1977, 22, N2, 143-145. - CA, 1977, 86, 197369. PA, 1977, 80, 40124.

Exp.

4.4.129. Ott, H. R., Andres, K., Wang, P. S., Wong, Y. H., Lüthi, B., Induced Jahn-Teller transition in praseodymium-copper ($PrCu_2$). - In: Cryst. Field Effects Met. & Alloys: Proc. 2nd Int. Conf. Zurich, 1976. New York; London, 1977, p. 84-88. - CA, 1977, 87, 175804.

Exp.; d^9.

4.4.130. Reinen, D., Propach, V., Friebel, C., Krause, S., Cu^{2+}-ions in tetragonally compressed octahedral coordination in equal ligands. - In: Proc. 17th Int. Conf. Coord. Chem. Hamburg, 1976. Oxford, etc., 1977, p. 282.

Exp.; E - e; O_h; d^9; 3.1.

4.4.131. Reinen, D., Steffens, F., Structure and bonding in transition metal compounds $M^{II}Me^{IV}F_6$. - In: Proc. 18th Int. Conf. Coord. Chem. Sao Paolo, 1977, p. 219.

Exp.; T - t_2; O_h; d^6.

4.4.132. Sams, J. R., Thompson, R. C., Tsin, T. B., Orbital ground state, crystal field splitting, and magnetic hyperfine interactions in iron(II) fluorosulphate. - Can. J. Chem., 1977, 55, N1, 115-121. - PA, 1977, 80, 48096.

Exp.; E - e; O_h; d^6.

4.4.133. Stalinski, B., Nowak, B., Magnetic properties of the dihydride phases of titanium-niobium and titanium-tantalum alloys. - Bull. Acad. Pol. Sci. Ser. Sci. Chim., 1977, 25, N1, 65-72. - CA, 1977, 86, 199026.

Exp.

4.4.134. Vekhter, B. G., Kaplan, M. D., The magnetic properties of crystals with the cooperative Jahn-Teller effect. [Russ.]. - In: Fiz. Mat. Metody Koord. Khim.: Tez. Dokl. 6 Vsesoyuz. Soveshch. Kishinev, 1977, p. 117-118.

Theor.; E - b_1; D_{4h}.

4.4.135. Wiegers, G. A., Van Bruggen, C. F., Magnetic and structural phase transitions in sodium intercalates Na_xVS_2 and Na_xVSe_2. - Physica B + C, 1977, 86-88, N2, 1009-1011. - CA, 1977, 86, 181980. PA, 1977, 80, 69558.

Exp.

4.4.136. Zelentsov, V. V., Shipilov, V. I., Magnetic phase transitions with alteration of the spin states in d-element complex compounds. [Russ.]. - In: Fiz. Mat. metody Koord. Khim.: Tez. Dokl. 6 Vsesoyuz. Soveshch. Kishinev, 1977, p. 5-6.

Theor.; d^5; d^6.

4.4.137. Zenchenko, V. P., Vekhter, B. G., Bersuker, I. B., The vibronic theory of the phase transition in magnetic fields: wide-gap ferroelectrics. [Russ.]. - In: Abstr. 4th Int. Meet. Ferroelectricity IMF-4. Leningrad, 1977, p. 109.

Theor.; 4.2.

4.4.138. Ziatdinov, A. M., Zaripov, M. M., Yablokov, Yu. V., Davidovich, R. L., Dynamics of the octahedral Jahn-Teller complexes $[Cu(H_2O)_6]^{2+}$ in crystals of the $ABF_6 \cdot 6H_2O$ type. [Russ.]. - In: Fiz. Mat. Metody Koord. Khim.: Tez. Dokl. 6 Vsesoyuz. Soveshch. Kishinev, 1977, p. 58-59.

Exp.; E - e; O_h; d^9; 3.1.

4.4.139. Zvezdin, A. K., Mukhin, A. A., Popov, A. I., Energy level crossing and instability of the magnetic structure in rare earth ferrite garnets. [Russ.]. - Zh. Eksp. & Teor. Fiz., 1977, 72, N3, 1097-1109.

Theor.; 3.1.

1978

4.4.140. Berger, M., Harley, R. T., Perry, C. H., Richter, W., Magnetic field dependence of the cooperative Jahn-Teller distortion in $DyVO_4$ and $TbVO_4$. - In: Proc. Int. Conf. Lattice Dynamics, 1977. Paris, 1978, p. 724-725. - CA, 1979, 90, 95774. PA, 1978, 81, 79733.

Exp.; 2.8; 4.3.

4.4.141. Dionne, G. F., Local site distortion model of magnetostriction. - 1978. - 28 p. - Rep. TR-532. ESD-TR-78-292. Order No. AD-A066250. - CA, 1979, 91, 167535.

Theor.; T - e; T - t; d^4; d^6.

4.4.142. Borukhovich, A. S., Zubkov, V. G., Bazuev, G. V., Phase transitions and Jahn-Teller distortions in rare earth orthovanadates. [Russ.]. - Fiz. Tverd. Tela, 1978, 20, N6, 1816-1821. - CA, 1978, 89, 95956. PA, 1979, 82, 28548. RZF, 1978, 9E687.

Exp.

4.4.143. Cooper, B. R., Search criterion for lattice internal rearrangement modes: Correlation between lattice structural and magnetic structural transitions. - Phys. Rev. B, 1978, 17, N1, 293-296.

Theor.

4.4.144. Eremin, M. V., Kalinenkov, V. N., The magnetic structure and the cooperative ordering of orbitals in $KCuF_3$ and $KCrF_3$. [Russ.]. - Fiz. Tverd. Tela, 1978, 20, N12, 3546-3552. PA, 1979, 82, 99687. RZF, 1979, 4E1443.

Theor.; d^4; d^9.

4.4.144a. Kawamura, H., Murase, K., Sugai, S., Takaoka, S., Nishikawa, S., Nishi, S., Katayama, S., Effect of electron-phonon interaction on ferroelectric transition in IV-VI compounds. - In: Proc. Int. Conf. Lattice Dynamics, 1977. Paris, 1978, 658-661. - CA, 1979, 90, 79724.

4.4.145. Kjems, J. K., Ott, H. R., Shapiro, S. M., Andres, K., Magnetic excitations and the cooperative Jahn-Teller transition in $PrCu_2$. - J. Phys. [France], 1978, 39, Colloq. N6, Pt. 2, 1010-1012. - CA, 1979, 90, 65724. PA, 1979, 82, 65192.

Exp.

4.4.146. Kovtun, N. M., Prokopenko, V. K., Prokhorenko, Yu. I., Shemyakov, A. A., Investigation of normal spinels $FeCr_2S_4$ by the method of NMR at low temperatures. [Russ.]. - Fiz. Nizk. Temp., 1978, 4, N11, 1477-1479. - PA, 1979, 82, 19785. RZF, 1979, 4D671.

Exp.; E - e; T_d; d^6; 3.3.

4.4.147. Kugel', K. I., Khomskiy, D. I., Two-spin systems in a magnetic field: Semiclassical approach. [Russ.]. - Fiz. Tverd. Tela, 1978, 20, 2660-2665. - CA, 1979, 90, 15481.

Theor.

4.4.148. Morin, P., Rouchy, J., Schmitt, D., Cooperative Jahn-Teller effect in thulium-zinc (TmZn). - Phys. Rev. B, 1978, 17, N9, 3684-3694. - CA, 1978, 89, 69780. PA, 1978, 81, 67998.

Exp.

4.4.149. Mulak, J., Zolnierek, Z., The cooperative Jahn-Teller effect in uranium(4+) hydroxysulfate. - Solid State Commun., 1978, 26, N4, 275-277. - CA, 1978, 89, 98774. PA, 1978, 81, 72121.

Exp.

4.4.150. Pebler, J., Reinen, D., Schmidt, K., Steffens, F., Study of the phase transition of iron hexafluorozirconate with Mössbauer spectroscopy. [Germ.]. - J. Solid State Chem., 1978, 25, N2, 107-114. - CA, 1978, 89, 82589.

Exp.

4.4.151. Picard, J., Hubsch, J., Le Gall, H., Guillot, M., Anisotropic behavior of the magnetization of dysprosium vanadate(V) in the basal plane under Jahn-Teller distortion. - J. Appl. Phys., 1978, 49, N3, 1386-1388. - CA, 1978, 88, 202069. PA, 1978, 81, 68222.

Exp.

4.4.152. Reinen, D., Köhler, P., Massa, W., Cooperative Jahn-Teller ordering and magnetic structure in Cr(II) and Mn(III)-compounds. - In: Proc. 19th Int. Conf. Coord. Chem. Prague, 1978, p. 153-155.

Exp.; d^4.

4.4.153. Van Bruggen, C. F., Bloembergen, J. R., Bos-Alberink, A. J. A., Wiegers, G. A., Magnetic and electrical properties related to structural transitions in Na_xVS_2 and Na_xVSe_2. - J. Less-Common Met., 1978, 60, N2, 259-282. - CA, 1978, 89, 18991. PA, 1978, 81, 91848.

Exp.; 4.2.

4.4.154. Yablokov, Yu. V., Radiospectroscopy of doped magnetic centers and peculiarities of their structure in crystals of coordination substances. - J. Mol. Struct., 1978, 46, 285-298. - RZF, 1978, 10D634.

Exp.; E - e; O_h; d^9; 3.1.

4.4.155. Zolnierek, Z., The cooperative Jahn-Teller effect and paramagnetic susceptibility of S = I systems. - Solid State Commun., 1978, 26, N5, 307-310. - CA, 1978, 89, 83899. PA, 1978, 81, 68134.

Theor.

1979

4.4.156. Abdulsabirov, R. Yu., Konov, I. S., Korableva, S. L., Lukin, S. N., Tagirov, M. S., Teplov, M. A., Magnetic resonance in the Van Vleck paramagnets $Tm(C_2H_5SO_4)_3 \cdot 9H_2O$ and $LiTmF_4$ under uniaxial and uniform compression. [Russ.]. - Zh. Eksp. & Teor. Phys., 1979, 76, N3, 1023-1027. - CA, 1979, 90, 178121. PA, 1979, 82, 61243.

Exp.; 3.3.

4.4.157. Ammeter, J., Buergi, H. B., Gamp, E., Meyer-Sandrin, V., Jensen, W. P., Static and dynamic Jahn-Teller distortions in CuN_6 complexes. Crystal structures and EPR spectra of complexes between copper(II) and rigid, tridentate cis,cis-1,3,5-triaminocyclohexane (tach): $Cu(tach)_2(ClO_4)_2$, $Cu(tach)_2(NO_3)_2$; Crystal structure of $Ni(tach)_2(NO_3)_2$. - Inorg. Chem., 1979, 18, N3, 733-750. - CA, 1979, 90, 131000.

Exp.; d^8; d^9; 3.1.

4.4.158. Belov, K. P., Zvezdin, A. K., Mukhin, A. A., The magnetic phase transition in the terbium orthoferrites. [Russ.]. - Zh. Eksp. & Teor. Phys., 1979, 76, N3, 1100-1110. - PA, 1979, 82, 61105.

Exp.

4.4.159. Belov, K. P., Zvezdin, A. K., Kadomtseva, A. M., Krynetskiy, I. B., Mukhin, A. A., Metamagnetic phase transitions and instability of the magnetic structure in the rare earth orthoferrites. [Russ.]. - Zh. Eksp. & Teor. Phys., 1979, 76, N4, 1421-1430. - CA, 1979, 90, 214343. PA, 1979, 82, 61107.

Exp.

4.4.160. Blaise, A., Lagnier, R., Mulak, J., Zolnierek, Z., Magnetic susceptibility and heat capacity anomalies of $[U(OH)_2SO_4]$ at 21 K. - J. Phys. [France], 1979, 40, Colloq. N4, 176-178. - CA, 1979, 90, 214262. PA, 1980, 83, 11151.

Exp.; 4.7.

4.4.161. Dionne, G. F., Origin of the magnetostriction effects from Mn^{3+}, Co^{2+}, and Fe^{2+} ions in ferrimagnetic spinels and garnets. - J. Appl. Phys., 1979, 50, N6, 4263-4272. - CA, 1979, 91, 101161. PA, 1979, 82, 91277.

Theor.; T - e; T - t; d^4; d^6.

4.4.162. Gerard, A., Grandjean, F., Charge transfer in chromium copper oxide ($CuCr_2O_4$): a Mössbauer study of iron-57 below the tetragonal-cubic transition. - J. Phys. C, 1979, 12, N21, 4601-4610. - CA, 1980, 92, 119087. PA, 1980, 83, 15836.

Exp.; d^5; d^6; d^9.

4.4.163. Kambara Takeshi, Theory of high-spin $\rightleftarrows$ low-spin transitions in transition metal compounds induced by the Jahn-Teller effect. - J. Chem. Phys., 1979, 70, N9, 4199-4206. - CA, 1979, 91, 29739. PA, 1979, 82, 69676.

Theor.

4.4.164. Lines, M. E., Elastic properties of magnetic materials. - Phys. Rep., 1979, 55, N2, 133-181. - PA, 1980, 83, 7142.

Rev.

4.4.164a. Litvinov, V. I., Volkov, V. L., Kovstyuk, K. D., The quantizing magnetic field effect on the ferroelectric phase transition in narrow-gap semiconductors. - Ferroelectrics, 1979, 22, 839-845.

Theor.

4.4.165. Morin, P., Rouchy, J., Schmitt, D., du Tremolet de Lacheisserie, E., Quadrupole interactions in cubic rare earth intermetallics. - J. Phys. [France], 1979, 40, Colloq. N5, 101-106. - PA, 1980, 83, 15489.

Theor.; 4.1.

4.4.166. Petrashen, V. E., Yablokov, Yu. V., Davidovich, R. L., The influence of the parameters of lattice structure on the configuration of Jahn-Teller centers. - In: Magn. Resonance and Related Phenomena: Proc. 20th Congr. AMPERE, Tallin, 1978. Berlin, etc., 1979, p. 270.

Exp.; E - e; O_h; d^9; 3.1.

4.4.166a. Takaoka, S., Murase, K., Anomalous resistivity near the ferroelectric phase transition in (Pb,Ge,Sn)Te alloy semiconductors. - Phys. Rev. B, 1979, 20, N7, 2823-2833.

Exp.; 4.2.

4.4.167. Zaripov, M. M., Ziatdinov, A. M., Yablokov, Yu. V., The nature and dynamics of distortions of Jahn-Teller centers under low symmetry perturbations. - In: Magn. Resonance and Related Phenomena: Proc. 20th Congr. AMPERE Tallin, 1978. Berlin, etc., 1979, p. 268.

Exp.; E - e; O_h; d^9; 3.1.

See also: 4.1.-25, 32, 40, 44, 60, 67, 72, 92, 102; 4.2.-12, 30, 60, 74, 79, 82, 95, 96, 97, 100, 103; 4.3.-19, 20, 21, 24, 25, 31, 35, 76; 4.5.-13, 25, 26, 28, 45; 4.6.-51, 70, 78, 141, 163, 184.

4.5. Band Jahn-Teller Effect and Peierls Transitions

In the majority of the above-mentioned investigations, systems where the widths of the electron bands are small and/or where the existence of bands (rather than the localized electronic states) is not important were examined. In this subsection a number of papers that are devoted to the study of the band Jahn-Teller effect are complied and systematized. Here the main attention is paid to the manifestation of electron-phonon interaction in the case of wide unfilled electronic bands. A special interest in this problem concerns attempts to develop a deeper understanding of superconductivity and, in particular, of the correlation between superconducting and structural transitions in A-15 compounds. The fundamentals of this approach have been formulated by Peierls [4.5.1], their further development being given in Refs. [4.5.-7, 8, 19, 16, 32, 37, 46] (see especially the review by Bulaevskii [4.5.26a]).

1930

4.5.1. Peierls, R. E., Theory of the electric and thermal conductivity of metals. [Germ.]. - Ann. Phys. [Germany], 1930, 4, N2, 121-148. - CA, 1930, 24, 2350.

Theor.

1962

4.5.2. Nesbet, R. K., Theory of superconductivity. I. Electron-lattice interaction. - Phys. Rev., 1962, 126, N6, 2014-2020. - RZF, 1963, 3E209.

Theor.

4.5.3. Nesbet, R. K., Theory of superconductivity. II. Properties of the modified electron lattice states of low energy. - Phys. Rev., 1962, 128, N1, 139-150. - RZF, 1963, 6E664.

Theor.

1964

4.5.4. Batterman, B. W., Barrett, C. S., Crystal structure of superconducting V_3Si. - Phys. Rev. Lett., 1964, 13, N13, 390-392. - CA, 1965, 62, 89b.

Exp.

4.5.5. Kristofel', N. N., On the configruation instability and its possible manifestation in the electron states of perfect crystals. [Russ.]. - Fiz. Tverd. Tela, 1964, 6, N11, 3266-3271.

Theor.

4.5.6. Kristofel', N. N., Possibility of the Jahn-Teller effect for band states of crystals. [Russ.]. - In: Tr. Inst. Fiz. Astron. Akad. Nauk Est.SSR, 1964, Vol. 27, 85-98. - CA, 1965, 63, 2471c.

1966

4.5.7. Labbe, J., Friedel, J., Electronic instability and change of crystalline phase in compounds of the V_3Si type at low temperature. [Fr.]. - J. Phys. [France], 1966, 27, N3/4, 153-165. - CA, 1966, 65, 12970c.

Theor.

4.5.8. Labbe, J., Friedel, J., Effect of temperature on the electronic instability and the crystalline phase change at low temperature of V_3Si-type compounds. [Fr.]. - J. Phys. [France], 1966, 27, N5/6, 303-308. - CA, 1967, 66, 32568.

Theor.

4.5.9. Labbe, J., Friedel, J., Stability of periodic distortion modes of a linear chain of transition atoms in a crystalline V_3Si-type structure. [Fr.]. - J. Phys. [France], 1966, 27, N11/12, 708-716. - CA, 1967, 67, 68597.

Theor.

4.5.10. Pratt, G. W., Jr., The materials game: An APW [augmented-plane wave] analysis of the lead salts. - In: Quantum Theory of Atoms, Molecules and Solid State. New York; London, 1966, p. 429-438. - CA, 1968, 68, 54447.

Theor.

4.5.11. Suzuki Haruhiko, Miyahara Syohei, The crystal distortion of vanadium. - J. Phys. Soc. Jpn., 1966, 21, N12, 2735-2736. - CA, 1967, 66, 119573.

Exp.

1967

4.5.12. Barisic, S., Labbe, J., The elastic constants of V_3Si-type compounds in the cubic phase. - J. Phys. & Chem. Solids, 1967, 28, N12, 2477-2486. - CA, 1968, 68, 62178.

Theor.

4.5.13. Labbe, J., Paramagnetic susceptibility in the V_3Si-type of compounds in the normal state. - Phys. Rev., 1967, 158, N3, 647-654. - CA, 1967, 67, 58495.

Theor.

4.5.14. Levanyuk, A. P., Suris, R. A., On some properties of superconductive compounds of the V_3Si type. [Russ.]. - Usp. Fiz. Nauk, 1967, 91, N1, 113-120. - CA, 1967, 66, 99509.

Rev.

4.5.15. Suzuki Haruhiko, Miyahara Syohei, Lattice distortion of vanadium metal by Jahn-Teller-like energy splitting. - J. Phys. Soc. Jpn., 1967, 23, N1, 135-136. - CA, 1968, 68, 24871.

Theor.

1968

4.5.16. Labbe, J., Relation between superconductivity and lattice instability in the β-tungsten compounds. - Phys. Rev., 1968, 172, N2, 451-455. - CA, 1968, 68, 55429.

Theor.

4.5.17. Labbe, J., Electron instability of the Jahn-Teller type for a metal with a three-dimensional narrow band structure: Case of periodic distortion modes. [Fr.]. - J. Phys. [France], 1968, 29, N2/3, 195-200. - CA, 1968, 68, 108048.

Theor.

1969

4.5.18. Cody, G. D., Cohen, R. W., Vieland, L. J., Characteristic parameters of β-W superconductors. - In: Proc. 11th Int. Conf. Low Temp. Phys., 1968. St. Andrews, 1969, Vol. 2, p. 1009-1016, 1049-1050. - CA, 1970, 73, 114272.

Exp.

1970

4.5.19. Pytte, E., Theory of the structural transition in Nb_3Sn and V_3Si. - Phys. Rev. Lett., 1970, 25, N17, 1176-1180.

Theor.

4.5.20. Weger, M., Some considerations regarding the application of a linear chain model to some compounds of the β-W structure. - J. Phys. & Chem. Solids, 1970, 31, N7, 1621-1639.

Theor.

1971

4.5.21. Pytte, E., Theory of phase transitions in the β-tungsten structure induced by the band Jahn-Teller effect. - Phys. Rev. B, 1971, 4, N4, 1094-1100. - CA, 1971, 75, 81194. PA, 1971, 74, 58741. RZF, 1972, 2E580.

Theor.

4.5.22. Shirane, G., Axe, J. D., Acoustic phonon instability and critical scattering in Nb_3Sb. - Phys. Rev. Lett., 1971, 27, N27, 1803-1806.

Exp.

4.5.23. Vieland, L. J., Cohen, R. W., Rehwald, W., Evidence for first-order structural transformation in Nb_3Sn. - Phys. Rev. Lett., 1971, 26, N7, 373-376.

Exp.; 4.6.

1973

4.5.24. Testardi, L. R., Elastic behavior and structural instability of high-temperature A-15 structure superconductors. - In: Phys. Acoustics. New York, 1973, Vol. 10, p. 193-296.

Rev.

4.2.25. Ting, C. S., Ganguly, A. K., Birman, J. L., Prediction of magnetic-field-induced structural phase transformation in V_3Si. - Phys. Rev. Lett., 1973, 30, N25, 1245-1248. - CA, 1973, 79, 47163.

Theor.; 4.4.

1974

4.5.26. Williamson, S. J., Ting, C. S., Fung, H. K., Influence of electronic lifetime on the lattice instability of V_3Si. - Phys. Rev. Lett., 1974, 32, N1, 9-12. - PA, 1974, 77, 13248.

Exp.; 4.4.

<u>1975</u>

4.5.26a. Bulaevskiy, L. N., Structural (Peierls) transition in quasi-one-dimensional crystals. [Russ.]. - Usp. Fiz. Nauk, 1975, <u>115</u>, N2, 263-300.

4.5.27. Hasegawa, A., Bremicker, B., Kübler, J., Energy bands of lanthanum-silver (LaAg) and lanthanum-cadmium (LaCd) and martensitic transformation. - Z. Phys. B, 1975, <u>22</u>, N3, 231-236. - CA, 1976, <u>84</u>, 24635. PA, 1976, <u>79</u>, 13917.

Theor.

4.5.28. Jardin, J. P., Labbe, J., Model for the electronic structure of manganese perovskite compounds. [Fr.]. - J. Phys. [France], 1975, <u>36</u>, N12, 1317-1326. - CA, 1976, <u>84</u>, 52366. PA, 1976, <u>79</u>, 35773.

Theor.

<u>1976</u>

4.5.29. Bhatt, R. N., McMillan, W. L., The Landau theory of the martensitic transition in A-15 compounds. - Phys. Rev. B, 1976, <u>14</u>, N3, 1007-1027. - CA, 1976, <u>85</u>, 152063.

Theor.

4.5.30. Egri, I., Theory of the one-dimensional Peierls-Hubbard model. - Z. Phys. B, 1976, <u>23</u>, N4, 381-387. - RZF, 1976, 9E144.

Theor.

4.5.31. Fukase, T., Tachiki, M., Toyota, N., Muto, Y., Anomalies in the longitudinal ultrasonic attenuation and the velocity variation in the mixed state of vanadium-silicon (V_3Si) single crystal. - Solid State Commun., 1976, <u>18</u>, N4, 505-508. - CA, 1976, <u>84</u>, 155854.

Exp.; 4.1.

4.5.32. Gor'kov, L. P., Dorokhov, O. N., On the theory of the structural properties of A-15 type materials. - J. Low Temp. Phys., 1976, <u>22</u>, N1/2, 1-26.

Theor.

4.5.33. Ihrig, H., Methfessel, S., Martensitic transformation in $CeAg_xIn_{1-x}$ and other lanthanide compounds. - Z. Phys. B, 1976, <u>24</u>, N4, 381-383. - PA, 1976, <u>79</u>, 79552.

Theor.

4.5.34. Ono Yoshiyuki, Peierls transition in non-half-filled tight-binding model. I. Phase diagram. - J. Phys. Soc. Jpn., 1976, 41, N3, 817-823.

Theor.

4.5.35. Sokoloff, J. B., Charge ordering in Fe_3O_4. - In: Magn. & Magn. Mater.: 21st AIP Annu. Conf., Philadelphia, 1975. New York, 1976, p. 381. - PA, 1976, 79, 87254. RZF, 1977, 7E1408.

Theor.

1977

4.5.36. Bhatt, R. N., Microscopic theory of the martensitic transition in A-15 compounds based on a three-dimensional band structure. - Phys. Rev. B, 1977, 16, N5, 1955-1932. - PA, 1978, 81, 19752.

Theor.

4.5.37. McMillan, W. L., Superconductivity and martensitic transformations in A-15 compounds. - In: Electron-Phonon Interactions and Phase Transitions. New York; London, 1977, p. 194-199. (NATO Adv. Study Inst. Ser., Ser. B, 1977; N29).

Theor.

1978

4.5.38. Bhatt, R. N., Structural transition in A-15 compounds: Possible Landau theory descriptions. - Phys. Rev. B, 1978, 17, N7, 2947-2955.

Theor.

4.5.39. Ghatak, S. K., Khanra, B. C., Ray, D. K., Effect of superconductivity on the cubic to tetragonal structural transition due to a two-fold degenerate electronic band. - Solid State Commun., 1978, 27, N8, 767-770.

Theor.

4.5.40. Kragler, R., Behavior of the elastic moduli of Nb_3Sn in the low-temperature phase. - Physica B + C, 1978, 93, N3, 314-326. - PA, 1978, 81, 39767.

Theor.; 4.1.

1979

4.5.41. Gupta, M., Electronically driven tetragonal distortion in TiH_2. - Solid State Commun., 1979, 29, N1, 47-51. - PA, 1979, 82, 37986.

Exp.

4.5.42. Haddon, R. C., Unified theory of resonance energies, ring currents, and aromatic character in the $(4n + 2)\pi$-electron annulenes. - J. Am. Chem. Soc., 1979, 101, N7, 1722-1728. - CA, 1979, 90, 203308.

Theor.

4.5.43. Kasai Hideaki, Miyazima Sasuke, Okiji Ayao, The influence of intra-atomic Coulomb interaction on band type of Jahn-Teller effect in V_3Si. - J. Phys. Soc. Jpn., 1979, 46, N6, 1955-1956. - CA, 1979, 91, 66535. PA, 1979, 82, 78057.

Theor.

4.5.44. Kragler, R., Multiple band electron-phonon transport theory in A-15 compounds. - In: Proc. Int. Conf. Quasi One-Dimensional Conductors. Dubrovnik, 1978. Berlin, 1979, Pt. 2, p. 234-237. - PA, 1980, 83, 24902.

Theor.

4.5.45. Lüthi, B., Sommer, R., Structural instabilities in rare earth intermetallic compounds with CsCl structure. - J. Phys. [France], 1979, 40, Colloq. N5, 139-140. PA, 1980, 83, 24587.

Theor.; 4.1; 4.4.

4.5.46. Matsushita, E., Matsubara, T., Relation between martensitic and superconducting phase transitions in A-15 type compounds. - Progr. Theor. Phys., 1979, 62, N4, 862-873. - PA, 1980, 83, 29497.

Theor.

4.5.47. Van Bruggen, C. F., Haas, C., Wiegers, G. A., Charge-density waves and electron localization in vanadium chalcogenides. - J. Solid State Chem., 1979, 27, N1, 9-18. - PA, 1979, 82, 47038.

Theor.

See also: 4.1.-30, 119; 4.2.-66; 4.4.-68a.

4.6. Crystal Chemistry

In this subsection more than 200 papers are compiled devoted to the investigation of the cooperative Jahn-Teller effect upon the crystal structure of concrete Jahn-Teller crystals.

1950

4.6.1. McMurdi, H. F., Sullivan, B. M., Mauer, F. A., High-temperature X-ray study of the system Fe_3O_4-Mn_3O_4. - J. Res. N. B. S., 1950, 45, 35-41. - CA, 1950, 44, 9243f.

Exp.

1957

4.6.2. Dunitz, J. D. Orgel, L. E., Electronic properties of transition-metal oxides. I. Distortions from cubic symmetry. - J. Phys. & Chem. Solids, 1957, 3, N1, 20-29.

Theor.

4.6.3. Orgel, L. E., Dunitz, J. D., Stereochemistry of cupric compounds. - Nature, 1957, 179, N4557, 462-465.

Theor.; E; T; O_h; T_d; d^1-d^9.

1958

4.6.4. Griffith, J. S., Orgel, L. E., Localized electrons in body-centered cubic metals. - Nature, 1958, 181, N4603, 170-172.

Rev.

1959

4.6.5. Bertaut, F., Forrat, F., Dulac, J., Rhodium spinels. - C.R. Hebd. Seances Acad. Sci., 1959, 249, N5, 726-728. - CA, 1960, 54, 2870f.

Exp.; d^8; d^9.

4.6.6. Knox, K., Structure of K_2CuF_4: New kind of distortion for octahedral copper(II). - J. Chem. Phys., 1959, 30, N4, 991-993.

Exp.; E - e; O_h; d^9.

4.6.7. Ohnishi Haruyuki, Teranishi Teruo, Miyahara Syohei, On the transition temperature of copper ferrite. - J. Phys. Soc. Jpn., 1959, 14, N1, 106. - RZF, 1960, N9, 23624.

Exp.

4.6.8. Okazaki Atsushi, Suemune Yasutaka, Fuchikami, T., The crystal structure of $KMnF_3$, $KFeF_3$, $KCoF_3$, $KNiF_3$, and $KCuF_3$. - J. Phys. Soc. Jpn., 1959, 14, 1823-1824.

Exp.

1960

4.6.9. Morosin, B., Lingafelter, E. C., The crystal structure of cesium tetrabromocuprate(II). - Acta Crystallogr., 1960, 13, N10, 807-809.

Exp.; T; T_d; d^9.

4.6.10. Shiratori Kiichi, Iida Shuichi, Anomalous temperature dependence of lattice constant of CrO_2. - J. Phys. Soc. Jpn., 1960, 15, 2362-2363. - CA, 1961, 55, 11005e.

Exp.; d^2.

1961

4.6.11. Aoki Ikuo, Imazeki Yoshihiro, Tateno Tomio, Sasaki Takayuki, Electrical and thermal properties in the phase transition of Co and Mn mixed ocides. - Chiba Daigaku Bunri Gakubu Kiyo, Shizen Kagaku, J. College Arts and Sciences Chiba Univ. Natural Sci. Ser., 1961, 3, 289-293. - CA, 1963, 58, 12024b.

Exp.

4.6.12. Cervinka, L., Krupicka, S., Synacek, V., To the existence of tetragonally distorted $Mn^{3+}O_6{}^{2-}$ octahedra in cubic $MnFe_2O_4$. - Phys. & Chem. Solids, 1961, 20, N1/2, 167-168. - RZF, 1962, 2E54.

Exp.; E - e; O_h; d^4.

4.61.13. Ohnishi Hazuyuki, Teranishi Teruo, Crystal distortion in copper ferrite-chromite series. - J. Phys. Soc. Jpn., 1961, 16, N5, 35-43. - RZF, 1961, 11E318.

Exp.

4.6.14. Okazaki Atsushi, Suemune Yasutaka, Electron distribution in $KMnF_3$, $KFeF_3$, $KCoF_3$, and $KNiF_3$. - J. Phys. Soc. Jpn., 1961, 16, 1474. - CA, 1962, 57, 4049i.

Exp.; d^9.

4.6.15. Tracy, J. W., Gregory, N. W., Lingafelter, E. C., Dunitz, J. D., Mez, H.-C., Rundle, R. E., Scheringer, C., Yakel, H. L., Jr., Wilkinson, M. K., The cyrstal structure of chromium(II) chloride. - Acta Crystallogr., 1961, 14, N9, 927-929.

Exp.; E; O_h; d^4.

1962

4.6.16. Goodenough, J. B., Spin-orbit vs Jahn-Teller deformation in chromium spinels. - J. Phys. Soc. Jpn., 1962, 17, Suppl. B-1, 185-188. - CA, 1962, 57, 15926d. RZF, 1963, 6E587.

Theor.; d^8.

4.6.17. Iberson, E., Gut, R., Gruen, D. M., Nickel chloride-cesium chloride phase diagram: Tetrahedral $NiCl_4^{--}$ ion in the new compound Cs_3NiCl_3. - J. Phys. Chem., 1962, 66, N1, 65-69. - CA, 1962, 56, 9481b.

Exp.; T; T_d; d^8.

4.6.18. Irani, K. S., Sinha, A. P. B., Biswas, A. B., Effect of temperature on the structure of manganites. - Phys. & Chem. Solids, 1962, 23, N6, 711-727. - CA, 1962, 57, 6708a.

Exp.; E; O_h; d^4.

4.6.19. Tsushima, T., Magnetic properties and crystal chemistry of nickel chromite single crystals. - J. Phys. Soc. Jpn., 1962, 17, Suppl. B-1, 189-191. - RZF, 1963, 5E564.

Exp.; d^8.

1963

4.6.20. Bergstein, A., The lattice constant of $Mn_xFe_{3-x}O_{4+\gamma}$ spinels. - Czech. J. Phys., 1963, 13, N8, 613-616. - CA, 1964, 60, 11443b.

Exp.

4.6.21. Goodenough, J. B., Magnetism and the Chemical Bond. - New York; London: Wiley Interscience, 1963. - XV, 393 p. - Bibliogr.: 722 ref. CA, 1963, 59, 2190a. RZF, 1964, 2E551K.

4.6.22. Goodenough, J. B., Electron ordering transitions. - In: Solid State Res. Lincoln Lab. Mass. Inst. Technol., 1963, Vol. 2, p. 37-39. - RZF, 1964, 9E116.

Theor.; E - e; T_d; d^6.

4.6.23. Rüdorff, W., Lincke, G., Babel, D., Investigation of the threefold fluorides. II. Cobalt(II)- and cuprum(II)-fluorides. [Germ.]. - Z. Anorg. Allg. Chem., 1963, 320, 150-170.

Exp.; E - e; O_h; d^7; d^9.

1964

4.6.24. Aoki Ikuo, Kameyama Ikuji, Omiya Fumiya, Hinata Noriyasu, Cation distribution in $CuMn_2O_4$. [Jap.]. - Chiba Daigaku Bunri Gakuba Kiyo, Shizen Kagaku, J. College Arts and Sciences Chiba Univ. Natural Sci. Ser., 1964, 4, N2, 17-22. - CA, 1966, 64, 11976h.

Exp.; d^9.

4.6.25. Arnott, R. J., Wold, A., Rogers, D. B., Electron ordering transitions in several chromium spinel systems. - J. Phys. & Chem. Solids, 1964, 25, N2, 161-166. - CA, 1964, 60, 10004g.

Exp.; T; T_d; d^7; d^8.

4.6.26. Goodenough, J. B., Jahn-Teller distortions induced by tetrahedral-site Fe^{2+} ions. - J. Phys. & Chem. Solids, 1964, 25, N2, 151-160. - CA, 1964, 60, 9991c. RZF, 1964, 9E117.

Theor.; E - e; T_d; d^6.

1965

4.6.27. Babel, D., Ternary fluorides. III. Structure of Na_2CuF_4. [Germ.]. - Z. Anorg. Allgem. Chem., 1965, 336, N3/4, 200-206. - CA, 1965, 63, 5049f.

Exp.; E - e; O_h; d^9.

4.6.28. Blasse, G., New compounds with perovskite-like structures. - J. Inorg. & Nucl. Chem., 1965, 27, N5, 993-1003. - CA, 1965, 62, 15450f.

Exp.

4.6.29. Kino Yoshihiro, Miyahara Shochei, Cubic-tetragonal transition and X-ray diffuse scattering in $NiCr_2O_4$. - J. Phys. Soc. Jpn., 1965, 20, N8, 1522. - CA, 1966, 64, 5840e.

Exp.; T; T_d; d^8.

4.6.30. Robbins, M., Baltzer, P. K., Cooperative Jahn-Teller distortions in the systems ZnM_2O_4-Cu_2GeO_4 and ZnM_2O_4-$ZnCuGeO_4$. - J. Appl. Phys., 1965, 36, N3, 1039-1040. - CA, 1965, 62, 13950f. RZF, 1966, 1E537.

Exp.

4.6.31. Robbrecht, G. G., Clerck, E. F., de, The temperature dependence of the tetragonal-to-cubic phase transformations in transition-metal spinels. I. Distrotions occurring only in the octahedral sites. - Physica, 1965, 31, N7, 1033-1045. - CA, 1965, 63, 6400d.

Theor.; O_h.

4.6.32. Robbrecht, G. G., Clerck, E. F., de, Temperature dependence of the tetragonal distortion of magnetic spinels with Jahn-Teller distortion of the tetrahedrally coordinated interstices. [Dut.]. - Bull. Soc. Belge Phys., 1965, N5, 340-346. - CA, 1966, 65, 1494d.

Theor.; T_d.

4.6.33. Robbrecht, G. G., Iserentant, C. M., LInear expansion and volume anomaly of Jahn-Teller disordered spinels with distortion of the tetrahedrally coordinated interstices in the spinels. [Dut.]. - Bull. Soc. Belge Phys., 1965, N5, 347-351. - CA, 1966, 64, 18428g.

Exp.; T; T_d; d^9.

4.6.34. Robbrecht, G. G., Iserentant, C. M., Clerck, E. F., de, Jahn-Teller deformed $CuFe_2O_4$. [Dut.]. - Bull. Soc. Belge Phys., 1965, N5, 352-357. - CA, 1966, 65, 1490d.

Exp.; E; O_h; d^9.

4.6.35. Rogers, D. B., Germann, R. W., Arnott, R. J., Effect of trivalent manganese on the crystal chemistry of some lithium spinels. - J. Appl. Phys., 1965, 36, N8, 2338-2342. - CA, 1965, 63, 9148g.

Exp.; T_d; d^4.

1966

4.6.36. Bateman, L. R., Blount, J. F., Dahl, L. F., Structural characterization of a new niobium subhalide Nb_6I_{11}, containing the first known $[M_6X_8]^n$ group with nonintegral metal oxidation state. - J. Am. Chem. Soc., 1966, 88, N5, 1082-1084. - CA, 1966, 64, 16745a.

Exp.; O_h.

4.6.37. Cervinka, L., Local Jahn-Teller distortions in the cubic spinel ferrite lattice and their effect on the cubic-to-tetragonal transition. - Acta Crystallogr., 1966, 21, N7, Suppl., A193. - PA, 1968, 71, 26408. RZF, 1967, 11E523.

Exp.; E; O_h; d^4.

4.6.38. Elliott, H., Hathaway, B. J., Slade, R. C., The properties of the hexanitro complexes of divalent iron, cobalt, nickel, and copper. - Inorg. Chem., 1966, 5, N4, 669-677. - CA, 1966, 64, 13731f.

Exp.; E; O_h; d^6; d^7; d^8; d^9.

4.6.39. Lukaszewicz, K., Crystal structure of α-$Cu_2P_2O_7$. - Bull. Acad. Pol. Sci. Ser. Sci. Chim., 1966, 14, N10, 725-729. - CA, 1967, 66, 89260.

Exp.; d^9.

4.6.40. Robbins, M., Darcy, L., Cooperative Jahn-Teller distortions and site preferences in Cu^{2+}-containing spinels. - J. Phys. & Chem. Solids, 1966, 27, N4, 741-743. - CA, 1966, 64, 18536c. RZF, 1966, 10E345.

Exp.; d^9.

4.6.41. Strens, R. G. J., The axial-ratio-inversion effect in Jahn-Teller distorted ML_6 octahedra in epidote and perovskite structures. - Mineral. Mag., 1966, 35, N273, 777-781. - CA, 1966, 65, 8106a.

Exp.

1967

4.6.42. Grenot, M., Huber, M., X-ray and neutron diffraction study of the tetragonal deformation due to the Jahn-Teller effect in spinels: $MgGa_{2-x}Mn_xO_4$ and $MgCr_{2-x}Mn_xO_4$. [Fr.]. - J. Phys. & Chem. Solids, 1967, 28, N12, 2441-2447. - CA, 1968, 68, 54236.

Exp.; E; O_h; d^4; d^7.

4.6.43. Grenot, M., Cation distribution and Jahn-Teller distortion in mixed spinels, $MgGa_{2-x}Mn_xC_4$, $MgCr_{2-x}O_4$, $MgIn_{2-x}MnO_4$. [Fr.]. - Bull. Soc. Chim. France, 1967, N6, 2043-2050. - CA, 1968, 68, 7159.

Exp.; E; O_h; d^4; d^7.

4.6.44. Leeuwen, P. W. N. M., van, Groeneveld, W. L., Complexes of ligands containing S = O. III. Tetramethylene sulfoxide and pentamethylene sulfoxide: Jahn-Teller effect in infrared spectra. - Rec. Trav. Chim. Pays-Bas, 1967, 86, N7, 721-730. - CA, 1967, 67, 69021.

Exp.; E; O_h; d^9.

4.6.45. Manaila, R., Pausescu, P., Structural changes in $MgMn_2O_4$ at high temperatures. - Phys. Status Solidi, 1967, 9, N2, 385-394.

Exp.

4.6.46. Scatturin, V., Crystal and magnetic structures of 3d-transition metal(II) double fluorides. [Ital.]. - In: Stereochim. Inorg.: 9 Corso Estivo Chim. Accad. Naz. Lincei, 1965. Rome, 1967, p. 355-384. - CA, 1969, 71, 117827.

Exp.

1968

4.6.47. Cervinka, L., Vetterkind, D., The influence of low temperature (293°K-4.5°K) on the structure of manganese-rich manganese ferrite. - J. Phys. & Chem. Solids, 1968, 29, N1, 171-180.

Exp.

4.6.48. Iserentant, C. M., Jahn-Teller distorted perovskite structures. [Hol.]. - Verh. Kon. Vlaam. Acad. Wetensch., Belg., Kl. Wetensch., 1968, 30, N102, 91 p. - CA, 1969, 70, 32256.

Exp.

4.6.49. Pausescu, P., Manaila, R., The phase transition in $NiCr_2O_4$. [Russ.]. - Kristallografiya, 1968, 13, N4, 627-630.

Exp.; T; T_d; d^8.

4.6.50. Reinen, D., Jahn-Teller effect of cupric ion in oxidic solids. I. Octahedral coordination. [Germ.]. - Z. Naturforsch. a, 1968, 23, N4, 521-529. - CA, 1968, 69, 31634.

Exp.; E - e; O_h; d^9.

1969

4.6.51. Iwamoto Nobuya, Adachi Akira, Formation mechanism of tetragonal chromite. - Trans. Iron and Steel Inst. Jpn., 1969, 9, N1, 59-65. - CA, 1969, 71, 24216.

Exp.; E - e; T_d; d^6; 3.4; 4.4.

4.6.52. Nakagawa Takehiko, Nomura Schoichiro, Crystal distortion in the system $Sr_2(CuW)O_6$-$SrCu_{1/3}Sb_{2/3}O_3$. - Jpn. J. Appl. Phys., 1969, 8, N11, 1354. - CA, 1970, 72, 71797. PA, 1970, 73, 20769.

Exp.; E - e; O_h; d^9.

4.6.53. Propach, V., Reinen, D., Jahn-Teller effect of cupric ions in oxidized solids. II. Crystal lattices of the trirutile, niobite, and perovskite types. [Germ.]. - Z. Anorg. Allg. Chem., 1969, 369, N3/6, 278-294. - CA, 1970, 72, 6963.

Exp.; E - e; O_h; d^9.

1970

4.6.54. Brostigen, G., Kjekshus, A., Compounds with the marcasite type crystal structure. V. Crystal structure of FeS_2, $FeTe_2$, and $CoTe_2$. - Acta Chem. Scand., 1970, 24, N6, 1925-1940. - CA, 1971, 74, 7462.

Exp.; E - e; T_d; d^6; d^7.

4.6.55. Cervinka, L., Hosemann, R., Vogel, W., Paracrystalline lattice distortions and microdomains in manganese ferrites near the cubic-to-tetragonal transitions. - Acta Crystallogr. A, 1970, 26, N2, 277-289. - CA, 1970, 73, 8058.

Exp.; E - e; O_h; d^4.

4.6.56. Iwamoto Nobuya, Adachi Akira, Formation behavior of inclusions in iron-chromium alloys. [Jap.]. - Tetsu To Hagane, J. Iron and Steel Inst. Jpn., 1970, 56, N13, 1661-1676. - CA, 1971, 74, 7449.

Exp.; d^4.

4.6.57. Kubo Takeji, Twin structure of $NiCr_2O_4$ single crystals due to cooperative Jahn-Teller distortion. - J. Phys. Soc. Jpn., 1970, 28, N2, 430-437. - CA, 1970, 72, 83721. PA,1970, 73, 30882.

Exp.; T; T_d; d^8.

4.6.58. Muyahara Syohei, Problems on the Jahn-Teller distortion of crystals. - Prog. Theor. Phys., Suppl., 1970, N46, 437-442.- CA, 1971, 75, 82101. PA,1971, 74, 69864.

Theor.; d^8; d^9.

4.6.59. Nogues, M., Poix, P., Octahedral deformation by Jahn-Teller effect: Evolution of the (metal-oxygen)$_6$ bond lengths and of the lattice constants in the $tZnMn_2O_4(1-t)Zn_2SO_4$ system. [Fr.]. - C.R. Hebd. Sean. Acad. Sci. Ser. C, 1970, 271, N16, 995-997. - CA, 1971, 74, 46418.

Exp.; T; O_h; d^4.

4.6.60. Portier, J., Tressaud, A., Dupin, J.-L., The perovskite fluorides $AgMeF_3$ (Me = Mg, Mn, Co, Ni, Cu, Zn). [Fr.]. - C.R. Hebd. Sean. Acad. Sci. Ser. C, 1970, 270, N2, 216-218. - PA, 1970, 73, 55826.

Exp.

4.6.61. Robbrecht, G. G., Henriet-Iserentant, C. M., On the lattice parameters and the tetragonal distortion of the copper and cadmium manganite systems. - Phys. Status Solidi, 1970, 41, N1, K43-K46. - PA, 1970, 73, 78472.

Exp.; T; T_d; d^9.

4.6.62. Vishnevskiy, I. I., Alapin, B. G., Skripak, V. N., Jahn-Teller effect in the $NiCr_2O_4$ spinel. [Russ.]. - Izv. Akad. Nauk SSSR. Neorg. Mater., 1970, 6, N2, 314-318. - CA, 1970, 72, 125927.

Exp.; T; T_d; d^8.

4.6.63. Zilber, R., Bertaut, E. F., Burlet, P., Order in $(CuO)_x(NiO)_{1-x}$ solid solutions. [Fr.]. - Solid State Commun., 1970, 8, N12, 935-941. - CA, 1970, 73, 49413. PA, 1970, 73, 50526.

Exp.; d^8; d^9.

1971

4.6.64. Bertrand, J. A., Carpenter, D. A., Kalyanaraman, A. R., Structure of $K_2BaCo(NO_2)_6$ at 233°K: Static Jahn-Teller distortion. - Inorg. Chim. Acta, 1971, 5, N1, 113-114. - CA, 1971, 75, 41637.

Exp.; O_h; d^7.

4.6.65. Forsyth, J. B., Sampson, C. F., Crystallographic distortion in $DyVO_4$ at 14°K. - Phys. Lett. A, 1971, 36, N3, 223-225.

Exp.

4.6.66. Kroese, C. J., Tindemans-van Eijndhoven, J. C. M., Maaskant, W. J. A., A phase transition in a compound with helical electric dipole structure: $CsCuCl_3$. - Solid State Commun., 1971, 9, N19, 1707-1709.

Exp.

4.6.67. Marais, A., Merceron, T., Porte, M., Effect of tetragonal distortion on the directional order in copper ferrite at high temperature. [Fr]. - J. Phys. [France], 1971, 32, Colloq. I, Pt. 2, 843-844. - CA, 1971, 75, 27321.

Exp.; d^9.

4.6.68. Reinen, D., Application of ligand field spectroscopy to problems of chemical bonding in solids. [Germ.]. - Angew. Chem., 1971, 83, N24, 991-999. - CA, 1972, 76, 65598. RZK, 1972, 11B644.

Exp.; E - e; O_h; d^9.

4.6.69. Reinen, D., Propach, V., Compounds with Cu^{2+} ions on the A-positions of ABO_3 perovskites. [Germ.]. - Inorg. & Nucl. Chem. Lett., 1971, 7, N7, 569-572. - CA, 1971, 75, 54787.

Exp.; E - e; O_h; d^9.

4.6.70. Sayetat, F., Boucherle, J. X., Belakhovsky, M., Kallel, A., Tcheou, F., Fuess, H., Experimental study of magnetic and crystallographic transitions in $DyVO_4$. - Phys. Lett. A, 1971, 34, N7, 361-362. - CA, 1971, 75, 55438.

Exp.; 3.4; 4.4.

4.6.71. Spooner, S., Lee, J. N., Moos, H. W., Configuration of moments in $TbPO_4$. - Solid State Commun., 1971, 9, N13, 1143-1145.

Exp.

1972

4.6.72. Amelinckx, S., Domain structures in crystals. - J. Less-Common Met., 1972, 28, N2, 325-355. - CA, 1972, 77, 53428. PA, 1972, 75, 68021.

Exp.

4.6.73. Baffier, N., Huber, M., X-ray and neutron diffraction study of relations between cation distribution and crystal distortion in ferromanganite spinels $xMn_3O_4 + (1-x)Cu(Fe,Cr)_2O_4$. [Fr.]. - J. Phys. & Chem. Solids, 1972, 33, N3, 737-747. - CA, 1972, 76, 104836.

Exp.; T_d; d^4.

4.6.74. Biagini, C. M., Chiesi, V. A., Guastini, C., Nardelli, M., Crystal and molecular structures of divalent metal complexes with pyridinecarboxylic acids. IV. Bis(hydrogen pyridine-2,6-dicarboxylato)copper(II) trihydrate. - Gazz. Chim. Ital., 1972, 102, N11, 1026-1033. - CA, 1973, 78, 129385.

Exp.; E - e; O_h; d^9.

4.6.75. Göbel, H., Will, G., Crystallographic low-temperature phase transformation in some compounds of the type $LnXO_4$ (Ln = rare earth element, X = V, As, P). [Germ.]. - Int. J. Magn., 1972, 3, N1/3, 123-128. - CA, 1973, 78, 34789. PA, 1972, 75, 77746.

Exp.

4.6.76. Göbel, H., Will, G., Low-temperature X-ray diffraction and phase transitions in $DyVO_4$ and $DyAsO_4$. - Phys. Status Solidi b, 1972, 50, N1, 147-154.

Exp.

4.6.77. Göbel, H., Will, G., Crystallographic phase transition in DyAsO$. - Phys. Lett. A, 1972, 39, N2, 79-80.

Exp.

4.6.78. Ioan, C., Maxim, G., Anisotropy induced in copper ferrite at the cubic-tetragonal phase transition [Jahn-Teller distortion]. [Fr.]. - Rev. Roum. Phys., 1972, 17, N10, 1213-1216. - PA, 1973, 76, 15382.

Theor.; T; T_d; d^9.

4.6.79. Pearson, W. B., The Crystal Chemistry and Physics of Metals and Alloys. - New York, etc.: Wiley-Intersci., 1972. - 806 p.

Ch. 5. §8.2.1. The Jahn-Teller Effect, p. 322-327.

Rev.

4.6.80. Sawaoka Akira, Saito, S., Inoue, K., Asada, T., Jahn-Teller type of crystal deformation of some oxides with spinel structure under high pressure. - Acta Crystallogr. A, 1972, 28, N4, Suppl., S 244. - PA, 1973, 76, 27118.

Exp.; T; T_d; d^8; d^9.

4.6.81. Sayetat, F., Experimental study of the crystallographic transitions of $DyVO_4$ and $TbVO_4$. [Fr.]. - Solid State Commun., 1972, 10, N9, 879-882. - RZF, 1972, 9E477.

Exp.

4.6.82. Shannon, R. D., Calvo, C., Crystal structure of a new form of copper vanadate ($Cu_3V_2O_8$). - Can. J. Chem., 1972, 50, N24, 3944-3949. - CA, 1973, 78, 103016.

Exp.; E - e; O_h; d^9.

4.6.83. Siebert, G., Hoppe, R., New fluoroperovskites with chromium(III) and manganese(III) $M(M^I_{0.5}Cr_{0.5})F_3$ and $M(M^I_{0.5}Mn_{0.5})F_3$; M, M^I = alkali metal. [Germ.]. - Z. Anorg. Allg. Chem., 1972, 391, N2, 117-125. - CA, 1972, 77, 119433.

Exp.; E - e; O_h; d^4; d^7.

4.6.84. Terauchi Hikaru, Mori Masahiro, Yamada Yasusada, X-ray critical scattering in $NiCr_2O_4$. - J. Phys. Soc. Jpn., 1972, 32, N4, 1049-1058. - CA, 1972, 76, 132642. PA, 1972, 75, 32777. RZF, 1972, 9E445.

Exp.; T; T_d; d^8.

4.6.85. Van Landuyt, J., De Ridder, R., Brabers, V. A. M., Amelinckx, S., Jahn-Teller domains in $Mn_xFe_{3-x}O_4$ as observed by electron microscopy. - Mater. Res. Bull., 1972, 7, N4, 327-337. - CA, 1972, 76, 145953. PA, 1972, 75, 43922. RZF, 1972, 10E1204.

Exp.; E - e; O_h; d^4.

4.6.86. Wieghardt, K., Weiss, J., Crystal structures of hexaamminechromium(III), hexafluoromanganate(III), and hexamminechromium(III) hexafluoroferrate(III). [Germ.]. - Acta Crystallogr. B, 1972, 28, N2, 529-534. - CA, 1972, 76, 77792.

Exp.; E - e; O_h; d^4.

4.6.87. Will, G., Göbel, H., Sampson, C. F., Forsyth, J. B., Crystallographic distortion in $TbVO_4$ at 32°K. - Phys. Lett. A, 1972, 38, N3, 207-208.

Exp.

1973

4.6.88. Besrest, F., Jaulmes, S., Crystal structure of chromium(II) iodide. [Fr.]. - Acta Crystallogr. B, 1973, 29, N8, 1560-1563. - CA, 1973, 79, 84327. PA, 1973, 76, 60692.

Exp.; E - e; O_h; d^4.

4.6.89. Calvo, C., Manolescu, D., Refinement of the structure of CuV_2O_6. - Acta Crystallogr. B, 1973, 29, N8, 1743-1745. PA, 1973, 76, 63994.

Exp.; E - e; O_h; d^9.

4.6.90. Casten, A., Becker, P. J., Influence of temperature and external magnetic fields on the shape of crystallographic domains. - Int. J. Magn., 1973, 5, N1/3, 157-160. - RZF, 1974, 4E863.

Exp.

4.6.91. Li Ting-i, Stucky, G. D., McPherson, G. L., The crystal structure of $CsMnCl_3$ and a summary of the structures of RMX_3 compounds. - Acta Crystallogr. B, 1973, 29, N6, 1330-1335.

Exp.; O_h; d^4; d^9.

4.6.92. Li Ting-i, Stucky, G. D., The effect of exchange coupling on the spectra of transition metal ions: The crystal structure and optical spectrum of $CsCrBr_3$. - Acta Crystallogr. B, 1973, 29, N7, 1529-1532.

Exp.; E - e; O_h; d^4.

4.6.93. Mansour, B., Baffier, N., Huber, M., Jahn-Teller type distortion and distribution of cations in the spinel oxides $CoMn_xCr_{2-x}O_4$ ($0 \leq x \leq 2$): X-ray and neutron diffraction studies. [Fr.]. - C.R. Hebd. Seances Acad. Sci. Ser. C, 1973, 277, N18, 867-869. - CA, 1974, 80, 75168. PA, 1974, 77, 32498.

Exp.

4.6.94. Nogues, M., Poix, P., Octahedral distortion of Mn_3O_4-Zn_2SnO_4 systems by the Jahn-Teller effect. [Fr.]. - C.R. Hebd. Seances Acad. Sci. Ser. C, 1973, 277, N21, 1117-1119. - CA, 1974, 80, 75104.

Exp.

4.6.95. Pollert, E., The cation distributions in the spinel phases of the mixed vanadium oxides $LiZn_xV_{2-x}O_4$ and $LiMg_xV_{2-x}O_4$. - Krist. & Tech., 1973, 8, N7, 859-866. - PA, 1973, 76, 75506.

Exp.; d^1.

4.6.96. Ust'yantsev, V. M., Mar'evich, V. P., X-ray diffraction study of the Jahn-Teller effect in a copper chromium oxide ($CuCr_2O_4$) spinel. [Russ.]. - Izv. Akad. Nauk SSSR, Neorg. Mater., 1973, 9, N2, 336-337. - CA, 1973, 78, 129271. PA, 1973, 76, 72048. RZF, 1973, 6E463.

Exp.; T; T_d; d^9.

4.6.97. Von Schnering, H. G., Brand, B. H., Structure and properties of the blue chromium(II) chloride tetrahydrate $CrCl_2\cdot$

$4H_2O$. [Germ.]. - Z. Anorg. Allg. Chem., 1973, 402, N2, 159-168. - CA, 1974, 80, 88202.

Exp.; E - e; O_h; d^4.

1974

4.6.98. Bhaduri, A., Keer, H. V., Biswas, A. B., Structural properties of some mixed oxidic spinels. - Indian J. Pure & Appl. Phys., 1974, 12, N11, 745-747. - PA, 1975, 78, 33139.

Exp.

4.6.99. Birgeneau, R. J., Kjems, J. K., Shirane, G., Van Uitert, L. G., Cooperative Jahn-Teller phase transition in praseodymium(III) aluminate. - Phys. Rev. B, 1974, 10, N6, 2512-2534. - CA, 1975, 82, 9443.

Exp.

4.6.100. Friebel, C., Spectroscopic studies of the Jahn-Teller effect of the copper(2+) ion compounds of type A_2CuL_6. I. Structural details of dibarium hexahydroxycuprate and distrontium hexahydroxycuprate. [Germ.]. - Z. Naturforsch. b, 1974, 29, N5/6, 295-303. - CA, 1974, 81, 143902.

Exp.; E - e; O_h; d^9.

4.6.101. Friebel, C., Ferrodistortively ordered tetragonally elongated silver(II) hexafluoride octahedron in $AgM^{IV}F_6$ [M^{IV} = tin, zirconium] compounds. [Germ.]. - Solid State Commun., 1974, 15, N3, 639-641. - CA, 1974, 81, 113282.

Exp.; E - e; O_h; d^9.

4.6.102. Hutchings, M. T., Gregson, A. K., Day, P., Leech, D. H., Neutron diffraction study of the crystal and magnetic structure of the ionic ferromagnet Cs_2CrCl_4. - Solid State Commun., 1974, 15, N2, 313-316.

Exp.; E - e; O_h; d^4.

4.6.103. Köhl, P., Reinen, D., Structural and spectroscopic investigations of barium copper tellurate Ba_2CuTeO_6. [Germ.]. - Z. Anorg. Allg. Chem., 1974, 409, N3, 257-272. - CA, 1975, 82, 37536.

Exp.; E - e; O_h; d^9; 4.3.

4.6.104. Kroese, C. J., Maaskant, W. J. A., Verschoor, G. C., The high-temperature structure of $CsCuCl_3$. - Acta Crystallogr. B, 1974, 30, N4, 1053-1056.

Exp.; E - e; O_h; d^9.

4.6.105. Nagata, Y., Ohta, K., Impurity effects on the phase transition of lithium ferrite. - J. Phys. Soc. Jpn., 1974, 36, N4, 1208. - PA, 1974, 77, 52889.

Exp.; 4.7.

4.6.106. Nogues, M., Poix, P., Cooperative Jahn-Teller effect in Mn_3O_4-Mn_2SnO_4 system. [Fr.]. - J. Solid State Chem., 1974, 9, N8, 330-335. - CA, 1974, 81, 7102. PA, 1974, 77, 45074. RZF, 1974, 9E1023.

Exp.; d^4.

4.6.107. Obi, Y., Crystal distortion in several spinel systems. - Phys. Status Solidi a, 1974, 25, N1, 293-299. - CA, 1974, 81, 161404. PA, 1974, 77, 79135. RZF, 1975, 5E749.

Exp.; E - e; O_h; d^9.

4.6.108. Segmüller, A., Melcher, R. L., Kinder, H., X-ray diffraction measurement of the Jahn-Teller distortion in $TmVO_4$. - Solid State Commun., 1974, 15, N1, 101-104. - CA, 1974, 81, 127868. PA, 1974, 77, 60586. RZF, 1974, 12E516.

Exp.; E - b_1; D_4h.

4.6.109. Smith, S. H., Wanklyn, B. M., Flux growth of rare earth vanadates and phosphates. - J. Cryst. Growth, 1974, 21, N1, 23-28. - CA, 1974, 80, 53063. PA,1974, 77, 21188.

Exp.

4.6.110. Walsh, D., Donnay, G., Donnay, J. D. H., Jahn-Teller effects in ferromagnesian minerals: Pyroxenes and olivines. - Bull. Soc. Fr. Mineral. Cristallogr., 1974, 97, N2/5, 170-183. - CA, 1975, 83, 30796. PA, 1975, 78, 77058.

Exp.; T; O_h; d^6.

1975

4.6.111. Ahsbahs, H., Dehnicke, K., Heger, G., Hellner, E., Helmboldt, R., Mullen, D., Reinen, D., Cooperative Jahn-Teller distortions in nitro-complexes $A^I_2PbCu(NO_2)_6$ (A = K, Rb, Cs, Tl). - Acta Crystallogr. A, 1975, 31, N3, Suppl., S190. - PA, 1976, 79, 13662.

Exp.; E - e; O_h; d^9.

4.6.112. Backes, G., Reinen, D., Jahn-Teller effect of low spin cobalt(2+) ions in nitro complexes $A^{I}_{2}M^{II}Co(NO_2)_6$. I. A^{I} = potassium, rubidium; M^{II} = strontium, barium. [Germ.]. - Z. Anorg. Allg. Chem., 1975, 418, N3, 217-228. - CA, 1976, 84, 81983.

Exp.; E - e; O_h; d^9.

4.6.113. Balarev, K., Karaivanova, V., Change in the crystal structure of zinc(II) sulfate heptahydrate and magnesium(II) sulfate heptahydrate due to isodimorphous substitution by copper(II), iron(II), and cobalt(II) ions. - Krist. & Tech., 1975, 10, N11, 1101-1110. - CA, 1975, 83, 200325.

Exp.; E - e; O_h; d^9.

4.6.114. Bersuker, I. B., Concerning the crystal chemistry of transition metal coordination compounds with degenerate or pseudo-degenerate electronic terms. [Russ.]. - Zh. Strukt. Khim., 1975, 16, N6, 935-943. - CA, 1976, 84, 158358. RZK, 1976, 13B587.

Theor.; E - e; O_h; d^9.

4.6.115. Bersuker, I. B., Vekhter, B. G., Jahn-Teller effect in the stereochemistry and crystal chemistry of coordination compounds. [Russ.]. - In: II Mendeleev. S'ezd. Obshch. i Prikl. Khim.: Ref. Dokl. i Soobshch. Moscow, 1975, p. 8. - CA, 1976, 84, 157391. RZK, 1976, 2B47.

Theor.

4.6.116. Chenavas, J., Sayetat, F., Collomb, A., Joubert, J. C., Marezio, M., Z-ray study of the low-temperature phase of sodium manganese manganate $[NaMn_3^{3+}](Mn_2^{3+}Mn_2^{4+})O_{12}$, a perovskite-like compound. - Solid State Commun., 1975, 16, N10/11, 1129-1132. - CA, 1975, 83, 51410. PA, 1975, 78, 65575. RZF, 1975, 11E662.

Exp.; d^3; d^4.

4.6.117. Chanavas, J., Capponi, J. J., Joubert, J. C., Marezio, M., High pressure synthesis of new perovskite-like compounds in the systems neodymium(III) oxide-(aluminum, chromium(III), gallium, or iron(III)) oxide-manganese(III) oxide. - In: Proc. 4th Int. Conf. High Pressure, 1974. Kyoto, 1975, p. 176-179. - CA, 1975, 83, 51416.

Exp.; T; O_h; d^4.

4.6.118. Dubler, E., Matthieu, J. P., Oswald, H. R., DTA and DSC investigation of phase transitions in solids with cooperative Jahn-Teller effects. - In: Therm. Anal.: Proc. 4th Int. Conf., 1974. London, 1975, Vol. 1, p. 377-386. - CA, 1977, 87, 32175.

Exp.; E - e; O_h; d^9.

4.6.119. Faber, J., Jr., Lander, G. H., Cooper, B. R., Neutron diffraction study of UO_2: observation of internal distortion. - Phys. Rev. Lett., 1975, 35, N26, 1770-1773. - PA, 1976, 79, 22587.

Exp.; T- t_2; O_h; f^2.

4.6.120. Ferraro, J. R., Long, G. J., Solid state pressure effects on stereochemically nonrigid structures. - Accounts Chem. Res., 1975, 8, N5, 171-179. - RZK, 1976, 3B763.

Theor.; 5.2.

4.6.121. Gazo, J., Bersuker, I. B., The property of "plasticity" of the Cu^{2+} coordination sphere and its manifestations. [Russ.]. - In: II Mendeleev. S'ezd. Obshch. i Prikl. Khim.: Ref. Dokl. i Soobshch. Moscow, 1975, p. 7-8. - RZK, 1976, 7B32.

Exp.; E - e; O_h; d^9.

4.6.122. Guen, L., Nguyen Huy Dung, Eholie, R., Flahaut, J., Chromium(II) iodide-iron(II) iodide system: Phase diagram, structural study, and cooperative Jahn-Teller effect. [Fr.]. - Ann. Chim. [France], 1975, 10, N1, 11-16. - CA, 1975, 83, 50971. RZK, 1975, 16B927.

Exp.; E - e; O_h; d^4.

4.6.123. Holba, P., Nevriva, M., Pollert, E., Tetragonal distortion of spinel solid solutions $MnCr_2O_4$-Mn_3O_4. - Mater. Res. Bull., 1975, 10, N8, 853-860. - CA, 1975, 83, 156005. PA, 1975, 78, 85478. RZK, 1976, 6B423.

Exp.; T_d; d^4.

4.6.124. Kjems, J. K., Hayes, W., Smith, S. H., Wave-vector depenedence of the Jahn-Teller interactions in $TmVO_4$. - Phys. Rev. Lett., 1975, 35, N16, 1089-1092. - PA, 1976, 79, 9445.

Exp.; E - b_1; D_{4h}.

4.6.125. Mullen, D., Heger, G., Reinen, D., Planar dynamic Jahn-Teller effects in nitro complexes: a single crystal neutron diffraction study of cesium lead nitrocuprate ($Cs_2PbCu(NO_2)_6$) at 293°K. - Solid State Commun., 1975, 17, N10, 1249-1252. - CA, 1976, 84, 52617. PA, 1976, 79, 9205.

Exp.; E - e; O_h; d^9.

4.6.126. Nishizawa Hitoshi, Koizumi Mitsue, Synthesis and infrared spectra of $Ca_3M_2Si_3O_{12}$ and $Cd_3B_2Si_3O_{12}$ (B = Al, Cr, Ga, V, Fe, Mn) garnets. - Am. Mineral., 1975, 60, N1/2, 84-87. - CA, 1975, 82, 114159. PA, 1975, 78, 43393.

Exp.; E - e; O_h; d^4; 2.7; 4.3.

4.6.127. Nogues, M., Poix, P., Cooperative Jahn-Teller effect and tetrahedral site affinity of manganese(2+) and zinc(2+) ions in the manganese oxide (Mn_3O_4)-zinc tin oxide (Zn_2SnO_4) system. [Fr.]. - J. Solid State Chem., 1975, 13, N3, 245-251. - CA, 1975, 83, 21014. PA, 1975, 78, 41707. RZF, 1975, 9E194.

Exp.; d^4.

4.6.128. Shannon, R. D., Gumerman, P. S., Chenavas, J., Effect of octahedral distortion on mean manganese(3+)-oxygen distances. - Am. Mineral., 1975, 60, N7/8, 714-716. - CA, 1975, 83, 124539.

Exp.; E - e; O_h; d^4.

1976

4.6.129. Gazo, J., Kabesova, M., Valakh, F., Melnik, M., The modifications of $M_2P_2O_7$ compounds and the problem of plasticity of coordination M(II) polyhedrons. [Russ.]. - Koord. Khim., 1976, 2, N5, 715-718. - CA, 1976, 85, 54987.

Exp.; E - e; O_h; d^9.

4.6.130. Grimes, N. W., The structure of the versatile spinels. - Spectrum, 1976, 14, N2, 27-30. - CA, 1977, 86, 75915.

Rev.

4.6.131. Guen, L., Nguyen Huy Dung, Manganese(II) chromium(II) diiodide: The β phase of (manganese,chromium) diiodide. - Acta Crystallogr. B, 1976, 32, N1, 311-312. - CA, 1976, 84, 98081. PA, 1976, 79, 21783.

Exp.; T; T_d; d^4.

4.6.132. Guen, L., Nguyen Huy Dung, Ehalie, R., Flahaut, J., Chromium(II) iodide-MI_2 systems (M = titanium, vanadium, manganese, iron, cobalt, nickel, zinc): Phase diagrams, structural study, and the cooperative Jahn-Teller effect. [Fr.]. - Ann. Chim. [France], 1976, 11, N1, 39-46. - CA, 1976, 85, 25919.

Exp.; E - e; O_h; d^4.

4.6.133. Helmbold, R., Mullen, D., Ahsbahs, H., Klopsch, A., Hellner, E., Heger, G., Studies on Jahn-Teller deformed structures of $A_2^IPbCu(NO_2)_6$ with A = K, Rb, Cs. [Germ.]. - Z. Kristallogr., Kristallgeom., Kristallphys., Kristallchem., 1976, 143, N1/6, 220-238. - CA, 1977, 86, 24597.

Exp.

4.6.134. Hutchings, M. T., Fair, M. J., Day, P., Walker, P. J., Neutron scattering measurement of spin-wave dispersion in Rb_2CrCl_4: A two-dimensional easy-plane ferromagnet. - J. Phys. C, 1976, 9, N3, L55-L60.

Exp.; E - e; O_h; d^4.

4.6.135. Jirak, Z., Vratislav, S., Zajicek, J., Oxygen parameters and Debye-Waller factors in $Mn_xCR_{3-x}O_4$ spinels. - Phys. Status Solidi a, 1976, 37, N1, K47-K51. - PA, 1976, 79, 90920.

Exp.; T; T_d; d^4.

4.6.136. Kazey, Z. A., Mill, B. V., Sokolov, V. I., Cooperative Jahn-Teller effect in $Ca_3M_2Ge_3O_{12}$. [Russ.]. - Pis'ma, Zh. Eksp. & Teor. Fiz., 1976, 24, N4, 229-232. - CA, 1976, 85, 152276. PA, 1977, 80, 1119.

Exp.

4.6.137. Morgenstern-Badarau, I., Jahn-Teller effect and crystal structure of copper(II) tin(IV) hydroxide. {Fr.]. - J. Solid State Chem., 1976, 17, N4, 399-406. - CA, 1976, 85, 54954. PA, 1976, 79, 61705.

Exp.; E - e; O_h; d^9.

4.6.138. Pollert, E., Jirak, Z., The temperature dependence of tetragonal distortion in some $Mn_xCr_{3-x}O_4$ (manganese chromium oxide) spinel. - Czech. J. Phys. B, 1976, 26, N4, 481-484. - CA, 1976, 85, 85675.

Exp.; T; T_d; d^4.

4.6.139. Reinen, D., Weitzel, H., The crystal structure of copper(2+)-containing oxidic elpasolites: Neutron diffraction studies of the crystal powders. [Germ.]. - Z. Anorg. Allg. Chem., 1976, 424, N1, 31-38. - CA, 1976, 85, 85862.

Exp.

4.6.140. Smith, D. W., Chlorocuprates(II). - Coord. Chem. Rev., 1976, 21, N2/3, 93-158. - Bibliogr.: 302 ref.

Rev.; E; T; O_h; T_d; d^9.

4.6.141. Suits, J. C., Structural instability in new magnetic Heusler compounds. - Solid State Commun., 1976, 18, N3, 423-425. - CA, 1976, 84, 129652. PA, 1976, 79, 35397.

Exp.

4.6.142. Switendick, A. C., Influence of the electronic structure on the titanium-vanadium-hydrogen phase diagram. - J. Less-Common Met., 1976, 49, 283-290. - CA, 1976, 85, 182957. PA, 1976, 79, 93564.

Theor.; 5.4.

4.6.143. Takagi Shozo, Joesten, M., Lenhert, L. P., Potassium strontium hexanitrocuprate(II) and potassium strontium hexanitronickelate(II). - Acta Crystallogr. B, 1976, 32, N8, 2524-2526. - CA, 1976, 85, 134952.

Exp.; E - e; O_h; d^9.

4.6.144. Tomas, A., Chevalier, R., Laruelle, P., Bachet, B., Crystal structure of $CrEr_2S_4$. - Acta Crystallogr. B, 1976, 32, N12, 3287-3289. - CA, 1977, 86, 49361. PA, 1977, 80, 15356.

Exp.; E - e; O_h; d^4.

4.6.145. Ust'yantsev, V. M., Mar'evich, V. P., X-ray investigation of the system $CuCr_2O_4$-$MgCr_2O_4$. [Russ.]. - Izv. Akad. Nauk SSSR. Neorg. Mater., 1976, 12, N3, 562-563. - PA, 1977, 80, 44224.

Exp.; E - e; O_h; d^9.

4.6.146. Zubkov, V. G., Bazuyev, G. V., Shveykin, B. P., Low-temperature neutron and X-ray diffraction studies of rare earth orthovanadates. [Russ.]. - Fiz. Tverd. Tela, 1976, 18, N7, 2002-2004. - CA, 1976, 85, 115806. PA, 1977, 80, 32169.

Exp.

1977

4.6.147. Bersuker, I. B., Structural manifestations of the Jahn-Teller effect. [Russ.]. - In: Tez. Dokl. 1 Vsesoyuz. Soveshch. Neorg. Kristallokhim. Zvenigorod, 1977, p. 4.

Theor.; 1.5; 4.2.

4.6.148. Gopalakrishnan, J., Colsmann, G., Reuter, B., Studies on the lanthanum strontium nickelate ($La_{2-x}Sr_xNiO_4$) ($0 \le x \le 1$) system. - J. Solid State Chem., 1977, 22, N2, 145-149. - CA, 1977, 87, 209746. PA, 1978, 81, 2088.

Exp.

4.6.149. Hughes, H. P., Structural distortion in titanium diselenide and related materials - a possible Jahn-Teller effect. - J. Phys. C, 1977, 10, N11, L319-L323. - CA, 1977, 87, 175756. PA, 1977, 80, 62875.

Theor.

4.6.150. Ihringer, J., Schmidbauer, E., Low temperature crystallographic phase transitions in the Fe^{2+}-Cr-Ti spinel system. - Solid State Commun., 1977, 21, N1, 129-131. - PA, 1977, 80, 15589.

Exp.; d^6.

4.6.151. Joesten, M. D., Takagi Shozo, Lenhert, P. G., Structure of potassium lead hexanitrocuprate(II) at 276 K: An apparent compressed tetragonal Jahn-Teller distortion. - Inorg. Chem., 1977, 16, N11, 2680-2685. - CA, 1977, 87, 175903.

Exp.; E - e; O_h; d^9.

4.6.152. Kjems, J. K., Neutron scattering studies of the cooperative Jahn-Teller effect. - In: Electron-Phonon Interactions and Phase Transitions. New York; London, 1977, p. 302-322. (NATO Adv. Study Inst. Ser., Ser. B, 1977, N29.) - CA, 1979, 90, 127650.

Ref.; 4.1.

4.6.153. Kjems, J. K., The cooperative Jahn-Teller effect studied by neutron scattering. - In: Cryst. Field Effects Met. & Alloys: Proc. 2nd Int. Conf. Zurich, 1976. New York; London, 1977, p. 174-183. - CA, 1977, 87, 159287.

Exp.; E - b_1; D_{4h}.

4.6.154. Köhl, P., Reinen, D., The crystal structure of the hexagonal elpasolites $Ba_3NiSb_2O_9$ and $Ba_3CuSb_2O_9$: x-ray and spectrsoscopic results. [Germ.]. - Z. Anorg. Allg. Chem., 1977, 433, N1, 81-93. - CA, 1977, 87, 175911.

Exp.; 4.3.

4.6.155. Long, F. G., Stager, C. V., Low temperature crystal structure of $TbAsO_4$ and $DyAsO_4$. - Can. J. Phys., 1977, 55, N18, 1633-1640. - PA, 1977, 80, 89411.

Exp.

4.6.156. Mollenbach, K., Kjems, J. K., Gamma-ray diffraction studies of the mosaic distribution in $TmAsO_4$ near the cooperative Jahn-Teller transition at 6 K. - In: Electron-Phonon Interactions and Phase Transitions. New York; London, 1977, p. 323-326. (NATO Adv. Study Inst. Ser., Ser. B, 1977, N29.) - CA, 1979, 90, 130934.

Exp.

4.6.157. Noda Yukio, Mori Masahiro, Yamada Yasusada, Incommensurate Jahn-Teller transition in dipotassium lead copper hexanitrite. - Solid State Commun., 1977, 23, N4, 247-248. - CA, 1977, 87, 109672. PA, 1977, 80, 69065.

Exp.; E - e; O_h; d^9.

4.6.158. O'Conner, C. J., Sinn, E., Carlin, R. L., Structural and magnetic properties of $[M(C_5H_5NO)_6]L_2$ (M = copper, zinc; L = perchlorate(1-), tetrafluoroborate(1-)). - Inorg. Chem., 1977, 16, N12, 3314-3320. - CA, 1977, 87, 192330.

Exp.; E - e; O_h; d^9.

4.6.159. Propach, V., Crystal structure of $Ca_{0.5}Cu_{1.5}Ti_2O_6$, $Cu_{1.5}TaTiO_6$ and $CuTa_2O_6$: The spectroscopic behavior of Cu^{2+} ions in cuboctahedral surrounding. [Germ.]. - Z. Anorg. Allg. Chem., 1977, 435, N1, 161-171. - CA, 1978, 88, 97716.

Exp.; d^9.

4.6.160. Reinen, D., Supplementary remarks to the paper: "Planar dynamic Jahn-Teller effects in nitrocomplexes." - Solid State Commun., 1977, 21, N1, 137-140. - CA, 1977, 86, 82995. PA, 1977, 80, 15804.

Theor.; E - e; O_h; d^9.

4.6.161. Reinen, D., Henke, W., Allmann, R., The structure of terpyridine complexes $M^{II}(terpy)_2X_2 \cdot nH_2O$ [$M^{II} = Ni^{2+}$, Cu^{2+}, low-spin Co^{2+}]: The steric influence of the Jahn-Teller effect. - In: Proc. 18th Int. Conf. Coord. Chem. Sao Paulo, 1977, p. 211.

Exp.; E - e; O_h; d^9.

4.6.162. Reinen, D., Weitzel, H., A refinement of the Ba_2CuF_6 structure by powder neutron diffraction. [Germ.]. - Z. Naturforsch. b, 1977, 32, N4, 476-478. - CA, 1977, 86, 163846.

Exp.; E - e; O_h; d^9.

4.6.163. Sayetat, F., Anisotropic magnetoelastic properties of terbium compounds with garnet and zircon structure. - Physica B + C, 1977, 86-88, N3, 1467-1468. - CA, 1977, 86, 198976. PA, 1977, 80, 69660.

Exp.

4.6.164. Tanner, B. K., Safa, M., Midgley, D., Cryogenic X-ray topography using synchrotron radiation. - J. Appl. Crystallogr., 1977, 10, N2, 91-99. - CA, 1977, 86, 131409. PA, 1977, 80, 39507.

Exp.

4.6.165. Zaripov, M. M., Ziatdinov, A. M., Yablokov, Yu. V., Davidovich, R. L., Correlation of distortions of the nearest-neighbor Jahn-Teller hexaaquacopper(2+) ($[Cu(H_2O)_6]^{2+}$) centers in a copper zinc hexafluorozirconate hexahydrate ($(Cu_xZn_{1-x})ZrF_6\cdot 6H_2O$) crystal. [Russ.]. - Fiz. Tverd. Tela, 1977, 19, N10, 3165-3167. - CA, 1978, 88, 41766. RZF, 1978, 5D671.

Exp.; E - e; O_h; d^9; 3.1; 4.4.

1978

4.6.166. Allmann, R., Henke, W., Reinen, D., Presence of a static Jahn-Teller distortion in copper(II) terpyridine complexes. I. Crystal structure of diterpyridinecopper(II) nitrate. - Inorg. Chem., 1978, 17, N2, 378-382. - CA, 1978, 88, 82096.

Exp.; E - e; O_h; d^9; 3.1.; 4.4.

4.6.167. Bagieu-Beucher, M., Crystal structure of trivalent manganese polyphosphate $Mn(PO_3)_3$. [Fr.]. - Acta Crystallogr. B, 1978, 34, N5, 1443-1446. - CA, 1978, 89, 51752. PA, 1978, 81, 55843.

Exp.; E - e; O_h; d^4.

4.6.168. Bertini, I., Dapporto, P., Gatteschi, D., Scozzafava, A., X-ray and ESR investigation of an elongated octahedral tris-(1,2-diaminoethane)copper(II) complex. - Solid State Commun., 1978, 26, N11, 749-751. - CA, 1978, 89, 190193. PA, 1978, 81, 80206.

Exp.; E - e; O_h; d^9.

4.6.169. Crama, W. J., Maaskant, W. J. A., Verschoor, G. C., The cooperative Jahn-Teller distorted structures of rubidium chromium(II) trichloride and cesium chromium(II) trichloride. - Acta Crystallogr. B, 1978, 34, N6, 1973-1974. - PA, 1978, 81, 63713.

Exp.; E - e; O_h; d^4.

4.6.170. Fackler, J. P., Jr., Chen, H. W., Coordination "plasticity": about Cu(II) in a series of bis(dipyridylamine) and related complexes. - In: Proc. 19th Int. Conf. Coord. Chem. Prague, 1978, p. 148.

Exp.

4.6.171. Glidewell, C., Skeletal structures of hexaorgano-substituted triatomics R_3XYZR_3: a rationalization using the second-order Jahn-Teller effect. - J. Organomet. Chem., 1978, 159, N1, 23-30. - CA, 1979, 90, 22173.

Exp.

4.6.172. Jirak, Z., Vratislav, S., Novak, P., Study of cubic and tetragonal structures in the system manganese chromium oxide ($Mn_xCr_{3-x}O_4$). - Phys. Status Solidi a, 1978, 50, N1, K21-K24. - CA, 1979, 90, 46870. PA, 1979, 82, 19369.

Exp.; E - e; O_h; d^4.

4.6.173. Klein, S., Reinen, D., The structure of the high-temperature α-modification of $CsPbCu(NO_2)_6$ and the Jahn-Teller induced α → β-phase transition - a neutron diffraction study. - J. Solid State Chem., 1978, 25, N4, 295-299. - CA, 1979, 90, 14807. PA, 1979, 82, 1498.

Exp.; E - e; O_h; d^9.

4.6.174. Köhl, P., Crystal structure of the hexagonal compounds $Ba_3{}^{II}M^{II}Sb_2{}^{V}O_9$. II. Barium copper antimonate $Ba_3CuSb_2O_9$. [Germ.]. - Z. Anorg. Allg. Chem., 1978, 442, 280-288. - CA, 1978, 89, 138687.

Exp.; E - e; O_h; d^9.

4.6.175. Kollewe, D., Gibson, W. M., Observation of a cooperative Jahn-Teller phase transition in nickel chromite by particle channeling. - Phys. Lett. A, 1978, 65, N3, 253-255. - CA, 1978, 88, 144510.

Exp.; T; T_d; d^8.

4.6.176. Mori Masahiro, Yamada Yasusada, Incommensurate phase transitions in $R_2PbCu(NO_2)_6$ (R: Rb, Cs). - In: Nippon Genshiryoku Kenkyusho, Annu. Rep. Neutron Scattering Stud. JAERI-M-8009, 1978, p. 17-18. - CA, 1979, 91, 115559.

Exp.; E - e; O_h; d^9.

4.6.177. Noda Yukio, Mori Masahiro, Yamada Yasusada, Successive Jahn-Teller phase transitions in dipotassium lead hexanitrocuprate-(II). - J. Phys. Soc. Jpn., 1978, 45, N3, 954-966. - CA, 1978, 89, 172025.

Exp.; E - e; O_h; d^9.

4.6.178. Noda Yukio, Mori Masahiro, Yamada Yasusada, Incommensurate Jahn-Teller phase transition in $K_2PbCu(NO_2)_6$. - In: Proc. Int. Conf. Lattice Dynamics, 1977. Paris, 1978, p. 568-570. - CA, 1979, 90, 79313. PA, 1978, 81, 79730.

Exp.; E - e; O_h; d^9.

4.6.179. Ott, H. R., Kjems, J. K., Andres, K., Crystal-field transitions in $PrCu_2$ near the cooperative Jahn-Teller transition at T_d = 8 K. - In: Rare Earth and Actinides. London, 1978, p. 149-150. - PA, 1978, 81, 84178.

Exp.; d^9.

4.6.180. Propach, V., Steffens, F., On the structures of $CuZrF_6$ modifications - neutron diffraction studies on crystal powders. [Germ.]. - Z. Naturforsch. b, 1978, 33, N3, 268-274. - CA, 1978, 88, 180787.

Exp.; E - e; O_h; d^9.

4.6.181. Serator, M., Langfelderova, H., Gazo, J., Stracelsky, J., Temperature-dependent stereochemical transformations of distortion isomers of $CuBr_2(NH_3)_2$. - Inorg. Chim. Acta, 1978, 30, N2, 267-270.

Exp.; E - e; O_h; d^9; 4.3.

4.6.182. Smith, S. R. P., Tanner, B. K., γ-Ray diffractometry of lattice distortions in $TbVO_4$ caused by the Jahn-Teller phase transition. - J. Phys. C, 1978, 11, N17, L717-L723. - CA, 1979, 90, 144468. PA, 1978, 81, 91205.

Exp.

4.6.183. Van Eijndhoven, J. C. M., On the relation between Jahn-Teller ordering and charge ordering. - 1978. - 159 p. - Report INIS-mf-4658. - INIS Atomindex, 1979, 10, N3, 424336. CA, 1979, 91, 12301.

Exp.; d^9.

4.6.184. Wintenberger, M., Srour, B., Meyer, C., Hartmann-Boutron, F., Gros, Y., Crystallographic and Mössbauer study of zinc blende type FeS. - J. Phys. [France], 1978, 39, N9, 965-979. - CA, 1978, 89, 172022. PA, 1978, 81, 87650.

Exp.; 3.4; 4.4.

1979

4.6.185. Abell, G. C., Quasimolecular Jahn-Teller resonance states in the BCC metallic hydrides of vanadium, niobium, and tantalum. - Phys. Rev. B, 1979, 20, N12, 4773-4788. - CA, 1980, 92, 135711. PA, 1980, 83, 44550.

Theor.

4.6.186. Bertini, I., Dapporto, P., Gatteschi, D., Scozzafava, A., Static-dynamic distortions of the tris(1,2-diaminoethane)copper-(II) cation $[Cu(en)_3]^{2+}$: Crystal structures of the salts $[Cu(en)_3]$ $[SO_4]$ at 120 K and of $[Cu(en)_3]Cl_2 \cdot 075$ en at 298 K. - J. Chem. Soc. Dalton Trans., 1979, N9, 1409-1414. - CA, 1979, 91, 202512.

Exp.; E - e; O_h; d^9; 3.1; 4.4.

4.6.187. Crama, W. J., Bakker, M., Verschoor, G. C., Maaskant, W. J. A., The cooperative Jahn-Teller distorted structure of γ-rubidium chromium(II) trichloride. - Acta Crystallogr. B, 1979, 35, N8, 1875-1877. - CA, 1979, 91, 132399. PA, 1979, 82, 90604.

Exp.; E - e; O_h; d^4.

4.6.188. Freyberg, D. P., Robbins, J. L., Raymond, K. N., Smart, J. C., Crystal and molecular structures of decamethylmanganocene and decamethylferrocene. - J. Am. Chem. Soc., 1979, 101, N4, 892-897. - CA, 1979, 90, 131124.

Exp.; E - e; D_{5d}.

4.6.189. Guen, L., Marchand, R., Jouini, N., Verbaere, A., Crystal structure of cesium triiodochromate(II). [Fr.]. - Acta Crystallogr. B, 1979, 35, N7, 1554-1557. - CA, 1979, 91, 81787.

Exp.; E - e; O_h; d^4.

4.6.190. Hidaka, M., Walker, P. J., A comment on the structure of K_2CuF_4. - Solid State Commun., 1979, 31, N5, 383-385. - RZF, 1980, 1E781.

Exp.; E - e; O_h; d^9.

4.6.191. Kemmler-Sack, S., Rother, H. J., The barium zinc copper uranium oxide ($Ba_2Zn_{1-x}Cu_xUO_6$) system: A vibrational spectroscopic proof of the Jahn-Teller effect. [Germ.]. - Z. Anorg. Allg. Chem., 1979, 448, 143-155. - CA, 1979, 90, 159386.

Exp.; E - e; O_h; d^9; 4.3.

4.6.192. Lentz, A., Fabian, W., Single crystal growth of triple nitrites in gels. - J. Cryst. Growth, 1979, 47, N1, 121-123. - CA, 1979, 91, 115424.

Exp.; E - e; O_h; d^9.

4.6.193. Mori Masahiro, Yamada Yasusada, Incommensurate Jahn-Teller phase transitions in $Tl_2PbCu(NO_2)_6$ and $Rb_2PbCu(NO_2)_6$. - J. Phys. Soc. Jpn., 1979, 46, N5, 1673-1674. - CA, 1979, 91, 30806. PA, 1979, 82, 65185.

Exp.; E - e; O_h; d^9.

4.6.194. Mori Masahiro, Watanabe, T., Yamada Yasusada, One-dimensional stacking order in $Cs_2PbCu(NO_2)_6$. - J. Phys. Soc. Jpn., 1979, 47, N6, 1948-1954. - PA, 1980, 83, 28941.

Exp.; E - e; O_h; d^9.

4.6.195. Mori Masahiro, Watanabe, T., Yamada Yasusada, Sequential stacking order in Jahn-Teller active dicesium lead copper hexanitrate. - Sci. Rep. Res. Inst. Tohoku Univ. Ser. A, 1979, 27, (Suppl. 1), 65-67. - CA, 1980, 92, 13833.

Exp.; E - e; O_h; d^9.

4.6.196. Mori Masahiro, Noda Yukio, Yamada Yasusada, Incommensurate Jahn-Teller phase transitions in $R_2PbCu(NO_2)_6$ (R = K, Rb, Cs, Tl). - In: Modulated Struct.: AIP Conf. Proc., 1979, Vol. 53, p. 214-216. - CA, 1979, 91, 132302. PA, 1979, 82, 95156.

Exp.; E - e; O_h; d^9.

4.6.197. Reinen, D., Krause, S., Cooperative Jahn-Teller ordering and magnetic structure - pyridine-N-oxide complexes of Cu^{2+}: $Cu(ONC_5H_5)_6X_2$ [X = ClO_4^-, BF_4^-]. - Solid State Commun., 1979, 29, N10, 691-699. - CA, 1979, 90, 212798. PA, 1979, 82, 60505.

Exp.; E - e; O_h; d^9; 4.4.

4.6.198. Reinen, D., The Jahn-Teller effect in solid state chemistry of transition metal compounds. - J. Solid State Chem., 1979, 27, N1, 71-85. - CA, 1979, 90, 194811. PA, 1979, 82, 47061. RZK, 1979, 16B581.

Rev.; E - e; O_h; d^9.

4.6.199. Stein, J., Fackler, J. P., Jr., McClune, G. J., Fee, J. A., Chan, L. T., Superoxide and manganese(III). Reactions of Mn-EDTA and Mn-CyDTA complexes with O_2^-: X-ray structure of KMn-EDTA·$2H_2O$. - Inorg. Chem., 1979, 18, N12, 3511-3519. - CA, 1979, 91, 221629.

Exp.; E - e; O_h; d^4.

4.6.200. Tanaka, K., Konishi, M., Marumo, F., Electron-density distribution in crystals of $KCuF_3$ with Jahn-Teller distortion. - Acta Crystallogr. B, 1979, 35, N6, 1303-1308. - CA, 1979, 91, 66692. PA, 1979, 82, 69061.

Exp.; d^9.

4.6.201. Vasudevan, S., Shaikh, A. M., Rao, C. N. R., Jahn-Teller effect induced phase transitions in $CsCuCl_3$. - Phys. Lett. A, 1979, 70, N1, 44-46. - CA, 1979, 90, 178361. PA, 1979, 82, 47439. RZF, 1979, 7E1032.

Exp.; E - e; O_h; d^9; 2.7.

4.6.202. Zandbergen, H. W., The crystal structure of α-thallium hexaiodochromate, α-Tl_4CrI_6. - Acta Crystallogr. B, 1979, 35, N12, 2852-2855. - CA, 1980, 92, 102573. RZK, 1980, 10B397.

Exp.; E - e; O_h; d^4.

See also: 3.4.-3; 4.4.-166, 167; 5.1.-52; 5.2.-48.

4.7. Other Manifestations

This subsection lists papers on cooperative Jahn-Teller effect problems that are not included in the previous, more specialized subsections.

1958

4.7.1. Inoue, T., Iida, S., Specific heats of copper ferrite.- J. Phys. Soc. Jpn., 1958, 13, N6, 656.

Exp.; d^9.

1962

4.7.2. Siratori Kiichi, Iida, Shuichi, Anomalous temperature variation of lattice parameters in CrO_2. - J. Phys. Soc. Jpn., 1962, 17, Suppl. B-1, 208-211. - CA, 1963, 58, 1963f.

Exp.; d^2.

1964

4.7.3. Masumoto Kanichi, Electrical conductivity of chromium selenides. - J. Sci. Hiroshima Univ. Ser. A-11, 1964, 27, N2/3, 87-91. - CA, 1964, 61, 7804a.

Exp.

1966

4.7.4. Vernon, M. W., Lovell, M. C., Anomalies in the electrical conductivity of nickel oxide above room temperature. - J. Phys. & Chem. Solids, 1966, 27, N6/7, 1125-1131. - CA, 1966, 65, 3130d.

Exp.; d^8.

1967

4.7.5. Springthorpe, A. J., The rhombohedral distortion in NiO. - Phys. Status Solidi, 1967, 24, N1, K3-K4. - Solid State Abstr., 1967, 8, 69451.

Exp.; d^8.

1968

4.7.6. Buyers, W. J. L., Dolling, G., Sakurai Junji, Cowley, R. A., Magnetic excitations in cobaltous oxide (CoO). - In: Neutron Inelastic Scattering: Proc. 4th Symp. Copenhagen, 1968. Vienna, 1968, Vol. 2, p. 123-131. - CA, 1969, 70, 91778.

Exp.; d^7.

1971

4.7.7. Cooke, A. H., Martin, D. M., Wells, M. R., The specific heat of dysprosium vanadate. - Solid State Commun., 1971, 9, N9, 519-523.

Exp.

4.7.8. Gehring, K. A., Rosenberg, H. M., Microscopic observation of domains of Jahn-Teller distortion in dysprosium vanadate. - Phys. Status Solidi b, 1971, 47, N2, K75-K77. - CA, 1971, 75, 155615. PA, 1971, 74, 76882.

Exp.

4.7.9. Will, G., Schäfer, W., Göbel, H., Neutron and X-ray diffraction study of the magnetic and crystallographic phase transition in $DyVO_4$ and $DyAsO_4$. - In: Rare Earth and Actinides. London, 1971, p. 226-229.

Exp.

1972

4.7.10. Colwell, J. H., Mangum, B. W., The heat capacity of $TmAsO_4$ near its Jahn-Teller transition: Evidence of a low-lying excited state. - Solid State Commun., 1972, 11, N1, 83-87. - CA, 1972, 77, 93696. PA, 1972, 75, 61105.

Exp.

4.7.11. Jirmanus, N. S., McCarthy, K. A., The low-temperature thermal conductivity of pressed crystalline solids and chalcogenide glass. - In: Phonon Scattering in Solids: Introduct. Abstr. Int. Conf., Paris. Saclay, 1972, p. 357-358. - PA, 1973, 76, 1848.

Exp.

4.7.12. Kino Yoshihiro, Lüthi, B., Mullen, M. E., Cooperative Jahn-Teller phase transiton in the nickel-zinc chromite system. - J. Phys. Soc. Jpn., 1972, 33, N3, 687-697. - CA, 1972, 77, 131675. PA, 1972, 75, 71693.

Exp.; T; T_d; d^8.

4.7.13. Parons, M. W. S., Rosenberg, H. M., The thermal conductivity of rare earth vanadates at low temperature. - In: Phonon Scattering in Solids: Introduct. Abstr. Int. Conf. Paris. Saclay, 1972, p. 326-329. - PA, 1973, 76, 1847.

Exp.

4.7.14. Wells, M. R., Worswick, R. D., Specific heat of terbium vanadate $TbVO_4$. - Phys. Lett. A, 1972, 42, N4, 269-270. - CA, 1973, 78, 63155. PA, 1973, 76, 11748.

Exp.

1973

4.7.15. Kjems, J. K., Shirane, G., Birgeneau, R. J., Van Uitert, L. G., Quadrupole exciton-phonon dynamics at the 151°K phase transition in praseodymium aluminate. - Phys. Rev. Lett., 1973, 31, N21, 1300-1303. - CA, 1974, 80, 88093.

Exp.

4.7.16. Lüthi, B., Mullen, M. E., Andres, K., Bucher, E., Maita, J. P., Experimental investigation of the cooperative Jahn-Teller effect in TmCd. - Phys. Rev. B, 1973, 8, N6, 2639-2648. - CA, 1974, 80, 8660.

Exp.

4.7.17. Yamashiro Tomonobu, Electrical switching and memory phenomena in iron ferrate ($CuFe_2O_4$). - Jpn. J. Appl. Phys., 1973, 12, N1, 148-149. - CA, 1973, 78, 103219.

Exp.; T; T_d; d^9.

1974

4.7.18. Golba, P., Nevriva, M., Shestak, Ya., Use of the DTA method to determine the heats of transition in the solid solutions Mn_3O_4-Mn_2CrO_4. [Russ.]. - Izv. Akad. Nauk SSSR, Neorg. Mater., 1974, 10, N11, 2097-2098. - PA, 1976, 79, 30792.

Exp.; d^4.

4.7.19. Kahle, H. G., Simon, A., Wüchner, W., Observation of a cooperative phase transition in $TbVO_4$ at 0.61 K. - Phys. Status Solidi b, 1974, 61, N2, K53-K54.

Exp.

4.7.20. Sawaoka Akira, Tomizuka, C. T., Effect of hydrostatic pressure on the cooperative Jahn-Teller phase transition temperature of $NiCr_2O_4$. - J. Phys. Soc. Jpn., 1974, 36, N3, 912. - CA, 1974, 81, 17810. PA, 1974, 77, 48753. RZF, 1974, 9E659.

Exp.; T; T_d; d^8.

4.7.21. Wun, M., Phillips, N. E., Low-temperature heat capacity of praseodymium-copper $(PrCu)_2$. - Phys. Lett. A, 1974, 50, N3, 195-196. - CA, 1975, 82, 130045. PA, 1975, 78, 15557. RZF, 1975, 5E369.

Exp.; d^1.

4.7.22. Wun, M., Colorimetric studies of singlet-ground-state systems. - 1974. - 55 p. - Report LBL 3177. - CA, 1975, 83, 34878.

Exp.; d^9.

1975

4.7.23. Fujii, Y., Shirane, G., Yamada, Y., Study of the 123°K phase transition of magnetite by critical neutron scattering. - Phys. Rev. B, 1975, 11, N5, 2036-2041. - CA, 1975, 83, 19364.

Exp.; O_h.

4.7.24. Hutchings, M. I., Schem, R., Smith, S. H., Smith, S. R. P., Inelastic neutron scattering studies of the Jahn-Teller phase transition in terbium orthovanadate. - J. Phys. C, 1975, 8, N19, L393-L396. - CA, 1975, 83, 211493. PA, 1976, 79, 1747.

Exp.

4.7.25. Takagi Shozo, Joesten, M. D., Lenhert, P. G., Common boundary in multiple lattices of $K_2MCu(NO_2)_6$, where M is calcium, barium, lead. - Chem. Phys. Lett., 1975, 34, N1, 92-94. - CA, 1975, 83, 140186. PA, 1975, 78, 76659.

Exp.; E - e; O_h; d^9.

4.7.26. Wüchner, W., The specific heat of terbium phosphate $TbPO_4$. - J. Magn. & Magn. Mater., 1975, 2, N1/3, 203-206. - PA, 1976, 79, 80031.

Exp.

1976

4.7.27. Gehring, G. A., Random strain fields on a molecular field system - dilute thulium vanadate. - Solid State Commun., 1976, 18, N1, 31-34. - PA, 1976, 79, 30827.

Exp.

4.7.28. Novak, B., Pislewski, N., Leszczynski, W., Proton spin-lattice relaxation time in the dihydride phase of the ternary Ti-Nb-H alloy. - Phys. Status Solidi a, 1976, 37, N2, 669-674. - PA, 1977, 80, 4576.

Exp.; 3.2.

1977

4.7.29. Ihrig, H., Lohmann, W., The structural phase transformation of $CeAg_xIn_{1-x}$ in the silver-rich concentration range [and the onset of magnetic ordering]. - J. Phys. F, 1977, 7, N9, 1957-1963. - PA, 1978, 81, 2252.

Exp.; 4.4.

4.7.30. Zubkov, V. G., Belousov, F. I., Bazuev, G. V., Shveykin, G. P., Neutron-diffraction investigation of praseodymium orthovanadate. [Russ.]. - Fiz. Tverd. Tela, 1977, 19, N2, 618-620. - PA, 1978, 81, 2685.

Exp.; d^2.

1978

4.7.31. Borsa, F., Corti, M., Rigamonti, A., Electronic spin-dynamics at a structural phase transition by cooperative Jahn-Teller effect: an aluminum-27 NMR study in praseodymium aluminate. - J. Appl. Phys., 1978, 49, N3, Pt. 2, 1383-1385. - CA, 1978, 88, 200551. PA, 1978, 81, 68403.

Exp.; 3.3.

4.7.32. Kollewe, D., Gibson, W. M., Application of particle channeling to the study of a cooperative Jahn-Teller phase transition in nickel chromite. - Nucl. Instrum. Methods, 1978, 149, N1/3, 411-415. - CA, 1978, 88, 180568. PA, 1979, 82, 4413.

Exp.; T; T_d; d^8.

4.7.33. Komarov, A. G., Korovin, L. I., Kudinov, E. K., Neutron scattering in crystals with orbital degeneracy. [Russ.]. - Fiz. Tverd. Tela, 1978, 20, N3, 830-836.

Theor.

1979

4.7.34. Bykov, A. M., Ganenko, V. E., Markovich, V. I., Shteingart, F. A., Tsintsadze, G. A., Heat capacities and thermodynamic properties of $Cu_xZn_{1-x}SiF_6 \cdot 6H_2O$ (x = 1, 0.06, and 0) from 14 to 300 K. - J. Chem. Thermodyn., 1979, 11, N11, 1065-1073. - PA, 1980, 83, 21314.

Exp.; E - e; O_h; d^9.

4.7.35. Day, P., Hutchings, M. T., Janke, E., Walker, P. J., Cooperative Jahn-Teller ordering in the crystal structure of Rb_2CrCl_4: a two-dimensional easy-plane ionic ferromagnet. - J. Chem. Soc. Chem. Commun., 1979, N16, 711-713. - CA, 1979, 91, 220660.

Exp.; O_h; 4.4.

4.7.36. Endoh, Y., Kataoka, M., Takei, F., Noda, Y., Ishikawa, Y., Neutron scattering study of the cooperative Jahn-Teller effect in Fe_2TiO_4. - J. Magn. & Magn. Mater., 1979, 14, N2/3, 191-193. - CA, 1980, 92, 139847. PA, 1980, 83, 49678. RZF, 1980, 6E1856.

Exp.

4.7.37. Kashida Shoji, Effect of uniaxial stress upon the successive phase transitions in copper complex [$K_2PbCu(NO_2)_6$]. - J. Phys. Soc. Jpn., 1979, 47, N4, 1134-1140. - CA, 1979, 91, 220571. PA, 1980, 83, 6512.

Exp.; E - e; O_h; d^9.

4.7.38. Loewenhaupt, M., Rainford, B. D., Steglich, F., Dynamic Jahn-Teller effect in a rare earth compound: $CeAl_2$. - Phys. Rev. Lett., 1979, 42, N25, 1709-1712. - CA, 1979, 91, 66121. PA, 1979, 82, 78060. RZF, 1980, 3E267.

Exp.; O_h.

4.7.39. Murase, K., Sugai, S., Raman scattering from soft TO-phonon in IV-VI compound semiconductors. - Solid State Commun., 1979, 32, N1, 89-93. - CA, 1979, 91, 165834.

Exp.

4.7.40. Wenzel, G., Transport in cooperative T - (e + t) Jahn-Teller system. I. Heat current operator. - Z. Phys. B, 1979, 36, N2, 133-140. - PA, 1980, 83, 34405.

Theor.; T - (e + t); O_h.

4.7.41. Wenzel, G., Transport in cooperative T - (e + t) Jahn-Teller system. II. Pseudo-spin thermal conductivity. - Z. Phys. B, 1979, 36, N2, 141-149. - CA, 1979, 92, 65667. PA, 1980, 83, 34406.

Theor.; T - (e + t); O_h.

5. VIBRONIC EFFECTS IN CHEMICAL PROBLEMS. QUANTUM-CHEMICAL CALCULATIONS*

Vibronic effects in chemistry are manifested in molecular stereochemistry, mechanisms of chemical reactions, reactivity, and activation by coordination and catalysis. For concrete systems the numerical calculation of adiabatic potentials and vibronic constants is important.

5.1. General. Vibronic Effects in Chemical Reactions. Activation in Catalysis

In the paper [5.1.2] the pseudo-Jahn-Teller effect was used to reveal mechanisms of mono- and bimolecular chemical reactions, assuming that the vibronic mixing with the low-lying excited states determines the reaction path. In the subsequent series of works [5.1.-14, 15, 16, 20, 21, 25, 41] this concept was developed to include symmetry considerations in a more explicit way, resulting in what is now called orbital symmetry rules in mechanisms of chemical reactions (analogous to Woodward-Hoffmann rules). The theory allows one to predict qualitatively the geometry of transition states and the reaction mechanisms for a number of systems including coordination compounds. A theoretical analysis of the orbital symmetry rules is given in Refs. [5.1.-19, 39]. Stability and stereochemistry with Jahn-Teller and pseudo-Jahn-Teller effects are discussed in Refs. [5.1.-12, 18, 35]. A new approach to the problem of the origin and mechanism of chemical activation by coordination based on vibronic interaction theory is developed in t he works [5.1.-43, 49, 50, 51, 52, 53, 59]. In papers [5.1.-42, 48, 56] the electron transfer and multipole mechanism of energy transfer between orbitally degenerate molecular systems are considered.

Nonadiabatic chemical reactions with high symmetry and orbitally degenerate intermediates are controlled by Jahn-Teller instability.

*The short introductions to this section and subsections were prepared by S. S. Budnikov.

The theory of inelastic scattering resulting, in particular, in triatomic intermediate states with an electronic E term is considered in Refs. [5.1.-28, 31]. The low electron transfer reaction rate in Jahn-Teller systems is attributed to the Franck-Condon factor [5.2.24] (see also [5.1.29]). The photochemical behavior of coordination compounds is considered in Ref. [5.1.55]. The redox reaction with chlorophyll is shown to be controlled by the pseudo-Jahn-Teller mixing of the ground state with the low-lying excited state [5.1.60]. References [5.1.-34, 37] are devoted to MO LCAO calculations of Jahn-Teller distortions in planar C_3H_3 and to the correlation of these distortions with reaction paths.

1932

5.1.1. Pelzer, H., Wigner, E., Velocity coefficient of interchange reactions. - Z. Phys. Chem. Frankfurt a.M. , 1932, B15, p. 445-471. - CA, 1932, 26, 2638.

Theor.; 1.2.

1962

5.1.2. Bader, R. F. W., Vibrationally induced perturbations in molecular electron distributions. - Can. J. Chem., 1962, 40, N6, 1164-1175. - CA, 1962, 57, 8068g. RZK, 1963, 7B32.

Theor.

1963

5.1.3. Dibeler, V. H., Rosenstock, H. M., Metastable transitions in mass spectra of methane and the deuteriomethanes. - J. Chem. Phys., 1963, 39, N5, 1326-1329. - CA, 1963, 59, 7065f.

Exp.; T; T_d.

5.1.4. Witkowski, A., Coupling of charge transfer with nuclear motion in aromatic free radicals. [Fr.]. - C.R. Hebd. Seances Acad. Sci., 1963, 256, N2, 419-421. - CA, 1963, 58, 7509g.

Theor.; 1.5.

1964

5.1.5. Cashion, J. K., Herschbach, D. R., Comment on the activation energy for the H + H_2 reaction. - J. Chem. Phys., 1964, 41, N7, 2199-2200. - PA, 1965, 68, 3001.

Theor.; 5.4.

5.1.6. Ellison, F. O., Potential-energy surface for the H + H_2 reaction. - J. Chem. Phys., 1964, 41, N7, 2198-2199. - PA, 1965, 68, 3000.

Theor.; 5.4.

5.1.7. Sharma, V. S., Mathur, H. B., Biswas, A. B., Jahn-Teller stabilization and entropy changes accompanying the formation of metal-amino acid complexes. - J. Inorg. & Nucl. Chem., 1964, 26, N2, 382-384. - CA, 1964, 60, 8712c.

Exp.; E - e; O_h; d^9.

1965

5.1.8. Merling, S., Application of crystal field theory to the study of the distribution of trace elements. [Ital.]. - Atti Soc. Toscana Sci. Nat. Pisa Proces. Verbali Mem. Ser. A, 1965, 72, N1, 174-185. - CA, 1967, 66, 22940.

Theor.

1966

5.1.9. Osherov, V. I., About the nonadiabatic transitions in a solid state. [Russ.]. - Fiz. Tverd. Tela, 1966, 8, N11, 3295-3298.

Theor.

1967

5.1.10. De Boer, E., Colpa, J. P., Signs of spin densities and vibronic interactions in 1 and 1,4 alkyl-substituted benzene anions. - J. Phys. Chem., 1967, 71, N1, 21-25.

Exp.; 1.5; 3.1; 3.3.

5.1.11. Kobayashi Hiroshi, Electronic structure of complex ions related to their chemical reactions. [Jap.]. - Shokubai, Catalyst, 1967, 9, N2, 113-126. - CA, 1968, 69, 14083.

Ref.; d^8.

1968

5.1.12. Bersuker, I. B., The coordination chemical bond and some aspects of catalysis. [Russ.]. - In: Probl. Kinetiki i Kataliza. T.13. Kompleksoobrazovanie v Katalize. Moscow, 1968, p. 7-35.

Theor.; 1.5.

1969

5.1.13. Kurz, G. E., OH^- dissociation and U_2 decomposition in hydroxide-doped potassium chloride and potassium bromide crystals. - Phys. Status Solidi, 1969, 31, N1, 93-106. - CA, 1969, 70, 82754.

Exp.; T.

5.1.14. Pearson, R. G., A symmetry rule for predicting molecular structure and reactivity. - J. Am. Chem. Soc.,1969, 91, N5, p. 1252-1254.

Theor.; 1.5; 5.2.

1970

5.1.15. Pearson, R. G., Symmetry rules for predicting the course of chemical reactions. - Theor. Chim. Acta, 1970, 16, N2, 107-110.

Theor.; 1.5.

5.1.16. Pearson, R. G., Molecular orbital symmetry rules. - Chem. Eng. News, 1970, 48, N41, 66-72. - CA, 1970, 73, 134330.

Theor.; 1.5.

1971

5.1.17. Barriol, J., Metallic ions and complexes. [Fr.]. - In: Interactions Cations Metal. Macromol. Biol.: Ecole Ete Roscoff, 1970. Paris, 1971, p. 5-14. - CA, 1972, 77, 118550.

Theor.

5.1.18. Bersuker, I. B., Nuclear dynamics and instability in coordination compounds. [Russ.]. - In: Primeneniya Noveyshikh. Fiz. Metodov Issled. Koord. Soedin.: Tez. Dokl. 4 Vsesoyuz. Soveshch. Kishinev, 1971, p. 7-9.

Theor.

5.1.19. George, T. F., Ross, J., Analysis of symmetry in chemical reactions. - J. Chem. Phys., 1971, 55, N8, 3851-3866. - CA, 1971, 75, 144100.

Theor.; 1.5.

5.1.20. Pearson, R. G., Orbital symmetry rules and the mechanism of inorganic reactions. - Pure & Appl. Chem., 1971, 27, N1/2, 145-160. - CA, 1972, 76, 63881.

Theor.; 1.5; 5.2.

5.1.21. Pearson, R. G., Symmetry rules for chemical reactions. - Accounts Chem. Res., 1971, 4, N4, 152-160. - CA, 1971, 74, 130550.

Theor.; 1.5.

1972

5.1.22. Murrell, J. N., Jahn-Teller theorem applied to transition states. - J. Chem. Soc. Chem. Commun., 1972, N18, 1044-1045. - CA, 1972, 77, 156482.

Theor.

5.1.23. Vorotyntsev, M. A., Granovskiy, A. A., Dogonadze, R. R., Kuznetsov, A. M., The possibility of nonadiabatic transition in the system of three terms. [Russ.]. - Vestn. Mosk. Univ. Fiz., Astron., 1972, N1, 59-65.

Theor.

1973

5.1.24. Glick, M. D., Kuszaj, J. M., Endicott, J. F., Structural evidence for Franck-Condon barrier to electron transfer between low-spin cobalt(II) and cobalt(III). - J. Am. Chem. Soc., 1973, 95, N15, 5097-5098. - RZF, 1973, 12D260.

Exp.

5.1.24a. Moss, D. B., Chin-tung Lin, Rorabacher, D. B., Kinetics of Aquocopper(II) ion reacting with polyamines and poly-(amino alcohols). Evidence for rapid Jahn-Teller inversion. - J. Amer. Chem. Soc., 1973, 95, 5179-5185.

5.1.25. Pearson, R. G., Orbital symmetry rules for inorganic reactions from perturbation theory. - Fortschr. Chem. Forsch., 1973, 41, 75-112. - CA, 1974, 80, 30952.

Theor.; 1.5.

5.1.26. Vorotyntsev, M. A., Dogonadze, R. R., Kuznetsov, A. M., On the theory of adiabatic and nonadiabatic transitions. I. The calculation of transition probability. [Russ.]. - Vestn. Mosk. Univ. Fiz., Astron., 1973, N2, 224-228.

Theor.

1974

5.1.27. Dewar, M. J. S., Kirschner, S., Kollmar, H. W., Wade, L., Orbital isomerism in biradical processes. - J. Am. Chem. Soc., 1974, 96, N16, 5242-5244. - CA, 1974, 81, 119742.

Theor.

5.1.28. Voronin, A. I., Osherov, V. I., Nonadiabatic transitions in triatomics. [Russ.]. - Zh. Eksp. & Teor. Fiz., 1974, 66, N1, 135-145.

Theor.; E - e; C_{3v}.

1975

5.1.29. Endicott, J. F., Ferraudi, G. J., Barber, J. R., Charge transfer spectroscopy, redox energetics, and photoredox behavior of transition metal ammine complexes: Critical comparison of observations with mechanisms and models. - J. Phys. Chem., 1975, 79, N6, 630-643. - CA, 1975, 82, 131481.

Exp.; D^6; 2.1.

5.1.30. Englman, R., Vibronic method in chemical reaction rate theory. - Collect. Phenom., 1975, 2, N1, 49-54. - CA, 1975, 82, 175792.

Theor.

5.1.31. Karkach, S. P., Osherov, V. I., Ushakov, V. G., Nonadiabatic transitions in triatomic systems. [Russ.]. - Zh. Eksp. & Teor. Fiz., 1975, 68, N2, 493-505. - CA, 1975, 82, 175365. RZF, 1975, 6D267.

Theor.; E - e; C_{3v}.

5.1.32. Mulak, J., Stevens, K. W. H., Electron transfer from Eu^{2+} to Eu^{3+}. - Z. Phys. B, 1975, 20, N1, 21-31. - PA, 1975, 78, 42390.

Theor.; 1.5.

5.1.33. Parkins, A. W., Slade, R. C., Model for catalysis by nickel(II) complexes. - J. Chem. Soc. Dolaton Trans., 1975, N13, 1352-1356. - CA, 1975, 83, 121311.

Theor.; T_d; d^8; 1.5.

1976

5.1.34. Beran, S., Zahradnik, R., MO study of the reactivity of cyclopropane, its radical and radical ions, and models of its transition metal complexes. - Collect. Czech. Chem. Commun., 1976, 41, N8, 2303-2319. - CA, 1976, 85, 192047.

Theor.; E - e; D_{3h}; 5.4.

5.1.35. Bersuker, I. B., Electronic effects in metalloenzyme catalysis. [Russ.]. - In: Tez. Dokl. Vyezdnoy Sessii Akad. Nauk SSSR. Krasnodar, 1976, p. 3.

Theor.; 1.5.

5.1.36. Burdett, J. K., Grzyboski, J. M., Peruts, R. N., Poliakoff, M., Turner, J. J., Turner, R. F., Photolysis and spectroscopy with polarized light: key to the photochemistry of $Cr(CO)_5$ and related species. - Inorg. Chem., 1976, 17, N1, 147-154.

Exp.; E; D_{3h}; 2.7.

5.1.37. Carsky, P., Zahradnik, R., Radicals: Their molecular orbitals, properties, and reactivity. - Acc. Chem. Res., 1976, 9, N11, 407-411. - CA, 1977, 86, 71012.

Theor.; E - e; D_{3h}.

5.1.38. Cayley, G., Cross, D., Knowles, P., Slow substitution reactions of the copper complex of 2,2',2"-triaminotriethylamine. - J. Chem. Soc. Chem. Commun., 1976, N20, 837-838. - CA, 1977, 86, 111592.

Exp.; E - e; O_h; d^9.

5.1.39. Chiu Ying-nan, Vibronic symmetry correlation in molecular combination and dissociation. - J. Chem. Phys., 1976, 64, N7, 2997-3002.

Theor.

5.1.40. Nguyen Dinh Phung, Tedenac, J. C., Crystallochemical study of complexes of the formula $M^{II}(bpy)_3AB_4 \cdot xH_2O$ ($M^{II} = Ni^{II}$ or Cu^{II}, $AB_4^{2-} = SO_4^{2-}$ or BeF_4^{2-}). [Fr.]. - C.R. Hebd. Seances Acad. Sci. Ser. C, 1976, 282, N5, 273-275. - CA, 1976, 84, 172378.

Exp.; E; D_3; 4.6.

5.1.41. Pearson, R. G., Symmetry Rules for Chemical Reactions: Orbital Topology and Elementary Processes. - New York: Wiley-Interscience, 1976. - 590 p.

5.1.42. Vekhter, B. G., Rafalovich, M. L., Special features of electron transport between degenerate states. [Russ.]. - Fiz. Tverd. Tela, 1976, 18, N7, 2093-2095. - PA, 1977, 80, 31987.

Theor.; E - e.

1977

5.1.43. Bersuker, I. B., To the theory of the elementary act in catalysis. [Russ.]. - Kinet. Katal., 1977, 18, N5, 1268-1282.

Theor.

5.1.44. Budnikov, S. S., Dimoglo, A. S., Electronic structure and orbital symmetry in substitution reactions of amine complexes of cobalt(III), rhodium(III), and iridium(III) taking into account the pseudo Jahn-Teller effect. [Russ.]. - Teor. & Eksp. Khim., 1977, 13, N4, 447-454. - CA, 1977, 87, 173130.

Theor.; 1.5.

5.1.45. Gordon, M. S., Excited states and photochemistry of saturated molecules: Extended basis calculations on the $1B_1(1T_2)$ state of methane. - Chem. Phys. Lett., 1977, 52, N1, 161-167.

Theor.; T; T_d.

5.1.46. Oksengendler, B. L., Yunusov, M. S., Study of atomic activation processes in nonmetals based on the Jahn-Teller effect. [Russ.]. - Izv. Akad. Nauk Uzb. SSR. Ser. Fiz. Mat. Nauk, 1977, N5, 65-70. - CA, 1978, 88, 55165.

Theor.

5.1.47. Osherov, V. I., Ushakov, V. G., Two-channel dissociation of $H + H_2 \rightarrow H + H + H$. [Russ.]. - Dokl. Akad. Nauk SSSR, 1977, 236, N1, 68-71. - CA, 1977, 87, 173042.

Theor.; E - e.

5.1.48. Vekhter, B. G., Rafalovich, M. L., Features of the electron transfer between degenerate states. - Chem. Phys., 1977, 21, N1, 21-25. - CA, 1977, 86, 195445. PA, 1977, 80, 38727.

Theor.; E; T.

1978

5.1.49. Bersuker, I. B., Vibronic mutual influence of weakly coordinated molecular groups in chemical reaction and catalysis. - Chem. Phys., 1978, 31, N1, 85-93.

Theor.

5.1.50. Bersuker, I. B., Vibronic factors in the catalytic fixation of nitrogen. [Russ.]. - Teor. & Eksp. Khim., 1978, N1, 3-12.

Theor.

5.1.51. Bersuker, I. B., The vibronic theory of elementary processes in coordination catalysis. - In: Proc. 19th Int. Conf. Coord. Chem. Prague, 1978, Vol. 2, p. 14.

Theor.; 1.5.

5.1.52. Bersuker, I. B., The significance of the Jahn-Teller effect in coordination chemistry. - In: Proc. 19th Int. Conf. Coord. Chem. Prague, 1978, Vol. 2, p. 145-147.

Theor.; 1.1; 1.5; 2.1; 4.6.

5.1.53. Bersuker, I. B., The Jahn-Teller effect and vibronic influence in structure and reactivity of coordination compounds. - In: Proc. 7th Symp. Coord. Chem. Bratislava, 1978, p. 19-20. - CA, 1980, 92, 87155.

5.1.54. Knyazev, S. P., Brattsev, V. A., Stanko, V. I., Manifestations of Jahn-Teller effects in rearrangements of polyhedral boron compounds: Chemically induced thermal rearrangements. [Russ.]. - Dokl. Akad. Nauk SSSR, 1978, 239, N5, 1136-1139. - CA, 1978, 88, 197904.

Exp.; 3.3.

5.1.55. Lehmann, G., Solid-state photochemistry - a method generating unusual valence states. - Angew. Chem., 1978, 17, N2, 89-96.

Exp.

5.1.56. Vekhter, B. G., Rafalovich, M. L., On the multipole mechanism of energy transfer between symmetrical biological compounds. [Russ.]. - Biofizika, 1978, 24, N6, 1097-1098.

Theor.

5.1.57. Volkov, S. V., Role of orbital and spin degeneracies and calculation of the Jahn-Teller effect in exchange reactions with electron transfer. [Russ.]. - Elektrokhimiya, 1978, 14, N10, 1528-1532. - CA, 1979, 90, 33314.

Theor.

1979

5.1.58. Bacci, M., Jahn-Teller effect in biomolecules. - Biophys. Chem., 1979, 11, N1, 39-47.

Rev.; $E - e$; $E - (b_1 + b_2)$; $T - e$; O_h; T_d; C_{4v}; d^6; d^9; 3.1; 3.4; 3.6; 5.3.

5.1.59. Bersuker, I. B., Modern aspects of structure/reactivity problems for coordination compounds. - In: Proc. 20th Int. Conf. Coord. Chem. Calcutta, 1979, p. 55-59.

Theor.; 1.1.

5.1.60. Davis, M. S., Forman, A., Fajer, J., Ligated chlorophyll cation radicals: their function in photosystem II of plant photosynthesis. - Proc. Natl. Acad. Sci. USA, 1979, 76, N9, 4170-4174. - RZBK, 1980, 7H397.

Exp.; D_{4h}; 1.5; 2.1.

5.1.61. Desouter-Lecomte, M., Galloy, C., Lorquet, J. C., Pires, M. V., Nonadiabatic interactions in unimolecular decay. V. Conical and Jahn-Teller intersections. - J. Chem. Phys., 1979, 71, N9, 3661-3672. - CA, 1980, 92, 31173. PA, 1980, 83, 25997.

Theor.

5.1.62. Gemmel, D. S., Kanter, E. P., Pietsch, W. J., Experimental confirmation of the Jahn-Teller distortion of methane(1+) ion. - 1979. - 8 p. - Report CONF - 790843-6. - CA, 1980, 92, 169655.

Exp.

5.1.83. Gordin, M. S., Caldwell, J. W., Excited states and photochemistry of saturated molecules. VII. Potential energy surfaces in excited singlet states of methane. - J. Chem. Phys., 1979, 70, N12, 5503-5514.

Theor.; T_d; 5.4.

5.1.64. Rosenberg, P. E., Foit, F. F., Jr., The stability of transition metal dolomites in carbonate system: a discussion. - Geochim. Cosmochim. Acta, 1979, 43, N7, 951-955. - CA, 1980, 92, 48205.

Theor.; O_h.

5.1.65. Sacher, E., A re-examination of the polyurethane reaction. - J. Macromol. Sci. B, 1979, 16, N4, 525-538.

Theor.; E - e; O_h; d^9.

5.1.66. Vogler, M., Observation of an electronically excited state of H_3 and determination of its vibrational level structure. - Phys. Rev. A, 1979, 19, N1, 1-5. - PA, 1979, 82, 41204.

Exp.

See also: 1.1.-15, 30, 68; 1.6.-20, 23.

5.2. Vibronic Effects in Stereochemistry

In papers [5.2.-3, 4, 5, 6, 8, 27, 48] the effect of the number of p and d electrons of the metal atom in coordination compounds on the Jahn-Teller distortions and their importance in the stereochemistry is discussed. Theoretical predictions of molecular geometry based on the Jahn-Teller effect are given in Refs. [5.2.-9, 21, 30, 36]. Work of the same nature for excited states appears in Ref. [5.2.43]. An interesting conclusion about the "plasticity" property of molecular systems having linear or two-dimensional troughs on the adiabatic potential surface, especially important in crystal chemistry (see Section 4), was made in papers [4.6.114, 5.2.72].

A number of works are devoted to the so-called mutual influence of ligands taking into account vibronic effects [5.2.-74, 75, 76, 83, 84, 94, 95, 96]. The pseudo-Jahn-Teller effect in metalloporphyrins and hemoproteins as a basis for studying metal stereochemistry, geometry of ligand coordination, cooperative effects, and photolysis is considered in Refs. [5.2.-77, 92, 97, 98, 99, 100, 103, 108]. Papers [5.2.-12, 34, 56, 70, 88, 105] are typical examples of the Jahn-Teller interpretation of stereochemical X-ray data for systems with degenerate electronic states.

1939

5.2.1. Penny, W. G., The theory of molecular structure. - Rep. Prog. Phys., 1939, 6, 212-237. - CA, 1941, 35, 3493.

Theor.; 1.2.

1957

5.2.2. Liehr, A. D., Adiabatic correlations for the formation of ions in degenerate electronic states. - J. Chem. Phys., 1957, 27, N2, 476-477.

Theor.; T - e; T_d; 1.2.

1960

5.2.3. Bersuker, I. B., On the problem of coordination compound symmetry. [Russ.]. - Dokl. Akad. Nauk SSSR, 1960, 132, N3, 587-590.

Theor.

1961

5.2.4. Bersuker, I. B., About the impossibility of a new type of isomerism based on internal configuration asymmetry in coordination compounds. [Russ.]. - Zh. Fiz. Khim., 1961, 35, N2, 471-472.

Theor.

5.2.5. Bersuker, I. B., About two conformations of inorganic transition metal complexes. [Russ.]. - Dokl. Akad. Nauk SSSR, 1961, 141 , N1, 87-89.

Theor.; 1.2.

5.2.6. Bersuker, I. B., Kroytoru, Yu. G., About the symmetry of ligand coordination around the central ion with unclosed shells. [Russ.]. - In: Tez. Dokl. 4 Soveshch. Kristallokhim. Kishinev, 1961, p. 33.

Theor.; O_h.

5.2.7. Coulson, C. A., Golebiewski, A., On perturbation calculation for the π-electrons and their application to bond length reconsiderations in aromatic hydrocarbons. - Proc. Phys. Soc. London, 1961, 78, N6, 1310-1320.

Theor.; 5.4.

1962

5.2.8. Bersuker, I. B., Titova, Yu. G., About the symmetry of ligand coordination around the central ion with unclosed p-shells. [Russ.]. - Izv. Akad. Nauk Mold.SSR, 1962, N10, 10-22.

Theor.; T - e; O_h.

5.2.9. Boer, D. H. W., den, Boer, P. C., den, Longuet-Higgins, H. C., The molecular symmetry of dibenzenechromium. - Mol. Pjys., 1962, 5, N4, 387-390. - RZK, 1963, 10B34.

Theor.; (A + B) - b; D_{6h}; 1.5.

5.2.10. Coulson, C. A., Golebiewski, A., The Jahn-Teller effect in aromatic ions and radicals. - Mol. Phys., 1962, 5, N1, 71-84. - CA, 1962, 57, 71 h.

Theor.

5.2.11. Coulson, C. A., Strauss, H. L., Static Jahn-Teller distortions in the small molecules: CH_4^+, CF_4^+, NH_3^+ (excited state) and NH_3 (excited state). - Proc. R. Soc. London, Ser. A, 1962, 269, N1339, 443-455. - CA, 1962, 57, 13304e.

Theor.; T_d; D_{3v}; D_{3h}.

1963

5.2.12. Albrecht, G., Chemical bonding in some tris(dipyridine) complexes of anomalous valency. - Z. Chem., 1963, 3, N1, 32-33. [Germ.]. - CA, 1963, 59, 4556b.

Exp.; 4.6.

1964

5.2.13. Herzberg, G., Determination of the structures of simple polyatomic molecules and radicals in electronically excited states: Twelfth Spiers Memorial Lecture. - Discuss. Faraday Soc., 1963, 35, 7-29. - RZF, 1964, 10D125.

Rev.

5.2.14. Morosin, B., Brathovde, J. R., The crystal structure and molecular configuration of trisacetylacetonatomanganese(III). - Acta Crystallogr., 1964, 17, 705-711. - CA, 1964, 61, 2564e.

Exp.; d^4; 4.6.

1965

5.2.15. Carrington, A., Longuet-Higgins, H. C., Todd, P. F., The tricyano-s-triazine anion: a permanent Jahn-Teller distortion? - Mol. Phys., 1965, 9, N3, 211-215. - CA, 1966, 64, 195d.

Exp.; 3.1; 5.4.

5.2.16. Pryce, M. H. L., Sinha, K. P., Tanabe, Y., On the tetragonal distortion of octahedral systems in an E_g electronic state. - Mol. Phys., 1965, 9, N1, 33-42. - CA, 1965, 63, 9070h. RZF, 1966, 1D117.

Theor.; E - e; O_h.

5.2.17. Sharnoff, M., Reimann, C. W., Intrinsic and lattice-induced distortion of the tetrachlorocuprate ion. - J. Chem. Phys., 1965, 43, N9, 2993-2996. - CA, 1965, 63, 17353a.

Exp.; T; T_d; d^9; 2.1; 3.1.

1966

5.2.18. Dunn, T. M., Spectrum and structure of C_6H_6. - In: Stud. Chem. Struct. and Reactivity. New York, 1966, p. 103-132. - CA, 1967, 66, 50414.

Rev.; D_{6h}; 2.6; 2.7; 2.8; 3.3.

5.2.19. Herzberg, G., Johns, J. W. C., The spectrum and structure of singlet CH_2. - Proc. R. Soc. London, Ser. A, 1966, 295, N1441, 107-128. - CA, 1966, 65, 19485d.

Exp.; 1.6; 2.1.

5.2.20. Russ, B. J., Wood, J. S., Molecular structure of bis-(trimethylene)titanium tribromide; an example of the Jahn-Teller effect in pentacoordinate molecules. - Chem. Commun., 1966, N20, 745-746. - CA, 1967, 66, 41359.

Exp.; E - e; D_{3h}; 2.5.

1967

5.2.21. Bartell, L. S., Evidence for pseudo-Jahn-Teller effect in XeF_6. - J. Chem. Phys., 1967, 46, N11, 4530-4531. - CA, 1967, 67, 48057.

Theor.; T - t; O_h; 1.5.

5.2.22. Collin, J. E., Delwiche, J., Ionization of methane and its electronic energy levels. - Can. J. Chem., 1967, 45, N16, 1875-1882. - CA, 1967, 67, 76426.

Exp.; T; T_d.

5.2.23. Herzberg, G., Johns, J. W. C., The spectrum and structure of the free BH_2 radical. - Proc. R. Soc. London, Ser. A, 1967, 298, N1453, 142-159. - CA, 1967, 66, 89882.

Exp.; 1.6.

5.2.23a. Bartell, L. S., Molecular geometry. Bonded versus nonbonded interactions. - J. Chem. Educ., 1968, 45, N12, 754-767.

1968

5.2.24. Buenker, R. J., Peyerimhoff, S. D., Study on the stability and geometry of cyclobutadiene. - J. Chem. Phys., 1968, 48, N1, 354-358.

Theor.; T.

5.2.25. Hathaway, B. J., Slade, R. C., The influence of cations on the properties of the hexanitrometallate anions in the $K_2M^{II}M^{II}$-$(NO_2)_6$ type of complex of the nickel(II) and copper(II) ions. - J. Chem. Soc. Ser. A, 1968, N1, 85-87. - CA, 1968, 68, 44402.

Exp.; E; O_h; d^9; 2.1; 2.7; 3.1; 3.6; 4.6.

5.2.26. Seip, H. M., Jahn-Teller effect: Experimental and theoretical studies in vanadium tetrachloride. [Norw.]. - Tidsskr. Kjemi Bergv. Met., 1968, 28, N8/9, 177-183. - CA, 1969, 70, 40668.

Theor.; T_d; 2.6.

1969

5.2.27. Bersuker, I. B., Vibronic interaction and stereochemistry of coordination compounds. - In: Proc. 2nd Conf. Coord. Chem. Smolenice. Bratislava, 1969, p. 23-24.

Theor.

5.2.28. Borden, W. T., Differences in expected stability of the triplet states in cyclic π-systems containing 4n electrons. - J. Chem. Soc. Ser. D, 1969, N15, 881-882. - CA, 1969, 71, 75725.

Theor.; 1.5.

5.2.29. Companion, A. L., Applications of diatomics-in-molecules theory. III. The Li_4 system. - J. Chem. Phys., 1969, 50, N3, 1165-1167. - PA, 1969, 72, 30034.

Theor.; T_d.

5.2.30. Pearson, R. G., A symmetry rule for predicting molecular structures. - J. Am. Chem. Soc., 1969, 91, N18, 4947-4955. - CA, 1969, 71, 74171. RZK, 1970, 6B72.

Theor.; 1.5.

5.2.31. Veidis, M. V., Schreiber, G. H., Gough, T. E., Palenik, G. J., Jahn-Teller distortions in octahedral copper(II) complexes. - J. Am. Chem. Soc., 1969, 91, N7, 1859-1860. - CA, 1969, 71, 16713.

Exp.; E - e; O_h; d^9; 2.5; 3.1.

1970

5.2.32. Bartlett, M., Palenik, G. J., The crystal structure of an octahedral high-spin Mn^{III} complex, acetylacetonatobis(N-phenylaminotroponiminato)manganese(III). - J. Chem. Soc. Ser. D, 1970, N7, 416.

Exp.; E - e; O_h; d^9; 4.6.

5.2.33. Burdett, J., Electron relaxation and molecular vibrations: triatomic systems of the first-row elements, the methyl radical, and diazomethane and ketene. - J. Chem. Phys., 1970, 52, N6, 2983-2992. - CA, 1970, 72, 103956.

Theor.; 1.5.

5.2.34. Cullen, D. L., Lingafelter, E. C., Redetermination of the crystal structure of tris-(ethylenediamine)copper(II) sulfate, $Cu(NH_2CH_2CH_2NH_2)_3SO_4$. - Inorg. Chem., 1970, 9, N8, 1858-1864. - CA, 1970, 73, 70876.

Exp.; d^3; 4.6.

5.2.35. Grimes, N. W., Hilleard, R. J., X-ray diffraction studies of the spinel series $Mg(Cr_x-Al_{2-x})O_4$. I. Lattice parameters and structure. - J. Phys. C, 1970, 3, N4, 866-871. - CA, 1970, 73, 49618.

Exp.; d^3.

5.2.36. Pearson, R. G., Symmetry rule for predicting the structure of molecules. II. Structure of X_2Y_n molecules. - J. Chem. Phys., 1970, 52, N5, 2167-2174. - CA, 1970, 72, 93510. PA, 1970, 73, 33346.

Theor.; 1.5.

5.2.37. Smith, D. W., Overlap considerations relevant to the geometry of copper(II). - J. Chem. Soc. Ser. A, 1970, N9, 1498-1503. - CA, 1970, 73, 29080.

Theor.; d^9.

5.2.38. Stockdale, J. A. D., Compton, R. N., Schweinler, H. C., Negative ion formation in selected hexafluoride molecules. - J. Chem. Phys., 1970, 53, N4, 1502-1507. - CA, 1970, 73, 69982. RZF, 1971, 1D355.

Exp.; O_h.

1971

5.2.39. Gergely, A., Nagypal, I., Kiraly, B., Equilibrium of α-amino acid complexes of transition metal ions. IV. Stability constants, enthalpy, and entropy changes of the alanine, phenylalanine, and tyrosine complexes. - Acta Chim. Acad. Sci. Hung., 1971, 68, N4, 285-296. - CA, 1971, 75, 41107.

Exp.; d^9.

5.2.40. Healy, P. C., White, A. H., Metal-ligand bond contraction and possible spin-orbit induced Jahn-Teller distortion in the iron(III) dialkyldithiocarbamate crossover system. - J. Chem. Soc. Ser. D, 1971, N22, 1446-1447. - CA, 1972, 76, 38463.

Exp.; d^9; 4.6.

5.2.41. Mingos, D. M., Geometries of coordinated ligands. - Nature. Phys. Sci., 1971, 229, N7, 193-198. - CA, 1971, 75, 14439. PA, 1971, 74, 32565.

Theor.; 1.5.

5.2.42. Paulic, N., Simeon, V., Bernik, B., Svigir, B., Fles, D., Stability and stereochemistry of cobalt(II) complexes with stereoisomeric 1,2-diaminocyclopentanetetraacetic acids. - J. Inorg. & Nucl. Chem., 1971, 33, N10, 3463-3469. - CA, 1972, 76, 30259.

Exp.; d^7; 2.1.

5.2.43. Pearson, R. G., The structures of molecules in excited electronic states. - Chem. Phys. Lett., 1971, 10, N1, 31-34. - PA, 1971, 74, 60979.

Theor.; 1.5.

1972

5.2.44. Fackler, J. P., Jr., Avdeef, A., Costamagna, J., Manganese(III) tropolonate: A Jahn-Teller molecule. - In: Proc. 14th Int. Conf. Coord. Chem., Toronto, 1972, p. 589.

Exp.; E; C_{3v}; d^4; 4.6.

5.2.45. Frisch, P. D., Dahl, L. F., Organometallic chalcogen complexes. XXV. Structural and magnetic studies of $Co_3(h^5-C_5H_5)_3$-(CO)(S), $Co_3(h^5-C_5H_5)_3S_2$, and the oxidized monocation, $[Co_3(h^5-C_5H_5)_3S_2]^+$: Sensitivity of the geometry of a triangular metal cluster system to antibonding electrons. - J. Am. Chem. Soc., 1972, 94, N14, 5082-5084. - CA, 1972, 77, 67252.

Exp.; E; D_{3h}; 3.6; 4.6.

5.2.46. Muller, E., Krausse, J., Schmiedeknecht, K., Structural investigation on organochromium compounds. III. X-ray structure analysis and spectroscopic investigation on sodium pentaphenylchromium compounds. III. X-ray structure analysis and spectroscopic investigation on sodium pentaphenylchromate. 3. Diethyl ether tetrahydrofuran. [Germ.]. - J. Organomet. Chem., 1972, 44, N1, 127-140. - CA, 1972, 77, 169888.

Exp.; D_{3h}; d^3; 4.6.

1973

5.2.47. Anderson, O. P., Crystal and molecular structure of tris(1,10-phenanthroline)copper(II) perchlorate. - J. Chem. Soc. Dalton Trans., 1973, N12, 1237-1241.

Exp.; O_h; d^9; 4.6.

5.2.48. Bersuker, I. B., Vekhter, B. G., Structural manifestations of the Jahn-Teller effect. [Russ.]. - In: Proc. 4th Conf. Coord. Chem. and 2nd Semin. Crystallochem. Coord. and Metallorganic Compounds. Bratislava, 1973, p. 132-143. - CA, 1974, 80, 112758. RZK, 1974, 18B93.

Theor.; 4.1.

5.2.49. Gouterman, M., Angular momentum, magnetic interactions, Jahn-Teller and environment effect in metalloporphyrin triplet states. - Ann. N.Y. Acad. Sci., 1973, 206, 70-83. - CA, 1974, 80, 92039. RZK, 1974, 9B33.

Theor.; T.

5.2.50. Miller, P. T., Lenhert, P. G., Joesten, M. D., Crystal and molecular structure of tris(tetraisopropylmethylenediphosphonate)copper(II) perchlorate with comments on the Jahn-Teller effect. - Inorg. Chem., 1973, 12, N1, 218-223. - CA, 1973, 78, 21177.

Exp.; E; O_h; d^9; 4.6.

5.2.51. Mingos, D. M. P., A general bonding model for linear and bent transition metal-nitrosyl complexes. - Inorg. Chem., 1973, 12, N5, 1209-1211.

Theor.; T_d; D_{4h}; 1.5.

5.2.52. Parikh, P. C., Bhattacharya, P. K., Mixed ligand complexes. II. - J. Indian Chem. Soc., 1973, 50, N12, 804-807. - CA, 1974, 81, 177681.

Exp.; d^9.

5.2.53. Pradilla-Sorzano, J., Fackler, J. P., Jr., Base adducts of β-ketoenolates. V. Crystal and molecular structure of cis-bis(1,1,1,6,6,6-hexafluoro-2,4-pentanedionato)bis(pyridine)-zinc(II) and copper(II). - Inorg. Chem., 1973, 12, N5, 1174-1189. - CA, 1973, 78, 152599. RZK, 1973, 23B341.

Exp.; E - e; O_h; d^9; 3.1.

1974

5.2.54. Avdeef, A., Costamagna, J. A., Fackler, J. P., Jr., Crystal and molecular structure of tris(tropolonato)-manganese(III), $Mn(O_2C_7H_5)_3$, a high-spin complex having structural features consistent with Jahn-Teller behavior for two distinct MnO_6 centers. - Inorg. Chem., 1974, 13, N8, 1854-1863. - CA, 1974, 81, 69577.

Exp.; E; C_{3v}; d^4; 4.6.

5.2.55. Burdett, J. K., Calculation of geometries of binary transition metal carbonyl and dinitrogen complexes. - J. Chem. Soc. Faraday Trans. II, 1974, 70, N9, 1599-1613. - CA, 1975, 82, 9440.

Theor.; d^5-d^{10}; 5.4.

5.2.56. Fackler, J. P., Jr., Avdeef, A., Crystal and molecular structure of tris(2,4-pentanedionato)manganese(III), $Mn(O_2C_5H_7)_3$, a distorted complex as predicted by Jahn-Teller arguments. - Inorg. Chem., 1974, 13, N8, 1864-1875. - CA, 1974, 81, 69578.

Exp.; E - e; O_h; d^4; 4.6.

5.2.57. Haegele, R., Babel, D., Redetermination of the dipotassium tetrafluorocuprate(II) structure. [Germ.]. - Z. Anorg. Allg. Chem., 1974, 409, N1, 11-22. - CA, 1975, 82, 10238.

Exp.; E - e; O_h; d^9; 4.6.

5.2.58. Peytavin, S., Philippot, E., Maurin, M., Structural study of the double dihydrate: $Na_2Cu(SeO_4)_2 \cdot 2H_2O$. [Fr.]. - J. Solid State Chem., 1974, 9, N1, 63-68. - PA, 1974, 77, 20269.

Exp.; O_h; 4.6.

5.2.59. Takagi, S., Lenhert, P. G., Joesten, M. D., Structure of potassium calcium hexanitrocuprate(II) at 295°K: Evidence for a static Jahn-Teller distortion. - J. Am. Chem. Soc., 1974, 96, N21, 6606-6609.

Exp.; E - e; 4.6.

5.2.60. Weissbluth, M., Hemoglobin: Cooperativity and Electronic Properties. - London: Chapman & Hall; Berlin: Springer, 1974. - VIII, 175 p.

1975

5.2.61. Avdeef, A., Fackler, J. P., Jr., Structural trends of tris(acetylacetonate)-, tris(tropolonate)-, and other tris(bidenatate ligand)-metal complexes: Additional comments on the trigonal twist and ligand-ligand repulsions in "octahedral" D_3 bidentate ligand-metal complexes. - Inorg. Chem., 1975, 14, N8, 2002-2006. - CA, 1975, 83, 87650.

Exp.; E - e; D_3; 2.7; 4.6.

5.2.62. Burdett, J. K., New method for the determination of the geometries of binary transition metal complexes. - Inorg. Chem., 1975, 14, N2, 375-382. - CA, 1975, 83, 85116.

Theor.; T_d.

5.2.63. Butcher, R. J., Sinn, E., Crystal and molecular structures of dichloromethane-solvate tris(morpholinocarbodithioato) complexes of chromium(III), manganese(III), and rhodium(III): Comparison of coordination spheres. - J. Chem. Soc. Dalton Trans., 1975, N23, 2517-2522. - CA, 1976, 84, 82832.

Exp.; d^3; d^4; 4.6.

5.2.64. Glidewell, C., Jahn-Teller effects in metallocenes and related species. - J. Organomet. Chem., 1975, 102, N3, 339-343. - CA, 1976, 84, 58068. RZK, 1976, 10B34.

Theor.; D_5; D_{5h}; D_{5d}; d^0-d^{10}.

5.2.65. Hawthone, F. C., Ferguson, R. B., Refinement of the crystal structure of krohnkite. - Acta Crystallogr. B, 1975, 31, N6, 1753-1755. - CA, 1975, 83, 89082. PA, 1975, 78, 54168.

Exp.; E - e; O_h; d^9; 4.6.

5.2.66. Ionova, G. V., Ionov, S. P., Lyubimov, V. S., Configurational instability and tunnel transfer of atoms along chemical bonds in complexes of platinum, palladium, gold, and tin. [Russ.]. - In: Tez. Dokl. 12 Vsesoyuz. Chugaevskogo Soveshch. Khim. Kompleks. Soedin. Novosibirsk, 1975, Vol. 2, p. 218-219. - CA, 1977, 86, 78909.

Theor.; 1.5.

5.2.67. Levin, C. C., Qualitative molecular orbital picture of electronnegativity effects on XH_3 inversion barriers. - J. Am. Chem. Soc., 1975, 97, N20, 5649-5655. - CA, 1975, 83, 183922.

Theor.; C_{3v}; 1.5; 5.4.

5.2.68. Pitzer, K. S., Bernstein, L. S., Molecular structure of xenon hexafluoride. - J. Chem. Phys., 1975, 63, N9, 3849-3856. - CA, 1975, 83, 210989.

Theor.; T - t; 1.5.

5.2.69. Schmidling, D. G., An electron-diffraction study of the molecular structure of vanadium hexacarbonyl. - J. Mol. Struct., 1975, 24, N1, 1-8. - CA, 1975, 83, 36153. PA, 1975, 78, 24712.

Exp.; 2.6.

5.2.70. Takagi, S., Joesten, M. D., The prototype structure is seen in crystals containing Jahn-Teller inactive complex such as $K_2PbNi(NO_2)_6$. - Acta Crystallogr. B, 1975, 31, N7, 1968-1970.

Exp.; O_h; d^7-d^9; 4.6.

1976

5.2.71. Cianchi, L., Gulisano, F., Mancini, M., Spina, G., The deoxygenation process in oxyhemoglobin and role of the Jahn-Teller effect. - Phys. Lett. A, 1976, 59, N3, 247-248. - CA, 1977, 86, 38928. PA, 1977, 80, 21474. RZK, 1977, 12B37.

Theor.; E - (b_1 + b_2); C_{4v}.

5.2.72. Gazo, J., Bersuker, I. B., Garaj, J., Kabesová, M., Kohout, J., Langfelderova, H., Melnik, M., Serator, M., Valach, F., Plasticity of the coordination sphere of copper(II) complexes, its manifestation and causes. - Coord. Chem. Rev., 1976, 19, N3, 253-297. - Bibliogr.: 220 ref. CA, 1976, 85, 136313.

Rev.; d^9; 4.6.

5.2.73. Grande, B., Müller-Buschbaum, H., Oxocuprates. XIV. Crystal chemistry of $Sr_2Cu_3O_4Cl_2$. [Germ.]. - Z. Naturforsch. b, 1976, 31, N4, 405-407. - CA, 1976, 85, 54921.

Exp.; O_h; 4.6.

5.2.74. Nefedov, V. I., Angular dependence of inductive effect and cis-trans influence in coordination compounds. - Chem. Phys., 1976, 14, N2, 241-254.

Theor.; 1.5.

5.2.75. Nefedov, V. I., Mutual influence of ligands in square complexes of transition and non-transition elements. - Chem. Phys., 1976, 14, N2, 255-260.

Exp.; D_{4h}; 1.5; 4.6.

5.2.76. Nefedov, V. I., Yarzhemsky, V. G., Tarasov, V. P., Nuclear spin-spin coupling constants and mutual influence of the ligands. - Chem. Phys., 1976, 18, N3/4, 417-430.

Theor.; O_h; 1.5.

1977

5.2.77. Bersuker, I. B., Stavrov, S. S., Vekhter, B. G., On the origin of the trigger mechanism of conformational transitions in hemoglobin oxygenation. [Russ.]. - In: Fiz. Mat. Metody Koord. Khim.: Tez. Dokl. 6 Vsesoyuz. Soveshch. Kishinev, 1977, p. 139.

Theor.; (E_g + E_u) - a_u; 1.5.

5.2.78. Chen Hsiang-Wen, Part I. Synthesis reactions, and crystal structures of arylxanthate and dithiophosphate complexes. Part II. Structural relationships in Jahn-Teller influenced copper(II) complexes. Ph.D Thesis. - 1977. - 409 p. - [Order N77-25149.] Diss. Abstr. Int. B, 1977, 38, N5, 2168. CA, 1978, 88, 31403.

Exp.; d^9; 4.6.

5.2.79. Glidewell, C., Jahn-Teller effects in some molecular polyhedra. - J. Organomet. Chem., 1977, 128, N1, 13-20. - CA, 1977, 86, 177756.

Theor.

5.2.80. Klyagina, A. P., Levin, A. A., Sign of distortion in the second-order Jahn-Teller effect and some aspects of stereochemistry of d^{10} compounds. [Russ.]. - In: Fiz. Mat. Metody Koord. Khim.: Tez. Dokl. 6 Vsesoyuz. Soveshch. Kishinev, 1977, p. 142.

Theor.; O_h; d^{10}; 5.3.

5.2.81. Koskenlinna, M., Valkonen, J., Jahn-Teller distortions in the structure of manganese(III) selenite trihydrate, $Mn_2(SeO_3)_3 \cdot 3H_2O$. - Acta Chem. Scand. A, 1977, 31, N8, 611-614. - CA, 1978, 88, 30643.

Exp.; O_h; 4.6.

5.2.82. Koskenlinna, M., Valkonen, J., Crystal structure of manganese(III) hydrogen selenite diselenite, $MnH(SeO_3)(Se_2O_5)$. - Acta Chem. Scand. A, 1977, 31, N8, 638-640. - CA, 1978, 88, 30645.

Exp.; O_h; 4.6.

5.2.83. Levin, A. A., Klyagina, A. P., MO theory of distortions in octahedral complexes within the framework of the perturbation theory. [Russ.]. - In: Proc. 3rd Int. Semin. Cryst. Chem. Coord. Organomet. Compound. Wroclaw, 1977, p. 145-148. - CA, 1980, 92, 221021.

Theor.; O_h; 1.5.

5.2.84. Levin, A. A., Klyagina, A. P., Sign of distortion in the second-order Jahn-Teller effect: Stereochemistry of compounds of d^{10}-metals. [Russ.]. - Koord. Khim., 1977, 3, N8, 1154-1160. - CA, 1978, 88, 96734.

Theor.; 1.5.

5.2.85. Mizuhashi Seiji, Out-of plane distortion of the iron in ferrohemoglobin due to pseudo Jahn-Teller effect. - J. Theor. Biol., 1977, 66, N1, 13-20. - CA, 1978, 87, 64301.

Theor.; 1.5.

5.2.86. Pluth, J. J., Smith, J. V., Mortier, W. J., Positions of cations and molecules in zeolite with the chabazite framework. IV. Hydrated and dehydrated Cu^{2+}-exchanged chabazite. - Mater. Res. Bull., 1977, 12, N10, 1001-1007. - CA, 1977, 87, 192391. PA, 1978, 81, 15709.

Exp.; T_d; d^9; 4.6.

5.2.87. Price, W. C., Potts, A. W., Gabriel, A., Williams, T. A., Jahn-Teller and substitutional splitting of degenerate orbitals - particularly in benzene and the fluorobenzenes. - In: Proc. 6th Conf. Mol. Spectrosc., 1976. London, 1977, p. 381-395. - CA, 1978, 89, 162588.

Rev.

5.2.88. Templeton, J. L., Jacobson, R. A., McCarley, R. E., Synthesis and structure of bis(tetrapropylammonium)tri-μ-bromo-hexabromoditungstate(2-): A novel odd-electron dimeric anion showing evidence of Jahn-Teller distortion. - Inorg. Chem., 1977, 16, N12, 3320-3328. - CA, 1977, 87, 192990.

Exp.; E; D_{3h}; 4.6.

5.2.89. Zhilinskii, B. I., Stepanov, N. F., Nuclear charge changes: influence on the energy of highly symmetrical molecules. - Chem. Phys. Lett., 1977, 45, N3, 589-591. - CA, 1977, 86, 130233.

Theor.

1978

5.2.90. Armand, M., Coic, L., Palvadeau, P., Rouxel, J., The M-O-X transition metal oxyhalides: a new class of lamellar cathode material. - J. Power Sources, 1978, 3, N2, 137-144. - CA, 1979, 90, 63472.

Exp.; 4.6.

5.2.91. Bamzai, A. S., Deb, B. M., A link between the second-order Jahn-Teller effect and the highest occupied molecular orbital postulate for molecular shapes. - Pramaña, A, J. Phys., 1978, 11, N2, 191-194. - CA, 1978, 90, 61618. PA, 1978, 81, 86780.

Theor.; 1.5.

5.2.92. Bersuker, I. B., Stavrov, S. S., Vekhter, B. G., Pseudo-Jahn-Teller effect and electron-conformational rearrangements in hemoglobin oxygenation. - In: Proc. 19th Int. Conf. Coord. Chem. Prague, 1978, Vol. 2, p. 14.

Theor.; $(E_u + E_g) - a_u$; D_{4h}; d^6; 1.5.

5.2.92a. Burdett, J. K., Meldola Medal Lectures. I. Molecular Chem. Soc. Rev., 1978, 7, N4, 507-526. - CA, 1979, 91, 181566.

5.2.93. Hoffmann, B. M., Ratner, M. A., Jahn-Teller effects in metalloporphyrins and other four-fold symmetric systems. - Mol. Phys., 1978, 35, N4, 901-925. - CA, 1978, 89, 204347. PA, 1978, 81, 58599. RZK, 1978, 23B83.

Theor.; E; 1.2; 3.2.

5.2.94. Levin, A. A., Klyagina, A. P., Dolin, S. P., Molecular-orbital theory of the influence of substituents, based on electronic energy analysis: Interatomic distances and mutual influence of ligands in octahedral complexes. [Russ.]. - Koord. Khim., 1978, 4, N3, 354-360.

Theor.; O_h; 1.5.

5.2.95. Levin, A. A., Klyagina, A. P., Dolin, S. P., Angular distortions in octahedral compounds of nontransition and d^{10} elements induced by substitutions and variation of bond lengths. [Russ.]. - Koord. Khim., 1978, 4, N4, 505-517.

Theor.; O_h; 1.5.

5.2.96. Levin, A. A., Dolin, S. P., Adiabatic potential surface method in the vibronic theory of the mutual effect of ligands. [Russ.]. - Koord. Khim., 1978, 4, N12, 1812-1817. - CA, 1979, 90, 92632.

Theor.; T - (e + t); 1.5.

5.2.97. Mizuhashi Seiji, R - T conformational change of ferrohemoglobin. - Denki Tsushin Daigaku Gakuho, Rep. Univ. Electro-Commun. (Tokyo), 1978, 29, N1, 73-78. - CA, 1979, 90, 163573. PA, 1979, 82, 70521.

Theor.; 1.5.

5.2.98. Mizuhashi Seiji, Out-of-plane vibration of iron-porphine compounds. - Denki Tsushin Daigaku Gakuho, Rep. Univ. Electro-Commun. (Tokyo), 1978, 29, N1, 79-83. - CA, 1979, 90, 147376.

Theor.; $(E_u + E_g) - a_u$; D_4h; d^6; 1.5.

5.2.99. Mizuhashi Seiji, The pseudo-Jahn-Teller Hamiltonian in heme. [Jap.]. - Denki Tsushin Daigaku Gakuho, Rep. Univ. Electro-Commun. (Tokyo), 1978, 29, N1, 85-88. - CA, 1979, 90, 147377.

Theor.; $(E_u + E_g) - a_u$; D_4h; d^6; 1.5.

5.2.100. Mizuhashi Seiji, Excited-state wave functions of ferrohemoglobin. - Denki Tsushin Daigaku Gakuho, Rep. Univ. Electro-Commun. (Tokyo), 1978, 29, N2, 311-312. - PA, 1979, 82, 8475.

Theor.; 1.5.

5.2.101. Mizuhashi Seiji, A_{2u} out-of-plane vibration of iron-porphin compounds. - J. Phys. Soc. Jpn., 1978, 45, N2, 612-618. - CA, 1978, 89, 137748. PA, 1978, 81, 82712. RZF, 1979, 1I423.

Theor.; D_4h; 1.5.

5.2.101a. Poliakoff, M., Meldola Medal Lectures. II. $Fe(CO)_4$. - Chem. Soc. Rev., 1978, 7, N4, 527-540. - CA, 1979, 91, 181566.

1979

5.2.102. Bellard, S., Rubinson, K. A., Sheldrick, G. M., Crystal and molecular structure of vanadium hexacarbonyl. - Acta Crystallogr. B, 1979, 35, N2, 271-274. - CA, 1979, 90, 113392. PA, 1979, 82, 32949.

Exp.; 4.6.

5.2.103. Bersuker, I. B., Stavrov, S. S., Vekhter, B. G., The pseudo-Jahn-Teller effect as a cause of the iron atom displacement from the plane of the heme group and the trigger mechanism of conformational changes induced by hemoglobin oxygenation. [Russ.]. - Biofizika, 1979, 24, N3, 412-418. - CA, 1979, 91, 51453.

Theor.; $(E_u + E_g) - a_u$; D_4h; C_4v; d^6.

5.2.104. Hoffman, B. M., Metal substitution in hemoglobin and myoglobin. - In: The Porphyrins. New York, 1979, Vol. 7, p. 403-444.

Rev.; $E - (b_1 + b_2)$; D_4h.

5.2.105. Jezowska-Trzebiatowska, B., Ostern, M., Structure and polarographic properties of the ruthenium(III) thiocyanate complex. - Electrochim. Acta, 1979, 24, N1, 47-49. - CA, 1979, 91, 46356.

Exp.; 4.6.

5.2.106. Levin, A. A., Dolin, S. P., Molecular-orbital theory of the effect of substituents based on analysis of electron energy: Interatomic distances and mutual effect in a complete σ-variation of the theory. [Russ.]. - Koord. Khim., 1979, 5, N3, 320-335. - CA, 1979, 90, 210362.

Theor.; O_h; 1.5.

5.2.107. Levin, A. A., Klyagina, A. P., Dolin, S. P., Electron-vibrational theory of the mutual effect of ligands in coordination compounds. [Russ.]. - Zh. Neorg. Khim., 1979, 24, N9, 2307-2316. - CA, 1979, 91, 181730.

Theor.; O_h; 1.5.

5.2.108. Mizuhashi Seiji, Pseudo-Jahn-Teller effect in ferrohemoglobin. - Denki Tsushin Daigaku Gakuho, Rep. Univ. Electro-Commun. (Tokyo), 1979, 30, N1, 95-99. - PA, 1980, 83, 55984.

Theor.; 1.5.

See also: 1.1.-15, 30, 68, 72, 76; 1.2.-3, 12, 14, 22, 24; 1.5.-21, 23, 53; 1.6.-26; 2.1.-219, 262, 270; 2.7.-34; 2.8.-50, 70; 4.6.-15, 188; 5.1.-20, 64; 5.4.-4, 11, 31.

5.3. Computation of Vibronic Constants

The very important problem of vibronic-constant calculation is not sufficiently developed. The first works [5.3.-1, 2, 4, 5, 6, 78] on the benzene molecule were based on MO LCAO calculation of vibronic coupling constants using a "floating basis." Further refinements were made by means of semiempirical calculations [5.3.-24, 26, 28, 29]. Other determinations of vibronic constants have also been reported [5.3.-9, 13, 14].

1956

5.3.1. Murrell, J. N., Pople, J. A., The intensities of the symmetry-forbidden electronic bands of benzene. - Proc. Phys. Soc. London. Ser. A, 1956, 69, N3, 245-252. - RZK, 1957, N10, 25726.

Theor.; D_{6h}; 1.5; 2.1.

1957

5.3.2. Liehr, A. D., Critical study of vibronic interaction calculations. - Can. J. Phys., 1957, 35, N9, 1123-1132. - RZF, 1958, N5, 10380.

Theor.; D_{6h}; 1.5; 2.1.

5.3.3. Pople, J. A., Sidman, J. W., Intensity of the symmetry-forbidden electronic absorption band of formaldehyde. - J. Chem. Phys., 1957, 27, N6, 1270-1277. - RZK, 1959, N4, 7171.

Theor.; (A + B) - b.

1958

5.3.4. Liehr, A. D., Interaction of the vibrational and electronic motions in some simple conjugated hydrocarbons. I. Exact calculation of the intensity of the $^1A_{1g} \rightarrow {^1B_{1u}}$, $^1B_{2u}$ vibronic transitions of benzene. - Z. Naturforsch. a, 1958, 13, N4, 311-335.

Theor.; D_{6h}; 1.5; 2.1.

5.3.5. Liehr, A. D., Interaction of the vibrational and electronic motions in some simple conjugated hydrocarbons. II. Algebraic evaluation of integrals. - Z. Naturforsch. a, 1958, 13, N6, 429-438. - RZF, 1959, N7, 16589.

Theor.; D_{6h}.

5.3.6. Liehr, A. D., Errata: Interaction of the vibrational and electronic motions in some simple conjugated hydrocarbons. I. - Z. Naturforsch. a, 1958, 13, 596-597.

Theor.; D_{6h}; 1.5; 2.1.

5.3.7. Liehr, A. D., Errata and some additional comments on critical study of vibronic interaction calculations. - Can. J. Phys., 1958, 36, N11, 1588-1589. - RZF, 1959, N7, 15121.

Theor.

1961

5.3.8. Liehr, A. D., Interaction of the vibrational and electronic motions in some simple conjugated hydrocarbons. III. A semiempirical formulation. - Z. Naturforsch. a, 1961, 16, N7, 641-668. - RZF, 1962, 2V63.

Theor.; (A + B) - b; E - e; (E + B) - e; D_{4h}; D_{5h}; 1.2; 1.5; 2.1.

1965

5.3.9. Ballhausen, C. J., Heer, J., de., Ligand field treatment of the Jahn-Teller distortion in VCl_4. - J. Chem. Phys., 1965, 43, N12, 4304-4314. - CA, 1966, 64, 4432d. RZF, 1966, 7D95.

Theor.; E; T_d.

5.3.10. Brown, T. L., Vibronic contribution to the infrared intensities of benzene. - J. Chem. Phys., 1965, 43, N8, 2780-2787. - PA, 1966, 69, 1312.

Theor.; 2.7.

5.3.11. Child, M. S., Roach, A. C., Jahn-Teller effect in ReF_6. - Mol. Phys., 1965, 9, N3, 281-285. - CA, 1966, 64, 166d.

Theor.; T - e; 4.1.

1966

5.3.12. Salem, L., The Molecular Orbital Theory of Conjugated Systems. - New York; Amsterdam: Benjamin, 1966. - XVI, 576 p.

Theor.; 1.5.

1969

5.3.13. Kristofel', N. N., Revision of the calculation of the Jahn-Teller effect in the $^3F_{1u}$ state of the KCl-Tl center. [Russ.]. - Izv. Akad. Nauk Est.SSR, Fiz. Mat., 1969, 18, N3, 353-355. - CA, 1969, 71, 130057.

Theor.; T - (e + t_2).

1970

5.3.14. Trinkler, M. F., Complex structure of the lower excited state of alkali halide crystals activated by thallium. [Russ.]. - In: Radiats. Fiz. Riga, 1970, Vol. 6, p. 3-47. - CA, 1971, 74, 117793. RZF, 1971, 2D525.

Theor.; T - (e + t_2).

1971

5.3.15. Larkins, F. P., Electronic structure of the isolated vacancy in silicon. - J. Phys. & Chem. Solids, 1971, 32, N5, 965-980. - CA, 1971, 75, 41381.

Theor.; T_d.

5.3.16. Larkins, F. P., Isolated single vacancy in diamond. II. Jahn-Teller effect. - J. Phys. & Chem. Solids, 1971, 32, N9, 2123-2128. - CA, 1971, 75, 135726. PA, 1971, 74, 73189. RZF, 1972, 3E815.

Theor.; T; E; T_d.

5.3.17. Tsukerblat, B. S., Brunstein, S., Effect of vibronic interaction on crystal field theory parameters. - Phys. Status Solidi b, 1971, 45, N2, K115-K117.

Theor.; T - (e + t_2).

1974

5.3.18. Wood, R. F., Wilson, T. M., Mostoller, M. E., Electronic structure of the F and F^+ centers in the alkaline-earth oxides. - In: Extended Abstr. Int. Conf. Color Centers in Ionic Crystals. Sendai, 1974, F101, 2 p. - PA, 1975, 78, 21808.

Theor.; T - e; T - t; 2.1.

1975

5.3.19. Borg, M. A., Milk, Y., Coefficients of the orbit-lattice Hamiltonian for D_{3h} and C_{3h} symmetries. [Fr.]. - J. Phys. [France], 1975, 36, N2, 175-189. - CA, 1975, 82, 116354.

Theor.; D_{3h}; C_{3h}.

5.3.20. Johnson, W. C., Weigang, O. E., Theory of vibronic interactions: the importance of floating basis set. - J. Chem. Phys., 1975, 63, N5, 2135-2143.

Theor.

1976

5.3.21. Merienne-Lafore, M. F., Trommsdorff, H. P., Vibronic mixing and vibronic coupling of nearby electronic states. II. CNDO calculations of vibronic mixing in p-benzoquinone. - J. Chem. Phys., 1976, 64, N9, 3791-3796. - CA, 1977, 86, 4684.

Theor.; 1.5.

5.3.22. Orlandi, G., The evaluation of vibronic coupling matrix elements. - Chem. Phys. Lett., 1976, 44, N2, 277-280.

Theor.

5.3.23. Singh, J., Polansky, O. E., On the floating basis set model for studying the vibronic interaction in polyatomic molecules. - Chem. Phys. Lett., 1976, 44, N2, 325-328.

Theor.

1978

5.3.24. Bacci, M., Jahn-Teller coupling constants in the framework of the angular overlap model. - Chem. Phys. Lett., 1978, 58, N4, 537-540. - CA, 1979, 90, 31223. PA, 1979, 82, 10043.

Theor.

5.3.25. Bersuker, I. B., Polinger, V. Z., Rafalovich, M. L., Slyusarenko, E. I., Excited state, vibronic constant, and instability of the planar configuration of the NH_3 molecule. [Russ.]. - In: Tez. Dokl. Vsesoyuz. Soveshch. Kvant. Khimii. Novosibirsk, 1978, p. 30.

Theor.; C_{3v}; 1.5; 5.4.

5.3.26. Obert, T., Bersuker, I. B., Determination of the vibronic constant of the distortion isomers α- and β-$CuBr_2(NH_3)_2$. - In: Proc. 19th Int. Conf. Coord. Chem. Prague, 1978, Vol. 2, p. 94.

Theor.; 1.5.

5.3.27. Vasil'ev, A. V., Natadze, A. L., Ryskin, A. I., Electron-vibrational interaction in zinc sulfide crystals activated by transition metal ions. [Russ.]. - In: Spektrosk. Kristallov. Leningrad, 1978, p. 138-148. - RZF, 1980, 8E1442.

Exp.; E - e; T - e; 2.4.

1979

5.3.28. Bacci, M., Jahn-Teller effect in transition metal ions: a parametrization method based on the angular overlap model. - Chem. Phys., 1979, 40, N1/2, 237-244. - CA, 1979, 91, 99259. PA, 1979, 84, 78052.

5.3.29. Bacci, M., Calculation of first-order and second-order Jahn-Teller coupling constants in $MgO:V^{2+}$ and $KMgF_3:V^{2+}$. - Phys. Status Solidi b, 1979, 92, N1, 193-198. - CA, 1979, 90, 159643. PA, 1979, 82, 65906.

Theor.; T; O_h; d^3.

See also: 1.2.-51; 1.5.-2; 1.7.-1, 40, 46; 5.1.-58; 5.4.-31.

5.4. Quantum-Chemical Calculations of Adiabatic Potential Surfaces

In this subsection works on adiabatic-potential calculations for molecular systems and impurity centers in crystals performed by different quantum-chemical methods, and their use for the prediction of Jahn-Teller distortions, equilibrium geometry, Jahn-Teller stabilization energy, and other vibronic parameters, are listed.

1956

5.4.1. Liehr, A. D., On the stability of cyclobutadiene, cyclopentadienyl radical, and benzene(+1): a comparison of the molecular orbital and valence bond predictions. - Z. Phys. Chem. [Frankfurt a.M.], 1956, 9, N5/6, 338-354.

Theor.; (A + B) - b; E - e; D_4h; D_5h; D_6h; 1.5.

1958

5.4.2. Clinton, W. L., Rice, B., Electronic structure of BH_3. - J. Chem. Phys., 1958, 29, N2, 445-446. - RZF, 1959, N5, 10331.

Theor.; E - e; D_3h.

1960

5.4.3. Hobey, W. D., McLachlan, A. D., Dynamical Jahn-Teller effect in hydrocarbon radicals. - J. Chem. Phys., 1960, 33, N6, 1695-1703. - CA, 1961, 55, 16126e. RZF, 1961, 10V24.

Theor.; E; D_6h.

5.4.4. Snyder, L. C., Jahn-Teller distortions in cyclobutadiene, cyclopentadienenyl radical, and benzene positive and negative ions. - J. Chem. Phys., 1960, 33, N2, 619-621. - CA, 1961, 55, 4139b. RZF, 1961, 3V36.

Theor.; D_4h; D_5h; D_6h.

1961

5.4.5. Bersuker, I. B., Internal asymmetry in coordination compounds. I. The method of calculations and main formulas in the crystal field theory approximation. [Russ.]. - Zh. Strukt. Khim., 1961, 2, N3, 350-360.

Theor.; O_h.

5.4.6. Bersuker, I. B., Internal asymmetry in coordination compounds. II. Some results and discussion. [Russ.]. - Zh. Strukt. Khim., 1961, 2, N6, 734-739.

Theor.; E.

5.4.7. Snyder, L. C., Jahn-Teller distortions of large aromatic radical anions: a molecular orbital description. - Bull. Am. Phys. Soc., 1961, 6, N2, 165-166.

Theor.

5.4.8. Zalewski, K., A note on the Jahn-Teller effect in the $C_6H_6^+$ ion. - Acta Phys. Pol., 1961, 20, N11, 937-939. - CA, 1963, 58, 2017g. RZF, 1962, 5V58.

Theor.; E - e; D_{6h}.

1962

5.4.9. Bersuker, I. B., Internal asymmetry in coordination compounds. IV. The strong field case, molecular orbital method. [Russ.]. - Zh. Strukt. Khim., 1962, 3, N5, 563-568.

Theor.; d^1-d^9.

5.4.10. Colpa, J. P., Calculation of the π-electron energy of aromatic ions with an odd number of electrons: Some remarks on the Jahn-Teller effect. - In: Proc. 4th Int. Meet. Mol. Spectr., Bologna, 1959. 1962, Vol. 1, p. 210-214. - CA, 1963, 59, 8249e.

Theor.

5.4.11. Jordan, P. C. H., Longuet-Higgins, H. C., The lower electronic levels of the radicals CH, CH_2, CH_3, NH, NH_2, BH, BH_2, and BH_3. - Mol. Phys., 1962, 5, N2, 121-138.

Theor.; D_{3h}.

5.4.12. Snyder, L. C., A simple molecular orbital study of aromatic molecules and ions having orbitally degenerate ground states. - J. Phys. Chem., 1962, 66, N12, 2299-2306. - CA, 1963, 58, 4024f. RZF, 1963, 10D54.

Theor.

5.4.13. Yamaguchi, T., Electronic states of single vacancies in diamond. - J. Phys. Soc. Jpn., 1962, 17, N9, 1359-1384. - RZF, 1963, 4E274.

Theor.; 2.1.; 3.1.

1963

5.4.14. Lohr, L. L., Jr., Lipscomb, W. N., An LCAO-MO study of static distortions of transition metal complexes. - Inorg. Chem., 1963, 2, N5, 911-917. - CA, 1963, 59, 14669f.

Theor.; T_d; O_h; d^9.

1964

5.4.15. Bersuker, I. B., Titova, Yu. G., Internal asymmetry of coordination compounds. V. Tetrahedral complexes. [Russ.]. - Zh. Strukt. Khim., 1964, 5, N1, 121-129.

Theor.; E; T; T_d.

5.4.16. Peacock, T. E., Wilkinson, P. T., Electronic structure and spectrum of sym-trinitrobenzene. - Proc. Phys. Soc. London, 1964, 83, N533, 355-359. - CA, 1964, 60, 10068c.

Theor.; E.

1965

5.4.17. Dewar, M. J. S., Gleicher, G. J., Self-consistent field molecular orbital calculations for cyclobutadiene. - J. Am. Chem. Soc., 1965, 87, N14, 3255-3256.

Theor.; E - (b_1 + b_2); D_{4h}.

5.4.18. Friedel, J., Brooks, H., Lidiard, A. B., March, N. H., Seeger, A., Watkins, G. D., Yamaguchi, T., Panel discussion on intrinsic point defects in the diamond structure. - In: Radiation Damage Semicond.: 7th Int. Conf. Phys. Semicond., Paris - Royaumont, 1964. Paris, 1965, Vol. 3, p. 335-338. - RZF, 1965, 8E367.

Theor.

5.4.19. Yamaguchi, T., Calculation of the electronic states of single defects in diamond. - In: Radiation Damage Semicond.: 7th Int. Conf. Phys. Semicond., Paris-Royaumont, 1964. Paris, 1965, Vol. 3, p. 323-333. - RZF, 1965, 12E972.

Theor.; 2.1; 3.1.

1966

5.4.20. Ballhausen, C. J., Johansen, H. A., MO calculation of the static Jahn-Teller effect in CuF_6^{-4}. - Mol. Phys., 1966, 10, N2, 183-189. - CA, 1966, 64, 18684a. RZF, 1966, 8D76.

Theor.; O_h; d^9.

5.4.21. Moccia, R., Randaccio, L., Potential function for the inversion process of ammonia. - J. Chem. Phys., 1966, 45, N11, 4303-4309. - PA, 1967, 70, 6685.

Theor.; D_{3h}; C_{3v}.

1967

5.4.22. Aminov, L. K., Malkin, B. Z., The Jahn-Teller effect in the V^{2+} ground state in CaF_2 taking into account covalency. [Russ.]. - Fiz. Tverd. Tela, 1967, 9, N5, 1316-1323. - CA, 1967, 67, 26886. PA, 1967, 70, 31950.

Theor.; T - e; d^3.

5.4.23. Borden, W. T., The electronic nature of trimethylenemethane. - Tetrahedron Lett., 1967, N3, 259-264. - CA, 1967, 66, 85262.

Theor.; E - e; D_{3h}.

5.4.24. Friedel, J., Lannoo, M., Leman, G., Jahn-Teller effect for a single vacancy in diamond-like covalent solids. - Phys. Rev., 1967, 164, N3, 1056-1069. - CA, 1968, 68, 33896.

Theor.; E; T; T_d.

5.4.25. Lannoo, M., Leman, G., Jahn-Teller effect distortion of a monovacancy in diamond. [Fr.]. - J. Phys. [France], 1967, 28, Colloq. N3, 168-173. - CA, 1968, 68, 33306.

Theor.

5.4.26. Lohr, L. L., Jr., A molecular orbital study of vibronic interactions in the hexachlorocuprate(II) complex. - Inorg. Chem., 1967, 6, N10, 1890-1900.

Theor.; O_h; d^9; 2.1; 3.1.

5.4.27. Pfeiffer, G. V., Huff, N. T., Greenawalt, E. M., Ellison, F. O., Method of diatomics in molecules. IV. Ground and excited states of H_3^+, H_4^+, H_5^+, and H_6^+. - J. Chem. Phys., 1967, 46, N2, 821-822. - CA, 1967, 66, 69117.

Theor.

5.4.28. Sen, P. N., Chakgrabarti, B., Bose, S. K., Hückel calculations on decacyclene. - Tetrahedron, 1967, 23, N10, 4177-4180. - CA, 1967, 67, 81745.

Theor.; E; C_{3v}.

1968

5.4.29. Fowler, W. B., Dexter, D. L., Electronic bubble states in liquid helium. - Phys. Rev., 1968, 176, N1, 337-343. - CA, 1969, 70, 6719. PA, 1969, 72, 34938.

Theor.; T.

5.4.30. Lannoo, M., Leman, G., Friedel, J., Jahn-Teller effect for a single vacancy in diamond-like covalent solids. - In: Radiat. Eff. Semicond.: Proc. Santa Fe Conf., 1967. New York; London, 1968, p. 37-42. - CA, 1968, 69, 69772. PA, 1971, 74, 21853.

Theor.; E - e; T - e.

5.4.31. Porter, R. N., Stevens, R. M., Karplus, M., Symmetric H_3: A semiempirical and ab initio study of a simple Jahn-Teller system. - J. Chem. Phys., 1968, 49, N11, 5163-5178. - CA, 1969, 70, 40887.

Theor.; E - e.

1969

5.4.32. Coulson, C. A., Deb, B. M., On statis Jahn-Teller distortion of the ground state of vanadium tetrachloride. - Mol. Phys., 1969, 16, N6, 545-552. - CA, 1969, 71, 53686. PA, 1969, 72, 30059. RZF, 1970, 1D128.

Theor.; E - e; T_d.

5.4.33. Dehlinger, U., Electronic structure of alloys. [Germ.]. - Z. Metallkd., 1969, 60, N1, 8-10. - CA, 1969, 70, 90102. PA, 1969, 72, 39180.

Theor.

5.4.34. Duvall, B., Celli, V., Electronic properties of negative ions in liquid helium. - Phys. Rev., 1969, 180, N1, 276-286. - CA, 1969, 70, 81073. PA, 1969, 72, 23560.

Theor.

1970

5.4.35. Allen, G. C., Clack, D. W., Molecular-orbital calculations on transition-metal complexes. Part I. Hexafluorometallate(II) ions. - J. Chem. Soc. Ser. A, 1970, N16, 2668-2672.

Theor.; O_h; d^9.

5.4.36. Arents, J., Allen, L. C., Ab initio study of the geometries of Jahn-Teller distortions and electronic charge distribution in the CH_4^+ ion. - J. Chem. Phys., 1970, 53, N1, 73-78. - CA, 1970, 73, 48643. PA, 1970, 73, 55359. RZF, 1971, 1D217.

Theor.; T_d.

5.4.37. Becker, C. A. L., Dahl, J. P., A CNDO-MO calculation of VCl_4. - Theor. Chim. Acta, 1970, 19, N2, 135-154. - CA, 1971, 74, 25081.

Theor.; T; T_d.

5.4.38. Haselbach, E., Jahn-Teller distortions in the radical cations of cyclopropane and allene. - Chem. Phys. Lett., 1970, 7, N4, 428-430. - CA, 1971, 74, 69773. RZF, 1971, 5D111.

Theor.; 2.5.

5.4.39. Messmer, R. P., Watkins, G. D., Linear combination of atomic orbital-molecular orbital treatment of the deep defect level in a semiconductor: nitrogen in diamond. - Phys. Rev. Lett., 1970, 25, N10, 656-659. - PA, 1970, 73, 75485.

Theor.

5.4.40. Olsen, J. F., Burnelle, L., Distortions in the trigonally symmetric radicals nitrogen and carbon trioxide. - J. Am. Chem. Soc., 1970, 92, N12, 3659-3664. - CA, 1970, 73, 70015.

Theor.; E - e; D_{3h}.

5.4.41. Nakayama Misunobu, I'Haya Yasumasa, J., Multi-configuration LCAO-MO study for complex unsaturated molecules. I. General theory and its application to the benzene anion. - Int. J. Quantum Chem., 1970, 4, N1, 21-42. - CA, 1970, 72, 93499.

Theor.; D_{6h}.

5.4.42. Nakayama Mitsunobu, I'Haya Yasumasa, J., Multi-configuration LCAO-MO study for complex unsaturated molecules. II. Application to the benzene cation. - Int. J. Quantum Chem., 1970, 4, N1, 45-55. - CA, 1970, 72, 93501.

Theor.; D_{6h}.

5.4.43. Nakayama Mitsunobu, Isomae, S., I'Haya Yasumasa, J., Jahn-Teller effect in the triphenylene mono-negative ion. [Jap.]. - Denki Tsushin Daigaku Gakuho, Rep. Univ. Electro-Commun. (Tokyo), 1970, N28, 101-104. - PA, 1971, 74, 47725. RZF, 1971, 4D166.

Theor.; D_{3h}.

5.4.44. Watkings, G. D., Messmer, R. P., LCAO-MO treatment for a defect level in a semiconductor. - In: Proc. 10th Int. Conf. Semicond. Springfield, 1970, p. 623-629. - CA, 1971, 74, 117607.

Theor.

1971

5.4.45. Bhattacharyya, B. D., Role of Jahn-Teller effect in the ligand field theory of octahedrally coordinated tris-acetylacetonatomanganese. - Phys. Status Solidi b, 1971, 43, N2, 495-503. - CA, 1971, 74, 92870. PA, 1971, 74, 28606. RZF, 1971, 7D202.

Theor.; E - e; d^4; 3.6.

5.4.46. Copeland, D. A., Ballhausen, C. J., An unrestricted CNDO-MO calculation of VCl_4. - Theor. Chim. Acta, 1971, 20, N4, 317-330. - RZF, 1971, 8D182.

Theor.; T_d.

5.4.47. Coulson, C. A., Larkins, F. P., Isolated single vacancy in diamond. I. Electronic structure. - J. Phys. & Chem. Solids, 1971, 32, N9, 2245-2257. - RZF, 1972, 3E814.

Theor.; E; T.

5.4.48. Dixon, R. N., Jahn-Teller distortions of CH_4^+. - Mol. Phys., 1971, 20, N1, 113-126. - CA, 1971, 74, 69784. PA, 1971, 74, 1174.

Theor.; T - t; T_d; 2.5.

5.4.49. I'Haya Yasumasa, J., Nakayama Mitsunobu, Iwabuchi, T., Static Jahn-Teller distortion in the triphenylene mononegative ion. - Int. J. Quantum Chem., Symp., 1971, N5, 227-232. - CA, 1972, 76, 50291. RZF, 1972, 6D203.

Theor.; D_{3h}; 2.1.

5.4.50. Larkins, F. P., Electronic structure of the isolated vacancy in silicon. - In: Int. Conf. Radiat. Eff. Semicond., Albany, 1970. London, 1971, p. 19-21. - PA, 1972, 75, 47935.

Theor.; 3.1.

5.4.51. Larkins, F. P., Electronic structure of the isolated vacancy in silicon. - Radiat. Eff., 1971, 9, N1/2, 5-7. - CA, 1971, 75, 12128. PA, 1971, 74, 52256.

Theor.; 3.1.

5.4.52. Messmer, R. P., Watkins, G. D., An LCAO-MO treatment of the vacancy in diamond. - In: Int. Conf. Radiat. Eff. Semicond., Albany, 1970. London, 1971, p. 23-28. - PA, 1972, 75, 47808.

Theor.

5.4.53. Messmer, R. P., Watkins, G. D., LCAO-MO treatment of vacancy in diamond. - Radiat. Eff., 1971, 9, N1/2, 9-14. - CA, 1971, 75, 26277. PA, 1971, 74, 52252.

Theor.

1972

5.4.54. Abulaffio, C., Irvine, J. M., Alpha-cluster ground state of oxygen-16. - Phys. Lett. B, 1972, 38, N7, 492-494. - CA, 1972, 77, 12433.

Theor.

5.4.55. Bhattacharyya, B. D., Lattice-ion interactions in the ligand field theory of trigonally distorted octahedral V^{3+} complexes. - Phys. Status Solidi b, 1972, 53, N2, 723-734. - CA, 1972, 77, 145503. PA, 1972, 75, 81205. RZF, 1973, 3E964.

Theor.; T - e; O_h; d^2; 3.1; 3.6.

5.4.56. Godoy Almiralli, J. J., The use of the CNDO method in (I) the determination of molecular equilibrium geometry of boron hydrides and (II) the study of the Jahn-Teller effect on CH_4^+, BH_4, and CF_4^+. - 1972. - 101 p. - [Order N72-21351.] Diss. Abstr. Int. B, 1972, 33, N2, 661. CA, 1973, 78, 34200.

Theor.; T_d.

5.4.57. Purins, D., Feeley, H. F., Effect of vibronic interaction on the singlet-triplet separation in cyclopentadienyl cations. - J. Chem. Phys., 1972, 57, N1, 221-232. - CA, 1972, 77, 54286.

Theor.; D_{5h}; 1.5.

5.4.58. Titova, Yu. G., Elektronnoye Stroyeniye i Svoystva Nekotorykh Koordinatsionnykh Soyedineniy Medi: Avtoref. Dis. ...Kand. Fiz.-Mat. Nauk. [Russ.]. - Kishinev, 1972. - 18 p. - Kishinev. Univ.

Electronic Structure and Properties of Some Copper Coordination Compounds.

Theor.; E; T; T_d.

1973

5.4.59. Bersuker, I. B., Calculations of electronic structure and spectra of transition metal coordination compounds and impurity centers in crystals. [Russ.]. - In: Spectrosk. Kristallov: Materialy 3 Simp. Spektrosk. Kristallov, Aktivir. Ionami Redkozem. i Perekhodnykh Metallov, 1970, Leningrad, 1973, p. 15-30.

Theor.; E; T; 2.1.

5.4.60. Cianchi, L., Mancini, M., Moretti, P., Influence of covalency in the dynamic Jahn-Teller effect in $Al_2O_3:Ti^{3+}$. - Phys. Rev. B, 1973, 7, N11, 5014-5017. - CA, 1973, 79, 11472. PA, 1973, 76, 41149. RZF, 1974, 3E989.

Theor.; E - e; d^1.

5.4.61. Coulson, C. A., Vacancy centers in the diamond lattice: a critique of current theoretical treatments. - In: Radiat. Damage and Defects in Semicond.: Proc. Int. Conf., Reading, Berks., 1972. London, 1973, p. 249-254. - PA, 1973, 76, 33839.

Theor.

5.4.62. Ha Tae-kyn, Graf, F., Guenthard, H. H., Ab initio study on the stability, geometry, and electronic structure of the cyclopropenyl system. - J. Mol. Struct., 1973, 15, N3, 335-350. - CA, 1973, 78, 102246.

Theor.; D_{3h}.

5.4.63. Jackson, J. L., Wyatt, R. E., H_3 potential surface in natural collision coordinates. - Chem. Phys. Lett., 1973, 18, N2, 161-165. - RZF, 1973, 7D137.

Theor.; E - e.

5.4.64. Telezhkin, V. A., Tolpygo, K. B., Electron structure of a donor-vacancy complex in diamond-type crystals. [Russ.]. - Fiz. Tverd. Tela, 1973, 15, N4, 1084-1089. - CA, 1973, 79, 24301. PA, 1974, 77, 13180. RZF, 1973, 8D638.

Theor.

5.4.65. Watkins, G. D., Vacancies and interstitials in semiconductors [Jahn-Teller distortions]. - In: Radiat. Damage and Defects in Semicond.: Proc. Int. Conf., Reading, Berks., 1972. London, 1973, p. 228-237. - PA, 1973, 76, 33837.

Theor.

1974

5.4.66. Spacu, P., Teodorescu, M., Lepadatu, C. I., The transition $^5E(t_2^3e^3) \rightarrow {}^5T(t_2^4e^2)$ in the iron(II) 1,10-phenanthroline and 2,2'-dipyridyl high-spin complexes [Fe(II)$phen_2X_2$] and [Fe(II)-$dipy_2X_2$]. - Z. Phys. Chem. [Frankfurt a. M.], 1974, 88, N1/6, 285-289. - CA, 1974, 81, 112969.

Theor.; E; T; d^6.

5.4.67. Watkins, G. D., Messmer, R. P., Many-electron effects for deep levels in solids: the lattice vacancy in diamond. - Phys. Rev. Lett., 1974, 32, N22, 1244-1247.

Theor.; E; T; T_d.

5.4.68. Yip Kwok Leung, The lattice vacancy in Si and Ge. - Phys. Status Solidi b, 1974, 66, N2, 619-626. - PA, 1975, 78, 18575.

Theor.; T - (e + t); T_d.

1975

5.4.69. Biernacki, S. W., LCAO treatment of scandium(2+) ion in zinc sulfide. - Phys. Status Solidi b, 1975, 68, N1, K77-K79. - CA, 1975, 82, 161970. PA, 1975, 78, 33719.

Theor.; E; 3.1.

5.4.70. Clack, D. W., Smith, W., On the low-spin behavior of the hexafluoronickelate(III) ion. - Mol. Phys., 1975, 29, N5, 1615-1618. - RZF, 1975, 10D253.

Theor.; E - e; O_h; d^6.

5.4.71. Glidewell, C., Structures of four-atom polyhedra. - Inorg. & Nucl. Chem. Lett., 1975, 11, N11, 761-764. - CA, 1976, 84, 11630.

Theor.; C_{3v}; D_{4h}; T_d; 1.5.

5.4.72. Moreno, M., Barriuso, M. T., Study of the spin-Hamiltonian of $[MX_4]^{2-}$ Jahn-Teller complexes on the basis of MO theory. - Solid State Commun., 1975, 17, N8, 1035-1037. - PA, 1976, 79, 5766. RZK, 1976, 8B30.

Theor.; D_{4h}; d^9.

5.4.73. Schoeller, W., Thermal rearrangement of methylene-cyclobutane MO study. [Germ.]. - Chem. Ber., 1975, 108, N4, 1285-1292. - CA, 1975, 82, 169763.

Theor.

5.4.74. Wood, R. F., Wilson, T. M., Electronic structure of the F-center in calcium oxide and magnesium oxide. - Solid State Commun., 1975, 16, N5, 545-548. - CA, 1975, 83, 19281. PA, 1975, 78, 50557. RZF, 1975, 8E57.

Theor.; T - e.

5.4.75. Yip Kwok Leung, Fowler, W. B., Electronic structure of E_1 centers in SiO_2. - Phys. Rev. B, 1975, 11, N6, 2327-2338. - PA, 1975, 78, 42190.

Theor.; E.

1976

5.4.76. Armstrong, D. R., Fortune, R., Perkins, P. G., A molecular-orbital study of the ground and excited-state properties of metallocenes of the first-row transition-metal ion. - J. Organomet. Chem., 1976, 111, N2, 197-213. - CA, 1976, 85, 77343.

Theor.; E; D_{5d}; d^1-d^9.

5.4.77. Calles, A., Jahn-Teller effect in the α-cluster model for ^{12}C. - Phys. Lett. B, 1976, 61, N1, 13-14.

Theor.; E - e; D_{3h}.

5.4.78. Harrison, W. A., Surface reconstruction on semiconductors. - Surf. Sci., 1976, 55, N1, 1-19. - CA, 1976, 84, 188434.

Theor.

5.4.79. Jackels, C. F., Davidson, E. R., The two lowest energy ^{2}A' states of nitrogen dioxide. - J. Chem. Phys., 1976, 64, N7, 2908-2917. - CA, 1976, 84, 171591. PA, 1976, 79, 57200.

Theor.

5.4.80. Louie, S. G., Schlüter, M., Chelikowsky, R., Cohen, M. L., Self consistent electronic states for reconstructed silicon vacancy models. - Phys. Rev. B, 1976, 13, N4, 1654-1663. - CA, 1976, 84, 143157.

Theor.; T.

5.4.81. Weigel, C., Corbett, J. W., Jahn-Teller distortions for carbon interstitial configurations in diamond. - Z. Phys. B, 1976, 23, N3, 233-238. - CA, 1976, 84, 128965. PA, 1976, 79, 44909.

Theor.

1977

5.4.82. Ammerlaan, C. A. J., Weigel, C., Extended-Huckel-theory calculation of hyperfine interactions of the positive divacancy in silicon. - In: Int. Conf. Radiat. Eff. Semicond., Dubrovnik, 1976. London, 1977, p. 448-457. - PA, 1977, 80, 89742.

Theor.; E; D_{3h}; C_{3v}.

5.4.83. Davidson, E. R., Global topology of triatomic potential surfaces. - J. Am. Chem. Soc., 1977, 99, N2, 397-402. - CA, 1977, 86, 96332. PA, 1977, 80, 50381.

Theor.; E; D_{3h}.

5.4.84. Davidson, E. R., Borden, W. T., The potential surface for planar cyclopropenyl radical and anion. - J. Chem. Phys., 1977, 67, N5, 2191-2195. - CA, 1978, 88, 37024. PA, 1978, 81, 5155.

Theor.; E.

5.4.85. Natadze, A. L., Ryskin, A. I., Chromium-doped zinc sulfide - the system with a "nearly dynamic" Jahn-Teller effect. - Solid State Commun., 1977, 24, N2, 147-150. - CA, 1977, 87, 208746. PA, 1978, 81, 2455.

Theor.; E - e; T - e; T - t; d^4; d^6-d^9.

5.4.86. Watkins, G. D., Lattice vacancies and interstitials in silicon. - Taipei, Chin. J. Phys., 1977, 15, N2, 92-101. - CA, 1978, 88, 67920.

Theor.

5.4.87. Weber, J., Bill, H., Jahn-Teller distortion and electronic structure of the atomic oxygen(-1) center in calcium fluoride: an MS X_α study. - Chem. Phys. Lett., 1977, 52, N3, 562-566. - CA, 1978, 88, 79243. PA, 1978, 81, 23057.

Theor.; E - e; O_h.

1978

5.4.88. Ammerlaan, C. A. J., Wolfrat, J. C., Extended-Hückel-theory calculations for the positive divacancy in silicon. - Phys. Status Solidi b, 1978, 89, N1, 85-94. - CA, 1978, 89, 169368. PA, 1978, 81, 87938.

Theor.

5.4.89. Borden, W. T., Davidson, E. R., Hart, P., The potential surfaces for the lowest singlet and triplet states of cyclobutadiene. - J. Am. Chem. Soc., 1978, 100, N2, 388-392. - CA, 1978, 88, 120422.

Theor.; D_{4h}; 1.5.

5.4.90. Gerber, W. H., Schumacher, E., The dynamic Jahn-Teller effect in the electronic ground state of triatomic lithium: An ab initio calculation of the BO hypersurface and the lowest vibronic states of triatomic lithium. - J. Chem. Phys., 1978, 69, N4, 1692-1703. - CA, 1978, 89, 169429. PA, 1978, 81, 94203. RZF, 1979, 3D106.

Theor.; E - e; D_{3h}.

5.4.91. Hinde, A. L., Poppinger, D., Radom, L., Ab initio study of the benzene radical anion. - J. Am. Chem. Soc., 1978, 100, N15, 4681-4685. - CA, 1978, 89, 90000.

Theor.; E - e; D_{6h}.

5.4.92. Köppel, H., Cederbaum, L. S., Domcke, W., Van Niessen, W., The Jahn-Teller effect in NH_3^+. - Mol. Phys., 1978, 35, N5, 1283-1299. - CA, 1978, 89, 206991. PA, 1978, 81, 62948. RZF, 1978, 11D267.

Theor.; T_d; 2.5; 5.3.

5.4.93. Mainwood, A., Relaxation about the vacancy in diamond. - J. Phys. C, 1978, 11, N13, 2703-2710. - CA, 1978, 89, 224121. PA, 1978, 81, 71722.

Theor.; E - e.

5.4.94. Martin, R. L., Davidson, E. R., Electronic structure of the sodium trimer. - Mol. Phys., 1978, 35, N6, 1713-1729. - CA, 1978, 89, 221122. PA, 1978, 81, 70996.

Theor.

5.4.95. Zunger, A., Englman, R., Self-consistent LCAO calculation of the electronic properties of graphite. II. Point vacancy in the two-dimensional crystal. - Phys. Rev. B, 1978, 17, N2, 642-661. - CA, 1978, 88, 141928.

Theor.; 2.1.

1979

5.4.96. Astier, M., Pottier, N., Bourgoin, J. C., Linear-combination-of-atomic-orbitals, self-consistent-field method for the determination of the electronic structure of deep levels in semiconductors. - Phys. Rev. B, 1979, 19, N10, 5265-5276. - PA, 1979, 82, 82555. RZF, 1979, 12E30.

Theor.

5.4.97. Bacci, M., Del Giallo, F., Pieralli, F., Ranfagni, A., A development of the molecular-orbital model for KCl:Tl. - Physica B + C, 1979, 97, N2/3, 165-171. - CA, 1980, 92, 28862. RZF, 1980, 4E88.

Theor.

5.4.98. Baraff, G. A., Kane, E. O., Schlüter, M., Silicon vacancy: A possible "Anderson-negative-V" system. - Phys. Rev. Lett., 1979, 43, N13, 956-959.

Theor.; T.

5.4.99. Baricic, P., Vibronic interaction in octahedral complexes with an electronically degenerate E term. [Slo.]. - In: Zb. Stud. Ved. Odb. Pr. Slov. Vys. Sk. Tech. Bratislave. Chemicko-technol. Fak., 1979, p. 40-41. - CA, 1980, 92, 101582.

Theor.; E - (a + e); 5.3.

5.4.100. Borden, W. T., Davidson, E. R., Potential surfaces for the planar cyclopentadienyl radical and cation. - J. Am. Chem. Soc., 1979, 101, N14, 3771-3775. - CA, 1979, 91, 74085.

Theor.; E - e; D_{5h}; 1.5.

5.4.101. King, H. F., Morokuma Keiji, Theory of the Rydberg spectrum of triatomic hydrogen. - J. Chem. Phys., 1979, 71, N8, 3213-3220. - CA, 1979, 91, 201660.

Theor.; E - e; D_{3h}.

5.4.102. Lipari, N. O., Bernhole, J., Pantelides, S. T., Electronic structure of the Jahn-Teller distorted vacancy in silicon. - Phys. Rev. Lett., 1979, 43, N18, 1354-1357. - CA, 1980, 92, 11378. PA, 1980, 83, 10869.

Theor.

5.4.103. Mainwood, A., Substitutional impurities in diamond. - J. Phys. C, 1979, 12, N13, 2543-2549. - CA, 1980, 92, 13012. PA, 1979, 82, 87056.

Theor.

5.4.104. Martin, R. L., Rydberg states of H_3. - J. Chem. Phys., 1979, 71, N8, 3541-3542. - PA, 1980, 83, 9494.

Theor.; D_{3h}.

5.4.105. Meyer, R., Graf, F., Ha Tae-kyn, Guenthard, H. H., Jahn-Teller effect in cyclopentadienyl radical: delocalized vibronic valence isomerization. - Chem. Phys. Lett., 1979, 66, N1, 65-71. - CA, 1980, 92, 128176. PA, 1979, 82, 97986.

Theor.; D_{5h}.

5.4.106. Tanaka Kiyoshi, Davidson, E. R., A theoretical study on the potential surfaces of the lower electronic states of formyl radical. - J. Chem. Phys., 1979, 70, N6, 2904-2913. - CA, 1979, 90, 194816.

Theor.; 1.5; 1.6.

5.4.107. Volland, W. V., Davidson, E. R., Borden, W. T., Effect of carbon atom pyramidalization on the bonding in ethylene. - J. Am. Chem. Soc., 1979, 101, N3, 533-537. - PA, 1979, 82, 93936.

Theor.; 1.5.

5.4.108. Yamatera Hideo, Jahn-Teller distortion in copper(II) complexes: Why tetragonal elongation is preferred to tetragonal compression. - Acta Chem. Scand. Ser. A, 1979, 33, N2, 107-111. - CA, 1979, 90, 192877.

Theor.; E - e; O_h.

See also: 1.6.-2; 5.1.-34; 5.3.-25.

AUTHOR INDEX

SUBJECT INDEX

FORMULA INDEX

SUBSTANCE INDEX

CHEMICAL ABBREVIATIONS

acac	acetylacetonate
bipy	bipyridine
cp	cyclopentadienyl
EDTA	ethylenediaminotetraacetic acid
en	ethylendiamine
Et	ethyl
halo	halogen
Me	methyl
mephen	5,6,-dimethyl-1,10-phenanthroline
Ph_3	triphenyl
phen	1,10-phenanthroline
terpy	terpyridine
ur	urea